T0189530

Developments in Mathematics

VOLUME 32

Series Editors:

Krishnaswami Alladi, *University of Florida*
Hershel M. Farkas, *Hebrew University of Jerusalem*

For further volumes:
http://www.springer.com/series/5834

Krishnaswami Alladi • Peter Paule
James Sellers • Ae Ja Yee
Editors

Combinatory Analysis

Dedicated to George Andrews

 Springer

Editors
Krishnaswami Alladi
Department of Mathematics
University of Florida
Gainesville, FL, USA

Peter Paule
Research Institute for Symbolic Computation
Johannes Kepler University Linz
Linz, Austria

James Sellers
Department of Mathematics
Penn State University
University Park, PA, USA

Ae Ja Yee
Department of Mathematics
Penn State University
University Park, PA, USA

ISSN 1389-2177
ISBN 978-1-4899-8835-5   ISBN 978-1-4614-7858-4 (eBook)
DOI 10.1007/978-1-4614-7858-4
Springer New York Heidelberg Dordrecht London

Mathematics Subject Classifications (2000): 11P82, 11P83, 33C05, 33E05, 11F11, 11R29, 11P81, 05A19, 05E10, 05E05, 05A15, 05A30, 33B10, 05A17, 05A19, 11F37, 33D15, 11E16, 11P85, 37E15, 11P21, 60G50, 33C67, 60E15

Preface

George Andrews is one of the most influential figures in number theory and combinatorics. In the theory of partitions and q-hypergeometric series and in the study of Ramanujan's work, he is the unquestioned leader. To suitably honor him during his 70th birthday year, an *International Conference on Combinatory Analysis* was held at The Pennsylvania State University during December 5–7, 2008. Three issues of the Ramanujan Journal comprising Volume 23 were published in 2010 as the refereed proceedings of that conference. The Ramanujan Journal was proud to bring out that volume honoring one of its Founding Editors. In view of the great interest that the mathematical community has in the influential work of Andrews, it was decided to republish Volume 23 of The Ramanujan Journal in this book form, so that the refereed proceedings are more readily available for those who do not subscribe to the journal but wish to possess this volume.

The conference was organized by James Sellers and Ae Ja Yee of The Pennsylvania State University with the assistance of Krishnaswami Alladi of the University of Florida and Peter Paule of the Research Institute of Symbolic Computation (RISC) at the Johannes Kepler University Linz, Austria. The generous support received from the National Science Foundation, the National Security Agency, and The Pennsylvania State University was crucial for the success of the conference. Right from the planning stage back in 2007, the head of the Penn State Mathematics Department, John Roe, and the Dean of the Eberly College of Science, Daniel Larson, gave their wholehearted support. For this we are most grateful.

The regard and affection that the international community of researchers has for George was demonstrated by the enthusiastic response to our invitations to speakers and participants. The conference had about 70 registered participants. There were 11 plenary talks including one by George Andrews himself, and more than 30 research presentations in 20 minute talks. The broad range of areas covered by these talks included the many facets of the theory of partitions, single and multi-variable q-hypergeometric series, various aspects of Ramanujan's mathematics, analytic number theory, and combinatorial number theory—areas which collectively were classified under the heading of *Combinatory Analysis* by MacMahon and his contemporaries in the early part of the twentieth century. The research conducted worldwide in the past several decades, especially that of George Andrews, and the talks of this conference, confirmed that Combinatory Analysis is still a vibrant area of research, and that Andrews continues to be one of its most influential leaders.

As a fitting tribute to George Andrews, many speakers from the conference contributed research papers to this volume which deals with a broad range of areas that signify the research interests of George Andrews. In reproducing Volume 23 of The Ramanujan Journal in this book form, we have included two papers—one by Hei-Chi Chan and Shaun Cooper, and another by Vyacheslav Spiridonov and Ole Warnaar—which were intended for Volume 23 of The Ramanujan Journal, but appeared in other issues.

The enormous productivity of George Andrews remains unabated in spite of the passage of time. His immensely fertile mind continues to pour forth seminal ideas year after year. Indeed, he has two research papers in this very volume! The only way to suitably recognize such an individual is to convene conferences in his honor every few years. There was one in Maratea, Italy in 1998 for his 60th birthday organized in the frame of the Lotharingien Seminar in Combinatorics, another at The Pennsylvania State University in 2003 for his 65th, and the conference on Combinatory Analysis for his 70th birthday in 2008. And now in 2013, we will be gathering at the Center for Combinatorics at Nankai University, China, to celebrate his 75th birthday with an *International Conference on the Combinatorics of q-Series and Partitions*. It is indeed appropriate that this book version of Volume 23 of The Ramanujan Journal is planned to be released at the conference in Nankai University, China. May his eternal youthfulness and his magnificent research output continue to inspire and influence researchers in the years ahead.

March 2013

Krishnaswami Alladi
(Editor-in-Chief, The Ramanujan Journal)

Peter Paule
(Editor, The Ramanujan Journal)

James Sellers
(Guest Editor)

Ae Ja Yee
(Editor, The Ramanujan Journal)

Contents

Ramanujan J (2010) 23: 3–15
DOI 10.1007/s11139-008-9143-z

Cranks—really, the final problem

**Bruce C. Berndt · Heng Huat Chan ·
Song Heng Chan · Wen-Chin Liaw**

Dedicated to our friend George Andrews on his 70th birthday

Received: 29 August 2007 / Accepted: 5 September 2008 / Published online: 1 October 2009
© Springer Science+Business Media, LLC 2009

Abstract A survey of Ramanujan's work on cranks in his lost notebook is given.
We give evidence that Ramanujan was concentrating on cranks when he died, that is
to say, the final problem on which Ramanujan worked was *cranks—not mock theta
functions.*

Keywords Crank · Partitions · Theta functions · Ramanujan's lost notebook · Rank

Mathematics Subject Classification (2000) Primary: 11P82 · Secondary: 11P83

B.C. Berndt's research is partially supported by grant H98230-07-1-0088. from the National Security
Agency.
H.H. Chan's research is partially supported by National University of Singapore Academic Research
Fund, Project Number R146000027112.
W.-C. Liaw's research is partially supported by grant NSC95-2115-M-194-012 from the National
Science Council of the Republic of China.

B.C. Berndt (✉)
Department of Mathematics, University of Illinois, 1409 West Green Street, Urbana, IL 61801, USA
e-mail: berndt@illinois.edu

H.H. Chan
Department of Mathematics, National University of Singapore, 2 Science Drive 2,
Singapore 117543, Republic of Singapore
e-mail: matchh@nus.edu.sg

S.H. Chan
Division of Mathematical Sciences, School of Physical and Mathematical Sciences, Nanyang
Technological University, 21 Nanyang Link, Singapore 637371, Republic of Singapore
e-mail: Chansh@ntu.edu.sg

W.-C. Liaw
Department of Mathematics, National Chung Cheng University, Min-Hsiung, Chia-Yi 62145,
Taiwan, Republic of China
e-mail: wcliaw@math.ccu.edu.tw

1 Introduction

Recall that in his last letter to G.H. Hardy [17, pp. xxxi–xxxii, 6, pp. 220–223], Ramanujan announced a new class of functions, which he called mock theta functions, and gave several examples and theorems in illustration. This letter was written on January 20, 1920, only slightly more than three months before his death on April 26, 1920. When G.N. Watson wrote his last two papers [20, 21] on Ramanujan's work in 1936 and 1937, respectively, it was natural to entitle one of them [20, 7, pp. 325–347], *The final problem: an account of the mock theta functions*, which was also the title of his retiring Presidential address to the London Mathematical Society. In conversations with one of us (BCB) in December, 1983, Ramanujan's widow, Janaki, told us that her husband worked on mathematics, or in her words, that he was "doing his sums," until four days before he died, when his pain became too great and he could no longer concentrate on mathematics. The wide ranging content of the lost notebook [19] gives evidence that Ramanujan probably derived theorems in several topics during the last three months of his life. Of course, we can only speculate on Ramanujan's focus as his life withered away, but one of the goals in this short survey is to provide evidence that Ramanujan devoted his very last days to cranks (not to mock theta functions as suggested by Watson), although, of course, Ramanujan would not have used this terminology.

First, we briefly relate the origin of cranks *after* Ramanujan had studied them in his lost notebook. Secondly, we describe the extensive material in the lost notebook pertaining to cranks. Thirdly, we draw some conclusions from this material and ask several questions, because we do not know the motivation for most of Ramanujan's calculations on cranks. Thus, one purpose in writing this paper is to elicit reader insights into Ramanujan's motivations.

We offer no proofs in this paper, but we indicate where readers can find proofs of all the theorems discussed here.

2 Dyson, ranks, and cranks

Recall Ramanujan's celebrated congruences for the partition function $p(n)$, namely,

$$p(5n + 4) \equiv 0 \pmod{5}, \tag{2.1}$$

$$p(7n + 5) \equiv 0 \pmod{7}, \tag{2.2}$$

$$p(11n + 6) \equiv 0 \pmod{11}. \tag{2.3}$$

In attempting to find combinatorial interpretations for (2.1)–(2.3), in 1944, F.J. Dyson [9] defined the *rank of a partition*.

Definition 2.1 The rank of a partition is equal to the largest part minus the number of parts.

Let $N(m, n)$ denote the number of partitions of n with rank m, and let $N(m, t, n)$ denote the number of partitions of n with rank congruent to m modulo t. Then Dyson conjectured that

$$N(k, 5, 5n + 4) = \frac{p(5n + 4)}{5}, \quad 0 \le k \le 4, \tag{2.4}$$

and

$$N(k, 7, 7n + 5) = \frac{p(7n + 5)}{7}, \quad 0 \le k \le 6. \tag{2.5}$$

Thus, if (2.4) and (2.5) were true, the partitions counted by $p(5n + 4)$ and $p(7n + 5)$ fall into five and seven equinumerous classes, respectively, thus providing a partial answer to Dyson's query. Dyson also conjectured that the generating function for $N(m, n)$ is given by

$$\sum_{m=-\infty}^{\infty} \sum_{n=0}^{\infty} N(m, n) a^m q^n = \sum_{n=0}^{\infty} \frac{q^{n^2}}{(aq; q)_n (q/a; q)_n}, \quad |q| < 1, \ |q| < |a| < 1/|q|. \tag{2.6}$$

These conjectures were first proved by A.O.L. Atkin and H.P.F. Swinnerton-Dyer [3] in 1954. Although Ramanujan did not record any statement equivalent to Definition 2.1, he recorded theorems on its generating function (2.6) in his lost notebook; in particular, see [19, p. 20].

The corresponding analogue to (2.4) and (2.5) does not hold for $p(11n + 6) \equiv 0$ (mod 11), and so Dyson conjectured the existence of a *crank* that would combinatorially explain this congruence. In his doctoral dissertation [10], F.G. Garvan defined vector partitions which became the forerunners of the crank. The *true crank* was discovered by Andrews and Garvan [1] during the afternoon of June 6, 1987 at Illinois Street Residence Hall, a student dormitory at the University of Illinois, following a meeting to commemorate the centenary of Ramanujan's birth.

Definition 2.2 For a partition π, let $\lambda(n)$ denote the largest part of π, let $\mu(\pi)$ denote the number of ones in π, and let $\nu(\pi)$ denote the number of parts of π larger than $\mu(\pi)$. The crank $c(\pi)$ is then defined to be [1]

$$c(\pi) = \begin{cases} \lambda(\pi), & \text{if } \mu(\pi) = 0, \\ \nu(\pi) - \mu(\pi), & \text{if } \mu(\pi) > 0. \end{cases}$$

For $n > 1$, let $M(m, n)$ denote the number of partitions of n with crank m, while for $n \le 1$ we set

$$M(0, 0) = 1 \quad \text{and} \quad M(m, 0) = 0, \quad \text{otherwise,}$$

$$M(0, 1) = -1, \quad M(1, 1) = M(-1, 1) = 1, \quad \text{and} \quad M(m, 1) = 0, \quad \text{otherwise.}$$

The generating function for $M(m, n)$ is given by

$$\sum_{m=-\infty}^{\infty} \sum_{n=0}^{\infty} M(m, n) a^m q^n = \frac{(q; q)_\infty}{(aq; q)_\infty (q/a; q)_\infty}. \tag{2.7}$$

The crank not only leads to a combinatorial interpretation of $p(11n + 6) \equiv 0$ (mod 11), as desired by Dyson, but also to similar interpretations for $p(5n + 4) \equiv 0$ (mod 5) and $p(7n + 5) \equiv 0$ (mod 7).

Let $M(m, t, n)$ denote the number of partitions of n with crank congruent to m modulo t.

Theorem 2.3 *With $M(m, t, n)$ defined above,*

$$M(k, 5, 5n + 4) = \frac{p(5n + 4)}{5}, \quad 0 \leq k \leq 4,$$

$$M(k, 7, 7n + 5) = \frac{p(7n + 5)}{7}, \quad 0 \leq k \leq 6,$$

$$M(k, 11, 11n + 6) = \frac{p(11n + 6)}{11}, \quad 0 \leq k \leq 10.$$

An excellent introduction to cranks can be found in Garvan's survey paper [11].

Remarkably, the generating function (2.7) for cranks can be found in Ramanujan's lost notebook [19, p. 179] in the form

$$F(q) := F_a(q) := \frac{(q; q)_\infty}{(aq; q)_\infty (q/a; q)_\infty} =: \sum_{n=0}^{\infty} \lambda_n q^n, \quad (2.8)$$

and so from (2.7),

$$\lambda_n = \sum_{m=-\infty}^{\infty} M(m, n) a^m.$$

Note that when $a = 1$, (2.7) and (2.8) reduce to the generating function for $p(n)$.

3 Dissections of the crank

Definition 3.1 If

$$P(q) := \sum_{n=0}^{\infty} a_n q^n$$

is any power series, then the m-dissection of $P(q)$ is given by

$$P(q) = \sum_{j=0}^{m-1} \sum_{n_j=0}^{\infty} a_{n_j m + j} q^{n_j m + j}.$$

We actually use below an extension of this definition to congruences in the ring of power series in the two variables a and q.

In his lost notebook [19], Ramanujan offers, in various guises, m-dissections for $F_a(q)$ for $m = 2, 3, 5, 7, 11$. In particular, on page 179 Ramanujan offers 2- and 3-dissections for $F_a(q)$ in the form of congruences. To give these dissections, we first need to define Ramanujan's theta function $f(a, b)$ by

$$f(a,b) := \sum_{n=-\infty}^{\infty} a^{n(n+1)/2} b^{n(n-1)/2}, \quad |ab| < 1. \tag{3.1}$$

Furthermore, Ramanujan sets

$$f(-q) := f(-q, -q^2) = (q; q)_\infty, \tag{3.2}$$

where the latter equality follows from the pentagonal number theorem. The two congruences are then given by

$$F(\sqrt{q}) \equiv \frac{f(-q^3, -q^5)}{(-q^2; q^2)_\infty} + \left(a - 1 + \frac{1}{a}\right) \sqrt{q} \frac{f(-q, -q^7)}{(-q^2; q^2)_\infty} \pmod{(a^2 + a^{-2})} \tag{3.3}$$

and

$$F(q^{1/3}) \equiv \frac{f(-q^2, -q^7) f(-q^4, -q^5)}{(q^9; q^9)_\infty}$$

$$+ \left(a - 1 + \frac{1}{a}\right) q^{1/3} \frac{f(-q, -q^8) f(-q^4, -q^5)}{(q^9; q^9)_\infty}$$

$$+ \left(a^2 + \frac{1}{a^2}\right) q^{2/3} \frac{f(-q, -q^8) f(-q^2, -q^7)}{(q^9; q^9)_\infty} \pmod{(a^3 + 1 + a^{-3})}. \tag{3.4}$$

Note that $\lambda_2 = a^2 + a^{-2}$, which trivially implies that $a^4 \equiv -1 \pmod{\lambda_2}$ and $a^8 \equiv 1 \pmod{\lambda_2}$. Thus, in (3.3), a behaves like a primitive 8th root of unity modulo λ_2. On the other hand, $\lambda_3 = a^3 + 1 + a^{-3}$, from which it follows that $a^9 \equiv -a^6 - a^3 \equiv 1 \pmod{\lambda_3}$. So in (3.4), a behaves like a primitive 9th root of unity modulo λ_3.

In contrast to (3.3) and (3.4), on page 20 in his lost notebook, Ramanujan gives a 5-dissection for $F_a(q)$ in terms of an *equality* instead of a congruence. For $a = e^{2\pi i n/5}$, with $n = 1$ or 2,

$$F_a(q^{1/5}) = \frac{f(-q^2, -q^3)}{f^2(-q, -q^4)} f^2(-q^5) - 4\cos^2(2n\pi/5) q^{1/5} \frac{f^2(-q^5)}{f(-q, -q^4)}$$

$$+ 2\cos(4n\pi/5) q^{2/5} \frac{f^2(-q^5)}{f(-q^2, -q^3)}$$

$$- 2\cos(2n\pi/5) q^{3/5} \frac{f(-q, -q^4)}{f^2(-q^2, -q^3)} f^2(-q^5). \tag{3.5}$$

Observe that (3.5) contains no terms involving $q^{4/5}$, which is a reflection of the fact that $p(5n+4) \equiv 0 \pmod 5$. In fact [4, p. 105], one can replace (3.5) by an equivalent 5-dissection that is a congruence, and conversely (3.3) and (3.4) can be replaced by equalities involving 8th and 9th roots of unity, respectively. Indeed, one can always convert a dissection from one involving roots of unity to one involving a congruence, and conversely [4, pp. 118–119]. One advantage of formulations in terms of congruences is that one can let $a = 1$ in these congruences and obtain congruences originally found by Atkin and Swinnerton-Dyer [3].

Ramanujan also derived the 7- and 11-dissections for $F_a(q)$, but he does not explicitly state them. Instead, on pages 71 and 70, respectively, in his lost notebook he vaguely provides the quotients of theta functions that appear in these representations. These dissections immediately yield the congruences $p(7n + 5) \equiv 0 \pmod{7}$ and $p(11n + 6) \equiv 0 \pmod{11}$, respectively. On pages 186 and 187 in the lost notebook [19], Ramanujan records modulo 11 the coefficients (in some cases up to 70) of the quotients of theta functions appearing in the 11-dissection of the crank generating function. Furthermore, on page 19 in his lost notebook, Ramanujan apparently intended to write down the 7-dissection of $F_a(q)$ when $a = e^{2\pi i n/7}$, but all he records is "$F_a(q)$." Did Ramanujan not record the 7-dissection for $F_a(q)$ because he was stymied by his illness? Or was the 7-dissection recounted on a *lost page* that is not contained in the extant lost notebook? If so, then possibly his seventh order mock theta functions, which are not found in the lost notebook, were also described on this possible missing page.

In versions involving roots of unity, Garvan first proved all five of Ramanujan's m-dissections [12, 13]. The present authors [4] devised uniform proofs of Ramanujan's dissections in the language of congruences in two distinct ways. One of the methods arises from obscure statements found on page 59 of the lost notebook. In the middle of that page are three (somewhat difficult to read) mathematical expressions, one of which is equal to $(q; q)_\infty$ times the generating function for the crank. Although not claimed by Ramanujan, the three expressions are equal, and it is one of these identities, namely,

$$\frac{(q; q)_\infty^2}{(aq; q)_\infty (q/a; q)_\infty} = \sum_{k=-\infty}^{\infty} (-1)^k \frac{(1 - a)q^{k(k+1)/2}}{1 - aq^k}, \tag{3.6}$$

that leads to one set of the aforementioned uniform proofs of Ramanujan's dissections [4]. See [5, pp. 94–96] for proofs and a history of earlier proofs.

Further references to other proofs of Ramanujan's five dissections for $F_a(q)$ and to proofs of certain other dissections of $F_a(q)$ can be found in [4]. The largest value of m for which a dissection of $F_a(q)$ has been given is $m = 11$.

On pages 179 and 180 in [19], Ramanujan recorded ten tables of values of n for which λ_n satisfies certain congruences. A rough preliminary version of this table can be found on page 61 in his lost notebook [19]. Although Ramanujan makes no claims about these tables, it is clear that he regarded them as complete. We provide below a list of the congruences and the number of values of n for which λ_n satisfies the given congruence. Following Ramanujan, we use the notation

$$a_n = a^n + a^{-n}. \tag{3.7}$$

For example, the 27 values of n with a_2 as a factor of λ_n are

2, 8, 9, 10, 11, 15, 19, 21, 22, 25, 26, 27, 28, 30, 31, 34, 40, 42, 45, 46, 47, 50, 55, 57, 58, 59, 62, 66, 70, 74, 75, 78, 79, 86, 94, 98, 106, 110, 122, 126, 130, 142, 154, 158, 170, 174, 206.

Table 1 Congruences for λ_n

$\lambda_n \equiv 0 \pmod{a_2}$	47	$\lambda_n \equiv 0 \pmod{a_1}$	3
$\lambda_n \equiv 1 \pmod{a_2}$	27	$\lambda_n \equiv 0 \pmod{a_1 - 1}$	19
$\lambda_n \equiv -1 \pmod{a_2}$	27	$\lambda_n \equiv 1 \pmod{a_1 - 1}$	26
$\lambda_n \equiv a_1 - 1 \pmod{a_2}$	22	$\lambda_n \equiv -1 \pmod{a_1 - 1}$	26
$\lambda_n \equiv -(a_1 - 1) \pmod{a_2}$	23	$\lambda_n \equiv 0 \pmod{a_1 + 1}$	2

For complete lists of all tables, see the authors' paper [5, pp. 85–88]. Remarkably, all ten tables are complete, except that in the third table Ramanujan missed one of the 27 values. For the first five tables, the appropriate values of n can be determined from examining the power series for the quotients of theta functions appearing in the 2-dissection (3.3) [5, pp. 85–86]. For example, consider the third congruence $\lambda_n \equiv -1 \pmod{a^2 + a^{-2}}$ listed in Table 1. To determine the values of n satisfying this congruence, from (3.3) with q replaced by q^2, we must determine the values of n for which the coefficient of q^n in

$$\frac{f(-q^6, -q^{10})}{(-q^4; q^4)_\infty} \tag{3.8}$$

is equal to -1 and simultaneously the coefficient of q^n in

$$q \frac{f(-q^2, -q^{14})}{(-q^4; q^4)_\infty} \tag{3.9}$$

is equal to 0. There are thus 27 such values.

For the remaining five congruences listed in Table 1, the appropriate values of n in Ramanujan's last five tables can be determined from congruences for the generating function (2.8) [5, pp. 86–88]. Thus, for the sixth congruence in Table 1, from the calculation

$$\frac{(q; q)_\infty}{(aq; q)_\infty (q/a; q)_\infty} \equiv \frac{(q; q)_\infty}{(-q^2; q^2)_\infty} = \frac{f(-q) f(-q^2)}{f(-q^4)} \pmod{a_1}, \tag{3.10}$$

where $f(-q)$ is defined by (3.2), we see that in his sixth table Ramanujan recorded the degree of q for the terms with zero coefficients in the power series expansion of $f(-q) f(-q^2)/f(-q^4)$ [5, p. 87].

Similarly [5, pp. 87–88],

$$\frac{(q; q)_\infty}{(aq; q)_\infty (q/a; q)_\infty} \equiv \frac{(q^2; q^2)_\infty}{(-q^3; q^3)_\infty} = \frac{f(-q^2) f(-q^3)}{f(-q^6)} \pmod{a_1 - 1} \tag{3.11}$$

and

$$\frac{(q; q)_\infty}{(aq; q)_\infty (q/a; q)_\infty} \equiv \frac{(q; q)_\infty^2}{(q^3; q^3)_\infty} = \frac{f^2(-q)}{f(-q^3)} \pmod{a_1 + 1}. \tag{3.12}$$

Thus, in his Tables 7–9, Ramanujan recorded the degree of q for the terms with coefficients 0, 1, and -1 respectively, in the theta quotient on the right side of (3.11), and in Table 10, he recorded the degree of q for the terms with zero coefficients in the theta quotient on the right-hand side of (3.12). In fact, the only values in this last table are $n = 14, 17$.

In order to demonstrate that Ramanujan's ten tables are complete, we need to examine the coefficients in the power series expansions of the quotients of theta functions in (3.8)–(3.12). In [5], using computer calculations, we conjectured that the coefficients in certain dissections of these products are monotonic for all n larger than certain specific values. If we could demonstrate that these conjectures are valid and then determine the values of λ_n below these certain thresholds, then we would have the means to show that Ramanujan's tables are complete. Using the Hardy–Ramanujan circle method, O.-Y. Chan [8] proved that the conjectured monotonicities did indeed hold. Upon checking the values of λ_n below the thresholds of monotonicity, Chan then completed the proof that all ten tables of values of n are complete (with the exception of one missing value noted earlier). It should be remarked that in the course of proving a conjecture of Andrews and R. Lewis [2] on ranks, D. Kane [14] also thereby proved the completeness of Ramanujan's Table 10.

The largest value of n found in Ramanujan's tables is $n = 302$ appearing in Table 3. Unless Ramanujan determined the values in his tables in a manner entirely different from how we have proceeded above, it seems that Ramanujan would have needed to calculate by hand the coefficients of five quotients of theta functions up to several hundred terms each. Even for Ramanujan, this appears to be a prodigious task, especially when he was dying.

Another question arises. Why did Ramanujan devote considerable energy in his last year to determining the values of λ_n satisfying these particular congruences in Table 1? Clearly, there was little hope of finding an infinite sequence of values of λ_n satisfying one of these congruences, like he discovered for $p(5n + 4)$, for example. Why were these particular congruences chosen? Was his goal to discover something about $p(n)$?

We now present evidence that, as surmised in the penultimate paragraph, Ramanujan did calculate the coefficients in quotients of certain power series. First define

$$L_{p,r}(q) := \frac{(q^p; q^p)_\infty}{(q^r; q^p)_\infty (q^{p-r}; q^p)_\infty} = (1 - q^r)^{-1} F_{q^r}(q^p).$$

The functions $L_{11,r}(q)$, $1 \le r \le 5$, appear in the 11-dissection of $F_a(q)$ when $a = e^{2\pi i n/11}$. On pages 63 and 64 in his lost notebook [19], Ramanujan calculates the power series of $L_{11,1}(q)$ and $L_{11,2}(q)$, respectively, the first up to q^{100} and the second up to q^{33}. All of the calculated coefficients in these series are nonnegative, and most (but not all) are increasing. Liaw [15, 16], possibly in concordance with Ramanujan's observations illustrated by these two examples, proved the following theorem.

Theorem 3.2 *Let p and r be positive integers with $p \ge 2$ and $r < p$. Write*

$$L_{p,r}(q) = \sum_{n=0}^{\infty} b_{p,r}(n)q^n. \tag{3.13}$$

Then $b_{p,r}(n) \geq 0$ for all n. Moreover, we let

$$L_{p,r}(q) + q^p := \sum_{n=0}^{\infty} c_{p,r}(n)q^n := \Sigma_0 + \Sigma_1 + \cdots + \Sigma_{r-1}, \qquad (3.14)$$

where the exponents in Σ_j are congruent to j modulo r, $0 \leq i \leq r - 1$, i.e.,

$$\Sigma_i = \sum_{n=0}^{\infty} c_{p,r}(nr + i)q^{nr+i}.$$

Then for each i the coefficient sequence $\{c_{p,r}(nr + i)\}_{n=0}^{\infty}$ is non-decreasing.

4 Factoring the crank

On page 58 in his lost notebook [19], Ramanujan wrote out completely the first 22 terms of the power series representation (2.8) for the crank. Recalling that a_n is defined in (3.7), we provide only a few of the terms

$$1 + q(a_1 - 1) + q^2 a_2 + q^3(a_3 + 1) + q^4(a_4 + a_2 + 1) + q^5(a_5 + a_3 + a_1 + 1) + \cdots$$

$$+ q^{20}(a_2 - a_1 + 1)(a_3 + 1)(a_5 + a_4 + a_3 + a_2 + a_1 + 1)$$

$$\times (a_{10} + a_6 + a_4 + a_3 + 2a_2 + 2a_1 + 3)$$

$$+ q^{21}a_1 a_2(a_3 + 1)(a_2 - a_1 + 1)(a_5 + a_4 + a_3 + a_2 + a_1 + 1)$$

$$\times (a_8 - a_6 + a_4 + a_1 + 2) + \cdots .$$

It should be noted that Ramanujan wrote the coefficients λ_n in factored form. On p. 59, Ramanujan wrote out some (but not all) of the factors of the coefficients of q^n, $13 \leq n \leq 26$. He gave no hint of his reasons for either recording the coefficients in detail, or for factoring them, or for examining only certain factors. It appears that Ramanujan was attempting to find some patterns in the coefficients of the crank generating function. Perhaps, on setting $a = 1$ he had then hoped to draw conclusions about the factors of $p(n)$.

On p. 181 in his lost notebook [19], Ramanujan returns to the coefficients λ_n in the generating function (2.8) of the crank. He factors λ_n, $1 \leq n \leq 21$, as before, but singles out nine particular factors by giving them special notation. The criterion that Ramanujan apparently uses is that of multiple occurrence, i.e., each of these nine factors appears more than once in the 21 factorizations, while other factors not favorably designated appear only once. The factors designated by Ramanujan are

$\rho_1 = a_1 - 1$	$\rho_5 = a_4 + a_2 + 1$
$\rho = a_2 - a_1 + 1$	$\rho_7 = a_3 + a_2 + a_1 + 1$
$\rho_2 = a_2$	$\rho_9 = (a_2 + 1)(a_3 + 1)$
$\rho_3 = a_3 + 1,$	$\rho_{11} = a_5 + a_4 + a_3 + a_2 + a_1 + 1$
$\rho_4 = a_1 a_2$	

At first glance, there does not appear to be any reasoning behind the choice of subscripts; note that there is no subscript for the second value. However, observe that in each case if we set $a = 1$, then the subscript n is equal to the right-hand side. The reason ρ does not have a subscript is that the value of n in this case would be $3 - 2 = 1$, which has been reserved for the first factor. In the table below, we record the content of page 181.

$$p(1) = 1, \qquad \lambda_1 = \rho_1,$$

$$p(2) = 2, \qquad \lambda_2 = \rho_2,$$

$$p(3) = 3, \qquad \lambda_3 = \rho_3,$$

$$p(4) = 5, \qquad \lambda_4 = \rho_5,$$

$$p(5) = 7, \qquad \lambda_5 = \rho_7 \rho,$$

$$p(6) = 11, \qquad \lambda_6 = \rho_1 \rho_{11},$$

$$p(7) = 15, \qquad \lambda_7 = \rho_3 \rho_5,$$

$$p(8) = 22, \qquad \lambda_8 = \rho_1 \rho_2 \rho_{11},$$

$$p(9) = 30, \qquad \lambda_9 = \rho_2 \rho_3 \rho_5,$$

$$p(10) = 42, \qquad \lambda_{10} = \rho \rho_2 \rho_3 \rho_7,$$

$$p(11) = 56, \qquad \lambda_{11} = \rho_4 \rho_7 (a_5 - a_4 + a_2),$$

$$p(12) = 77, \qquad \lambda_{12} = \rho_7 \rho_{11} (a_4 - 2a_3 + 2a_2 - a_1 + 1),$$

$$p(13) = 101, \qquad \lambda_{13} = \rho \rho_1 \, (a_{10} + 2a_9 + 2a_8 + 2a_7 + 3a_6$$
$$+ 4a_5 + 6a_4 + 8a_3 + 9a_2 + 9a_1 + 9),$$

$$p(14) = 135, \qquad \lambda_{14} = \rho_5 \rho_9 (a_5 - a_3 + a_1 + 1),$$

$$p(15) = 176, \qquad \lambda_{15} = \rho_4 \rho_{11} (a_7 - a_6 + a_4 + a_1),$$

$$p(16) = 231, \qquad \lambda_{16} = \rho_3 \rho_7 \rho_{11} (a_5 - 2a_4 + 2a_3 - 2a_2 + 3a_1 - 3),$$

$$p(17) = 297, \qquad \lambda_{17} = \rho_9 \rho_{11} (a_7 - a_6 + a_3 + a_1 - 1),$$

$$p(18) = 385, \qquad \lambda_{18} = \rho_5 \rho_7 \rho_{11} (a_6 - 2a_5 + a_4 + a_3 - a_2 + 1),$$

$$p(19) = 490, \qquad \lambda_{19} = \rho_1 \rho_2 \rho_5 \rho_7 (a_9 - a_7 + a_4 + 2a_3 + a_2 - 1),$$

$$p(20) = 627, \qquad \lambda_{20} = \rho \rho_3 \rho_{11} (a_{10} + a_6 + a_4 + a_3 + 2a_2 + 2a_1 + 3),$$

$$p(21) = 792, \qquad \lambda_{21} = \rho \rho_3 \rho_4 \rho_{11} (a_8 - a_6 + a_4 + a_1 + 2).$$

These factors lead to the rapid calculation of values for $p(n)$. For example, since $\lambda_{10} = \rho \rho_2 \rho_3 \rho_7$, then $p(10) = 1 \cdot 2 \cdot 3 \cdot 7 = 42$.

Ramanujan evidently was searching for some general principles or theorems on the factorization of λ_n so that he could not only compute $p(n)$ but make deductions about the divisibility of $p(n)$. No theorems are stated by Ramanujan. Is it possible to determine that certain factors appear in some precisely described infinite family of values of λ_n? What are the motivations that led Ramanujan to make these factorizations?

5 The evidence: summary

In the foregoing discussions of the crank, pp. 20, 58–59, 61, 63–64, 70–71, and 179–181 in the lost notebook were cited. Ramanujan's 5-dissection for the crank is given on p. 20. *All* of the material on the remaining ten pages is devoted to the crank. In fact, all of pp. 58–89 contain either incipient material or scratch work. Needless to say, it is very difficult to determine what Ramanujan was contemplating in most of his scratch work, but it is likely that all these pages are related to cranks. Pages 65 (which is the same as p. 73), 66, 72, 77, 80–81, and 83–85 are almost certainly related to crank calculations, while we are unable to determine conclusively if pp. 60, 62, 67–69, 74–76, 78–79, 82, and 86–89 pertain to cranks.

6 Ramanujan's final days

In December, 1983, one of the present authors (BCB) asked Janaki about the extent of the papers left behind by her late husband. In particular, Berndt mentioned to her that Ramanujan's lost notebook contains 138 pages. She claimed that Ramanujan had many more than 138 pages in his possession at his death. She related that as her husband "did his sums," he would deposit his papers in a large leather trunk beneath his bed. She told Berndt that during her husband's funeral, certain people, whom she named but whom we do not name here, came to her home and stole most of her husband's papers and never returned them. She later donated the un-stolen papers of Ramanujan to the University of Madras. These papers certainly contain or possibly entirely comprise the lost notebook.

It is our contention that Ramanujan kept at least two stacks of papers while doing mathematics in his last year. In one stack, he put primarily the pages he wanted to save, i.e., those containing the statements of his theorems, and in another stack or stacks he put papers containing his calculations and proofs. The one stack of papers containing the lost notebook was likely in a different place and missed by those taking his other papers. Of course, it is certainly possible that more than one stack of papers contained statements of results that Ramanujan wanted to save. Undoubtedly, Ramanujan produced scores of pages containing calculations, scratch work, and proofs, but, remarkably, the extant scratch work belongs almost exclusively to one topic, ***cranks***. Our guess is that when Ramanujan ceased research four days prior to his death, he was thinking about cranks. His power series expansions, factorizations, preliminary tables, and scratch work were part of his deliberations and had not yet been put in a secondary pile of papers. Thus, these sheets were found with the papers that were set aside containing his final theorems and thus unofficially became part of his lost notebook. In particular, pp. 58–89 likely include some pages that Ramanujan intended to keep in his principal stack, but most of this work probably would have been relegated to a secondary stack if Ramanujan had lived longer.

7 Did Ramanujan think combinatorially?

Since Definitions 2.1 and 2.2 of rank and crank, respectively, are important combinatorial concepts, it is natural to ask if Ramanujan knew the combinatorial interpretations of the generating functions (2.6) for the rank and (2.8) for the crank. His
notebooks [18] have little discourse, and his lost notebook [19] contains almost no
words at all. Since Ramanujan had so many ideas and so little time in which to record
his findings, doubtless, he was not going to use the small amount of valuable time
that remained in the last year of his life to record definitions or combinatorial interpretations. To him, such interpretations might have been obvious with no need to
record them, and especially since both the "ordinary" notebooks and lost notebook
were never intended for publication but were simply his own personal compilation of
what he had discovered, the need for Ramanujan to record such interpretations was
probably close to nil. Some of his published papers are testimony that Ramanujan
was an excellent combinatorial thinker. Thus, in conclusion, one can only speculate
about how much or how little Ramanujan thought combinatorially, especially about
ranks and cranks, but Ramanujan probably discovered more combinatorially than he
has been given credit for.

Acknowledgements   The authors gratefully acknowledge perspicacious observations about dissections
supplied by Frank Garvan.

References

1. Andrews, G.E., Garvan, F.G.: Dyson's crank of a partition. Bull. Am. Math. Soc. (N.S.) **18**, 167–171
    (1988)
2. Andrews, G.E., Lewis, R.: The ranks and cranks of partitions moduli 2, 3, and 4. J. Number Theory
    85, 74–84 (2000)
3. Atkin, A.O.L., Swinnerton-Dyer, H.P.F.: Some properties of partitions. Proc. Lond. Math. Soc. (3) **4**,
    84–106 (1954)
4. Berndt, B.C., Chan, H.H., Chan, S.H., Liaw, W.-C.: Cranks and dissections in Ramanujan's lost note-
    book. J. Comb. Theory (A) **109**, 91–120 (2005)
5. Berndt, B.C., Chan, H.H., Chan, S.H., Liaw, W.-C.: Ramanujan and cranks. In: Ismail, M.E.H.,
    Koelink, E. (eds.) Theory and Applications of Special Functions. A Volume Dedicated to Mizan Rah-
    man, pp. 77–98. Springer, Dordrecht (2005)
6. Berndt, B.C., Rankin, R.A.: Ramanujan: Letters and Commentary. American Mathematical Society,
    Providence (1995). London Mathematical Society, London (1995)
7. Berndt, B.C., Rankin, R.A.: Ramanujan: Essays and Surveys. American Mathematical Society, Prov-
    idence (2001). London Mathematical Society, London (2001)
8. Chan, O.-Y.: Some asymptotics for cranks. Acta Arith. **120**, 107–143 (2005)
9. Dyson, F.J.: Some guesses in the theory of partitions. Eureka (Cambridge) **8**, 10–15 (1944)
10. Garvan, F.G.: Generalizations of Dyson's rank. Ph. D. Thesis, Pennsylvania State University, Univer-
    sity Park, PA (1986)
11. Garvan, F.G.: Combinatorial interpretations of Ramanujan's partition congruences. In: Andrews, G.E.,
    Askey, R.A., Berndt, B.C., Ramanathan, K.G., Rankin, R.A. (eds.) Ramanujan Revisited, pp. 29–45.
    Academic Press, Boston (1988)
12. Garvan, F.G.: New combinatorial interpretations of Ramanujan's partition congruences mod 5, 7 and
    11. Trans. Am. Math. Soc. **305**, 47–77 (1988)
13. Garvan, F.G.: The crank of partitions mod 8, 9 and 10. Trans. Am. Math. Soc. **322**, 79–94 (1990)
14. Kane, D.M.: Resolution of a conjecture of Andrews and Lewis involving cranks of partitions. Proc.
    Am. Math. Soc. **132**, 2247–2256 (2004)

15. Liaw, W.-C.: Contributions to Ramanujan's theories of modular equations, Ramanujan-type series for $1/\pi$, and partitions. Ph. D. Thesis, University of Illinois at Urbana-Champaign, Urbana (1999)
16. Liaw, W.-C.: Note on the monotonicity of coefficients for some q-series arising from Ramanujan's lost notebook. Ramanujan J. **3**, 385–388 (1999)
17. Ramanujan, S.: Collected Papers. Cambridge University Press, Cambridge (1927). reprinted by Chelsea, New York (1962); reprinted by the American Mathematical Society, Providence (2000)
18. Ramanujan, S.: Notebooks. Tata Institute of Fundamental Research, Bombay (1957) (2 volumes)
19. Ramanujan, S.: The Lost Notebook and Other Unpublished Papers. Narosa, New Delhi (1988)
20. Watson, G.N.: The final problem: an account of the mock theta functions. J. Lond. Math. Soc. **11**, 55–80 (1936)
21. Watson, G.N.: The mock theta functions (2). Proc. Lond. Math. Soc. **42**, 274–304 (1937)

Ramanujan J (2010) 23: 17–44
DOI 10.1007/s11139-008-9155-8

Eisenstein series and Ramanujan-type series for $1/\pi$

Nayandeep Deka Baruah · Bruce C. Berndt

Dedicated to George Andrews on the occasion of his 70-th birthday

Received: 27 January 2007 / Accepted: 19 December 2008 / Published online: 5 November 2009
© Springer Science+Business Media, LLC 2009

Abstract Using certain representations for Eisenstein series, we uniformly derive several Ramanujan-type series for $1/\pi$.

Keywords Eisenstein series · Theta-functions · Modular equations

Mathematics Subject Classification (2000) 33C05 · 33E05 · 11F11 · 11R29

1 Introduction

In his famous paper [36], [37, pp. 36–38], Ramanujan recorded 17 hypergeometric-like series representations for $1/\pi$. Proofs of the first three series representations were briefly sketched by Ramanujan [37, p. 36]. These three series belong to the classical theory of elliptic functions, while the latter fourteen series depend on Ramanujan's alternative theories of elliptic functions. The mathematical community seems to have forgotten that in 1928 S. Chowla [24, 25], [26, pp. 87–91, 116–119] gave the first published proof of a general series representation for $1/\pi$ and used it to derive the first of Ramanujan's series for $1/\pi$ [36, Eq. (28)]. It was not until 1987 that proofs of all 17 formulas were found by J.M. and P.B. Borwein [13]. These authors subsequently discovered many further series for $1/\pi$ [14–18], where [18] is coauthored

N.D. Baruah research partially supported by BOYSCAST Fellowship grant SR/BY/M-03/05 from DST, Govt. of India.
B.C. Berndt research partially supported by NSA grant MSPF-03IG-124.

N.D. Baruah
Department of Mathematical Sciences, Tezpur University, Napaam-784028, Sonitpur, Assam, India
e-mail: nayan@tezu.ernet.in

B.C. Berndt (✉)
Department of Mathematics, University of Illinois, 1409 West Green St., Urbana, IL 61801, USA
e-mail: berndt@illinois.edu

with D.H. Bailey. D.V. Chudnovsky and G.V. Chudnovsky [27] independently proved several of Ramanujan's series representations for $1/\pi$ and established new ones as well. Further particular series representations for $1/\pi$ as well as some general formulas have subsequently been derived by Berndt and H.H. Chan [9], Berndt, Chan, and W.-C. Liaw [12], H.H. Chan, S.H. Chan, and Z. Liu [23], H.H. Chan and Liaw [19], H.H. Chan and K.P. Loo [20], H.H. Chan, Liaw, and V. Tan [22], and H.H. Chan and H. Verrill [21]. J. Guillera [28–32] discovered some beautiful series for $1/\pi$ as well as for $1/\pi^2$. Further work has been accomplished by W. Zudilin [40–42].

The purpose of this paper is to return to Ramanujan's ideas expressed in Sect. 13 of his fundamental paper [36], [37, p. 36] and use them in conjunction with twelve identities for Eisenstein series recorded without proofs in Sect. 10 of [36], [37, pp. 33–34] and with further identities of this type to reprove 13 of Ramanujan's 17 identities from [36] as well as to establish many new series representations for $1/\pi$. In particular, we rely on Ramanujan's initial ideas more so than previous authors. For example, the Borweins employ Legendre's relation between elliptic integrals; in our derivations, we do not need knowledge of elliptic integrals. However, in contrast to Ramanujan's proposed derivations, "from these [alternative] theories we can deduce further series for $1/\pi$," *we do not appeal to Ramanujan's alternative theories in this paper.* In another paper [3], we employ Ramanujan's ideas once again, but now with his alternative theories, to derive several new series representations for $1/\pi$.

The formulas from Sect. 10 to which we alluded above are also found in Ramanujan's second notebook [38] and were first proved by the second author in [6, Chap. 21]. Berndt's proofs of some of the formulas follow a hint given by Ramanujan at the beginning of Chap. 21, but unfortunately Berndt was not able to use Ramanujan's idea, or any idea with which Ramanujan might have been familiar, to prove most of the identities. Thus, Berndt resorted to the theory of modular forms to prove most of Ramanujan's formulas. It would be of enormous interest to discover how Ramanujan might have proved all twelve identities.

The authors are extremely indebted to Heng Huat Chan who provided this direction for us. At the request of the second author, Chan provided notes more fully developing Ramanujan's ideas in Sect. 13 of [36] and used them to give complete proofs of Ramanujan's formulas (28)–(30) in [36], [37, pp. 36–37]. These three series representations are, in fact, published with Ramanujan's lost notebook [39] but are clearly *not* from the last year of Ramanujan's life, but instead are from his earlier days in India or more likely from his first year at Cambridge.

2 Preliminary definitions and results

We use the standard shifted or rising factorial notation

$$(a)_0 := 1, \qquad (a)_n := a(a+1)(a+2)\cdots(a+n-1), \quad n \geq 1.$$

The hypergeometric functions $_pF_{p-1}$, $p \geq 1$, are defined by

$$_pF_{p-1}\left(a_1,\ldots,a_p; b_1,\ldots,b_{p-1}; x\right) := \sum_{n=0}^{\infty} \frac{(a_1)_n \cdots (a_p)_n}{(b_1)_n \cdots (b_{p-1})_n} \frac{x^n}{n!}, \quad |x| < 1.$$

If

$$q = \exp\left(-\pi \frac{{}_2F_1(\frac{1}{2}, \frac{1}{2}; 1; 1-x)}{{}_2F_1(\frac{1}{2}, \frac{1}{2}; 1; x)}\right),$$

then one of the fundamental results in the theory of elliptic functions [6, p. 101, Entry 6] is given by

$$\phi^2(q) = {}_2F_1\left(\frac{1}{2}, \frac{1}{2}; 1; x\right), \tag{2.1}$$

where here, and for the sequel,

$$\phi(q) := \sum_{n=-\infty}^{\infty} q^{n^2} \quad \text{and} \quad \psi(q) = \sum_{n=0}^{\infty} q^{n(n+1)/2}, \quad |q| < 1. \tag{2.2}$$

We also need Ramanujan's function

$$f(-q) := \sum_{n=-\infty}^{\infty} (-1)^n q^{n(3n-1)/2} = (q; q)_\infty, \tag{2.3}$$

where the latter identity is Euler's pentagonal number theorem, which is easily derived from Jacobi's triple product identity. Following Ramanujan, define

$$z := z(q) := {}_2F_1\left(\frac{1}{2}, \frac{1}{2}; 1; x\right) = \phi^2(q). \tag{2.4}$$

In the sequel, we often emphasize that x is also a function of q when writing $x = x(q)$.

The identity (2.1) enables one to derive formulas for Ramanujan's functions ϕ, ψ, and f at different arguments in terms of x, q, and z. In particular, Ramanujan recorded the following identities in his second notebook [38], [6, pp. 122–124].

Lemma 2.1 *We have*

$$\phi(q) = \sqrt{z}, \tag{2.5}$$

$$\phi(-q^2) = \sqrt{z}(1-x)^{1/8}, \tag{2.6}$$

$$\phi(q^2) = \sqrt{z}\left(\frac{1}{2}(1+\sqrt{1-x})\right)^{1/2}, \tag{2.7}$$

$$\psi(q) = \sqrt{\frac{z}{2}}\left(\frac{x}{q}\right)^{1/8}, \tag{2.8}$$

$$\psi(q^2) = \frac{\sqrt{z}}{2}\left(\frac{x}{q}\right)^{1/4}, \tag{2.9}$$

$$f(-q^2) = \frac{\sqrt{z}\{x(1-x)\}^{1/12}}{2^{1/3}q^{1/12}}. \tag{2.10}$$

We lastly define a modular equation as understood by Ramanujan [34, p. 214]. Suppose that the equality

$$n\frac{{}_2F_1(\frac{1}{2},\frac{1}{2};1;1-k^2)}{{}_2F_1(\frac{1}{2},\frac{1}{2};1;k^2)} = \frac{{}_2F_1(\frac{1}{2},\frac{1}{2};1;1-\ell^2)}{{}_2F_1(\frac{1}{2},\frac{1}{2};1;\ell^2)} \tag{2.11}$$

holds for some positive integer n. Then a modular equation of degree n is a relation between the moduli k and ℓ that is implied by (2.11). Ramanujan recorded his modular equations in terms of α and β, where $\alpha = k^2$ and $\beta = l^2$. We say that β has degree n over α.

3 The development of Ramanujan's ideas

Ramanujan's series representations for $1/\pi$ depend upon Clausen's product formulas for hypergeometric series and Ramanujan's Eisenstein series

$$P(q) := 1 - 24\sum_{k=1}^{\infty}\frac{kq^k}{1-q^k}, \quad |q| < 1. \tag{3.1}$$

More precisely, but briefly, by combining two different relations between $P(q)$ and $P(q^n)$, for certain positive integers n, along with a Clausen formula, we can obtain series representations for $1/\pi$.

As with the Borweins in their proofs, we begin with Clausen's formulas. We quote Theorems 5.7(i)–(vi) and Formula (5.5.9) in [13, pp. 180–181]. Let

$$A_k := \frac{(\frac{1}{2})_k^3}{k!^3}, \quad B_k := \frac{(\frac{1}{4})_k(\frac{1}{2})_k(\frac{3}{4})_k}{k!^3}, \quad C_k := \frac{(\frac{1}{6})_k(\frac{1}{2})_k(\frac{5}{6})_k}{k!^3}, \tag{3.2}$$

and

$$X := 4x(1-x), \quad Y := \frac{4x}{(1-x)^2}, \quad U := \frac{x^2}{4(1-x)}, \quad V := \frac{4\sqrt{x}(1-x)}{(1+x)^2},$$

$$W := \frac{2\sqrt{X}}{1-X}, \quad L := \frac{27X^2}{(4-X)^3}, \quad \text{and} \quad M := \frac{27X}{(1-4X)^3}.$$

Then

$$z^2 = {}_3F_2\left(\frac{1}{2},\frac{1}{2},\frac{1}{2};1,1;X\right) = \sum_{k=0}^{\infty}A_kX^k,$$

$$0 \le x \le \frac{1}{2}, \tag{3.3}$$

$$= \frac{1}{1-x}\,{}_3F_2\left(\frac{1}{2},\frac{1}{2},\frac{1}{2};1,1;-Y\right) = \frac{1}{1-x}\sum_{k=0}^{\infty}(-1)^kA_kY^k,$$

$$0 \le x \le 3 - 2\sqrt{2}, \tag{3.4}$$

$$= \frac{1}{\sqrt{1-x}} \quad {}_3F_2\left(\frac{1}{2}, \frac{1}{2}, \frac{1}{2}; 1, 1; -U\right) = \frac{1}{\sqrt{1-x}} \sum_{k=0}^{\infty} (-1)^k A_k U^k,$$

$$0 \le x \le 2\sqrt{2} - 2, \tag{3.5}$$

$$= \frac{1}{1+x} \quad {}_3F_2\left(\frac{1}{4}, \frac{1}{2}, \frac{3}{4}; 1, 1; V^2\right) = \frac{1}{1+x} \sum_{k=0}^{\infty} B_k V^{2k},$$

$$0 \le x \le 3 - 2\sqrt{2}, \tag{3.6}$$

$$= \frac{1}{1-2x} \quad {}_3F_2\left(\frac{1}{4}, \frac{1}{2}, \frac{3}{4}; 1, 1; -W^2\right) = \frac{1}{1-2x} \sum_{k=0}^{\infty} (-1)^k B_k W^{2k},$$

$$0 \le x \le \frac{1}{2}(1 - 2^{1/4}\sqrt{2 - \sqrt{2}}), \tag{3.7}$$

$$= \frac{2}{\sqrt{4-X}} \quad {}_3F_2\left(\frac{1}{6}, \frac{1}{2}, \frac{5}{6}; 1, 1; L\right) = \frac{2}{\sqrt{4-X}} \sum_{k=0}^{\infty} C_k L^k,$$

$$0 \le x \le \frac{1}{2}, \tag{3.8}$$

$$= \frac{1}{\sqrt{1-4X}} \quad {}_3F_2\left(\frac{1}{6}, \frac{1}{2}, \frac{5}{6}; 1, 1; -M\right) = \frac{1}{\sqrt{1-4X}} \sum_{k=0}^{\infty} (-1)^k C_k M^k,$$

$$0 \le x \le \frac{1}{2}. \tag{3.9}$$

Now, differentiating (3.3) with respect to x, we find that

$$2z\frac{dz}{dx} = \sum_{k=0}^{\infty} A_k k X^{k-1} \cdot 4(1-2x). \tag{3.10}$$

Next, recall the representation for $P(q^2)$ given in [6, p. 120, Entry 9(iv)], namely,

$$P(q^2) = (1-2x)z^2 + 6x(1-x)z\frac{dz}{dx}. \tag{3.11}$$

Employing (3.3) and (3.10) in (3.11), we find that

$$P(q^2) = (1-2x) \sum_{k=0}^{\infty} (3k+1) A_k X^k. \tag{3.12}$$

Now set

$$x_n := x(e^{-\pi\sqrt{n}}) \quad \text{and} \quad z_n := z(e^{-\pi\sqrt{n}}). \tag{3.13}$$

The numbers x_n are *singular moduli*. In his notebooks [38], Ramanujan calculated the values of many singular moduli, and in the sequel, we frequently appeal to

Ramanujan's values, as recorded and proved in [7]. It also can be easily shown that [1, Chap. 15]

$$1 - x_n = x_{1/n} \quad \text{and} \quad z_{1/n} = \sqrt{n} z_n. \tag{3.14}$$

Setting $q = e^{-\pi \sqrt{n}}$ in (3.12), we deduce that

$$P(e^{-2\pi \sqrt{n}}) = (1 - 2x_n) \sum_{k=0}^{\infty} (3k + 1) A_k X_n^k, \tag{3.15}$$

where

$$X_n = 4x_n(1 - x_n).$$

Similarly, differentiating each of (3.4)–(3.9) with respect to x, and proceeding as above, we can find that

$$P(e^{-2\pi \sqrt{n}}) = \frac{1 + x_n}{1 - x_n} \sum_{k=0}^{\infty} (3k + 1)(-1)^k A_k Y_n^k, \tag{3.16}$$

$$= \frac{2 - x_n}{2\sqrt{1 - x_n}} \sum_{k=0}^{\infty} (6k + 1)(-1)^k A_k U_n^k, \tag{3.17}$$

$$= \sum_{k=0}^{\infty} \frac{3k(1 - 6x_n + x_n^2) + 1 - 4x_n + x_n^2}{1 + X_n} B_k V_n^{2k}, \tag{3.18}$$

$$= \sum_{k=0}^{\infty} (-1)^k \frac{3k(1 + X_n) + 1 + X_n/2}{1 - X_n} B_k W_n^{2k}, \tag{3.19}$$

$$= \frac{\sqrt{1 - X_n}(X_n + 8)}{(4 - X_n)^{3/2}} \sum_{k=0}^{\infty} (6k + 1) C_k L_n^k, \tag{3.20}$$

$$= \frac{\sqrt{1 - X_n}}{(1 - 4X_n)^{3/2}} \sum_{k=0}^{\infty} (-1)^k \{3(1 + 8X_n)k + 1 + 2X_n\} C_k M_n^k, \tag{3.21}$$

where

$$Y_n := \frac{4x_n}{(1 - x_n)^2}, \quad U_n := \frac{x_n^2}{4(1 - x_n)}, \quad V_n := \frac{4\sqrt{x_n}(1 - x_n)}{(1 + x_n)^2},$$

$$W_n := \frac{2\sqrt{X_n}}{1 - X_n}, \quad L_n := \frac{27X_n^2}{(4 - X_n)^3}, \quad \text{and} \quad M_n := \frac{27X_n}{(1 - 4X_n)^3}.$$

Next, we determine a transformation formula for $P(q)$. Recall the transformation formula for Ramanujan's function $f(-q)$ [6, p. 43, Entry 27(iii)]. If $\alpha, \beta > 1$ with

$\alpha\beta = \pi^2$, then

$$e^{-\alpha/12}\alpha^{1/4}f(-e^{-2\alpha}) = e^{-\beta/12}\beta^{1/4}f(-e^{-2\beta}). \tag{3.22}$$

Taking the logarithm of both sides of (3.22), we obtain

$$-\frac{\alpha}{12} + \frac{1}{4}\log\alpha + \sum_{k=1}^{\infty}\log(1 - e^{-2k\alpha})$$

$$= -\frac{\beta}{12} + \frac{1}{4}\log\beta + \sum_{k=1}^{\infty}\log(1 - e^{-2k\beta}). \tag{3.23}$$

Differentiating both sides of (3.23) with respect to α, we find that

$$-\frac{1}{12} + \frac{1}{4\alpha} + \sum_{k=1}^{\infty}\frac{2ke^{-2k\alpha}}{1 - e^{-2k\alpha}} = \frac{\beta}{12\alpha} - \frac{1}{4\alpha} - \sum_{k=1}^{\infty}\frac{(2k\beta/\alpha)e^{-2k\beta}}{1 - e^{-2k\beta}}. \tag{3.24}$$

Multiplying both sides of (3.24) by 12α, rearranging, and then employing the definition of $P(q)$ from (3.1), we deduce that

$$6 - \alpha P(e^{-2\alpha}) = \beta P(e^{-2\beta}). \tag{3.25}$$

Setting $\alpha = \pi/\sqrt{n}$, so that $\beta = \pi\sqrt{n}$, in (3.25), we arrive at

$$n P(e^{-2\pi\sqrt{n}}) + P(e^{-2\pi/\sqrt{n}}) = \frac{6\sqrt{n}}{\pi}. \tag{3.26}$$

If we set $n = 1$ in (3.26), we deduce that

$$P(e^{-2\pi}) = \frac{3}{\pi}, \tag{3.27}$$

which has been established many, many times in the literature, including by Ramanujan in Sect. 8 of Chap. 14 of his second notebook [38]. See [5, p. 256] for references to several proofs.

In his paper [36], Ramanujan recorded twelve representations for

$$f_n(q) := n P(q^{2n}) - P(q^2), \tag{3.28}$$

corresponding to twelve values of n, namely, $n = 2, 3, 4, 5, 7, 11, 15, 17, 19, 23, 31,$ and 35. He also recorded the representations for $n = 2$ and 4 in Chap. 17 and for the remaining ten values and for $n = 9$ and $n = 25$ in Chap. 21 of his second notebook [38]. These representations for $q = e^{-\pi/\sqrt{n}}$ combined with (3.26) and the identities (3.15)–(3.21) are the primary ingredients in our derivations of series representations for $1/\pi$ in the following sections.

In our derivations which follow, for each value of n, there are potentially seven series representations for $1/\pi$, each arising from one of the formulas (3.3)–(3.9). However, it may happen that one or more of these series diverge. Also, we have needed to make decisions about the elegance or inelegance of series representations, and consequently we have not listed certain representations which we think are inelegant.

4 Example: $n = 2$

Theorem 4.1 *If A_k and C_k, $k \geq 0$, are defined by (3.2), then*

$$\frac{1}{\pi} = \sum_{k=0}^{\infty} \{(8 - 5\sqrt{2})k + 3 - 2\sqrt{2}\} A_k (2\sqrt{2} - 2)^{3k}, \tag{4.1}$$

$$\frac{2}{\pi} = \sum_{k=0}^{\infty} (-1)^k (4k + 1) A_k, \tag{4.2}$$

$$\frac{2\sqrt{\sqrt{2} - 1}}{\pi} = \sum_{k=0}^{\infty} (-1)^k \{(4\sqrt{2} - 2)k + \sqrt{2} - 1\} A_k \left(\frac{\sqrt{2} - 1}{2}\right)^{3k}, \tag{4.3}$$

$$\frac{5\sqrt{5}}{\pi} = \sum_{k=0}^{\infty} (28k + 3) C_k \left(\frac{3}{5}\right)^{3k}. \tag{4.4}$$

The identity (4.2) was first proved by G. Bauer in 1859 [4]. Many years later, the formula (4.2) was communicated by Ramanujan in his first letter to G.H. Hardy [37, p. xxvi], [10, p. 25]. Ramanujan also recorded (4.2) as Example 14 in Sect. 7 of Chap. 10 in his second notebook [38], [5, pp. 23–24]. Ramanujan [38], W.N. Bailey [2, p. 96], and Hardy [33], [35, pp. 517–518] all observed that (4.2) is a limiting case of Dougall's theorem. See also [13, p. 184]. The identity (4.4) is due to the Borweins [14]. The remaining two identities are new.

Proof of (4.1) From, for example, [7, p. 281, Theorem 9.2], $x_2 = (\sqrt{2} - 1)^2$, and from [6, p. 127, Entry 13(ix)],

$$f_2(q) = 2P(q^4) - P(q^2) = z^2(q)\left(1 - \frac{x(q)}{2}\right). \tag{4.5}$$

Setting $q = e^{-\pi/\sqrt{2}}$ and using (3.14), we find that $z(e^{-\pi/\sqrt{2}}) = \sqrt{2}z(e^{-\pi\sqrt{2}}) = \sqrt{2}z_2$ and $x(e^{-\pi/\sqrt{2}}) = 1 - x(e^{-\pi\sqrt{2}}) = 1 - x_2$. Hence, we deduce that

$$f_2(e^{-\pi/\sqrt{2}}) = 2P(e^{-2\pi\sqrt{2}}) - P(e^{-2\pi/\sqrt{2}}) = z_2^2(1 + x_2) = (4 - 2\sqrt{2})z_2^2. \tag{4.6}$$

Setting $n = 2$ in (3.26), we find that

$$2P(e^{-2\pi\sqrt{2}}) + P(e^{-2\pi/\sqrt{2}}) = \frac{6\sqrt{2}}{\pi}. \tag{4.7}$$

From (4.6) and (4.7), we deduce that

$$P(e^{-2\pi\sqrt{2}}) = \frac{3}{\pi\sqrt{2}} + \frac{\sqrt{2} - 1}{\sqrt{2}} z_2^2. \tag{4.8}$$

With the help of (3.3), we can rewrite (4.8) in the form

$$P(e^{-2\pi\sqrt{2}}) = \frac{3}{\pi\sqrt{2}} + \frac{\sqrt{2}-1}{\sqrt{2}} \sum_{k=0}^{\infty} A_k X_2^k, \qquad (4.9)$$

where $X_2 = 4x_2(1 - x_2) = 8(\sqrt{2} - 1)^3$.
   Now, setting $n = 2$ in (3.15), we find that

$$P(e^{-2\pi\sqrt{2}}) = (1 - 2x_2) \sum_{k=0}^{\infty} (3k + 1) A_k X_2^k = (4\sqrt{2} - 5) \sum_{k=0}^{\infty} (3k + 1) A_k X_2^k. \quad (4.10)$$

From (4.9) and (4.10), we readily deduce (4.1). $\qquad\square$

Proof of (4.2) With the help of (3.4), we can rewrite (4.8) as

$$P(e^{-2\pi\sqrt{2}}) = \frac{3}{\pi\sqrt{2}} + \frac{\sqrt{2}-1}{\sqrt{2}(1-x_2)} \sum_{k=0}^{\infty} (-1)^k A_k Y_2^k$$

$$= \frac{3}{\pi\sqrt{2}} + \frac{1}{2\sqrt{2}} \sum_{k=0}^{\infty} (-1)^k A_k, \qquad (4.11)$$

where we also used the fact that $Y_2 = 1$.
   Setting $n = 2$ in (3.16), we find that

$$P(e^{-2\pi\sqrt{2}}) = \frac{1+x_2}{1-x_2} \sum_{k=0}^{\infty} (3k + 1) A_k = \sqrt{2} \sum_{k=0}^{\infty} (3k + 1) A_k. \qquad (4.12)$$

From (4.11) and (4.12), we easily arrive at (4.2). $\qquad\square$

Proof of (4.3) Employing (3.5) in (4.8), we obtain the equality

$$P(e^{-2\pi\sqrt{2}}) = \frac{3}{\pi\sqrt{2}} + \frac{\sqrt{2}-1}{\sqrt{2}\sqrt{1-x_2}} \sum_{k=0}^{\infty} (-1)^k A_k U_2^k$$

$$= \frac{3}{\pi\sqrt{2}} + \frac{\sqrt{\sqrt{2}-1}}{2} \sum_{k=0}^{\infty} (-1)^k A_k U_2^k, \qquad (4.13)$$

where $U_2 = \{(\sqrt{2} - 1)/2\}^3$.
   Also, setting $n = 2$ in (3.17), we deduce that

$$P(e^{-2\pi\sqrt{2}}) = \frac{2-x_2}{2\sqrt{1-x_2}} \sum_{k=0}^{\infty} (6k + 1)(-1)^k A_k U_2^k$$

$$= \frac{2\sqrt{2}-1}{2\sqrt{2}\sqrt{\sqrt{2}-1}} \sum_{k=0}^{\infty} (6k + 1)(-1)^k A_k U_2^k. \qquad (4.14)$$

From (4.13) and (4.14), we deduce (4.3). $\qquad\square$

Proof of (4.4) Using (3.8) in (4.8), we find that

$$P(e^{-2\pi\sqrt{2}}) = \frac{3}{\pi\sqrt{2}} + \frac{2(\sqrt{2}-1)}{\sqrt{2}\sqrt{4-X_2}} \sum_{k=0}^{\infty} C_k L_2^k$$

$$= \frac{3}{\pi\sqrt{2}} + \frac{1}{\sqrt{10}} \sum_{k=0}^{\infty} C_k L_2^k, \qquad (4.15)$$

where $L_2 = (3/5)^3$.

Also, setting $n = 2$ in (3.20), we find that

$$P(e^{-2\pi\sqrt{2}}) = \frac{\sqrt{1-X_2}(X_2+8)}{(4-X_2)^{3/2}} \sum_{k=0}^{\infty} (6k+1) C_k L_2^k$$

$$= \frac{7\sqrt{2}}{5\sqrt{5}} \sum_{k=0}^{\infty} (6k+1) C_k L_2^k. \qquad (4.16)$$

From (4.15) and (4.16), we easily deduce (4.4).   □

5 Example: $n = 4$

Theorem 5.1 *If A_k, B_k, and C_k, $k \geq 0$, are defined by (3.2), then*

$$\frac{1}{\pi} = \sum_{k=0}^{\infty} \{(48\sqrt{2}-66)k + 20\sqrt{2} - 28\} A_k (1584\sqrt{2} - 2240)^k, \qquad (5.1)$$

$$\frac{2\sqrt{2}}{\pi} = \sum_{k=0}^{\infty} (-1)^k (6k+1) A_k \left(\frac{1}{8}\right)^k, \qquad (5.2)$$

$$\frac{2 \cdot 2^{1/4}}{\pi} = \sum_{k=0}^{\infty} (-1)^k \{(18 - 6\sqrt{2})k + 5 - 3\sqrt{2}\} A_k \left(\frac{(\sqrt{2}-1)^6}{16\sqrt{2}}\right)^k, \qquad (5.3)$$

$$\frac{9}{2\pi} = \sum_{k=0}^{\infty} (7k+1) B_k \left(\frac{32}{81}\right)^k, \qquad (5.4)$$

$$\frac{\sqrt{33}}{\pi} = \sum_{k=0}^{\infty} (126k + 10) C_k \left(\frac{2}{11}\right)^{3k+1}. \qquad (5.5)$$

The identities (5.4) and (5.5) are due to Berndt, Chan, and Liaw [12] and the Borweins [14], respectively. Guillera [30] proved the identity (5.2) by the WZ-method. The remaining two identities seem to be new.

Proof of (5.1) We note from [7, p. 284] that $x_4 = (\sqrt{2} - 1)^4 = 17 - 12\sqrt{2}$. We set $q = e^{-\pi}$ in (4.5) and then use (3.27) to obtain

$$2P(e^{-4\pi}) - \frac{3}{\pi} = z^2(e^{-\pi})\left(1 - \frac{x(e^{-\pi})}{2}\right). \tag{5.6}$$

Upon using the trivial evaluation $x(e^{-\pi}) = x_1 = \frac{1}{2}$ in (5.6), we deduce that

$$P(e^{-4\pi}) = \frac{3}{2\pi} + \frac{3}{8}z_1^2. \tag{5.7}$$

Now, from (2.5) and (2.7) in Lemma 2.1, we find that

$$\phi^4(q) = \frac{4}{(1 + \sqrt{1 - x(q)})^2}\phi^4(q^2). \tag{5.8}$$

Setting $q = e^{-\pi}$ in (5.8), we find that

$$z_1^2 = \frac{4}{(1 + \sqrt{1 - x_1})^2}z_4^2 = 8(3 - 2\sqrt{2})z_4^2. \tag{5.9}$$

Employing (5.9) in (5.7), we arrive at

$$P(e^{-4\pi}) = \frac{3}{2\pi} + 3(3 - 2\sqrt{2})z_4^2. \tag{5.10}$$

Now, using (3.3) in (5.10), we find that

$$P(e^{-4\pi}) = \frac{3}{2\pi} + 3(3 - 2\sqrt{2})\sum_{k=0}^{\infty} A_k X_4^k, \tag{5.11}$$

where

$$X_4 = 4x_4(1 - x_4) = 16\sqrt{2}(\sqrt{2} - 1)^6. \tag{5.12}$$

Setting $n = 4$ in (3.15), we obtain

$$P(e^{-4\pi}) = (1 - 2x_4)\sum_{k=0}^{\infty}(3k + 1)A_k X_4^k$$

$$= (24\sqrt{2} - 33)\sum_{k=0}^{\infty}(3k + 1)A_k X_4^k. \tag{5.13}$$

From (5.11) and (5.13), we easily deduce (5.1). □

Proof of (5.2) Employing (3.4) in (5.10), we find that

$$P(e^{-4\pi}) = \frac{3}{2\pi} + \frac{3(3 - 2\sqrt{2})}{1 - x_4}\sum_{k=0}^{\infty}(-1)^k A_k Y_4^k$$

$$= \frac{3}{2\pi} + \frac{3}{4\sqrt{2}} \sum_{k=0}^{\infty} (-1)^k A_k Y_4^k, \tag{5.14}$$

where

$$Y_4 = \frac{4x_4}{(1-x_4)^2} = \frac{1}{8}.$$

Setting $n = 4$ in (3.16), we find that

$$P(e^{-4\pi}) = \frac{1+x_4}{1-x_4} \sum_{k=0}^{\infty} (-1)^k (3k+1) A_k Y_4^k$$

$$= \frac{3}{2\sqrt{2}} \sum_{k=0}^{\infty} (-1)^k (3k+1) A_k Y_4^k. \tag{5.15}$$

From (5.14) and (5.15), we easily arrive at (5.2). $\qquad\square$

Proof of (5.3) We employ (3.4) to rewrite (5.10) as

$$P(e^{-4\pi}) = \frac{3}{2\pi} + \frac{3(3-2\sqrt{2})}{\sqrt{1-x}} \sum_{k=0}^{\infty} (-1)^k A_k U_4^k$$

$$= \frac{3}{2\pi} + \frac{3(\sqrt{2}-1)}{2 \cdot 2^{1/4}} \sum_{k=0}^{\infty} (-1)^k A_k U_4^k, \tag{5.16}$$

where

$$U_4 = \frac{x_4^2}{4(1-x_4)} = \frac{(\sqrt{2}-1)^6}{16\sqrt{2}}.$$

Setting $n = 4$ in (3.17), we find that

$$P(e^{-4\pi}) = \frac{2-x_4}{2\sqrt{1-x_4}} \sum_{k=0}^{\infty} (-1)^k (6k+1) A_k U_4^k$$

$$= \frac{3(4\sqrt{2}-5)}{4 \cdot 2^{1/4}(\sqrt{2}-1)} \sum_{k=0}^{\infty} (-1)^k (6k+1) A_k U_4^k. \tag{5.17}$$

From (5.16) and (5.17), we readily deduce (5.3). $\qquad\square$

Proof of (5.4) Using (3.6) in (5.10), we find that

$$P(e^{-4\pi}) = \frac{3}{2\pi} + \frac{3(3-2\sqrt{2})}{1+x} \sum_{k=0}^{\infty} B_k V_4^{2k}$$

$$= \frac{3}{2\pi} + \frac{1}{2} \sum_{k=0}^{\infty} B_k V_4^{2k}, \tag{5.18}$$

where

$$V_4 = \frac{4\sqrt{x_4}(1 - x_4)}{(1 + x_4)^2} = \frac{4\sqrt{2}}{9}.$$

Next, setting $n = 4$ in (3.18), we find that

$$P(e^{-4\pi}) = \sum_{k=0}^{\infty} \sum_{k=0}^{\infty} \frac{3k(1 - 6x_4 + x_4^2) + 1 - X_4}{1 + X_4} B_k V_4^{2k}$$

$$= \sum_{k=0}^{\infty} (-1)^k \left(\frac{7k}{3} + 56 \right) B_k V_4^{2k}, \tag{5.19}$$

where we have also used (5.12). From (5.18) and (5.19), we arrive at (5.4). □

Proof of (5.5) With the aid of (3.6), we can rewrite (5.10) as

$$P(e^{-4\pi}) = \frac{3}{2\pi} + \frac{6(3 - 2\sqrt{2})}{\sqrt{4 - X_4}} \sum_{k=0}^{\infty} C_k L_4^k$$

$$= \frac{3}{2\pi} + \frac{3}{\sqrt{33}} \sum_{k=0}^{\infty} C_k L_4^k, \tag{5.20}$$

where

$$L_4 = \frac{27 X_4^2}{(4 - X_4)^3} = \left(\frac{2}{11} \right)^3.$$

Next, setting $n = 4$ in (3.20), we find that

$$P(e^{-4\pi}) = \frac{\sqrt{1 - X_4}(8 + X_4)}{(4 - X_4)^{3/2}} \sum_{k=0}^{\infty} \sum_{k=0}^{\infty} (6k + 1) C_k L_4^k$$

$$= \frac{63}{11\sqrt{33}} \sum_{k=0}^{\infty} (-1)^k (6k + 1) C_k L_4^k, \tag{5.21}$$

where we have also used (5.12). From (5.20) and (5.21), we readily deduce (5.5). □

6 Example: $n = 6$

Theorem 6.1 *If A_k and B_k, $k \geq 0$, are defined by (3.2), then*

$$\frac{\sqrt{6} + \sqrt{2} + 1}{\pi} = \sum_{k=0}^{\infty} \{(6\sqrt{3} + 3\sqrt{6} - 6)k + 2\sqrt{3} + \sqrt{6} - 3 - \sqrt{2}\}$$

$$\times A_k \{8(\sqrt{2} + 1)^2(\sqrt{3} - \sqrt{2})^3(2 - \sqrt{3})^3\}^k, \tag{6.1}$$

$$\frac{2}{\pi} = \sum_{k=0}^{\infty} (-1)^k \{(12\sqrt{2} - 12)k + 4\sqrt{2} - 5\} A_k (\sqrt{2} - 1)^{4k}, \quad (6.2)$$

$$\frac{2\sqrt{2\sqrt{3} + 2\sqrt{2}}}{\pi} = \sum_{k=0}^{\infty} \{(4\sqrt{6} + 6\sqrt{3} - 6)k + 3\sqrt{2} - 3\sqrt{3} + 5 - \sqrt{6}\}$$

$$\times (-1)^k A_k \{8(\sqrt{2} + 1)^2 (\sqrt{3} + \sqrt{2})^3 (2 + \sqrt{3})^3\}^{-k}, \quad (6.3)$$

$$\frac{2\sqrt{3}}{\pi} = \sum_{k=0}^{\infty} (8k + 1) B_k \left(\frac{1}{9}\right)^k. \quad (6.4)$$

The last identity was recorded by Ramanujan in his paper [36, Eq. (40)], [37, p. 38]. To the best of our knowledge, the remaining identities are new.

Proof First of all, we derive an expression for $f_6(e^{-\pi/\sqrt{6}}) = 6P(e^{-2\pi\sqrt{6}}) - P(e^{-2\pi/\sqrt{6}})$. To this end, from Entry 3(ii) in Chap. 21 of Ramanujan's second notebook [38], [6, p. 460], we record that

$$1 + 12 \sum_{k=1}^{\infty} \frac{kq^{2k}}{1 - q^{2k}} - 36 \sum_{k=1}^{\infty} \frac{kq^{6k}}{1 - q^{6k}} = \phi^2(q)\phi^2(q^3) - 4q\psi^2(-q)\psi^2(-q^3), \quad (6.5)$$

where $\phi(q)$ and $\psi(q)$ are defined by (2.2). Now, replacing q by $-q^2$ in (6.5) and then employing the definition of P in (3.1), we find that

$$3P(q^{12}) - P(q^4) = 2\phi^2(-q^2)\phi^2(-q^6) + 4q^2\psi^2(q^2)\psi^2(q^6). \quad (6.6)$$

Transcribing (6.6) with the aid of (2.5), (2.6), (2.8), and (2.9) of Lemma 2.1, we deduce that

$$3P(q^{12}) - P(q^4) = z(q)z(q^6)\{2(1 - x(q))^{1/4}(1 - x(q^6))^{1/2} + x^{1/2}(q)x^{1/4}(q^6)\}. \quad (6.7)$$

Multiplying (6.7) by 2 and adding the resulting equality to (4.5), we find that

$$6P(q^{12}) - P(q^2) = z^2(q)\left(1 - \frac{x(q)}{2}\right) + 2z(q)z(q^6)$$

$$\times \{2(1 - x(q))^{1/4}(1 - x(q^6))^{1/2} + x^{1/2}(q)x^{1/4}(q^6)\}. \quad (6.8)$$

Now set $q = e^{-\pi/\sqrt{6}}$, so that, by (3.14), $x(q) = x(e^{-\pi/\sqrt{6}}) = 1 - x(e^{-\pi\sqrt{6}}) = 1 - x_6$, and $z(q) = z(e^{-\pi/\sqrt{6}}) = \sqrt{6}z(e^{-\pi\sqrt{6}}) = \sqrt{6}z_6$. We therefore deduce that

$$6P(e^{-2\pi\sqrt{6}}) - P(e^{-2\pi/\sqrt{6}}) = \{3(1 + x_6) + 6\sqrt{6}x_6^{1/4}(1 - x_6)^{1/2}\}z_6^2. \quad (6.9)$$

Now, the singular modulus x_6 is given by [7, p. 282]

$$x_6 = (2 - \sqrt{3})^2(\sqrt{3} - \sqrt{2})^2 = \frac{\sqrt{6} - \sqrt{2} - 1}{\sqrt{6} + \sqrt{2} - 1},$$

so that

$$1 - x_6 = 2(\sqrt{2}+1)^2(2-\sqrt{3})(\sqrt{3}-\sqrt{2}) \tag{6.10}$$

and

$$1 + x_6 = 1 - x_6 + 2x_6 = 2\sqrt{6}(\sqrt{2}+1)(2-\sqrt{3})(\sqrt{3}-\sqrt{2}). \tag{6.11}$$

Thus, from (6.9), we deduce that

$$6P(e^{-2\pi\sqrt{6}}) - P(e^{-2\pi/\sqrt{6}}) = 6\sqrt{6}(\sqrt{2}+1)^2(2-\sqrt{3})(\sqrt{3}-\sqrt{2})z_6^2. \tag{6.12}$$

Next, setting $n = 6$ in (3.26), we find that

$$6P(e^{-2\pi\sqrt{6}}) + P(e^{-2\pi/\sqrt{6}}) = \frac{6\sqrt{6}}{\pi}. \tag{6.13}$$

Adding (6.12) and (6.13), we obtain the identity

$$P(e^{-2\pi\sqrt{6}}) = \frac{\sqrt{3}}{\pi\sqrt{2}} + \frac{\sqrt{3}}{\sqrt{2}}(\sqrt{2}+1)^2(2-\sqrt{3})(\sqrt{3}-\sqrt{2})z_6^2. \tag{6.14}$$

Now we are ready to prove (6.1). Using (3.3) in (6.14), we find that

$$P(e^{-2\pi\sqrt{6}}) = \frac{\sqrt{3}}{\pi\sqrt{2}} + \frac{\sqrt{3}}{\sqrt{2}}(\sqrt{2}+1)^2(2-\sqrt{3})(\sqrt{3}-\sqrt{2})\sum_{k=0}^{\infty} A_k X_6^k, \tag{6.15}$$

where $X_6 = 4x_6(1-x_6) = 8(\sqrt{2}+1)^2(2-\sqrt{3})^3(\sqrt{3}-\sqrt{2})^3$.
Next, setting $n = 6$ in (3.15), we find that

$$P(e^{-2\pi\sqrt{6}}) = (1 - 2x_6)\sum_{k=0}^{\infty}(3k+1)A_k X_6^k$$

$$= (\sqrt{2}+1)(2-\sqrt{3})(\sqrt{3}-\sqrt{2})(3\sqrt{2}+3-\sqrt{6})\sum_{k=0}^{\infty}(3k+1)A_k X_6^k. \tag{6.16}$$

From (6.15) and (6.16), we deduce (6.1).

Similarly, employing (3.4), (3.5), and (3.6) in (6.14) and setting $n = 6$ in each of (3.16), (3.17), and (3.18), we can derive the series identities in (6.2)–(6.4). $\qquad\square$

7 Example: $n = 5$

Theorem 7.1 *If A_k and B_k, $k \geq 0$, are defined by (3.2), then*

$$\frac{2}{\pi} = \sum_{k=0}^{\infty}\{4\sqrt{5}k + \sqrt{5} - 1\}A_k(\sqrt{5}-2)^{2k+1/2}, \tag{7.1}$$

$$\frac{8}{\pi} = \sum_{k=0}^{\infty} (-1)^k \left[2 \left\{ (15 + 5\sqrt{5})\sqrt{\sqrt{5}+1} - 7\sqrt{10} - 5\sqrt{2} \right\} k + (9 + 3\sqrt{5})\sqrt{\sqrt{5}+1} \right.$$

$$\left. - 7\sqrt{2} - 5\sqrt{10} \right] A_k \left(\frac{\sqrt{5}-1}{4} \right)^{3k} \left(\frac{\sqrt{5}+1}{2} - \sqrt{\frac{\sqrt{5}+1}{2}} \right)^{6k}, \tag{7.2}$$

$$\frac{8}{\pi} = \sum_{k=0}^{\infty} (-1)^k (20k + 3) B_k \frac{1}{4^k}. \tag{7.3}$$

The last identity was recorded by Ramanujan [36, Eq. (35)], [37, p. 38]. The other identities appear to be new.

Proof of (7.1) From Entry 4(iii) in Chap. 21 of Ramanujan's second notebook [38], [6, p. 464], we see that

$$1 + 6 \sum_{k=1}^{\infty} \frac{kq^{2k}}{1 - q^{2k}} - 30 \sum_{k=1}^{\infty} \frac{kq^{10k}}{1 - q^{10k}}$$

$$= \frac{1}{4\sqrt{2}} \phi^2(q) \phi^2(q^5) \left\{ 3 + \sqrt{x(q)x(q^5)} + \sqrt{(1 - x(q))(1 - x(q^5))} \right\}$$

$$\times \left\{ 1 + \sqrt{x(q)x(q^5)} + \sqrt{(1 - x(q))(1 - x(q^5))} \right\}^{1/2}. \tag{7.4}$$

With the help of (3.1) we can rewrite (7.4) in the form

$$5P(q^{10}) - P(q^2)$$

$$= \frac{1}{\sqrt{2}} \phi^2(q) \phi^2(q^5) \left\{ 3 + \sqrt{x(q)x(q^5)} + \sqrt{(1 - x(q))(1 - x(q^5))} \right\}$$

$$\times \left\{ 1 + \sqrt{x(q)x(q^5)} + \sqrt{(1 - x(q))(1 - x(q^5))} \right\}^{1/2}. \tag{7.5}$$

Now we set $q = e^{-\pi/\sqrt{5}}$ and use (3.14) and (2.4) to deduce that $x(q) = x_{1/5} = 1 - x_5$, $x(q^5) = x_5$, and $\phi^2(e^{-\pi/\sqrt{5}}) = \sqrt{5}\phi^2(e^{-\pi\sqrt{5}}) = \sqrt{5}z_5$. Thus, from (7.5), we find that

$$5P(e^{-2\pi\sqrt{5}}) - P(e^{-2\pi/\sqrt{5}})$$

$$= \sqrt{5}z_5^2 \{ 3 + 2(x_5(1 - x_5))^{1/2} \} \left\{ \frac{1}{2}(1 + 2(x_5(1 - x_5))^{1/2}) \right\}^{1/2}$$

$$= \sqrt{5}z_5^2 \{ 3 + \sqrt{X_5} \} \left\{ \frac{1}{2}(1 + \sqrt{X_5}) \right\}^{1/2}, \tag{7.6}$$

where $X_5 = 4x_5(1 - x_5)$.

But, by [8], the singular modulus x_5 is given by

$$x_5 = \frac{1}{2} - \left(\frac{\sqrt{5}-1}{2}\right)^{3/2},$$

so that

$$X_5 = 9 - 4\sqrt{5} \quad \text{and} \quad \sqrt{X_5} = \sqrt{5} - 2. \tag{7.7}$$

Thus, from (7.6), we find that

$$5P(e^{-2\pi\sqrt{5}}) - P(e^{-2\pi/\sqrt{5}}) = \sqrt{5}(\sqrt{5}+1)\left(\frac{\sqrt{5}-1}{2}\right)^{1/2} z_5^2. \tag{7.8}$$

Next, setting $n = 5$ in (3.26), we find that

$$5P(e^{-2\pi\sqrt{5}}) + P(e^{-2\pi/\sqrt{5}}) = \frac{6\sqrt{5}}{\pi}. \tag{7.9}$$

Adding (7.8) and (7.9), we deduce that

$$P(e^{-2\pi\sqrt{5}}) = \frac{3}{\pi\sqrt{5}} + \frac{\sqrt{5}+1}{\sqrt{5}} \cdot \left(\frac{\sqrt{5}-1}{2}\right)^{1/2} z_5^2. \tag{7.10}$$

Now, employing (3.3) in (7.10), we deduce the identity

$$P(e^{-2\pi\sqrt{5}}) = \frac{3}{\pi\sqrt{5}} + \frac{\sqrt{5}+1}{\sqrt{5}} \cdot \left(\frac{\sqrt{5}-1}{2}\right)^{1/2} \sum_{k=0}^{\infty} A_k X_5^k. \tag{7.11}$$

Next, setting $n = 5$ in (3.15), we find that

$$P(e^{-2\pi\sqrt{5}}) = (1 - 2x_5) \sum_{k=0}^{\infty} (3k+1) A_k X_5^k$$

$$= 2(\sqrt{5}-2)^{1/2} \sum_{k=0}^{\infty} (3k+1) A_k X_5^k. \tag{7.12}$$

Using (7.11) and (7.12), we arrive at (7.1). $\qquad\square$

Proof of (7.3) Employing (3.7) in (7.10), we find that

$$P(e^{-2\pi\sqrt{5}}) = \frac{3}{\pi\sqrt{5}} + \frac{\sqrt{5}+1}{\sqrt{5}(1-2x_5)} \cdot \left(\frac{\sqrt{5}-1}{2}\right)^{1/2} \sum_{k=0}^{\infty} (-1)^k B_k W_5^{2k}$$

$$= \frac{3}{\pi\sqrt{5}} + \frac{3+\sqrt{5}}{4\sqrt{5}} \sum_{k=0}^{\infty} (-1)^k B_k W_5^{2k}, \tag{7.13}$$

where

$$W_5 = \frac{2\sqrt{X_5}}{1 - X_5} = \frac{1}{2}.$$

Next, setting $n = 5$ in (3.19), we find that

$$P(e^{-2\pi\sqrt{5}}) = \sum_{k=0}^{\infty} \frac{3k(1 + X_n) + 1 + X_n/2}{1 - X_n} B_k (-1)^k W_n^{2k}$$

$$= \sum_{k=0}^{\infty} \left(\frac{3\sqrt{5}}{2}k + \frac{3\sqrt{5} + 2}{8} \right) (-1)^k B_k W_5^{2k}. \qquad (7.14)$$

From (7.13) and (7.14), we readily arrive at (7.3). Thus, we complete the proof. The proof of (7.2) is similar. ☐

8 Example: $n = 3$

Theorem 8.1 *If A_k, B_k, and C_k, $k \geq 0$, are defined by (3.2), then*

$$\frac{4}{\pi} = \sum_{k=0}^{\infty} (6k + 1) A_k \frac{1}{4^k}, \qquad (8.1)$$

$$\frac{1}{\pi} = \sum_{k=0}^{\infty} (-1)^k \{(15\sqrt{3} - 24)k + 6\sqrt{3} - 10\} A_k 2^k (\sqrt{3} - 1)^{6k}, \qquad (8.2)$$

$$\frac{4\sqrt{2}}{\pi} = \sum_{k=0}^{\infty} (-1)^k \{(30 - 6\sqrt{3})k + 7 - 3\sqrt{3}\} A_k \frac{(2 - \sqrt{3})^{3k}}{2^{4k}}, \qquad (8.3)$$

$$\frac{8\sqrt{2}}{\pi} = \sum_{k=0}^{\infty} \{(85\sqrt{3} - 135)k + 8\sqrt{3} - 12\} B_k \left(\frac{8\sqrt{2}}{51\sqrt{3} - 75} \right)^{2k+1}, \qquad (8.4)$$

$$\frac{5\sqrt{5}}{2\pi\sqrt{3}} = \sum_{k=0}^{\infty} (11k + 1) C_k \left(\frac{4}{125} \right)^k. \qquad (8.5)$$

The identities (8.1) and (8.5) are due to Ramanujan [36, Eqs. (28), (33), resp.], [37, pp. 36–37]. The remaining identities are new. Because the proofs are similar to those in previous sections, we do not give them. Our proofs use the identity

$$f_3(e^{-\pi/\sqrt{3}}) = 3P(e^{-2\pi\sqrt{3}}) - P(e^{-2\pi/\sqrt{3}}) = \frac{3\sqrt{3}}{2} z_3^2.$$

9 Example: $n = 7$

Theorem 9.1 *If A_k, B_k, and C_k, $k \geq 0$, are defined by (3.2), then*

$$\frac{16}{\pi} = \sum_{k=0}^{\infty}(42k+5)A_k\frac{1}{2^{6k}}, \tag{9.1}$$

$$\frac{1}{\pi} = \sum_{k=0}^{\infty}(-1)^k\{(255\sqrt{7}-672)k+112\sqrt{7}-296\}A_k(32-12\sqrt{7})^{3k}, \tag{9.2}$$

$$\frac{8\sqrt{2}}{\pi} = \sum_{k=0}^{\infty}\{(102\sqrt{7}-210)k+35\sqrt{7}-89\}(-1)^k A_k\left(\frac{8-3\sqrt{7}}{4}\right)^{3k}, \tag{9.3}$$

$$\frac{29241}{\pi} = \sum_{k=0}^{\infty}\{(76160-455\sqrt{7})k+784\sqrt{7}+6728\}$$

$$\times B_k\left(\frac{8\sqrt{2}(325+119\sqrt{7})}{29241}\right)^{2k}, \tag{9.4}$$

$$\frac{9\sqrt{7}}{\pi} = \sum_{k=0}^{\infty}(65k+8)(-1)^k B_k\left(\frac{16}{63}\right)^{2k}, \tag{9.5}$$

$$\frac{85\sqrt{85}}{18\pi\sqrt{3}} = \sum_{k=0}^{\infty}(133k+8)C_k\left(\frac{4}{85}\right)^{3k}, \tag{9.6}$$

$$\frac{5\sqrt{15}}{\pi} = \sum_{k=0}^{\infty}(63k+8)(-1)^k C_k\left(\frac{4}{5}\right)^{3k}. \tag{9.7}$$

The identities (9.1) and (9.6) are due to Ramanujan [36, Eqs. (29), (34), resp.], [37, pp. 36–37], and (9.5) is due to Berndt, Chan, and Liaw [12]. The other four identities seem to be new. Because the proofs are similar to those in previous sections, we do not record them, but we do note that we need the calculation

$$f_7(e^{-\pi/\sqrt{7}}) = 7P(e^{-2\pi\sqrt{7}}) - P(e^{-2\pi/\sqrt{7}}) = \frac{27\sqrt{7}}{8}z_7^2.$$

10 Example: $n = 9$

Theorem 10.1 *If A_k and B_k, $k \geq 0$, are defined by (3.2), then*

$$\frac{12^{1/4}}{\pi} = \sum_{k=0}^{\infty}\{(24\sqrt{3}-36)k+9\sqrt{3}-15\}A_k(2-\sqrt{3})^{4k}, \tag{10.1}$$

$$\frac{4}{\pi} = \sum_{k=0}^{\infty} [6\{9\sqrt{2} + 7\sqrt{6} - 3^{1/4}(5\sqrt{3} + 11)\}k$$

$$+ 15\sqrt{6} + 21\sqrt{2} - 3^{1/4}(13\sqrt{3} + 27)]$$

$$\times (-1)^k A_k \left(\frac{(\sqrt{3} + 1)^{1/3}(\sqrt{2} - 3^{1/4})}{2^{2/3}} \right)^{6k}, \tag{10.2}$$

$$\frac{16}{\pi\sqrt{3}} = \sum_{k=0}^{\infty} (-1)^k (28k + 3) B_k \frac{1}{48^k}. \tag{10.3}$$

The identity (10.1) is due to J.M. and P.B. Borwein [13, 14]; (10.3) is due to Ramanujan [36, Eq. (36)], [37, p. 38]; and (10.2) is new. The proofs are similar to those in previous sections. Our proofs depend on the calculation

$$f_9(e^{-\pi/3}) = 9P(e^{-6\pi}) - P(e^{-2\pi/3}) = 9\sqrt{2} \cdot 3^{1/4}(\sqrt{3} - 1) z_9^2.$$

11 Example: $n = 10$

Theorem 11.1 *If A_k and B_k, $k \geq 0$, are defined by (3.2), then*

$$\frac{3\sqrt{2} + \sqrt{5} + 2}{\pi} = \sum_{k=0}^{\infty} \{(15\sqrt{2} + 6\sqrt{10} - 6\sqrt{5})k + 2\sqrt{10} - 3\sqrt{5} + 5\sqrt{2} - 4\}$$

$$\times A_k \{(3 + \sqrt{5})(2 + \sqrt{5})(3\sqrt{2} - \sqrt{5} - 2)\}^{3k}, \tag{11.1}$$

$$\frac{2}{\pi} = \sum_{k=0}^{\infty} (-1)^k \{(60 - 24\sqrt{5})k + 23 - 10\sqrt{5}\} A_k (\sqrt{5} - 2)^{4k}, \tag{11.2}$$

$$\frac{9}{2\pi\sqrt{2}} = \sum_{k=0}^{\infty} (10k + 1) B_k \frac{1}{92^k}, \tag{11.3}$$

$$\frac{6}{\pi} = \sum_{k=0}^{\infty} (-1)^k [6\{a(90\sqrt{2} + 12\sqrt{10} - 30 - 36\sqrt{5})$$

$$+ b(90 + 12\sqrt{5} - 30\sqrt{2} - 36\sqrt{10})\}k$$

$$+ a(27\sqrt{2} - 29 + 5\sqrt{10} - 12\sqrt{5})$$

$$+ b(39 - 9\sqrt{10} + 8\sqrt{5} - 17\sqrt{2})]$$

$$\times A_k \left(\frac{(\sqrt{10} - 3)(3 - \sqrt{5})(3 - 2\sqrt{2})}{4} \right)^{3k}, \tag{11.4}$$

where

$$a = \sqrt{\frac{\sqrt{10}+1}{2}} \quad and \quad b = \sqrt{\frac{\sqrt{10}-1}{2}}.$$

The identity (11.3) is due to Ramanujan [36, Eq. (41)], [37, p. 38]. The remaining three are new. As in the foregoing sections, we do not give the proofs. In our proofs we need the identity

$$f_{10}(e^{-\pi/\sqrt{10}}) = 10P(e^{-2\pi\sqrt{10}}) - P(e^{-2\pi/\sqrt{10}})$$
$$= (4\sqrt{5}+5)\left(1 + (\sqrt{10}-3)^2(3-2\sqrt{2})^2\right) z_{10}^2.$$

12 Example: $n = 13$

Theorem 12.1 *If A_k and B_k, $k \geq 0$, are defined by (3.2), then*

$$\frac{\sqrt{5\sqrt{13}-17}}{\pi} = \sum_{k=0}^{\infty} \{(273\sqrt{2}-75\sqrt{26})k + 112\sqrt{2}-31\sqrt{26}\}$$
$$\times A_k(5\sqrt{13}-18)^{2k}, \tag{12.1}$$

$$\frac{12}{\pi} = \sum_{k=0}^{\infty} (-1)^k \{12\sqrt{13}(1+2(c-d)^4)k$$
$$+ 3\sqrt{13}-7 + (9\sqrt{13}+7)(c-d)^4\}$$
$$\times A_k(\sqrt{13}+3)^{3k} \left(\sqrt{\frac{19+5\sqrt{13}}{2}} - \sqrt{\frac{17+5\sqrt{13}}{2}}\right)^{6k}, \tag{12.2}$$

$$\frac{72}{\pi} = \sum_{k=0}^{\infty} (-1)^k (260k+23)B_k \frac{1}{18^{2k}}, \tag{12.3}$$

where

$$c = \sqrt{\frac{19+5\sqrt{13}}{2}} \quad and \quad d = \sqrt{\frac{17+5\sqrt{13}}{2}}.$$

The identity (12.3) is due to Ramanujan [36, Eq. (37)], [37, p. 38]. The remaining identities are new.

Proof Ramanujan did not record any expression for $f_{13}(q) = 13P(q^{26}) - P(q^2)$. So our first task is to find an expression for $f_{13}(e^{-\pi/\sqrt{13}})$.

From (2.3) and (2.10) of Lemma 2.1, we find that

$$\frac{q^{p/12}f(-q^{2p})}{q^{1/12}f(-q^2)} = \frac{1}{\sqrt{m}}\left(\frac{\beta(1-\beta)}{\alpha(1-\alpha)}\right)^{1/12}, \tag{12.4}$$

where β has degree p over α, and $m = \phi^2(q)/\phi^2(q^p)$. Taking logarithms on both sides of (12.4), differentiating with respect to q, and then using the definition of $P(q)$ from (3.1), we arrive at

$$\frac{1}{12q}\{pP(q^{2p}) - P(q^2)\} = \frac{1}{12\beta(1-\beta)} \cdot \frac{d}{dq}(\beta(1-\beta))$$

$$- \frac{1}{12\alpha(1-\alpha)} \cdot \frac{d}{dq}(\alpha(1-\alpha)) - \frac{1}{2m}\frac{dm}{dq} \quad (12.5)$$

$$= \frac{1-2\beta}{12\beta(1-\beta)} \cdot \frac{d\beta}{d\alpha} \cdot \frac{d\alpha}{dq} - \frac{1-2\alpha}{12\alpha(1-\alpha)} \cdot \frac{d\alpha}{dq}$$

$$- \frac{1}{2m}\frac{dm}{d\alpha} \cdot \frac{d\alpha}{dq}. \quad (12.6)$$

Now, from Entry 9(i) of Chap. 17 and Entry 24(vi) in Chap. 18 in Ramanujan's second notebook [38], [6, pp. 120, 217], we find that

$$\frac{d\alpha}{dq} = \frac{\alpha(1-\alpha)\phi^4(q)}{q} \quad \text{and} \quad \frac{d\beta}{d\alpha} = \frac{\beta(1-\beta)}{\alpha(1-\alpha)} \cdot \frac{p}{m^2}. \quad (12.7)$$

Employing (12.7) in (12.5) and then simplifying, we find that

$$pP(q^{2p}) - P(q^2) = \phi^2(q)\phi^2(q^p)\left\{\frac{p(1-2\beta)}{m} - (1-2\alpha)m - 6\alpha(1-\alpha)\frac{dm}{d\alpha}\right\}. \quad (12.8)$$

Next, we record the following modular equation of degree 13 from Chap. 20 of Ramanujan's second notebook [38], [6, p. 376, Entry 8(iii)]. If β has degree 13 over α, then

$$m = \left(\frac{\beta}{\alpha}\right)^{1/4} + \left(\frac{1-\beta}{1-\alpha}\right)^{1/4} - \left(\frac{\beta(1-\beta)}{\alpha(1-\alpha)}\right)^{1/4} - 4\left(\frac{\beta(1-\beta)}{\alpha(1-\alpha)}\right)^{1/6}. \quad (12.9)$$

Differentiating (12.9) with respect to α, we find that

$$\frac{dm}{d\alpha} = \frac{1}{\sqrt{\alpha}}\left\{\frac{\alpha^{1/4}}{4\beta^{3/4}} \cdot \frac{d\beta}{d\alpha} - \frac{\beta^{1/4}}{4\alpha^{3/4}}\right\} + \frac{1}{\sqrt{1-\alpha}}\left\{\frac{-(1-\alpha)^{1/4}}{4(1-\beta)^{3/4}} \cdot \frac{d\beta}{d\alpha} + \frac{(1-\beta)^{1/4}}{4(1-\alpha)^{3/4}}\right\}$$

$$- \frac{1}{\sqrt{\alpha(1-\alpha)}}\left\{\frac{(\alpha(1-\alpha))^{1/4}(1-2\beta)}{4(\beta(1-\beta))^{3/4}} \cdot \frac{d\beta}{d\alpha} - \frac{(\beta(1-\beta))^{1/4}(1-2\alpha)}{4(\alpha(1-\alpha))^{3/4}}\right\}$$

$$- \frac{4}{(\alpha(1-\alpha))^{1/3}}\left\{\frac{(\alpha(1-\alpha))^{1/6}(1-2\beta)}{6(\beta(1-\beta))^{5/6}} \cdot \frac{d\beta}{d\alpha} - \frac{(\beta(1-\beta))^{1/6}(1-2\alpha)}{6(\alpha(1-\alpha))^{5/6}}\right\}. \quad (12.10)$$

Next, if $q = e^{-\pi/\sqrt{13}}$, then, by (3.14), $\alpha = 1 - \beta = 1 - x_{13}$, $m = \sqrt{13}$, and, by (12.7),

$$\left[\frac{d\beta}{d\alpha}\right]_{q=e^{-\pi/\sqrt{13}}} = 1.$$

Thus, setting $q = e^{-\pi/\sqrt{13}}$ in (12.10), we deduce that

$$\left[\frac{dm}{d\alpha}\right]_{q=e^{-\pi/\sqrt{13}}} = \frac{1 - 2x_{13}}{4x_{13}(1 - x_{13})}\left\{\left(\frac{x_{13}}{1 - x_{13}}\right)^{1/4} + \left(\frac{1 - x_{13}}{x_{13}}\right)^{1/4}\right\}$$
$$- \frac{11(1 - 2x_{13})}{6x_{13}(1 - x_{13})}. \tag{12.11}$$

Now, setting $p = 13$ and $q = e^{-\pi/\sqrt{13}}$ in (12.8), and then using (12.11), we find that

$$f_{13}(e^{-\pi/\sqrt{13}}) = 13P(e^{-2\pi\sqrt{13}}) - P(e^{-2\pi/\sqrt{13}})$$
$$= \sqrt{13}(1 - 2x_{13})\Big[(11 + 2\sqrt{13})$$
$$- \frac{3}{2}\left\{\left(\frac{x_{13}}{1 - x_{13}}\right)^{1/4} + \left(\frac{1 - x_{13}}{x_{13}}\right)^{1/4}\right\}\Big]z_{13}^2. \tag{12.12}$$

From [11], we note that

$$x_{13} = \frac{1}{2}\left(\frac{\sqrt{13} - 3}{2}\right)^3 \left(\sqrt{\frac{7 + \sqrt{13}}{4}} - \sqrt{\frac{3 + \sqrt{13}}{4}}\right)^4,$$

so that

$$1 - x_{13} = \frac{1}{2}\left(\frac{\sqrt{13} - 3}{2}\right)^3 \left(\sqrt{\frac{7 + \sqrt{13}}{4}} + \sqrt{\frac{3 + \sqrt{13}}{4}}\right)^4,$$

$$1 - 2x_{13} = 1 - x_{13} - x_{13}$$
$$= \frac{1}{2}\left(\frac{\sqrt{13} - 3}{2}\right)^3$$
$$\times \left\{\left(\sqrt{\frac{7 + \sqrt{13}}{4}} + \sqrt{\frac{3 + \sqrt{13}}{4}}\right)^4 - \left(\sqrt{\frac{7 + \sqrt{13}}{4}} - \sqrt{\frac{3 + \sqrt{13}}{4}}\right)^4\right\}$$
$$= (7\sqrt{13} - 25)\sqrt{\frac{17 + 5\sqrt{13}}{2}}, \tag{12.13}$$

and

$$\left(\frac{x_{13}}{1 - x_{13}}\right)^{1/4} + \left(\frac{1 - x_{13}}{x_{13}}\right)^{1/4} = 5 + \sqrt{13}. \tag{12.14}$$

With the help of (12.13) and (12.14), we can rewrite (12.12) in the form

$$f_{13}(e^{-\pi/\sqrt{13}}) = 13P(e^{-2\pi\sqrt{13}}) - P(e^{-2\pi/\sqrt{13}})$$

$$= (7\sqrt{13} - 25)\frac{13 + 7\sqrt{13}}{2} \cdot \sqrt{\frac{17 + 5\sqrt{13}}{2}} z_{13}^2. \qquad (12.15)$$

Now, we prove (12.1). Setting $n = 13$ in (3.26), we find that

$$13P(e^{-2\pi\sqrt{13}}) + P(e^{-2\pi/\sqrt{13}}) = \frac{6\sqrt{13}}{\pi}. \qquad (12.16)$$

Adding (12.15) and (12.16), we obtain

$$P(e^{-2\pi\sqrt{13}}) = \frac{3}{\pi\sqrt{13}} + (7\sqrt{13} - 25)\frac{7 + \sqrt{13}}{4\sqrt{13}} \cdot \sqrt{\frac{17 + 5\sqrt{13}}{2}} z_{13}^2. \qquad (12.17)$$

Employing (3.3) in (12.17), we find that

$$P(e^{-2\pi\sqrt{13}}) = \frac{3}{\pi\sqrt{13}} + (7\sqrt{13} - 25)\frac{7 + \sqrt{13}}{4\sqrt{13}} \cdot \sqrt{\frac{17 + 5\sqrt{13}}{2}} \sum_{k=0}^{\infty} A_k X_{13}^k, \qquad (12.18)$$

where $X_{13} = 4x_{13}(1 - x_{13}) = \{(\sqrt{13} - 3)/2\}^6$.

Next, setting $n = 13$ in (3.15), we obtain

$$P(e^{-2\pi\sqrt{13}}) = (1 - 2x_{13}) \sum_{k=0}^{\infty} (3k + 1)A_k X_{13}^k$$

$$= (7\sqrt{13} - 25)\sqrt{\frac{17 + 5\sqrt{13}}{2}} \sum_{k=0}^{\infty} (3k + 1)A_k X_{13}^k. \qquad (12.19)$$

From (12.18) and (12.19), we readily arrive at (12.1).

Similarly, we can prove (12.2) and (12.3). □

Because the proofs in the closing Sects. 13–18 follow along the same lines as those in previous sections, we omit the proofs.

13 Example: $n = 14$

Theorem 13.1 *If B_k, $k \geq 0$, is defined by (3.2), then*

$$\frac{1}{\pi\sqrt{14}\sqrt{44\sqrt{2} - 50}} = \sum_{k=0}^{\infty} \left\{ \frac{2}{7}k + \frac{3}{196}(3 - \sqrt{2}) \right\} B_k \left(\frac{1}{11 + 8\sqrt{2}} \right)^{2k}. \qquad (13.1)$$

This identity is due to Berndt, Chan, and Liaw [12]. Our proof uses the identity

$$f_{14}(e^{-\pi/\sqrt{14}}) = 14P(e^{-2\pi\sqrt{14}}) - P(e^{-2\pi/\sqrt{14}})$$

$$= \{7(1+x_{14}) + 6\sqrt{14}\left(3+2\sqrt{2}\right)x_{14}^{1/4}(1-x_{14})^{1/2}\}z_{14}^2,$$

where

$$x_{14} = 995 + 704\sqrt{2} + 8\sqrt{30926 + 21868\sqrt{2}}$$

$$-2\left\{990130 + 700128\sqrt{2} + 4(6218652 + 4397251\sqrt{2})\sqrt{1562\sqrt{2} - 2209}\right\}^{1/2}$$

We have calculated the singular modulus x_{14} with the aid of [7, p. 301, Theorem 9.14] and [6, p. 314, Entry 19(i)].

14 Example: $n = 15$

Theorem 14.1 *If A_k and B_k, $k \geq 0$, are defined by (3.2), then*

$$\frac{32}{\pi} = \sum_{k=0}^{\infty}\{(42\sqrt{5} + 30)k + 5\sqrt{5} - 1\}A_k\frac{1}{2^{6k}}\left(\frac{\sqrt{5} - 1}{2}\right)^{8k}, \tag{14.1}$$

$$\frac{121\sqrt{3}}{\pi} = \sum_{k=0}^{\infty}(-1)^k\{(1365\sqrt{5} - 2240)k + 300\sqrt{5} - 604\}B_k\left(\frac{4}{45}\right)^k. \tag{14.2}$$

The first identity is due to Ramanujan [36, Eq. (30)], [37, p. 37], and the other is new. In our proof we use the identity

$$f_{15}(e^{-\pi/\sqrt{15}}) = 15P(e^{-2\pi\sqrt{15}}) - P(e^{-2\pi/\sqrt{15}}) = \frac{3\sqrt{15}(11 + 9\sqrt{5})}{16}z_{15}^2.$$

15 Example: $n = 17$

Theorem 15.1 *If B_k, $k \geq 0$, is defined by (3.2), then*

$$\frac{16\sqrt{38 + 10\sqrt{17}}}{\pi} = \sum_{k=0}^{\infty}(-1)^k\{(60\sqrt{17} + 340)k + 33 + 3\sqrt{17}\}$$

$$\times B_k\left(\frac{25\sqrt{17} - 103}{128}\right)^k. \tag{15.1}$$

The representation (15.1) is new. Our proof depends on the identity

$$f_{17}(e^{-\pi/\sqrt{17}}) = 17P(e^{-2\pi\sqrt{17}}) - P(e^{-2\pi/\sqrt{17}})$$

$$= \frac{\sqrt{17}}{2}\left(\sqrt{17} + 3\sqrt{4+\sqrt{17}}\right)(1-2x_{17})\,z_{17}^2,$$

where, from [7, p. 290, Theorem 9.9],

$$x_{17} = \frac{1}{2}\left(\sqrt{\frac{7+\sqrt{17}}{4}} - \sqrt{\frac{3+\sqrt{17}}{4}}\right)^4 \left(\sqrt{\frac{3+\sqrt{4+\sqrt{17}}}{4}} - \sqrt{\frac{\sqrt{4+\sqrt{17}}-1}{4}}\right)^8.$$

16 Example: $n = 18$

Theorem 16.1 *If A_k and B_k, $k \geq 0$, are defined by (3.2), then*

$$\frac{2}{\pi} = \sum_{k=0}^{\infty}(-1)^k(84k+21-6\sqrt{6})A_k(\sqrt{3}-\sqrt{2})^{8k+2}, \tag{16.1}$$

$$\frac{2\sqrt{\sqrt{2}-1}}{3\pi} = \sum_{k=0}^{\infty}(-1)^k\{2(106+45\sqrt{6}-65\sqrt{3}-72\sqrt{2})k+90+38\sqrt{6}$$

$$-55\sqrt{3}-62\sqrt{2}\}A_k\left(\frac{(5\sqrt{2}-7)^3(7-4\sqrt{3})^3(5-2\sqrt{6})^2}{8}\right)^k, \tag{16.2}$$

$$\frac{1}{3\pi\sqrt{3}} = \sum_{k=0}^{\infty}(40k+3)B_k\frac{1}{492^k}. \tag{16.3}$$

The identity (16.1) is due to the Borwein brothers [13, 14], and (16.3) is due to Ramanujan [36, Eq. (42)], [37, p. 38]. Identity (16.2) is new. In our proof we employ the identity

$$f_{18}(e^{-\pi/3\sqrt{2}}) = 18P(e^{-6\pi\sqrt{2}}) - P(e^{-2\pi/3\sqrt{2}})$$

$$= 18\sqrt{2}(5+2\sqrt{6})(7+6\sqrt{6})(5\sqrt{2}-7)(7-4\sqrt{3})\,z_{18}^2.$$

17 Example: $n = 22$

Theorem 17.1 *If A_k and B_k, $k \geq 0$, are defined by (3.2), then*

$$\frac{2}{\pi} = \sum_{k=0}^{\infty}(-1)^k\{(660\sqrt{2}-924)k+284\sqrt{2}-401\}A_k(\sqrt{2}-1)^{12k}, \tag{17.1}$$

$$\frac{2}{\pi\sqrt{11}} = \sum_{k=0}^{\infty}(280k + 19)B_k\frac{1}{99^{2k+1}}. \tag{17.2}$$

Identity (17.2) is due to Ramanujan [36, Eq. (43)], [37, p. 38], while (17.1) is new. Our proof depends upon the identity

$$f_{22}\left(e^{-\pi/\sqrt{22}}\right) = 22P(e^{-2\pi\sqrt{22}}) - P(e^{-2\pi/\sqrt{22}})$$

$$= 6\sqrt{22}(11 + 17\sqrt{2})(10 - 3\sqrt{11})(3\sqrt{11} - 7\sqrt{2})(5\sqrt{2} + 7)\,z_{22}^2.$$

18 Example: $n = 25$

Theorem 18.1 *If A_k and B_k, $k \geq 0$, are defined by (3.2), then*

$$\frac{5^{1/4}}{\pi} = \sum_{k=0}^{\infty}\{(540\sqrt{5} - 1200)k + 235\sqrt{5} - 525\}A_k(\sqrt{5} - 2)^{8k}, \tag{18.1}$$

$$\frac{288}{\pi\sqrt{5}} = \sum_{k=0}^{\infty}(-1)^k(644k + 41)B_k\frac{1}{(72\sqrt{5})^{2k}}. \tag{18.2}$$

Identity (18.2) is due to Ramanujan [36, Eq. (38)]; the identity (18.1) is new. Our proof uses the identity

$$f_{25}(e^{-\pi/5}) = 25P(e^{-10\pi}) - P(e^{-2\pi/5}) = 300 \cdot 5^{1/4}(\sqrt{5} - 2)^{5/3}\,z_{25}^2.$$

References

1. Andrews, G.E., Berndt, B.C.: Ramanujan's Lost Notebook, Part II. Springer, New York (2009)
2. Bailey, W.N.: Generalized Hypergeometric Series. Cambridge University Press, London (1935)
3. Baruah, N.D., Berndt, B.C.: Ramanujan's series for $1/\pi$ arising from his cubic and quartic theories of elliptic functions. J. Math. Anal. Appl. **341**, 357–371 (2008)
4. Bauer, G.: Von den Coefficienten der Reihen von Kugelfunctionen einer Variabeln. J. Reine Angew. Math. **56**, 101–121 (1859)
5. Berndt, B.C.: Ramanujan's Notebooks, Part II. Springer, New York (1989)
6. Berndt, B.C.: Ramanujan's Notebooks, Part III. Springer, New York (1991)
7. Berndt, B.C.: Ramanujan's Notebooks, Part V. Springer, New York (1998)
8. Berndt, B.C., Chan, H.H.: Notes on Ramanujan's singular moduli. In: Gupta, R., Williams, K.S. (eds.) Number Theory, pp. 7–16. Fifth Conference of the Canadian Number Theory Association. American Mathematical Society, Providence (1999)
9. Berndt, B.C., Chan, H.H.: Eisenstein series and approximations to π. Ill. J. Math. **45**, 75–90 (2001)
10. Berndt, B.C., Rankin, R.A.: Ramanujan: Letters and Commentary. American Mathematical Society/London Mathematical Society, Providence/London (1995)
11. Berndt, B.C., Chan, H.H., Zhang, L.-C.: Ramanujan's singular moduli. Ramanujan J. **1**, 53–74 (1997)
12. Berndt, B.C., Chan, H.H., Liaw, W.-C.: On Ramanujan's quartic theory of elliptic functions. J. Number Theory **88**, 129–156 (2001)
13. Borwein, J.M., Borwein, P.B.: Pi and the AGM; A Study in Analytic Number Theory and Computational Complexity. Wiley, New York (1987)
14. Borwein, J.M., Borwein, P.B.: Ramanujan's rational and algebraic series for $1/\pi$. J. Indian Math. Soc. **51**, 147–160 (1987)

15. Borwein, J.M., Borwein, P.B.: More Ramanujan-type series for $1/\pi$. In: Andrews, G.E., Askey, R.A., Berndt, B.C., Ramanathan, K.G., Rankin, R.A. (eds.) Ramanujan Revisited, pp. 359–374. Academic Press, Boston (1988)
16. Borwein, J.M., Borwein, P.B.: Some observations on computer aided analysis. Not. Am. Math. Soc. **39**, 825–829 (1992)
17. Borwein, J.M., Borwein, P.B.: Class number three Ramanujan type series for $1/\pi$. J. Comput. Appl. Math. **46**, 281–290 (1993)
18. Borwein, J.M., Borwein, P.B., Bailey, D.H.: Ramanujan, modular equations, and approximations to pi or how to compute one billion digits of pi. Am. Math. Mon. **96**, 201–219 (1989)
19. Chan, H.H., Liaw, W.-C.: Cubic modular equations and new Ramanujan-type series for $1/\pi$. Pac. J. Math. **192**, 219–238 (2000)
20. Chan, H.H., Loo, K.P.: Ramanujan's cubic continued fraction revisited. Acta Arith. **126**, 305–313 (2007)
21. Chan, H.H., Verrill, H.: The Apéry numbers, the Almkvist-Zudilin numbers and new series for $1/\pi$. Math. Res. Lett. **16**, 405–420 (2009)
22. Chan, H.H., Liaw, W.-C., Tan, V.: Ramanujan's class invariant λ_n and a new class of series for $1/\pi$. J. Lond. Math. Soc. **64**(2), 93–106 (2001)
23. Chan, H.H., Chan, S.H., Liu, Z.: Domb's numbers and Ramanujan-Sato type series for $1/\pi$. Adv. Math. **186**, 396–410 (2004)
24. Chowla, S.: Series for $1/K$ and $1/K^2$. J. Lond. Math. Soc. **3**, 9–12 (1928)
25. Chowla, S.: On the sum of a certain infinite series. Tôhoku Math. J. **29**, 291–295 (1928)
26. Chowla, S.: The Collected Papers of Sarvadaman Chowla, vol. 1. Les Publications Centre de Recherches Mathématiques, Montreal (1999)
27. Chudnovsky, D.V., Chudnovsky, G.V.: Approximation and complex multiplication according to Ramanujan. In: Andrews, G.E., Askey, R.A., Berndt, B.C., Ramanathan, K.G., Rankin, R.A. (eds.) Ramanujan Revisited, pp. 375–472. Academic Press, Boston (1988)
28. Guillera, J.: Some binomial series obtained by the WZ-method. Adv. Appl. Math. **29**, 599–603 (2002)
29. Guillera, J.: About a new kind of Ramanujan-type series. Exp. Math. **12**, 507–510 (2003)
30. Guillera, J.: Generators of some Ramanujan formulas. Ramanujan J. **11**, 41–48 (2006)
31. Guillera, J.: A new method to obtain series for $1/\pi$ and $1/\pi^2$. Exp. Math. **15**, 83–89 (2006)
32. Guillera, J.: Hypergeometric identities for 10 extended Ramanujan type series. Ramanujan J. **15**, 219–234 (2008)
33. Hardy, G.H.: Some formulae of Ramanujan. Proc. Lond. Math. Soc. **22**(2), xii–xiii (1924)
34. Hardy, G.H.: Ramanujan. Cambridge University Press, Cambridge (1940); reprinted by Chelsea, New York (1960); reprinted by the American Mathematical Society, Providence (1999)
35. Hardy, G.H.: Collected Papers, vol. 4. Clarendon Press, Oxford (1969)
36. Ramanujan, S.: Modular equations and approximations to π. Quart. J. Math. (Oxford) **45**, 350–372 (1914)
37. Ramanujan, S.: Collected Papers. Cambridge University Press, Cambridge (1927); reprinted by Chelsea, New York (1962); reprinted by the American Mathematical Society, Providence (2000)
38. Ramanujan, S.: Notebooks (2 volumes). Tata Institute of Fundamental Research, Bombay (1957)
39. Ramanujan, S.: The Lost Notebook and Other Unpublished Papers. Narosa, New Delhi (1988)
40. Zudilin, W.: Ramanujan-type formulae and irrationality measures of certain multiples of π. Mat. Sb. **196**, 51–66 (2005)
41. Zudilin, W.: Quadratic transformations and Guillera's formulae for $1/\pi^2$. Math. Zametki **81**, 335–340 (2007) (Russian); Math. Notes **81**, 297–301 (2007)
42. Zudilin, W.: Ramanujan-type formulae for $1/\pi$: A second wind? In: Yui, N., Verrill, H., Doran, C.F. (eds.) Modular Forms and String Duality. Fields Institute Communications, vol. 54, pp. 179–188. American Mathematical Society & The Fields Institute for Research in Mathematical Sciences, Providence (2008)

Ramanujan J (2010) 23: 45–90
DOI 10.1007/s11139-008-9150-0

Parity in partition identities

George E. Andrews

Received: 11 November 2008 / Accepted: 26 November 2008 / Published online: 17 April 2010
© Springer Science+Business Media, LLC 2010

Abstract This paper considers a variety of parity questions connected with classical partition identities of Euler, Rogers, Ramanujan and Gordon. We begin by restricting the partitions in the Rogers-Ramanujan-Gordon identities to those wherein even parts appear an even number of times. We then take up questions involving sequences of alternating parity in the parts of partitions. This latter study leads to: (1) a bi-basic q-binomial theorem and q-binomial series, (2) a new interpretation of the Rogers-Ramanujan identities, and (3) a new natural interpretation of the fifth-order mock theta functions $f_0(q)$ along with a new proof of the Hecke-type series representation.

Keywords Partitions · Rogers-Ramanujan · Parity index

Mathematics Subject Classification (2000) Primary 11P83 · 11P81 · 05A19 · 05A17

1 Introduction

Parity has played a role in partition identities from the beginning. For example [7, p. 5].

Euler's partition identity The number of partitions of any positive integer n into distinct parts equals the number of partitions of n into *odd* parts.

Partially supported by National Science Foundation Grant DMS 0200097.

G.E. Andrews (✉)
Department of Mathematics, The Pennsylvania State University, University Park, PA 16802, USA
e-mail: andrews@math.psu.edu

Equivalently in terms of generating functions: for $|q| < 1$, [7, p. 5, eq. (1.2.5)]

$$\prod_{n=1}^{\infty} (1 + q^n) = \prod_{n=1}^{\infty} \frac{1}{1 - q^{2n-1}}. \tag{1.1}$$

Now the Rogers-Ramanujan identities [7, p. 109] do not immediately involve parity.

First Rogers-Ramanujan identity The number of partitions of any positive integer n into distinct non-consecutive parts equals the number of partitions of n into parts congruent to ± 1 (mod 5).

Equivalently in terms of generating functions: for $|q| < 1$, [7, p. 113, Cor. 7.9]

$$1 + \sum_{n=1}^{\infty} \frac{q^{n^2}}{(1 - q)(1 - q^2) \cdots (1 - q^n)} = \prod_{n=1}^{\infty} \frac{1}{(1 - q^{5n-1})(1 - q^{5n-4})}. \tag{1.2}$$

However B. Gordon [15, 17] and H. Göllnitz [13, 14] independently introduced parity considerations as follows:

First Göllnitz-Gordon identity The number of partitions of n into distinct non-consecutive parts with no even parts differing by exactly 2 equals the number of partitions of n into parts $\equiv 1, 4$, or 7 (mod 8).

Equivalently in terms of generating functions: for $|q| < 1$,

$$1 + \sum_{n=1}^{\infty} \frac{q^{n^2}(1 + q)(1 + q^3) \cdots (1 + q^{2n-1})}{(1 - q^2)(1 - q^4) \cdots (1 - q^{2n})}$$

$$= \prod_{n=1}^{\infty} \frac{1}{(1 - q^{8n-1})(1 - q^{8n-4})(1 - q^{8n-7})}. \tag{1.3}$$

There are several results of this sort related to the Rogers-Ramanujan identities [14]. In addition, in his Lost Notebook [8, 10] Ramanujan found q series identities such as [8, p. 57, eq. $(1.10)_R$]

$$\sum_{n=1}^{\infty} \frac{(1 - q^n)}{(1 + q^n)} \left(1 + \sum_{m=1}^{\infty} \frac{q^{m(m+1)/2}}{(1 - q^2)(1 - q^4) \cdots (1 - q^{2m})} \right)$$

$$+ \sum_{n=0}^{\infty} \frac{(-1)^n q^{n(n+1)/2}}{(1 + q)^2 (1 + q^2)^2 \cdots (1 + q^n)^2} = 2 \sum_{n=0}^{\infty} \frac{(-q)^{n(n+1)/2}}{(1 + q^2)(1 + q^4) \cdots (1 + q^{2n})}. \tag{1.4}$$

It was noted in [8, p. 55], that

$$\sum_{n=0}^{\infty} OE(n)q^n := 1 + \sum_{n=1}^{\infty} \frac{q^{n(n+1)/2}}{(1-q^2)(1-q^4)\cdots(1-q^{2n})} \qquad (1.5)$$

is the generating function for $OE(n)$, the number of partitions of n into distinct parts in which the parity of parts alternates and the smallest part is odd.

These examples from the literature foreshadow the deeper examination of parity in partition identities that will be undertaken here.

Our first major exploration will concern parity in the celebrated Rogers-Ramanujan-Gordon identities [1, 16]:

Rogers-Ramanujan-Gordon identities Let $B_{k,a}(n)$ denote the number of partitions of n for the form

$$b_1 + b_2 + \cdots + b_j,$$

where $b_i \geq b_{j+1}$, $b_i - b_{i+k-1} \geq 2$ and at most $a - 1$ of the b_i are equal to 1 and $1 \leq a \leq k$. Let $A_{k,a}(n)$ denote the number of partitions of n into parts $\not\equiv 0, \pm a \pmod{2k+1}$. Then for all $n \geq 0$,

$$A_{k,a}(n) = B_{k,a}(n).$$

After Gordon's proof of this theorem in 1961 [16], there was subsequently discovered a generating function version in 1974 [6]: for $|q| < 1$

$$\sum_{n_1,\ldots,n_{k-1} \geq 0} \frac{q^{N_1^2+N_2^2+\cdots+N_{k-1}^2+N_a+N_{a+1}+\cdots+N_{k-1}}}{(q;q)_{n_1}(q;q)_{n_2}\cdots(q;q)_{n_{k-1}}} = \prod_{\substack{n=1 \\ n \not\equiv 0, \pm a \pmod{2k+1}}}^{\infty} \frac{1}{1-q^n}, \qquad (1.6)$$

where

$$N_j = n_j + n_{j+1} + \cdots + n_{k-1},$$

and

$$(A;q)_n = (A)_n = (1-A)(1-Aq)\cdots(1-Aq^{n-1}).$$

We now involve parity restrictions:

Theorem 1 *Suppose $k \geq a \geq 1$ are integers with $k \equiv a \pmod 2$. Let $\mathcal{W}_{k,a}(n)$ denote the number of those partitions enumerated by $B_{k,a}(n)$ with the added restriction that* even *parts appear an* even *number of times. If k and a are both even, let $G_{k,a}(n)$ denote the number of partitions of n in which no odd part is repeated and no even part $\equiv 0, \pm a \pmod{2k+2}$. If k and a are both odd, let $G_{k,a}(n)$ denote the number of partitions of n into parts that are neither $\equiv 2 \pmod 4$ nor $\equiv 0, \pm a \pmod{2k+2}$. Then for all $n \geq 0$,*

$$\mathcal{W}_{k,a}(n) = G_{k,a}(n).$$

Theorem 2 *Suppose $k \geq a \geq 1$ with k odd and a even. Let $\overline{W}_{k,a}(n)$ denote the number of those partitions enumerated by $B_{k,a}(n)$ with added restriction that* odd *parts appear an even number of times. Then*

$$\sum_{n \geq 0} \overline{W}_{k,a}(n)q^n = \prod_{m=1}^{\infty} \frac{1}{1+q^{2m-1}} \prod_{\substack{n=1 \\ n \not\equiv 0,\pm a \pmod{2k+2}}}^{\infty} \frac{1}{1-q^n}. \tag{1.7}$$

In analogy with (1.4), we shall prove equivalent generating function identities.

Theorem 3 *For $k \geq a \geq 1$, $k \equiv a \pmod 2$,*

$$\sum_{n_1,n_2,\ldots,n_{k-1}} \frac{q^{N_1^2+N_2^2+\cdots+N_{k-1}^2+2N_a+2N_{a+2}+\cdots+2N_{k-2}}}{(q^2;q^2)_{n_1}(q^2;q^2)_{n_2}\cdots(q^2;q^2)_{n_{k-1}}}$$

$$= \sum_{n \geq 0} \mathcal{W}_{k,a}(n)q^n$$

$$= \sum_{n \geq 0} G_{k,a}(n)q^n$$

$$= \frac{(-q;q^2)_\infty(q^a;q^{2k+2})_\infty(q^{2k+2-a};q^{2k+2})_\infty(q^{2k+2};q^{2k+2})_\infty}{(q^2;q^2)_\infty} \tag{1.8}$$

and if $k \geq a \geq 2$, k odd, a even, then

$$\sum_{n_1,\ldots,n_{k-1} \geq 0} \frac{q^{N_1^2+\cdots+N_{k-1}^2+n_1+n_3+\cdots+n_{a-3}+N_{a-1}+N_a+\cdots+N_{k-1}}}{(q^2;q^2)_{n_1}(q^2;q^2)_{n_2}\cdots(q^2;q^2)_{n_{k-1}}}$$

$$= \frac{(-q^2;q^2)_\infty(q^a;q^{2k+2})_\infty(q^{2k+2-a};q^{2k+2})_\infty(q^{2k+2};q^{2k+2})_\infty}{(q^2;q^2)_\infty}$$

$$= \frac{(q^a;q^{2k+2})_\infty(q^{2k+2-a};q^{2k+2})_\infty(q^{2k+2};q^{2k+2})_\infty}{(-q;q^2)_\infty(q;q)_\infty}. \tag{1.9}$$

The second part of our investigation concerns the alternating parity of parts related to partition functions like $OE(n)$ from (1.5). There will be numerous results derived on this topic. We sample the flavor of these results with the following unexpected variation on the first Rogers-Ramanujan identity. To do this we require the following definition:

The *upper even index* of a partition $\lambda_1 + \lambda_2 + \cdots + \lambda_s$ is the number of terms in the longest decreasing subsequence of parts λ_i beginning with an even part and alternating in parity.

Corollary 15 *The number of partitions of n into distinct parts each larger than upper even index equals the number of partitions of n into parts $\equiv \pm 1 \pmod 5$.*

Beyond this variation on the first Rogers-Ramanujan identity, we are led naturally to Ramanujan's mock theta functions [9].

Let us define RUE partitions of n to be those described in the first part of Corollary 15.

Theorem 17 *Let $\mathrm{ru}_e(n)$ (resp. $\mathrm{ru}_o(n)$) denote the number of RUE partitions of n with an even (resp. odd) upper even index. Then for the fifth order mock theta function $f_0(q)$, we have*

$$f_0(q) := \sum_{n=0}^{\infty} \frac{q^{n^2}}{(-q;q)_n} = \sum_{n=0}^{\infty} \left(\mathrm{ru}_e(n) - \mathrm{ru}_o(n)\right)q^n.$$

We shall, in addition, look at a variety of alternating parity questions, and shall be led inexorably to improving the development of the Hecke type series for $f_0(q)$. Surprisingly there is a nice relationship of $f_0(q)$ with the little q-Jacobi polynomials [12, p. 27].

It should be emphasized that there is more in common than just parity considerations between the refinements of Gordon's Theorem (Sects. 2–5) and parity indices (Sects. 6–12). Indeed comparison of Theorem 4 at $k = 2$ and Theorem 18 reveals that we are considering instances of

$$\sum_{n=0}^{\infty} \frac{q^{n^2} x^n f_n(y;q)}{(q^2;q^2)_n}$$

where we want $f_n(0;q) = 1$ and $f_n(1;q) = (-q;q)_n$.

For such polynomials we see that $y = 0$ leads to Euler's series

$$\sum_{n=0}^{\infty} \frac{q^{n^2} x^n}{(q^2;q^2)_n}$$

and $y = 1$ leads to the Rogers-Ramanujan series

$$\sum_{n=0}^{\infty} \frac{q^{n^2} x^n}{(q;q)_n}.$$

This entire project began with the observation that there are two obvious choices for f, namely H_n, the Rogers-Szegö polynomial (defined in Sect. 5) and $(-yq;q)_n$. The first choice leads naturally to Theorem 4 and subsequently to everything in Sects. 2–5; the latter choice leads to the material in Sects. 6–12.

Finally we note that K. Kursungoz has made an extensive combinatorial study of many of the theorems in this paper and has obtained corresponding bijective or sieve-theoretic proofs. We will catalog his achievements in Sect. 13.

2 Background

In subsequent sections, we require a number of results and techniques from the literature. We shall collect all those elements in this section.

Our proof of Theorems 1 and 2 will require the following. As in [1], we define

$$C_{k,i}(x; q) = \sum_{n=0}^{\infty} (-1)^n x^{kn} q^{\frac{1}{2}(2k+1)n(n+1)-in} \frac{(1 - x^i q^{(2n+1)i})(xq; q)_n}{(q; q)_n}, \quad (2.1)$$

$$Q_{k,i}(x; q) = \frac{C_{k,i}(x; q)}{(xq; q)_{\infty}}. \quad (2.2)$$

It was shown in [1] and [3], that

$$Q_{k,0}(x; q) = 0, \quad (2.3)$$

$$Q_{k,k+1}(x; q) = Q_{k,k}(x; q), \quad (2.4)$$

$$Q_{k,-i}(x; q) = -(xq)^{-i} Q_{k,i}(x; q), \quad (2.5)$$

$$Q_{k,i}(x; q) - Q_{k,i-1}(x; q) = x^{i-1} q^{i-1} Q_{k,k-i+1}(xq; q). \quad (2.6)$$

The functions $Q_{k,i}(x; q)$ play a vital role in the proof of the Rogers-Ramanujan-Gordon identities given in the introduction. Namely, if we let $b_{k,a}(m, n)$ denote the number of partitions enumerated by $B_{k,a}(n)$ that have exactly m parts, then [1]

$$Q_{k,a}(x; q) = \sum_{m,n \geq 0} b_{k,a}(m, n) x^m q^n. \quad (2.7)$$

In addition (cf. [1] or [3])

$$Q_{k,a}(1; q) = \frac{(q^a; q^{2k+1})_{\infty}(q^{2k+1-a}; q^{2k+1})_{\infty}(q^{2k+1}; q^{2k+1})_{\infty}}{(q; q)_{\infty}}. \quad (2.8)$$

In [6], it was shown that

$$Q_{k,i}(x; q) = \sum_{n_1,\dots,n_k \geq 0} \frac{q^{N_1^2+N_2^2+\cdots+N_{k-1}^2+N_i+N_{i+1}+\cdots+N_{k-1}} x^{N_1+N_2+\cdots+N_{k-1}}}{(q; q)_{n_1}(q; q)_{n_2}\cdots(q; q)_{n_{k-1}}}. \quad (2.9)$$

We also require the q-binomial coefficients

$$\begin{bmatrix} N \\ M \end{bmatrix}_k = \begin{cases} 0 & \text{if } M < 0 \text{ or } M > N, \\ \frac{(q^k; q^k)_N}{(q^k; q^k)_M (q^k; q^k)_{N-M}} & \text{otherwise.} \end{cases} \quad (2.10)$$

These satisfy two basic recurrences [7, p. 35, eqs. (3.3.4) and (3.3.3)]

$$\begin{bmatrix} N \\ M \end{bmatrix}_k = \begin{bmatrix} N - 1 \\ M - 1 \end{bmatrix}_k + q^{kM} \begin{bmatrix} N - 1 \\ M \end{bmatrix}_k, \quad (2.11)$$

$$\begin{bmatrix} N \\ M \end{bmatrix}_k = \begin{bmatrix} N - 1 \\ M \end{bmatrix}_k + q^{k(N-M)} \begin{bmatrix} N - 1 \\ M - 1 \end{bmatrix}_k. \quad (2.12)$$

We conclude this section with some remarks about standard methods used for partition generating functions.

Often we will refer to $f(x; q)$ as the generating function for all partitions subject to certain constraints C. By this we mean that if $P_C(m, n)$ is the number of partitions of n into m summands subject to the constraint C, then

$$f(x; q) = \sum_{m,n \geq 0} P_C(m, n) x^m q^n.$$

Next we shall often use the *Shift Rule*. Namely

$$f(xq^j; q)$$

is the generating function for partitions subject to C wherein each part has had j added to it.

Additionally we often shall consider that the partitions subject to constraint C must also have all their summands $\leq N$. The generating function will now be denoted by $f(N, x; q)$.

Often we will obtain recurrences for $f(N, x; q)$ utilizing the *Largest Part Decomposition Principle* as follows:

$$f(N, x; q) = f(N - 1, x; q)$$

$$+ \text{generating function for those partitions subject to } C \text{ and}$$

$$\text{having } N \text{ as a summand.}$$

Next let us suppose that adding a given j to each part of every partition constrained by C produces exactly all those partitions constrained by C whose parts are $> j$. Under these circumstances we may have a *Smallest Parts Decomposition Principle* as follows:

$$f(x; q) = f(xq^j; q)$$

$$+ \text{generating function for those partitions subject to } C \text{ and}$$

$$\text{having at least one summand } \leq j.$$

Finally suppose that we have two sets of functions $f_i(x, q)$, $1 \leq i \leq r$ and $g_i(x, q)$, $1 \leq i \leq r$ which are analytic in x and q for $|q| < 1$ and $|x| < |q|^{-1}$. Furthermore suppose that for each i, $f_i(0, q) = g_i(0, q)$,

$$f_i(x, q) = \sum_{j=1}^{r} h_{i,j}(x, q) f_j\left(xq^{e(i,j)}, q\right) \tag{2.13}$$

and

$$g_i(x, q) = \sum_{j=1}^{r} h_{i,j}(x, q) g_j\left(xq^{e(i,j)}, q\right), \tag{2.14}$$

where the $e(i, j)$ are all positive integers and the $h_{i,j}(x, q)$ are polynomials in x and q. Then it follows by a double induction on the double power series coefficients that

$$f_i(x, q) = g_i(x, q), \quad 1 \leqq i \leqq r.$$

We shall refer to this process of identifying two sets of functions as the *Defining q-Difference Equations Principle*.

When we invoke this principle, it may be the case that linear combinations of the given functional equations are necessary to fulfill precisely (2.13) and (2.14).

3 Theorems 1 and 2

Suppose we let $w_{k,a}(m, n)$ (resp. $\overline{w}_{k,a}(m, n)$) denote the number of partitions of the type enumerated by $W_{k,a}(n)$ (resp. $\overline{W}_{k,a}(n)$) that have exactly m parts. The related generating functions are

$$\mathcal{W}_{k,a}(x; q) := \sum_{m,n \geq 0} w_{k,a}(m, n) x^m q^n \tag{3.1}$$

and

$$\overline{\mathcal{W}}_{k,a}(x; q) := \sum_{m,n \geq 0} \overline{w}_{k,a}(m, n) x^m q^n. \tag{3.2}$$

Our first object will be to show that

$$\mathcal{W}_{k,a}(x; q) = (-xq; q^2)_\infty Q_{\frac{k}{2}, \frac{a}{2}}(x^2; q^2) \tag{3.3}$$

provided $k \equiv a \pmod 2$, and when k is odd and a even,

$$\overline{\mathcal{W}}_{k,a}(x; q) = (-xq; q^2)_\infty Q_{\frac{k}{2}, \frac{a}{2}}(x^2; q^2). \tag{3.4}$$

We begin with k even (and thus only consider (3.3)).

Case 1. The even case.   Here we replace k by $2k$ and a by $2a$, and consequently we need (3.2) to show that

$$\mathcal{W}_{2k,2a}(x; q) := (-xq; q^2)_\infty Q_{k,a}(x^2; q^2). \tag{3.5}$$

We start off by adding together instances of (2.6) for $1 \leq i \leq a$ and noting $Q_{k,0}(x^2; q^2) = 0$ by (2.3), thus by (2.6)

$$Q_{k,a}(x^2; q^2) = \sum_{i=1}^{a} (x^2 q^2)^{i-1} Q_{k,k-i+1}(x^2 q^2; q^2)$$

$$= \sum_{i=1}^{a} (x^2 q^2)^{i-1} \sum_{h=1}^{k-i+1} (x^2 q^4)^{h-1} Q_{k,k-h+1}(x^2 q^4; q^2). \tag{3.6}$$

Now defining

$$V_{2k,2a}(x;q) = (-xq;q^2)_\infty Q_{k,a}(x^2;q^2), \qquad (3.7)$$

we see that (3.6) implies

$$V_{2k,2a}(x;q) = (1+xq)\sum_{i=1}^a x^{2(i-1)}q^{\overbrace{1+1+\cdots+1}^{2i-2 \text{ times}}}$$

$$\times \sum_{h=1}^{k-i+1} x^{2(h-1)}q^{\overbrace{2+2+\cdots+2}^{2h-2 \text{ times}}}V_{2k,2k-2h+2}(xq^2;q). \qquad (3.8)$$

However, we may establish combinatorially that

$$\mathcal{W}_{2k,2a}(x;q) = \sum_{i=1}^a \left(x^{2(i-1)}q^{\overbrace{1+1+\cdots+1}^{2i-2 \text{ times}}} + x^{2i-1}q^{\overbrace{1+1+\cdots+1}^{2i-1 \text{ times}}} \right)$$

$$\times \sum_{h=1}^{k-i+1} x^{2(h-1)}q^{\overbrace{2+2+\cdots+2}^{2h-2 \text{ times}}}\mathcal{W}_{2k,2a}(xq^2;q). \qquad (3.9)$$

To see this we note that if in the partitions enumerated by $w_{k,a}(m,n)$ 1 appears $2i-2$ times and 2 appears $2h-2$ times, then 3 can appear at most $2k-(2h-2)-1 = (2k-2h+2)-1$ times. Thus by the Shift Rule, this particular set of partitions is generated by

$$x^{2i-2}q^{2i-2}x^{2h-2}q^{4h-4}\mathcal{W}_{2k,2k-2h+2}(xq^2;q).$$

In precisely the same way, those partitions enumerated by $w_{k,a}(m,n)$ in which 1 appears $2i-1$ times and 2 appears $2h-2$ times have as their generating function

$$x^{2i-1}q^{2i-1}x^{2h-2}q^{4h-4}\mathcal{W}_{2k,2k-2h+2}(xq^2;q).$$

Now summing over $1 \le i \le a$ and $1 \le h \le k-i+1$, we obtain (3.9). Thus by Defining q-Difference Equations Principle

$$V_{2k,2a}(x;q) = \mathcal{W}_{2k,2a}(x;q) \qquad (3.10)$$

which establishes (3.3) in the even case.

Case 2. The odd case. This case is somewhat less intricate because we can combine the proof of (3.3) and (3.4) in this case. Namely replacing k by $2k+1$ and a by $2a+1$ in (3.3) and by $2a$ in (3.4), we see that we need to prove

$$\mathcal{W}_{2k+1,2a+1}(x;q) = (-xq;q^2)_\infty Q_{k+\frac{1}{2},a+\frac{1}{2}}(x^2;q^2) \qquad (3.11)$$

and

$$\overline{\mathcal{W}}_{2k+1,2a}(x;q) = (-xq^2;q^2)_\infty Q_{k+\frac{1}{2},a}(x^2;q^2). \qquad (3.12)$$

Following the example of the previous case, we define

$$V_{2k+1,2a+1}(x;q) = (-xq;q^2)_\infty Q_{k+\frac{1}{2},a+\frac{1}{2}}(x^2;q^2) \tag{3.13}$$

and

$$\overline{V}_{2k+1,2a}(x;q) = (-xq^2;q^2)_\infty Q_{k+\frac{1}{2},a}(x^2;q^2). \tag{3.14}$$

We shall next derive a defining set of q-difference equations and initial conditions for the V's and $\overline{V}$'s. First by (2.3)

$$\overline{V}_{2k+1,0}(x;q) = 0 \tag{3.15}$$

and otherwise

$$1 = V_{2k+1,2a+1}(0;q) = V_{2k+1,2a+1}(x;0) = \overline{V}_{2k+1,2a}(0;q) = \overline{V}_{2k+1,2a}(x;0). \tag{3.16}$$

As for the q-difference equations

$$V_{2k+1}(x;q) = (-xq;q^2)Q_{k+\frac{1}{2},\frac{1}{2}}(x^2;q^2),$$

and by (2.5) and (2.6),

$$Q_{k+\frac{1}{2},\frac{1}{2}}(x^2;q^2)\left(1+(x^2q^2)^{-\frac{1}{2}}\right) = (x^2q^2)^{-\frac{1}{2}}Q_{k+\frac{1}{2},k+1}(x^2q^2;q^2);$$

so

$$V_{2k+1,1}(x;q) = \overline{V}_{2k+1,2k+2}(xq;q). \tag{3.17}$$

For $0 < a \leq k$

$$V_{2k+1,2a+1}(x;q) - V_{2k+1,2a-1}(x;q)$$
$$= (-xq;q^2)_\infty\left(Q_{k+\frac{1}{2},a+\frac{1}{2}}(x^2;q^2) - Q_{k+\frac{1}{2},a-\frac{1}{2}}(x^2;q^2)\right)$$
$$= (-xq;q^2)_\infty Q_{k+\frac{1}{2},k-a+1}(x^2q^2;q^2)(x^2q^2)^{a-\frac{1}{2}}$$
$$= (xq)^{2a-1}(1+xq)\overline{V}_{2k+1,2k-2a+2}(xq;q) \tag{3.18}$$

and

$$\overline{V}_{2k+1,2a+2}(x;q) - \overline{V}_{2k+1,2a}(x;q)$$
$$= (-xq^2;q^2)_\infty\left(Q_{k+\frac{1}{2},a+1}(x^2;q^2) - Q_{k+\frac{1}{2},a}(x^2;q^2)\right)$$
$$= (-xq^2;q^2)_\infty(x^2q^2)^a Q_{k+\frac{1}{2},k-a+\frac{1}{2}}(x^2q^2;q^2)$$
$$= (xq)^{2a} V_{2k+1,2k-2a+1}(xq;q). \tag{3.19}$$

The initial conditions (3.15) and (3.16) plus the q-difference equations (3.17) and (3.18) will allow us to identify the V's with the W's provided we prove that the W's satisfy the same conditions.

Clearly

$$\overline{W}_{2k+1,0}(x;q) = 0 \tag{3.20}$$

because there are no partitions with ≤ -1 appearances of 1.

Also

$$1 = W_{2k+1,2a+1}(0;q) = W_{2k+1,2a+1}(x;0)$$
$$= \overline{W}_{2k+1,2a}(0;q) = \overline{W}_{2k+1,2a}(x;0) \tag{3.21}$$

because in each case the only partition allowed is the empty partition of 0.

Now

$$w_{2k+1,2a+1}(m,n) - w_{2k+1,2a-1}(m,n)$$

counts those partitions enumerated by $w_{2k+1,2a+1}(m,n)$ with the added condition that 1 appears either $2a$ times or $2a - 1$ times. We transform these partitions by deleting all the 1's and subtracting 1 from each of the remaining summands. If there were initially $2a$ ones, then there were initially at most $(2k - 2a + 1) - 1$ twos (remember two cannot appear an odd number of times), and subtracting 1 from each part changes the parity of each part. Hence this first class of partitions is enumerated by

$$\overline{w}_{2k+1,2k-2a+2}(m - 2a, n - m).$$

If there were initially $2a - 1$ ones, then there were at most $(2k - 2a + 2) - 1$ twos. Hence this second class of partitions is enumerated by

$$\overline{w}_{2k+1,2k-2a+2}(m - 2a + 1, n - m).$$

Putting this all together, we see that

$$w_{2k+1,2a+1}(m,n) - w_{2k+1,2a-1}(m,n)$$
$$= \overline{w}_{2k+1,2k-2a+2}(m - 2a, n - m) + \overline{w}_{2k+1,2k-2a+2}(m - 2a + 1, n - m) \tag{3.22}$$

and (3.22) may be translated to the related generating functions:

$$W_{2k+1,2a+1}(x;q) - W_{2k+1,2a-1}(x;q)$$
$$= (xq)^{2a-1}(1 + xq)W_{2k+1,2k-2a+2}(xq;q), \tag{3.23}$$

which is (3.18) for the W's. Finally

$$\overline{w}_{2k+1,2a+2}(m,n) - \overline{w}_{2k+1,2a}(m,n)$$

counts those partitions enumerated by $\overline{w}_{2k+1,2a+2}(m,n)$ with the added condition that 1 appear $2a$ times (keep in mind that 1, being odd, cannot appear $2a + 1$ times). We transform these partitions by deleting the 1's and subtracting 1 from each of the remaining summands. Since 1 appears $2a$ times, then there were initially at most $(2k + 1 - 2a) - 1$ twos. As before, subtracting 1 from each part changes the parity. Hence the transformed partitions are enumerated by

$$w_{2k+1,2k-2a+1}(m-2a, n-m).$$

Therefore

$$\overline{w}_{2k+1,2a+2}(m,n) - \overline{w}_{2k+1,2a}(m,n) = w_{2k+1,2k-2a+1}(m-2a, n-m), \quad (3.24)$$

and (3.24) may be translated to the related generating functions

$$\overline{W}_{2k+1,2a+1}(x;q) - \overline{W}_{2k+1,2a-1}(x;q) = (xq)^{2a} W_{2k+1,2k-2a+1}(xq;q) \quad (3.25)$$

which is (3.19) for the W's.

Hence by the Defining q-Difference Equations Principle, the W's and the V's are identical. I.e. (3.3) and (3.4) are valid.

The proofs of Theorem 1 and 2 are now quite straight forward.

$$\sum_{n=0}^{\infty} W_{k,a}(n)q^n$$

$$= \sum_{n,m\geq 0} w_{k,a}(m,n)q^n$$

$$= W_{k,a}(1,q) = (-q;q^2)_\infty Q_{\frac{k}{2},\frac{a}{2}}(1;q^2)$$

$$= \frac{(-q;q^2)_\infty (q^a;q^{2k+2})_\infty (q^{2k+2-a};q^{2k+2})_\infty (q^{2k+2};q^{2k+2})_\infty}{(q^2;q^2)_\infty} \quad \text{(by (2.8))}$$

$$= \frac{(q^a;q^{2k+2})_\infty (q^{2k+2-a};a^{2k+2})_\infty (q^{2k+2};q^{2k+2})_\infty}{(q;q^2)_\infty (q^4;q^4)_\infty}$$

$$= \sum_{n=0}^{\infty} G_{k,a}(n)q^n. \quad (3.26)$$

The last line follows from the antepenultimate line if k an a are both even and from the penultimate line if k and a are both odd. Comparing coefficients in the extremes of (3.26), we see that Theorem 1 is proved.

Finally for k odd and a even,

$$\sum_{n=0}^{\infty} \overline{W}_{k,a}(n)q^n = \sum_{n,m\geq 0} \overline{w}_{k,a}(n)q^n$$

$$= \overline{W}_{k,a}(1,q) = (-q^2;q^2)_\infty Q_{\frac{k}{2},\frac{a}{2}}(1;q^2)$$

$$= \frac{(q^a;q^{2k+2})_\infty (q^{2k+2-a};q^{2k+2})_\infty (q^{2k+2};q^{2k+2})_\infty}{(-q;q^2)_\infty (q;q)_\infty}, \quad (3.27)$$

by (2.8) as asserted in Theorem 2.

While there is no really clean classical partition-theoretic interpretation of (3.27), it can be interpreted via overpartitions (cf. [11, 18]). Also the equivalent assertion

$$\left(1+2\sum_{n=1}^{\infty}(-1)^n q^{2n^2}\right)\sum_{n=0}^{\infty}\overline{\mathcal{W}}_{k,a}(n)q^n = \sum_{n=-\infty}^{\infty}(-1)^n q^{(k+1)n(n-1)+an}$$

yields a very nice recurrence for $\overline{\mathcal{W}}_{k,a}(n)$.

4 Theorem 3

In light of (3.2) and Theorem 1 we shall first prove that

$$(-q;q^2)_\infty Q_{\frac{k}{2},\frac{a}{2}}(1;q^2)$$

$$= \sum_{n_1,\ldots,n_{k-1}\geq 0}\frac{q^{N_1^2+N_2^2+\cdots+N_{k-1}^2+2N_a+2N_{a+2}+\cdots+2N_{k-2}}}{(q^2;q^2)_{n_1}(q^2;q^2)_{n_2}\cdots(q^2;q^2)_{n_{k-1}}},\tag{4.1}$$

where, as noted earlier, $N_i = n_i + n_{i+1} + \cdots + n_{k-1}$.

Case 1. The even case. As before we replace k by $2k$ and a by $2a$. We shall prove that

$$(-xq;q^2)_\infty Q_{k,a}(x^2;q^2)$$

$$= \sum_{n_1,\ldots,n_{2k-1}=-\infty}^{\infty}\frac{q^{N_1^2+N_2^2+\cdots+N_{k-1}^2+2N_{2a}+2N_{2a+2}+\cdots+2N_{2k-2}}x^{N_1+N_2+\cdots+N_{2k-1}}}{(q^2;q^2)_{n_1}(q^2;q^2)_{n_2}\cdots(q^2;q^2)_{n_{2k-1}}}.\tag{4.2}$$

Note that we sum from $-\infty$ to ∞ for convenience; indeed the extended definition $(q;q)_n = \prod_{j=1}^{\infty}(1-q^j)/(1-q^{n+j})$, means that $1/(q;q)_n = 0$ if $n < 0$. First we replace each n_{2i} by $n_{2i} - n_{2i+1}$ for $1 \leq i \leq k-1$. With this change of indices

$$N_{2i-1} = n_{2i-1} + n_{2i} + n_{2i+2} + \cdots + n_{2k-2}$$

and

$$N_{2i} = n_{2i} + n_{2i+2} + \cdots + n_{2k-2}.$$

Thus we may rewrite the left-hand side of (4.2) as

$$\sum_{n_2,n_4,\ldots,n_{2k-2}=-\infty}^{\infty}\frac{q^{2N_2^2+2N_4^2+\cdots+2N_{2k-2}^2+2N_{2a}+2N_{2a+2}+\cdots+2N_{2k-2}}x^{2N_2+2N_4+\cdots+2N_{2k-2}}}{(q^2;q^2)_{n_2}(q^2;q^2)_{n_4}\cdots(q^2;q^2)_{n_{2k-2}}}$$

$$\times\sum_{n_1,n_3,\ldots,n_{2k-1}\geq 0}q^{n_1^2+2n_1N_2+n_3^2+2n_3N_4+\cdots+n_{2k-3}^2+2n_{2k-3}N_{2k-2}+n_{2k-1}^2}x^{n_1+n_3+\cdots+n_{2k-1}}$$

$$\times\frac{1}{(q^2;q^2)_{n_1}}\begin{bmatrix}n_2\\n_3\end{bmatrix}_2\begin{bmatrix}n_4\\n_5\end{bmatrix}_2\cdots\begin{bmatrix}n_{2k-2}\\n_{2k-1}\end{bmatrix}_2$$

$$= \sum_{n_2,n_4,\ldots,n_{2k-2}=-\infty} \frac{q^{2N_2^2+2N_4^2+\cdots+2N_{2k-2}^2+2N_{2a}+\cdots+2N_{2k-2}} x^{2N_2+\cdots+2N_{2k-2}}}{(q^2;q^2)_{n_2}(q^2;q^2)_{n_4}\cdots(q^2;q^2)_{n_{2k-2}}}$$

$$\times \left(-xq^{1+2N_2};q^2\right)_\infty\left(-xq^{1+2N_4};q^2\right)_{n_2}\left(-xq^{1+2N_6};q^2\right)_{n_4}\cdots\left(-xq;q^2\right)_{n_{2k-2}}$$

$$= (-xq;q^2)_\infty Q_{k,a}(x^2;q^2) \quad \text{(by [10, p. 36, eq. (3.3.6)]),}$$

where we have combined the inner numerator factors into the single infinite product $(-xq;q^2)_\infty$ and have invoked (2.9).

Finally setting $x = 1$ in (4.2) yields (4.1) in the even case.

Case 2. The odd case. Now we replace k by $2k+1$ and a by $2a+1$. We shall prove that

$$(-xq;q^2)_\infty Q_{k+\frac{1}{2},a+\frac{1}{2}}(x^2;q^2)$$

$$= \sum_{n_1,\ldots,n_{2k}\geq 0} \frac{q^{N_1^2+N_2^2+\cdots+N_{2k}^2+2N_{2a+1}+2N_{2a+3}+\cdots+2N_{2k-1}} x^{N_1+N_2+\cdots+N_{2k}}}{(q^2;q^2)_{n_1}(q^2;q^2)_{n_2}\cdots(q^2;q^2)_{n_{2k}}}. \tag{4.3}$$

We now define

$$R_{2k+1,2a+1}(x) = (-xq;q^2)_\infty Q_{k+\frac{1}{2},a+\frac{1}{2}}(x^2;q^2) \tag{4.4}$$

and

$$R_{2k+1,2a+2}(x) = (-xq^2;q^2)_\infty Q_{k+\frac{1}{2},a+1}(x^2;q^2). \tag{4.5}$$

Hence for $k+1 \geq a > 0$,

$$R_{2k+1,2a+1}(x) - R_{2k+1,2a-1}(x)$$

$$= (-xq;q^2)_\infty \left(Q_{k+\frac{1}{2},a+\frac{1}{2}}(x^2;q^2) - Q_{k+\frac{1}{2},a-\frac{1}{2}}(x^2;q^2)\right)$$

$$= (-xq;q^2)_\infty (x^2q^2)^{a-\frac{1}{2}} Q_{k+\frac{1}{2},k-a+1}(x^2q^2;q^2)$$

$$= (xq)^{2a-1}(1+xq) R_{2k+1,2k-2a+2}(xq) \tag{4.6}$$

(note that for $a = k+1$, this implies $R_{2k+1,2k+3}(x) = R_{2k+1,2k+1}(x)$), for $k \geq a \geq 0$

$$R_{2k+1,2a+2}(x) - R_{2k+1,2a}(x)$$

$$= (-xq^2;q^2)_\infty \left(Q_{k+\frac{1}{2},a+1}(x^2;q^2) - Q_{k+\frac{1}{2},a}(x^2;q^2)\right)$$

$$= (xq)^{2a}(-xq^2;q^2)_\infty Q_{k+\frac{1}{2},k-a+\frac{1}{2}}(x^2q^2;q^2)$$

$$= (xq)^{2a} R_{2k+1,2k-2a+1}(xq), \tag{4.7}$$

and for $a = 0$

$$R_{2k+1,1}(x)(1 + x^{-1}q^{-1}) = R_{2k+1,1}(x) - R_{2k+1,-1}(x)$$
$$= (xq)^{-1}(1 + xq)R_{2k+1,2k+2}(xq),$$

or

$$R_{2k+1,1}(x) = R_{2k+1,2k+2}(xq). \tag{4.8}$$

Now (4.6)–(4.8) are a defining set of q-difference equations once we note the obvious initial conditions for $x = 0$ and $q = 0$.

Furthermore, when $k = 0$, the unique solutions are

$$R_{1,1}(x) = R_{1,2}(x) = 1, \qquad R_{1,0}(x) = 0.$$

We shall now prove that if for $0 \leq a \leq k$

$$R_{2k+1,2a+1}(x) = \sum_{n \geq 0} \frac{(-xq; q^2)_n q^{(2k-1)n^2 + 2(k-a)n} x^{(2k-1)n} R_{2k-1,2a+1}(xq^{2n})}{(q^2; q^2)_n}$$

$$\tag{4.9}$$

and for $a = k + 1$

$$\mathcal{R}_{2k+1,2k+3}(x) = \sum_{n \geq 0} \frac{(-xq; q^2)_n q^{(2k-1)n^2} x^{(2n-1)n} R_{2k-1,2k-1}(xq^{2n})}{(q^2; q^2)_n}$$

$$= \mathcal{R}_{2k+1,2k+1}(x), \tag{4.10}$$

and if for $0 \leq a \leq k$

$$R_{2k+1,2a}(x) = \sum_{n \geq 0} \frac{(-xq^2; q^2)_n q^{(2k-1)n^2 + (2k-2a+1)n} x^{(2k-1)n} R_{2k-1,2a}(xq^{2n})}{(q^2; q^2)_n},$$

$$\tag{4.11}$$

while

$$\mathcal{R}_{2k+1,2k+2}(x) = \sum_{n \geq 0} \frac{(-x; q^2)_n q^{(2k-1)n^2 + n} x^{(2k-1)n} R_{2k-1,2k}(xq^{2n})}{(q^2; q^2)_n}, \tag{4.12}$$

then for $0 \leq A \leq 2k + 3, k > 0$,

$$\mathcal{R}_{2k+1,A}(x) = R_{2k+1,A}(x). \tag{4.13}$$

We now proceed to the base case, $k = 1$, in order to establish (4.13) by mathematical induction on k. The asserted expansions for the $\mathcal{R}$'s are:

$$R_{3,1}(x) = \sum_{n \geq 0} \frac{(-xq; q^2)_n q^{n^2 + 2n} x^n}{(q^2; q^2)_n}, \tag{4.14}$$

$$\mathcal{R}_{3,2}(x) = \sum_{n \geq 0} \frac{(-xq^2; q^2)_n q^{n^2+n} x^n}{(q^2; q^2)_n}, \tag{4.15}$$

$$\mathcal{R}_{3,5}(x) = \mathcal{R}_{3,3}(x) = \sum_{n \geq 0} \frac{(-xq; q^2)_n q^{n^2} x^n}{(q^2; q^2)_n}, \tag{4.16}$$

$$\mathcal{R}_{3,4}(x) = \sum_{n \geq 0} \frac{(-x; q^2)_n q^{n^2+n} x^n}{(q^2; q^2)_n}. \tag{4.17}$$

By inspection we see that,

$$\mathcal{R}_{3,5}(x) - \mathcal{R}_{3,3}(x) = 0 = x^3 (1 + xq) \mathcal{R}_{3,0}(xq), \tag{4.18}$$

$$\mathcal{R}_{3,1}(x) = \mathcal{R}_{3,4}(xq), \tag{4.19}$$

$$\mathcal{R}_{3,2}(x) = \mathcal{R}_{3,3}(xq). \tag{4.20}$$

Also

$$\mathcal{R}_{3,3}(x) - \mathcal{R}_{3,1}(x) = \sum_{n \geq 1} \frac{(-xq; q^2)_n q^{n^2} x^n}{(q^2; q^2)_{n-1}}$$

$$= xq(1 + xq) \sum_{n \geq 0} \frac{(-(xq)q^2; q^2))_n q^{n^2+n} (xq)^n}{(q^2; q^2)_n}$$

$$= xq(1 + xq) \mathcal{R}_{3,2}(xq), \tag{4.21}$$

and

$$\mathcal{R}_{3,4}(x) - \mathcal{R}_{3,2}(x) = \sum_{n \geq 0} \frac{(-xq^2; q^2)_{n-1} q^{n^2+n} x^n ((1 + x) - (1 + xq^{2n}))}{(q^2; q^2)_n}$$

$$= x \sum_{n \geq 1} \frac{(-xq^2; q^2)_{n-1} q^{n^2+n} x^n}{(q^2; q^2)_{n-1}}$$

$$= x^2 q^2 \sum_{n \geq 0} \frac{(-(xq)q; q^2)_n q^{n^2+2n} (xq)^n}{(q^2; q^2)_n}$$

$$= x^2 q^2 \mathcal{R}_{3,1}(xq). \tag{4.22}$$

Now (4.18)–(4.22) are just (4.6), (4.7) and (4.8) in the case $k = 1$. Thus, noting the concurrent initial conditions, we see that by the Defining q-Difference Equation Principle, (4.13) is valid for $k = 1$.

We now assume that (4.13) is valid up to but not including a fixed k. Hence

$$\mathcal{R}_{2k+1,1}(x) = \sum_{n\geq 0} \frac{(-xq;q^2)_n q^{(2k-1)n^2+2kn} x^{(2k-1)n} R_{2k-1,1}(xq^{2n})}{(q^2;q^2)_n}$$

$$= \sum_{n\geq 0} \frac{(-xq;q^2)_n q^{(2k-1)n^2+n}(xq)^{(2k-1)n} R_{2k-1,2k}(xq^{2n+1})}{(q^2;q^2)_n} \quad \text{(by (4.8))}$$

$$= \mathcal{R}_{2k+1,2k+2}(xq). \tag{4.23}$$

So (4.8) is valid with the R's replaced by $\mathcal{R}$'s.

Next for $1 \leq a < k$,

$$\mathcal{R}_{2k+1,2a+1}(x) - \mathcal{R}_{2k+1,2a-1}(x)$$

$$= \sum_{n\geq 0} \frac{(-xq;q^2)_n q^{(2k-1)n^2+2(k-a)n} x^{(2k-1)n}}{(q^2;q^2)_n}$$

$$\times \left(R_{2k-1,2a+1}(xq^{2n}) - q^{2n} R_{2k-1,2a-1}(xq^{2n}) \right)$$

$$= \sum_{n\geq 0} \frac{(-xq;q^2)_n q^{(2k-1)n^2+2(k-a)n} x^{(2k-1)n}}{(q^2;q^2)_n}$$

$$\times \left(R_{2k-1,2a-1}(xq^{2n}) + (xq^{2n+1})^{2a-1}(1+xq^{2n+1}) R_{2k-1,2k-2a}(xq^{2n+1}) \right.$$

$$\left. - q^{2n} R_{2k-1,2a-1}(xq^{2n}) \right) \quad \text{(by (4.6))}$$

$$= \sum_{n\geq 1} \frac{(-xq;q^2)_n q^{(2k-1)n^2+2(k-a)n} x^{(2k-1)n} R_{2k-1,2a-1}(xq^{2n})}{(q^2;q^2)_{n-1}}$$

$$+ (xq)^{2a-1}(1+xq)$$

$$\times \sum_{n\geq 0} \frac{(-(xq)q^2;q^2)_n q^{(2k-1)n^2+(2a-1)n} x^{(2k-1)n} R_{2k-1,2k-2a}(xq^{2n})}{(q^2;q^2)_n}$$

$$= \sum_{n\geq 1} \frac{(-xq;q^2)_{n+1} q^{(2k-1)(n+1)^2+2(k-a)(n+1)} x^{(2k-1)(n+1)} R_{2k-1,2a-1}(xq^{2n+2})}{(q^2;q^2)_n}$$

$$+ (xq)^{2a-1}(1+xq) \sum_{n\geq 0} \frac{(-xq^3;q^2)_n q^{(2k-1)n^2+(2a-1)n}(xq)^{(2k-1)n}}{(q^2;q^2)_n}$$

$$\times \left(R_{2k-1,2k-2a+2}(xq^{2n+1}) - (xq^{2n+2})^{2(k-a)} R_{2k-1,2a-1}(xq^{2n+2}) \right) \quad \text{(by (4.7))}$$

$$= (xq)^{2a-1}(1+xq)\mathcal{R}_{2k+1,2k-2a+2}(xq), \tag{4.24}$$

where the sums involving $R_{2k-1,2a-1}(xq^{2n+1})$ cancel each other. Hence we have (4.6) proved with $\mathcal{R}$'s replacing R's as long as $1 \leq a \leq k$. At $a = k+1$

$$\mathcal{R}_{2k+1,2k+3}(x) - \mathcal{R}_{2k+1,2k+1}(x) = 0 \quad \text{(by (4.10))}$$
$$= (xq)^{2k+1}(1+xq)\mathcal{R}_{2k+1,0}(xq) \quad (4.25)$$

(by (4.5) because $Q_{k+\frac{1}{2},0}(x^2; q^2) = 0$). Hence (4.6) is proved in all cases with $\mathcal{R}$'s replacing R's.

Next for $0 \leq a < k$

$$\mathcal{R}_{2k+1,2a+2}(x) - \mathcal{R}_{2k+1,2a}(x)$$

$$= \sum_{n \geq 0} \frac{(-xq^2; q^2)_n q^{(2k-1)n^2+(2k-2a-1)n} x^{(2k-1)n}}{(q^2; q^2)_n}$$

$$\times \left(R_{2k-1,2a+2}(xq^{2n}) - q^{2n} R_{2k-1,2a}(xq^{2n}) \right)$$

$$= \sum_{n \geq 0} \frac{(-xq^2; q^2)_n q^{(2k-1)n^2+(2k-2a-1)n} x^{(2k-1)n}}{(q^2; q^2)_n}$$

$$\times \left(R_{2k-1,2a}(xq^{2n}) + (xq^{2n+1})^{2a} R_{2k-1,2k-2a-1}(xq^{2n+1}) \right.$$

$$\left. - q^{2n} R_{2k-1,2a}(xq^{2n}) \right) \quad \text{(by (4.6))}$$

$$= \sum_{n \geq 1} \frac{(-xq^2; q^2)_n q^{(2k-1)n^2+(2k-2a-1)n} x^{(2k-1)n} R_{2k-1,2a}(xq^{2n})}{(q^2; q^2)_{n-1}}$$

$$+ (xq)^{2a} \sum_{n \geq 0} \frac{(-xq^2; q^2)_n q^{(2k-1)n^2+(2k-2a-1)n+4na} x^{(2k-1)n}}{(q^2; q^2)_n}$$

$$\times \left(R_{2k-1,2k-2a+1}(xq^{2n+1}) \right.$$

$$\left. - (xq^{2n+2})^{2k-2a-1}(1 + xq^{2n+2}) R_{2k-1,2a}(xq^{2n+2}) \right) \quad \text{(by (4.7))}$$

$$= (xq)^{2a} \mathcal{R}_{2k+1,2k-2a+1}(xq),$$

where the two sums involving $R_{2k-1,2a}$ cancel each other out once n has been replaced by $n+1$ in the first of the two. Hence we have (4.7) proved for $0 \leq a < k$.

Finally for $a = k$

$$\mathcal{R}_{2k+1,2k+2}(x) - \mathcal{R}_{2k+1,2k}(x)$$

$$= \sum_{n \geq 1} \frac{(-xq^2; q^2)_{n-1} q^{(2k-1)n^2+n} x^{(2k-1)n} R_{2k-1,2k}(xq^{2n})}{(q^2; q^2)_n} \left((1+x) - (1 + xq^{2n}) \right)$$

$$= x^{2k}q^{2k} \sum_{n \geq 0} \frac{(-xq^2; q^2)_n q^{(2k-1)n^2+2kn}(xq)^{(2k-1)n} R_{2k-1,2k}(xq^{2n+2})}{(q^2; q^2)_n}$$

$$= x^{2k}q^{2k} \sum_{n \geq 0} \frac{(-(xq)q; q^2)_n q^{(2k-1)n^2+2kn}(xq)^{(2k-1)n} R_{2k-1,1}(xqq^{2n})}{(q^2; q^2)_n} \quad \text{(by (4.8))}$$

$$= (xq)^{2k} \mathcal{R}_{2k+1,1}(xq)$$

which establishes (4.7) with the R's replaced by $\mathcal{R}$'s. Hence by the Defining q-Difference Equations Principle (4.13) is true.

Hence (4.9)–(4.13) are valid if $\mathcal{R}$ is replaced by R.

I now claim that for $0 \leq a \leq k$

$$R_{2k+1,2a+1}(x) = \sum_{n_1,\ldots,n_{2k} \geq 0} \frac{q^{N_1^2+\cdots+N_{2k}^2+2\sum_{j=a+1}^{k} N_{2j-1}} x^{N_1+\cdots+N_{2k}}}{(q^2; q^2)_{N_1-N_2} \cdots (q^2; q^2)_{N_{2k-1}-N_{2k}}(q^2; q^2)_{N_{2k}}},$$

(4.26)

and

$$R_{2k+1,2a}(x)$$

$$= \sum_{n_1,\ldots,n_{2k} \geq 0} \frac{q^{N_1^2+\cdots+N_{2k}^2+N_{2a-1}+N_{2a}+\cdots+N_{2k}+n_1+n_3+n_{2a-3}} x^{N_1+\cdots+N_{2k}}}{(q^2; q^2)_{N_1-N_2} \cdots (q^2; q^2)_{N_{2k-2}-N_{2k-1}}(q^2; q^2)_{N_{2k-1}-N_{2k}}(q^2; q^2)_{N_{2k}}},$$

(4.27)

where $N_i = n_i + n_{i+1} + \cdots + n_{2k}$ and empty sums are 0.

If we denote the right hand side of (4.26) by $\rho_{2k+1,2a+1}(x)$, we see that by summing the n_{2k} series using the q-binomial theorem [7, p. 36, eq. (3.3.6)] and letting $n = n_{2k-1}$ we obtain

$$\rho_{2k+1,2a+1}(x) = \sum_{n \geq 0} \frac{(-xq; q^2)_n q^{(2k-1)n^2+2(k-a)n} x^{(2k-1)n}}{(q^2; q^2)_n} \rho_{2k-1,2a+1}(xq^{2n}),$$

an exact replica of (4.9) where we add the caveat that $\rho_{2k+1,2k+3}(x) = \rho_{2k+1,2k+1}(x)$ (i.e. (4.10)).

If we denote the right hand side of (4.7) by $\rho_{2k+1,2a}(x)$, we see that by summing the n_{2k} series using the q-binomial theorem [7, p. 36, eq. (3.36)] and letting $n = n_{2k-1}$, we obtain

$$\rho_{2k+1,2a}(x) = \sum_{n \geq 0} \frac{(-xq^2; q^2)_n q^{(2k-1)n^2+(2k-2a+1)n} x^{(2k-1)n}}{(q^2; q^2)_n} \rho_{2k-1,2a}(xq^{2n})$$

and

$$\rho_{2k+1,2k+2}(x) = \sum_{n \geq 0} \frac{(-x; q^2)_n q^{(2k-1)n^2+n} x^{(2k-1)n}}{(q^2; q^2)_n} \rho_{2k-1,2k}(xq^{2n}).$$

Thus we have exactly the same functional equations for the ρ's as for the R's and exactly the same argument shows that

$$\rho_{2k+1,A}(x) = R_{2k+1,A}(x) \tag{4.28}$$

for $0 \leq A \leq 2k + 3$.

Now (4.4), (4.26) and (4.28) prove (4.2) in the odd case and consequently we have proved (4.1).

Finally to prove (1.9), the final assertion in Theorem 3, we need only set $x = 1$ in (4.5) and invoke (2.8).

5 Evens appearing Oddly in Gordon's partitions

Our object in this section is to prove the following result:

Theorem 4 *If $k = 2$ or 3, then the coefficient of $q^n x^m y^j$ in*

$$\sum_{n_1,n_2,\ldots,n_{k-1} \geq 0} \frac{q^{N_1^2+N_2^2+\cdots+N_{k-1}^2} x^{N_1+\cdots+N_{k-1}} H_{n_1} H_{n_2} \cdots H_{n_{k-1}}}{(q^2;q^2)_{n_1} (q^2;q^2)_{n_2} \cdots (q^2;q^2)_{n_{k-1}}} \tag{5.1}$$

is the number of partitions enumerated by $B_{k,k}(n)$ with exactly m parts and exactly j different even parts that appear an odd number of times. In (5.1),

$$H_n = \sum_{j=0}^{n} y^j q^j \begin{bmatrix} n \\ j \end{bmatrix}_2,$$

and as before

$$N_i = n_i + n_{i+1} + \cdots + n_{k-1}.$$

Noting that $H_n = 1$ when $y = 0$ and $H_n = (-q;q)_n$ when $y = 1$ [4, p. 49, Ex. 5], we see that the theorem is true for all k when $y = 1$ (it is the case $i = k$ of (2.9)) and when $y = 0$ (it is the case $i = k$ of Theorem 3).

Remark K. Kursungoz has found a combinatorial proof of (5.1) valid for *all* $k \geq 2$. The proof here is purely analytic. The possibility of a fully analytic proof will be considered in Sect. 13.

To set up our proof we require a couple of lemmas. First we define

$$S_k(a_{k-1}, a_{k-2}, \ldots, a_1; x; q)$$

$$= \sum_{n_1,\ldots,n_{k-1} \geq 0} \frac{q^{N_1^2+N_2^2+\cdots+N_{k-1}^2+2a_1 n_1+\cdots+2a_{k-1} n_{k-1}} x^{N_1+\cdots+N_{k-1}} H_{n_1} H_{n_2} \cdots H_{n_{k-1}}}{(q^2;q^2)_{n_1} (q^2;q^2)_{n_2} \cdots (q^2;q^2)_{n_{k-2}}} \tag{5.2}$$

$$= \sum_{n \geq 0} \frac{q^{(k-1)n^2+2a_{k-1}n} x^{(k-1)n} H_n}{(q^2;q^2)_n} S_{k-1}(a_{k-2}, \ldots, a_1; xq^{2n}; q). \tag{5.3}$$

Lemma 5 *For* $1 \leq i \leq k - 1, k \geq 2,$

$$S_k(a_{k-1}, a_{k-2}, \ldots, a_{k-i}, \ldots, a_1; x; q)$$
$$- S_k(a_{k-1}, a_{k-2}, \ldots, a_{k-i} + 1, \ldots, a_1; x; q)$$
$$= x^{k-i}(1 + yq)q^{(k-i)+2a_{k-i}}$$
$$\times S_k(a_{k-1} - i + 1, a_{k-2} - i + 2, \ldots, a_{k-i+1} - 1, a_{k-i}, \ldots, a_1; xq^2; q)$$
$$- x^{2(k-i)}yq^{4(a_{k-i}+k-i)+1}$$
$$\times S_k(a_{k-1} - 2i + 2, a_{k-2} - 2i + 4, \ldots, a_{k-i+1} - 2, a_{k-i}, \ldots, a_1; xq^4; q). \tag{5.4}$$

Proof We proceed by mathematical induction on k. First $k = 2$. Consequently $i = 1$. In the proof, we need the following recurrence for the H's [7, p. 49, Ex. 6]

$$H_{n+1} = (1 + yq)H_n - (1 - q^{2n})yq H_{n-1}. \tag{5.5}$$

Hence

$$S_2(a_1; x; q) - S_2(a_1 + 1; x; q)$$

$$= \sum_{n \geq 0} \frac{q^{n^2 + 2a_1 n} x^n H_n (1 - q^{2n})}{(q^2; q^2)_n}$$

$$= \sum_{n \geq 0} \frac{q^{(n+1)^2 + 2a_1(n+1)} x^{n+1} H_{n+1}}{(q^2; q^2)_n}$$

$$= \sum_{n \geq 0} \frac{q^{(n+1)^2 + 2a_1(n+1)} x^{n+1} ((1 + yq)H_n - (1 - q^{2n})yq H_{n-1})}{(q^2; q^2)_n}$$

$$= xq^{2a_1+1}(1 + yq)S_2(a_1; xq^2; q) - yq \sum_{n \geq 0} \frac{q^{(n+2)^2 + 2a_1(n+2)} x^{n+2} H_n}{(q^2; q^2)_n}$$

$$= x(1 + yq)q^{1+2a_1} S_2(a_1; xq^2; q) - x^2 yq^{4(a_1+1)+1} S_2(a_1; xq^4; q).$$

Thus the case $k = 2$ is established.

We now assume the theorem is true up to but not including a specific k. When $i = 1$

$$S_k(a_{k-1}, a_{k-2}, \ldots, a_1; x; q) - S_k(a_{k-1} + 1, a_{k-2}, \ldots, a_1; x; q)$$

$$= \sum_{n \geq 0} \frac{q^{(k-1)n^2 + 2a_{k-1}n} x^{(k-1)n} H_n (1 - q^{2n}) S_{k-1}(a_{k-2}, \ldots, a_1; xq^{2n}; q)}{(q^2; q^2)_n}$$

$$= \sum_{n \geq 0} \frac{q^{(k-1)(n+1)^2 + 2a_{k-1}(n+1)} x^{(k-1)(n+1)} S_{k-1}(a_{k-2}, \ldots, a_1; xq^{2n+2}; q)}{(q^2; q^2)_n}$$

$$\times ((1 + yq)H_n - yq(1 - q^{2n})H_{n-1})$$

$$= x^{k-1}(1+yq)q^{k-1+2a_{k-1}} S_k(a_{k-1}, a_{k-2}, \ldots, a_1; xq^2; q)$$

$$- yq \sum_{n \geq 0} \frac{q^{(k-1)(n+2)^2 + 2a_{k-1}(n+2)} x^{(k-1)(n+2)} H_n}{(q^2; q^2)_n}$$

$$\times S_{k-1}(a_{k-2}, \ldots, a_1; xq^{2n+4}; q)$$

$$= x^{k-1}(1+yq)q^{k-1+2a_{k-1}} S_k(a_{k-1}, \ldots, a_1; xq^2; q)$$

$$- x^{2(k-1)} yq^{1+4(k-1)+4a_{k-1}} S_k(a_{k-1}, \ldots, a_1; xq^4; q)$$

which is the desired result at k with $i = 1$. Now assume $k - 1 \geq i > 1$,

$$S_k(a_{k-1}, \ldots, a_{k-i}, \ldots, a_1; x; q) - S_k(a_{k-1}, \ldots, a_{k-i} + 1, \ldots, a_1; x; q)$$

$$= \sum_{n \geq 0} \frac{q^{(k-1)n^2 + 2a_{k-1}n} x^{(k-1)n} H_n}{(q^2; q^2)_n}$$

$$\times \Big\{ S_{k-1}(a_{k-2}, \ldots, a_{k-1-(i-1)}, \ldots, a_1; xq^{2n}; q)$$

$$- S_{k-1}(a_{k-2}, \ldots, a_{k-1-(i-1)} + 1, \ldots, a_1; xq^{2n}; q) \Big\}$$

$$= \sum_{n \geq 0} \frac{q^{(k-1)n^2 + 2a_{k-1}n} x^{(k-1)n} H_n}{(q^2; q^2)_n}$$

$$\times \Big((xq^{2n})^{k-i}(1+yq)q^{k-i+2a_{k-i}}$$

$$\times S_{k-1}(a_{k-2} - i + 2, a_{k-3} - i + 3, \ldots, a_{k-i}, \ldots, a_1; xq^{2n+2}; q)$$

$$- (xq^{2n})^{2(k-i)} yq^{4(k-i)+4a_{k-i}+1}$$

$$\times S_{k-1}(a_{k-2} - 2i + 4, a_{k-3} - 2i + 6, \ldots, a_{k-i}, \ldots, a_1; xq^{2n+4}; q) \Big)$$

(by the induction hypothesis)

$$= x^{k-i}(1+yq)q^{k-i+2a_{k-i}}$$

$$\times \sum_{n \geq 0} \frac{q^{(k-1)n^2 + 2(a_{k-1}-i+1)n}(xq^2)^{(k-1)n} H_n S_{k-1}(a_{k-2} - i + 2, \ldots; xq^{2n+2}; q)}{(q^2; q^2)_n}$$

$$- x^{2(k-i)} yq^{4(k-i)+4a_{k-i}+1}$$

$$\times \sum_{n \geq 0} q^{(k-1)n^2 + 2(a_{k-1}-2i+2)n}(xq^4)^{(k-1)n} H_n S_{k-1}(a_{k-2} - 2i + 4, \ldots; xq^{2n+4}; q)$$

$$= x^{k-i}(1+yq)q^{k-i+2a_{k-i}}$$

$$\times S_k(a_{k-1} - i + 1, a_{k-2} - i + 2, \ldots, a_{k-i}, \ldots, a_1; xq^2; q)$$

$$- x^{2(k-i)} yq^{4(k-i)+4a_{k-i}+1}$$

$$\times S_k(a_{k-1} - 2i + 2, a_{k-2} - 2i + 4, \ldots, a_{k-i}, \ldots, a_1; xq^4; q)$$

and so we have proved every instance at k thus proving the lemma. $\square$

Next,

Lemma 6

$$S_k(a_{k-1} + k - 1, a_{k-2} + k - 2, \ldots, a_1 + 1; x; q)$$
$$= S_k(a_{k-1}, a_{k-2}, \ldots, a_1; xq^2; q). \tag{5.6}$$

Proof This follows immediately from inspection of (5.2). $\qquad\square$

Finally,

Proof of Theorem 4 We begin with $k = 2$. Let us define $G_k(x, y, q)$ to be the generating function for the partitions named in the statement of the theorem. It is easy to see that

$$G_2(x, y, q) = (1 + xq + xyq^2)G_2(xq^2, y, q) - x^2yq^{2+3}G_2(xq^4, y, q). \tag{5.7}$$

The first term on the right exhibits the allowable partitions accordingly as there are no parts smaller than 3 (the "1"), 1 is a part (the "xq"), or 2 is a part (the "xyq^2"). However partitions with smallest parts $2 + 3$ have been inadmissably allowed. Thus the second term (with "x^2yq^{2+3}") removes these partitions.

On the other hand, by Lemmas 5 and 6,

$$S_2(0; x; q) = S_2(1; x; q) + x(1 + yq)q\,S_2(0; xq^2; q) - x^2yq^5 S_2(0; xq^4; q)$$
$$= (1 + xq + xyq^2)S_2(0; xq^2; q^2) - x^2yq^5 S_2(0; xq^4; q). \tag{5.8}$$

Comparing (5.8) and (5.7), we see by the Defining q-Difference Equation Principle that

$$G_2(x, y, q) = S_2(0; x; q).$$

This proves the case $k = 2$.

We now consider $k = 3$. We note that by Lemma 5

$$S_3(0, 0; x; q) - S_3(1, 0; x; q)$$
$$= x^2q^2(1 + yq)S_3(0, 0; xq^2; q) - x^4yq^9 S_3(0, 0; xq^4; q), \tag{5.9}$$
$$S_3(1, 0; x; q) - S_3(2, 0; x; q)$$
$$= x^2q^4(1 + yq)S_3(1, 0; xq^2; q) - x^4yq^{13}S_3(1, 0; xq^4; q), \tag{5.10}$$
$$S_3(2, 0; x; q) - S_3(2, 1; x; q)$$
$$= xq(1 + yq)S_3(1, 0; xq^2; q) - x^2yq^5 S_3(0, 0; xq^4; q), \tag{5.11}$$

and by Lemma 6:

$$S_3(2, 1; x; q) = S_3(0, 0; xq^2; q). \tag{5.12}$$

We may obtain $S_3(1, 0; x; q)$ in terms of instances of $S_3(0, 0; xq^{2i}; q)$ from (5.9). Then the same can be done for $S_3(2, 0; x; q)$ from (5.10). Finally all of these results plus (5.12) can be substituted in (5.11) to obtain the following q-difference equation for $S_3(0, 0; x; q)$:

$$S_3(0, 0; x; q) = \left(1 + qx(1 + qy)(1 + xq + xq^3)\right)S_3(0, 0; xq^2; q)$$
$$+ \left(x^2 yq^5 + x^3 q^7(1 + qy)^2\right.$$
$$+ x^4 q^9(q + y + 2q^2 y + yq^4 + y^2 q^3)\right)S_3(0, 0; xq^4; q)$$
$$- x^5 q^{18} y(1 + yq)(1 + xq^3 + xq^5)S_3(0, 0; xq^6; q)$$
$$+ x^8 y^2 q^{38} S_3(0, 0; xq^8; q). \tag{5.13}$$

Now we need a q-difference equation for $G_3(x, y, q)$. This may be obtained as follows: $G_3(x, 1, q)$ is the generating function for the original Gordon partition $b_{3,3}(m, n)$. Following [3, p. 443, eqs. (5.6)–(5.9)], we see that the Smallest Parts Decomposition Principle may be invoked to yield

$$G_3(x, 1, q) = (1 + xq)G_3(xq, 1, q) - x^2 q^{1+1} G_3(xq^2, 1, q)$$
$$+ x^3 q^{2+2+1} G_3(xq^3, 1, q). \tag{5.14}$$

But in instances of $G_3(xq^{2j+1}, y, q)$ we see that y would be counting instances of different *odd* parts appearing an odd number of times. Hence we must do several applications of the shift rule in order to eliminate instances of $G_3(xq^{2j+1}, 1, q)$. Thus with $G_3(x) = G_3(x, 1, q)$, we have that

$$G_3(xq) = (1 + xq^2)G_3(xq^2) - x^2 q^{2+2} G_3(xq^3) + x^3 q^{3+3+2} G_3(xq^4),$$
$$G_3(xq^2) = (1 + xq^3)G_3(xq^3) - x^2 q^{3+3} G_3(xq^4) + x^3 q^{4+4+3} G_3(xq^4),$$
$$G_3(xq^3) = (1 + xq^4)G_3(xq^4) - x^2 q^{4+4} G_3(xq^5) + x^4 q^{5+5+4} G_3(xq^6).$$

We now have four linear equations which can be solved for $G_3(x)$, $G_3(xq)$, $G_3(xq^3)$ and $G_3(xq^5)$ in terms of $G_3(xq^2)$, $G_3(xq^4)$ and $G_3(xq^6)$. As a result

$$G_3(x) = (1 + xq + x^2 q^{1+1} + xq^2 + x^2 q^{2+1} + x^2 q^{2+2})G_3(xq^2)$$
$$- \left(x^3 q^{3+2+2} + x^3 q^{3+3+2} + x^4 q^{3+3+2+2}\right.$$
$$+ x^4 q^{3+3+2+1} + x^4 q^{4+3+2+2}\right)G_3(xq^4)$$
$$+ x^6 q^{5+5+4+3+2+2} G_3(xq^6). \tag{5.15}$$

Now equation (5.15) may be viewed as an application of the Smallest Parts Decomposition Principle considering the partitions starting with parts ≤ 2. Thus to extend (5.15) to the full $G_3(x, y, q)$ we need only insert y's to account for appearances of evens an odd number of times. Hence

$$G_3(x, y, q) = (1 + xq + x^2 q^{1+1} + xyq^2 + x^2 yq^{2+1} + x^2 q^{2+2})G_3(xq^2, y, q)$$

$$- (x^3 q^{3+2+2} + x^3 y q^{3+3+2} + x^4 q^{3+3+2+2}$$

$$+ x^4 y q^{3+3+2+1} + x^4 y q^{4+3+2+2}) G_3(xq^4, y, q)$$

$$+ x^6 y q^{5+5+4+3+2+2} G_3(xq^6, y, q). \tag{5.16}$$

So if we define $E_3(x, q)$ to be the left-hand side of (5.16) minus the right-hand side, then we see that $E_3(x, q)$ is identically equal to 0. Therefore

$$0 = E_3(x, q) - x^2 y q^5 E_3(xq^2, q)$$

$$= G_3(x, y, q) - (1 + qx(1 + qy)(1 + xq + xq^3)) G_3(xq^2, y, q)$$

$$+ (x^2 y q^5 + x^3 q^7 (1 + qy)^2$$

$$+ x^4 q^9 (q + y + 2q^2 y + yq^4 + y^2 q^3)) G_3(xq^4, y, q)$$

$$- x^5 q^{18} y (1 + yq)(1 + xq^3 + xq^5) G_3(xq^6, y, q) + x^8 y^2 q^{38} G_3(xq^8, y, q). \tag{5.17}$$

Comparing (5.17) with (5.13) we see that the Defining q-Difference Equations Principle implies that

$$G_3(x, y, q) = S_3(0, 0; x; q). \qquad \square$$

6 The parity indices

The next few sections will be devoted to considerations of partitions subject to constraints on the following partition parameters.

Definition Let λ be a partition $\lambda: \lambda_1 + \lambda_2 + \cdots + \lambda_j$ where $\lambda_1 \geq \lambda_2 \geq \cdots \geq \lambda_j$. We define $I_{UE}(\lambda)$ (resp. I_{UO}) to be the maximum length of *nonincreasing* subsequences of $\{\lambda_1, \lambda_2, \ldots, \lambda_j\}$ whose terms alternate in parity starting with an even (resp. odd) λ_i.

Definition Let λ be a partition $\lambda: \lambda_1 + \lambda_2 + \cdots + \lambda_j$ where $\lambda_1 \geq \lambda_2 \geq \cdots \geq \lambda_j$. We define $I_{LE}(\lambda)$ (resp. I_{LO}) to be the maximum length of *nondecreasing* subsequences of $\{\lambda_1, \lambda_2, \ldots, \lambda_j\}$ whose terms alternate in parity starting with an even (resp. odd) λ_i.

It is immediate from these definitions that if λ_1 is even, then $I_{UO}(\lambda) = I_{UE}(\lambda) - 1$, and if λ_1 is odd, then $I_{UE}(\lambda) = I_{UO}(\lambda) - 1$. In the same way, if λ_j is even, then $I_{LO}(\lambda) = I_{LE}(\lambda) - 1$, and if λ_j is odd, then $I_{LE}(\lambda) = I_{LO}(\lambda) - 1$.

For example, if $\lambda = 7 + 7 + 5 + 4 + 4 + 3 + 2 + 1 + 1$, then $I_{UE}(\lambda) = 4$, $I_{UO}(\lambda) = 5$, $I_{LE}(\lambda) = 4$, $I_{LO}(\lambda) = 5$.

7 Lower parity indices in partitions with distinct parts

We begin with notation for the relevant partition functions: $p_e(r, m, n)$ (resp. $p_o(r, m, n)$) is to denote the number of partitions of n into m distinct parts with lower even (resp. odd) parity index equal to r.

Next,

$$P_e(y, x; q) = \sum_{r,m,n \geq 0} p_e(r, m, n) y^r x^m q^n, \tag{7.1}$$

and

$$P_o(y, x; q) = \sum_{r,m,n \geq 0} p_o(r, m, n) y^r x^m q^n. \tag{7.2}$$

Theorem 7

$$P_e(y, x; q) = (1 + xq) P_o(y, xq; q), \tag{7.3}$$

$$P_o(y, x; q) = \sum_{n \geq 0} \frac{x^n y^n q^{n(n+1)/2} (-q/y)_n}{(q^2; q^2)_n} \tag{7.4}$$

$$= (-xq)_\infty \sum_{n \geq 0} \frac{(y)_n (-xq)^n}{(q^2; q^2)_n}. \tag{7.5}$$

Proof By the Smallest Parts Decomposition Principle

$$P_e(y, x; q) = P_o(y, xq; q) + xq P_o(y, xq; q), \tag{7.6}$$

$$P_o(y, x; q) = P_e(y, xq; q) + xqy P_o(y, xq; q). \tag{7.7}$$

Equation (7.6) immediately establishes (7.3). We now substitute (7.6) into (7.7) to obtain

$$P_o(y, x; q) = (1 + xq^2) P_o(y, xq^2; q) + xqy P_o(y, xq; q). \tag{7.8}$$

We now expand $P_o(y, x; q)$ in a power series in x:

$$P_o(y, x; q) = \sum_{n \geq 0} A_n x^n. \tag{7.9}$$

Substituting (7.9) into (7.8) and noting that $A_0 = 1$, we see that

$$A_n = q^{2n} A_n + q^{2n} A_{n-1} + yq^n A_{n-1},$$

or

$$A_n = \frac{yq^n (1 + q^n/y)}{(1 - q^{2n})} A_{n-1}.$$

Iteration then reveals that

$$A_n = \frac{y^n q^{n(n+1)/2}(-q/y)_n}{(q^2; q^2)_n}.$$ (7.10)

Combining (7.10) with (7.9), we see that (7.4) is established.

Finally by (7.4) using the basic hypergeometric series notation [12, p. 4]

$$P_o(y, x; q) = \lim_{\tau \to 0} {}_2\phi_1 \left(\begin{matrix} -q/\tau, -q/y; q, xy\tau \\ -q \end{matrix} \right)$$

$$= \lim_{\tau \to 0} \frac{(-xq)_\infty}{(xy\tau)_\infty} {}_2\phi_1 \left(\begin{matrix} \tau, y; q, -xq \\ -q \end{matrix} \right) \quad \text{(by [12, p. 10, eq. (1.4.6)])}$$

$$= (-xq)_\infty \sum_{n=0}^{\infty} \frac{(y)_n(-xq)^n}{(q^2; q^2)_n},$$

which proves (7.5). □

8 Lower parity indices in unrestricted partitions

Again we start with the relevant notation:

Let $u_e(r, m, n)$ (resp. $u_o(r, m, n)$) denote the number of partitions of n into m parts with lower even (resp. odd) parity index equal to r.

Next

$$U_e(y, x; q) = \sum_{r,m,n \geq 0} u_e(r, m, n) y^r x^m q^n,$$ (8.1)

and

$$U_o(y, x; q) = \sum_{r,m,n \geq 0} u_o(r, m, n) y^r x^m q^n.$$ (8.2)

Theorem 8

$$U_e(y, x; q) = \frac{U_o(y, xq; q)}{1 - xq},$$ (8.3)

$$U_o(y, x; q) = \frac{1}{(xq)_\infty} \sum_{n \geq 0} \frac{x^n y^n q^{n(n+1)/2}(1/y)_n}{(q^2; q^2)_n}$$ (8.4)

$$= (1 - x) \sum_{n \geq 0} \frac{(-qy)_n x^n}{(q^2; q^2)_n}.$$ (8.5)

Proof By the Smallest Parts Decomposition Principle

$$U_e(y, x; q) = U_o(y, xq; q) + \frac{xq}{1 - xq} U_o(y, xq; q),$$ (8.6)

$$U_o(y, x; q) = U_e(y, xq; q) + \frac{xqy}{1 - xq} U_o(y, xq; q).$$ (8.7)

Equation (8.6) immediately establishes (8.3). We now substitute (8.6) into (8.7) to obtain

$$U_o(y, x; q) = \frac{U_o(y, xq^2; q)}{1 - xq^2} + \frac{xqy}{1 - xq} U_o(y, xq; q). \tag{8.8}$$

Let us now define

$$v(y, x; q) = (xq)_\infty U_o(y, x; q). \tag{8.9}$$

Substituting (8.9) into (8.8), we see that

$$v(y, x; q) = (1 - xq)v(y, xq^2; q) + xqyv(y, xq; q). \tag{8.10}$$

We now expand $v(y, x; q)$ in a power series in x:

$$v(y, x; q) = \sum_{n \geq 0} \alpha_n x^n. \tag{8.11}$$

Substituting (8.11) in (8.10) and noting that $\alpha_0 = 1$, we see that

$$\alpha_n = \alpha_n q^{2n} - \alpha_{n-1} q^{2n-1} + \alpha_{n-1} yq^n,$$

or

$$\alpha_n = \frac{yq^n(1 - q^{n-1}/y)}{(1 - q^{2n})} \alpha_{n-1}.$$

Iteration then reveals that

$$\alpha_n = \frac{y^n q^{n(n+1)/2}(1/y)_n}{(q^2; q^2)_n}. \tag{8.12}$$

Combining (8.12) with (8.11) and (8.9), we see that (8.4) is established.

Finally by (8.4) using the notation of [12, p. 4]

$$
\begin{aligned}
U_o(y, x; q) &= \frac{1}{(xq)_\infty} \lim_{\tau \to 0} {}_2\phi_1\left(\begin{matrix} -q/\tau, 1/y; q, \tau xy \\ -q \end{matrix} \right) \\
&= \frac{1}{(xq)_\infty} \lim_{\tau \to 0} (x)_\infty {}_2\phi_1\left(\begin{matrix} \tau, -qy; q, x \\ -q \end{matrix} \right) \quad \text{(by [9, p. 10, eq. (1.4.6)])} \\
&= (1 - x) \sum_{n=0}^{\infty} \frac{(-qy)_n x^n}{(q^2; q^2)_n}.
\end{aligned}
$$

$\square$

Corollary 9

$$U_o(y, x; q) = \frac{1}{(xq)_\infty} P_o(-yq, -x/q; q). \tag{8.13}$$

Proof This follows immediately by comparison of (8.4) and (7.4). $\square$

9 Upper parity indices in partitions with distinct parts

As before, we begin with notation for the relevant partition functions:

$\delta_e(N, r, m, n)$ (resp. $\delta_o(N, r, m, n)$) denotes the number of partitions of n into m distinct parts each $\leq N$ and with upper even (resp. odd) parity index equal to r.

Next

$$D_e(N, y, x; q) := D_e(N) = \sum_{r,m,n \geq 0} \delta_e(N, r, m, n) y^r x^m q^n, \tag{9.1}$$

$$D_o(N, y, x; q) := D_o(N) = \sum_{r,m,n \geq 0} \delta_o(N, r, m, n) y^r x^m q^n. \tag{9.2}$$

Theorem 10

$$D_o(2n-1) = \sum_{i,j \geq 0} x^i y^{2j} q^{(i-j)^2+j^2+i+j} \begin{bmatrix} n+j-1 \\ i \end{bmatrix}_2 \begin{bmatrix} i \\ 2j \end{bmatrix}_1$$

$$+ \sum_{i,j \geq 0} x^i y^{2j-1} q^{(i-j)^2+j^2+i-j} \begin{bmatrix} n+j-1 \\ i \end{bmatrix}_2 \begin{bmatrix} i \\ 2j-1 \end{bmatrix}_1, \tag{9.3}$$

$$D_e(2n) = \sum_{i,j \geq 0} x^i y^{2j} q^{(i-j)^2+j^2+j} \begin{bmatrix} n+j \\ i \end{bmatrix}_2 \begin{bmatrix} i \\ 2j \end{bmatrix}_1$$

$$+ \sum_{i,j \geq 0} x^i y^{2j+1} q^{(i-j)^2+j^2+3j+1} \begin{bmatrix} n+j \\ i \end{bmatrix}_2 \begin{bmatrix} i \\ 2j+1 \end{bmatrix}_1. \tag{9.4}$$

Proof By the Largest Part Decomposition Principle, we see that

$$D_e(2N) = D_e(2N-1) + xyq^{2N} D_o(2N-1), \tag{9.5}$$

$$D_o(2N) = (1 + xq^{2N}) D_o(2N-1), \tag{9.6}$$

$$D_e(2N-1) = (1 + xq^{2N-1}) D_e(2N-2), \tag{9.7}$$

$$D_o(2N-1) = D_o(2N-2) + xyq^{2N-1} D_e(2N-2). \tag{9.8}$$

In addition, we note the initial values $D_e(0) = 1$, $D_o(1) = 1 + xyq$.

In order to prove our result we must reduce (9.5)–(9.8) to two recurrences only involving $D_o(2N-1)$ and $D_e(2N)$. By (9.5) and (9.7)

$$D_e(2N) = (1 + xq^{2N-1}) D_e(2N-2) + xyq^{2N} D_o(2N-1), \tag{9.9}$$

and by (9.6) and (9.8)

$$D_o(2N-1) = (1 + xq^{2N-2}) D_o(2N-3) + xyq^{2N-1} D_e(2N-2). \tag{9.10}$$

Clearly (9.9) and (9.10) plus the initial values completely define $D_e(2N)$ and $D_o(2N-1)$. To complete the proof of the theorem we need only show that the right-hand sides of (9.3) and (9.4) satisfy the same recurrences. The initial conditions are clear by inspection.

Let us define $d_o(2n-1)$ as the right-hand side of (9.3) and $d_e(2n)$ as the right-hand side of (9.4).

First we note that

$$d_e(2n) - d_e(2n-2)$$

$$= \sum_{i,j \geq 0} x^i y^{2j} q^{(i-j)^2 + j(j+1)} q^{2(n+j-i)} \begin{bmatrix} n+j-1 \\ i-1 \end{bmatrix}_2 \begin{bmatrix} i \\ 2j \end{bmatrix}_1$$

$$+ \sum_{i,j \geq 0} x^i y^{2j+1} q^{(i-j)^2 + (j+1)(j+2)-1} q^{2(n+j-i)} \begin{bmatrix} n+j-1 \\ i-1 \end{bmatrix}_2 \begin{bmatrix} i \\ 2j+1 \end{bmatrix}_1$$

$$= x q^{2n-1} \left\{ \sum_{i,j \geq 0} x^i y^{2j} q^{(i-j)^2 + 2(i-j) + j(j+1) + 2j - 2i} \begin{bmatrix} n+j-1 \\ i \end{bmatrix}_2 \begin{bmatrix} i+1 \\ 2j \end{bmatrix}_1 \right.$$

$$\left. + \sum_{i,j \geq 0} x^i y^{2j+1} q^{(i-j)^2 + 2(i-j) + (j+1)(j+2) - 1 + 2j - 2i} \begin{bmatrix} n+j-1 \\ i \end{bmatrix}_2 \begin{bmatrix} i+1 \\ 2j+1 \end{bmatrix}_1 \right\}.$$

$$(9.11)$$

So to complete the proof of (9.9) with D replaced by d, we need only identify $d_e(2n-2) + y q d_o(2n-1)$ with the expression inside the above curly brackets.

Now

$$d_e(2n-2) - y q d_o(2n-1)$$

$$= \sum_{i,j \geq 0} x^i y^{2j} q^{(i-j)^2 + j(j+1)} \begin{bmatrix} n+j-1 \\ i \end{bmatrix}_2 \begin{bmatrix} i \\ 2j \end{bmatrix}_1$$

$$+ \sum_{i,j \geq 0} x^i y^{2j+1} q^{(i-j)^2 + (j+1)(j+2)-1} \begin{bmatrix} n+j-1 \\ i \end{bmatrix}_2 \begin{bmatrix} i \\ 2j+1 \end{bmatrix}_1$$

$$+ \sum_{i,j \geq 0} x^i y^{2j+1} q^{i^2 - (2j-1)i + \binom{2j+1}{2} + 1} \begin{bmatrix} n+j-1 \\ i \end{bmatrix}_2 \begin{bmatrix} i \\ 2j \end{bmatrix}_1$$

$$+ \sum_{i,j \geq 0} x^i y^{2j} q^{i^2 - (2j-1)i + \binom{2j}{2} + 1} \begin{bmatrix} n+j-1 \\ i \end{bmatrix}_2 \begin{bmatrix} i \\ 2j-1 \end{bmatrix}_1$$

$$= \sum_{i,j \geq 0} x^i y^{2j} q^{i^2 - 2ij + 2j^2 + j} \begin{bmatrix} n+j-1 \\ i \end{bmatrix}_2 \left(\begin{bmatrix} i \\ 2j \end{bmatrix}_1 + q^{i-2j+1} \begin{bmatrix} i \\ 2j-1 \end{bmatrix}_1 \right)$$

$$+ \sum_{i,j \geq 0} x^i y^{2j+1} q^{i^2 - 2ij + 2j^2 + 3j + 1} \begin{bmatrix} n+j-1 \\ i \end{bmatrix}_2 \left(\begin{bmatrix} i \\ 2j+1 \end{bmatrix}_1 + q^{i-2j} \begin{bmatrix} i \\ 2j \end{bmatrix}_1 \right)$$

(where we have combined the first with the fourth sum and the second with the third)

(by (2.12))

$$= \sum_{i,j\geq 0} x^i y^{2j} q^{i^2-2ij+2j^2+j} \begin{bmatrix} n+j-1 \\ i \end{bmatrix}_2 \begin{bmatrix} i+1 \\ 2j \end{bmatrix}_1$$

$$+ \sum_{i,j\geq 0} x^i y^{2j+1} q^{i^2-2ij+2j^2+3j+1} \begin{bmatrix} n+j-1 \\ i \end{bmatrix}_2 \begin{bmatrix} i+1 \\ 2j+1 \end{bmatrix}_1 . \tag{9.12}$$

Now inspection reveals that this last entry in (9.12) is identical with the final expression in curly brackets in (9.11). Hence (9.9) has been established with the D's replaced by the d's.

To conclude we must do the same for (9.10). Now we see that

$$d_o(2n-1) - d_o(2n-3)$$

$$= \sum_{i,j\geq 0} x^i y^{2j} q^{i^2-(2j-1)i+\binom{2j+1}{2}} q^{2(n+j-1-i)} \begin{bmatrix} n+j-2 \\ i-1 \end{bmatrix}_2 \begin{bmatrix} i \\ 2j \end{bmatrix}_1$$

$$+ \sum_{i,j\geq 0} x^i y^{2j-1} q^{i^2-(2j-1)i+\binom{2j}{2}} q^{2(n+j-1-i)} \begin{bmatrix} n+j-2 \\ i-1 \end{bmatrix}_2 \begin{bmatrix} i \\ 2j-1 \end{bmatrix}_1$$

(by (2.11))

$$= xq^{2n-2}\left\{ \sum_{i,j\geq 0} x^i y^{2j} q^{i^2-(2j-1)i+\binom{2j+1}{2}} \begin{bmatrix} n+j-2 \\ i \end{bmatrix}_2 \begin{bmatrix} i+1 \\ 2j \end{bmatrix}_1 \right.$$

$$\left. + \sum_{i,j\geq 0} x^i y^{2j-1} q^{i^2-(2j-1)i+\binom{2j}{2}} \begin{bmatrix} n+j-2 \\ i \end{bmatrix}_2 \begin{bmatrix} i+1 \\ 2j-1 \end{bmatrix}_1 \right\}, \tag{9.13}$$

where in the final line we have shifted i to $i+1$ in each sum.

So to finish the proof of (9.10) with D replaced by d, we need only identify $d_o(2n-3) + yqd_e(2n-2)$ with the above expression contained in the curly brackets.

$$d_o(2n-3) + yqd_e(2n-2)$$

$$= \sum_{i,j\geq 0} x^i y^{2j} q^{i^2-(2j-1)i+\binom{2j+1}{2}} \begin{bmatrix} n+j-2 \\ i \end{bmatrix}_2 \begin{bmatrix} i \\ 2j \end{bmatrix}_1$$

$$+ \sum_{i,j\geq 0} x^i y^{2j-1} q^{i^2-(2j-1)i+\binom{2j}{2}} \begin{bmatrix} n+j-2 \\ i \end{bmatrix}_2 \begin{bmatrix} i \\ 2j-1 \end{bmatrix}_1$$

$$+ \sum_{i,j\geq 0} x^i y^{2j+1} q^{(i-j)^2+j(j+1)+1} \begin{bmatrix} n+j-1 \\ i \end{bmatrix}_2 \begin{bmatrix} i \\ 2j \end{bmatrix}_1$$

$$+ \sum_{i,j \geq 0} x^i y^{2j+2} q^{(i-j)^2+(j+1)(j+2)} \begin{bmatrix} n+j-1 \\ i \end{bmatrix}_2 \begin{bmatrix} i \\ 2j+1 \end{bmatrix}_1,$$

$$= \sum_{i,j \geq 0} x^i y^{2j} q^{i^2-(2j-1)i+\binom{2j+1}{2}}$$

$$\times \begin{bmatrix} n+j-2 \\ i \end{bmatrix}_2 \left(\begin{bmatrix} i \\ 2j \end{bmatrix}_1 + q^{i-2j+1} \begin{bmatrix} i \\ 2j-1 \end{bmatrix}_1 \right)$$

$$+ \sum_{i,j \geq 0} x^i y^{2j-1} q^{i^2-(2j-1)+\binom{2j}{2}} \begin{bmatrix} n+j-2 \\ i \end{bmatrix}_2$$

$$\times \left(\begin{bmatrix} i \\ 2j-1 \end{bmatrix}_1 + q^{i-2j+2} \begin{bmatrix} i \\ 2j-2 \end{bmatrix}_1 \right)$$

(where we replaced j by $j-1$ in the third and fourth sums and then combined sum 1 with sum 4 and sum 2 with sum 3)

(by (2.12))

$$= \sum_{i,j \geq 0} x^i y^{2j} q^{i^2-(2j-1)i+\binom{2j+1}{2}} \begin{bmatrix} n+j-2 \\ i \end{bmatrix}_2 \begin{bmatrix} i+1 \\ 2j \end{bmatrix}_1$$

$$+ \sum_{i,j \geq 0} x^i y^{2j-1} q^{i^2-(2j-1)i+\binom{2j}{2}} \begin{bmatrix} n+j-2 \\ i \end{bmatrix}_2 \begin{bmatrix} i+1 \\ 2j-1 \end{bmatrix}_1. \tag{9.14}$$

As before, the last entry in (9.14) is identical with the final expression in curly brackets in (9.13). Hence (9.10) is established with D replaced by d. Thus our theorem is proved. □

Corollary 11

$$D_o(\infty) = \sum_{i,j \geq 0} \frac{x^i y^{2j} q^{(i-j)^2+j^2+i+j}}{(-q;q)_i (q)_{2j} (q)_{i-2j}} + \sum_{i,j \geq 0} \frac{x^i y^{2j-1} q^{(i-j)^2+j^2+i-j}}{(-q;q)_i (q)_{2j-1} (q)_{i-2j+1}}, \tag{9.15}$$

$$D_e(\infty) = \sum_{i,j \geq 0} \frac{x^i y^{2j} q^{(i-j)^2+j^2+j}}{(-q;q)_i (q)_{2j} (q)_{i-2j}} + \sum_{i,j \geq 0} \frac{x^i y^{2j+1} q^{(i-j)^2+j^2+3j+1}}{(-q;q)_i (q)_{2j+1} (q)_{i-2j-1}}. \tag{9.16}$$

Proof Let $N \to \infty$ in the theorem and then algebraically simplify each term. □

We note that if we set $y = 0$ in (9.16), then $D_e(\infty)$ must be the generating function for partitions into distinct odd parts, and in this instance the right-hand side of (9.16) reduces to the familiar sum in Euler's identity

$$\sum_{i \geq 0} \frac{x^i q^{i^2}}{(q^2; q^2)_i} = (-xq; q^2)_\infty \tag{9.17}$$

(cf. [12, p. 9, eq. (1.3.16)]).

Also if we set $y = 0$ in (9.15) then $D_o(\infty)$ must be the generating function for partitions into distinct even parts, and in this instance the right-hand side of (9.15) reduces to a change of x to xq in (9.17).

However, setting $y = 1$ in either $D_o(\infty)$ or $D_e(\infty)$ must produce the generating function for all partitions into distinct parts. Consequently

$$(-xq; q)_\infty = \sum_{i,j \geq 0} \frac{x^i q^{(i-j)^2 + j^2 + i + j}}{(-q; q)_i (q)_{2j} (q)_{i-2j}} + \sum_{i,j \geq 0} \frac{x^i q^{(i-j)^2 + j^2 + i - j}}{(-q; q)_i (q)_{2j-1} (q)_{i-2j+1}}$$

$$= \sum_{i,j \geq 0} \frac{x^i q^{(i-j)^2 + j^2 + j}}{(-q; q)_i (q)_{2j} (q)_{i-2j}} + \sum_{i,j \geq 0} \frac{x^i q^{(i-j)^2 + j^2 + 3j + 1}}{(-q; q)_i (q)_{2j+1} (q)_{i-2j-1}},$$

(9.18)

which appear to be new.

We note that similar finite results can be obtained by setting $y = 0$ and $y = 1$ in Theorem 10. The $y = 0$ case is the classical q-binomial theorem [7, p. 36, eq. (3.3.6)]. The $y = 1$ case also appears to be new.

10 Upper parity indices with unrestricted parts

Again we begin with notation for the relevant partition functions:

$\phi_e(N, r, m, n)$ (resp. $\phi_o(N, r, m, n)$) denotes the number of partitions of n into m parts each $\leq N$ and with upper even (resp. odd) parity index equal to r.

Next

$$F_e(N, y, x; q) := F_e(N) = \sum_{r,m,n \geq 0} \phi_e(N, r, m, n) y^r x^m q^n, \qquad (10.1)$$

and

$$F_o(N, y, x; q) := F_o(N) = \sum_{r,m,n \geq 0} \phi_o(N, r, m, n) y^r x^m q^n. \qquad (10.2)$$

Theorem 12

$$F_e(2n - 1) = \sum_{i,j \geq 0} x^i y^{2j} q^{\binom{2j}{2} + i} \begin{bmatrix} n + i - j - 1 \\ i \end{bmatrix}_2 \begin{bmatrix} i \\ 2j \end{bmatrix}_1$$

$$+ \sum_{i,j \geq 0} x^i y^{2j+1} q^{\binom{2j+1}{2} + 2i} \begin{bmatrix} n + i - j - 2 \\ i \end{bmatrix}_2 \begin{bmatrix} i \\ 2j + 1 \end{bmatrix}_1, \qquad (10.3)$$

$$F_o(2n) = \sum_{i,j \geq 0} x^i y^{2j} q^{\binom{2j}{2} + 2i} \begin{bmatrix} n + i - j - 1 \\ i \end{bmatrix}_2 \begin{bmatrix} i \\ 2j \end{bmatrix}_1$$

$$+ \sum_{i,j \geq 0} x^i y^{2j+1} q^{\binom{2j+1}{2} + i} \begin{bmatrix} n + i - j - 1 \\ i \end{bmatrix}_2 \begin{bmatrix} i \\ 2j + 1 \end{bmatrix}_1. \qquad (10.4)$$

Proof By the Largest Part Decomposition Principle, we see that

$$F_e(2n) = F_e(2n-1) + xyq^{2n}F_o(2n), \tag{10.5}$$

$$F_o(2n) = F_o(2n-1) + xq^{2n}F_o(2n), \tag{10.6}$$

$$F_e(2n-1) = F_e(2n-2) + xq^{2n-1}F_e(2n-1), \tag{10.7}$$

$$F_o(2n-1) = F_o(2n-2) + xyq^{2n-1}F_e(2n-1). \tag{10.8}$$

Now we eliminate $F_o(2n-1)$ and $F_e(2n)$ from (10.5) – (10.8). This yields

$$F_e(2n-1) = F_e(2n-3) + xyq^{2n-2}F_o(2n-2) + xq^{2n-1}F_e(2n-1), \tag{10.9}$$

and

$$F_o(2n) = F_o(2n-2) + xyq^{2n-1}F_e(2n-1) + xq^{2n}F_o(2n). \tag{10.10}$$

At first glance, it would appear that these are not defining recurrences for the F's in that the left-hand side of each equation also appears on the right. However, the offending expression on the right can easily be moved to the left to overcome this objection. Of course, this means that the $F's$ will be rational functions not polynomials as were the D's of Sect. 9.

Thus (10.9) and (10.10) together with the initial values $F_o(0) = 1$ and $F_e(1) = \frac{1}{1-xq}$ uniquely determine $F_o(2n)$ and $F_e(2n-1)$. Inspection reveals that for $n = 0$, the right-hand side of (10.4) $= 1$, and for $n = 1$, the right-hand side of (10.3) is

$$\sum_{i \geq 0} x^i q^i = \frac{1}{1-xq}.$$

Consequently in order to complete the proof of our theorem we need only show that the right-hand sides of (10.3) and (10.4) satisfy (10.9) and (10.10).

Let us define $f_e(2n-1)$ as the right-hand side of (10.3) and $f_o(2n)$ as the right-hand side of (10.4).

We treat (10.9) first noting that

$$f_e(2n-1) - f_e(2n-3)$$

$$= \sum_{i,j \geq 0} x^i y^{2j} q^{\binom{2j}{2}+i} \begin{bmatrix} n+i-j-2 \\ i-1 \end{bmatrix}_2 q^{2(n-j-1)} \begin{bmatrix} i \\ 2j \end{bmatrix}_1$$

$$+ \sum_{i,j \geq 0} x^i y^{2j+1} q^{\binom{2j+1}{2}+2i} \begin{bmatrix} n+i-j-3 \\ i-1 \end{bmatrix}_2 q^{2(n-j-2)} \begin{bmatrix} i \\ 2j+1 \end{bmatrix}_1 \quad \text{(by (2.12))}$$

$$= xq^{2n-2} \left\{ \sum_{i,j \geq 0} x^i y^{2j} q^{\binom{2j-1}{2}+i} \begin{bmatrix} n+i-j-1 \\ i \end{bmatrix}_2 \begin{bmatrix} i+1 \\ 2j \end{bmatrix}_1 \right.$$

$$+ \left. \sum_{i,j \geq 0} x^i y^{2j+1} q^{\binom{2j}{2}+2i} \begin{bmatrix} n+i-j-2 \\ i \end{bmatrix}_2 \begin{bmatrix} i+1 \\ 2j+1 \end{bmatrix}_1 \right\}. \tag{10.11}$$

Therefore to conclude our treatment of (10.9) we must show that $yf_o(2n-2) + qf_e(2n-1)$ is equal to this last expression in curly brackets.

$$
yf_o(2n-2) + qf_e(2n-1)
$$

$$
= \sum_{i,j\geq 0} x^i y^{2j+1} \begin{bmatrix} n+i-j-2 \\ i \end{bmatrix}_2 \begin{bmatrix} i \\ 2j \end{bmatrix}_1 q^{2i+\binom{2j}{2}}
$$

$$
+ \sum_{i,j\geq 0} x^i y^{2j+2} \begin{bmatrix} n+i-j-2 \\ i \end{bmatrix}_2 \begin{bmatrix} i \\ 2j+1 \end{bmatrix}_1 q^{i+\binom{2j+1}{2}}
$$

$$
+ \sum_{i,j\geq 0} x^i y^{2j} q^{\binom{2j}{2}+i+1} \begin{bmatrix} n+i-j-1 \\ i \end{bmatrix}_2 \begin{bmatrix} i \\ 2j \end{bmatrix}_1
$$

$$
+ \sum_{i,j\geq 0} x^i y^{2j+1} q^{\binom{2j+1}{2}+2i+1} \begin{bmatrix} n+i-j-2 \\ i \end{bmatrix}_2 \begin{bmatrix} i \\ 2j+1 \end{bmatrix}_1
$$

$$
= \sum_{i,j\geq 0} x^i y^{2j} \begin{bmatrix} n+i-j-1 \\ i \end{bmatrix}_2 q^{i+\binom{2j-1}{2}} \left(\begin{bmatrix} i \\ 2j-1 \end{bmatrix}_1 + q^{2j} \begin{bmatrix} i \\ 2j \end{bmatrix}_1 \right)
$$

$$
+ \sum_{i,j\geq 0} x^i y^{2j+1} \begin{bmatrix} n+i-j-2 \\ i \end{bmatrix}_2 q^{2i+\binom{2j}{2}} \left(\begin{bmatrix} i \\ 2j \end{bmatrix}_1 + q^{2j+1} \begin{bmatrix} i \\ 2j+1 \end{bmatrix}_1 \right)
$$

(where we have combined the second and third sums into the new first sum and the first and fourth in to the new second sum)

(by (2.11))

$$
= \sum_{i,j\geq 0} x^i y^{2j} q^{i+\binom{2j-1}{2}} \begin{bmatrix} n+i-j-1 \\ i \end{bmatrix}_2 \begin{bmatrix} i+1 \\ 2j \end{bmatrix}_1
$$

$$
+ \sum_{i,j\geq 0} x^i y^{2j+1} q^{2i+\binom{2j}{2}} \begin{bmatrix} n+i-j-2 \\ i \end{bmatrix}_2 \begin{bmatrix} i \\ 2j+1 \end{bmatrix}_1, \tag{10.12}
$$

which is precisely the expression inside the curly brackets in the final entry in (10.11). We now move to (10.10) for the f's noting that

$$
f_o(2n) - f_o(2n-2)
$$

$$
= \sum_{i,j\geq 0} x^i y^{2j} \begin{bmatrix} n+i-j-2 \\ i-1 \end{bmatrix}_2 q^{2(n-j-1)} \begin{bmatrix} i \\ 2j \end{bmatrix}_1 q^{2i+\binom{2j}{2}}
$$

$$
+ \sum_{i,j\geq 0} x^i y^{2j+1} \begin{bmatrix} n+i-j-2 \\ i-1 \end{bmatrix}_2 q^{2(n-j-1)} \begin{bmatrix} i \\ 2j+1 \end{bmatrix}_1 q^{i+\binom{2j+1}{2}} \quad \text{(by (2.12))}
$$

$$= xq^{2n-1} \left\{ \sum_{i,j \geq 0} x^i y^{2j} \begin{bmatrix} n+i-j-1 \\ i \end{bmatrix}_2 \begin{bmatrix} i+1 \\ 2j \end{bmatrix}_1 q^{\binom{2j-1}{2}+2i} \right.$$

$$\left. + \sum_{i,j \geq 0} x^i y^{2j+1} \begin{bmatrix} n+i-j-1 \\ i \end{bmatrix}_2 \begin{bmatrix} i+1 \\ 2j+1 \end{bmatrix}_1 q^{\binom{2j}{2}+i} \right\}. \tag{10.13}$$

In order to complete (10.10) for the f's, we must identify this last expression in curly brackets with $y f_e(2n-1) + q f_o(2n)$.

$$y f_e(2n-1) + q f_o(2n)$$

$$= \sum_{i,j \geq 0} x^i y^{2j+1} q^{\binom{2j}{2}+i} \begin{bmatrix} n+i-j-1 \\ i \end{bmatrix}_2 \begin{bmatrix} i \\ 2j \end{bmatrix}_1$$

$$+ \sum_{i,j \geq 0} x^i y^{2j+2} q^{\binom{2j+1}{2}+2i} \begin{bmatrix} n+i-j-2 \\ i \end{bmatrix}_2 \begin{bmatrix} i \\ 2j+1 \end{bmatrix}_1$$

$$+ \sum_{i,j \geq 0} x^i y^{2j} q^{2i+\binom{2j}{2}+1} \begin{bmatrix} n+i-j-1 \\ i \end{bmatrix}_2 \begin{bmatrix} i \\ 2j \end{bmatrix}_1$$

$$+ \sum_{i,j \geq 0} x^i y^{2j+1} q^{i+\binom{2j+1}{2}+1} \begin{bmatrix} n+i-j-1 \\ i \end{bmatrix}_2 \begin{bmatrix} i \\ 2j+1 \end{bmatrix}_1$$

$$= \sum_{i,j \geq 0} x^i y^{2j} q^{\binom{2j-1}{2}+2i} \begin{bmatrix} n+i-j-1 \\ i \end{bmatrix}_2 \left(\begin{bmatrix} i \\ 2j-1 \end{bmatrix}_1 + q^{2j} \begin{bmatrix} i \\ 2j \end{bmatrix}_1 \right)$$

$$+ \sum_{i,j \geq 0} x^i y^{2j+1} q^{\binom{2j}{2}+i} \begin{bmatrix} n+i-j-1 \\ i \end{bmatrix}_2 \left(\begin{bmatrix} i \\ 2j \end{bmatrix}_1 + q^{2j+1} \begin{bmatrix} i \\ 2j+1 \end{bmatrix}_1 \right)$$

(combining the second and third sums into the first new sum, and the first and fourth into the second new sum)

(by (2.11))

$$= \sum_{i,j \geq 0} x^i y^{2j} q^{\binom{2j-1}{2}+2i} \begin{bmatrix} n+i-j-1 \\ i \end{bmatrix}_2 \begin{bmatrix} i+1 \\ 2j \end{bmatrix}_1$$

$$+ \sum_{i,j \geq 0} x^i y^{2j+1} q^{\binom{2j}{2}+i} \begin{bmatrix} n+i-j-1 \\ i \end{bmatrix}_2 \begin{bmatrix} i+1 \\ 2j+1 \end{bmatrix}_1, \tag{10.14}$$

and this is indeed the expression inside curly brackets in (10.13). So (10.10) is proved for the f's, and, as a result, our theorem is proved.   □

Corollary 13

$$F_e(\infty) = \sum_{i,j \geq 0} \frac{x^i y^{2j} q^{\binom{2j}{2}+i}}{(-q;q)_i (q)_{2j} (q)_{i-2j}} + \sum_{i,j \geq 0} \frac{x^i y^{2j+1} q^{\binom{2j+1}{2}+2i}}{(-q;q)_i (q)_{2j+1} (q)_{i-2j-1}}, \quad (10.15)$$

$$F_o(\infty) = \sum_{i,j \geq 0} \frac{x^i y^{2j} q^{\binom{2j}{2}+2i}}{(-q;q)_i (q)_{2j} (q)_{i-2j}} + \sum_{i,j \geq 0} \frac{x^i y^{2j+1} q^{\binom{2j+1}{2}+i}}{(-q;q)_i (q)_{2j+1} (q)_{i-2j-1}}. \quad (10.16)$$

Proof Let $n \to \infty$ in (10.3) and (10.4) respectively. $\qquad\square$

We note that setting $y = 0$ in (10.15) yields the generating function for partitions into odd parts (cf. [12, p. 9, eq. (1.3.15)])

$$\sum_{i \geq 0} \frac{x^i q^i}{(q^2;q^2)_i} = \frac{1}{(xq;q^2)_\infty}, \quad (10.17)$$

while setting $y = 0$ in (10.16) yields the generating function for partitions into even parts

$$\sum_{i \geq 0} \frac{x^i q^{2i}}{(q^2;q^2)_i} = \frac{1}{(xq^2;q^2)_\infty}. \quad (10.18)$$

Identities (10.17) and (10.18) both go back to Euler.

However setting $y = 1$ in either $F_e(\infty)$ or $F_o(\infty)$ must produce the generating function for all partitions.

$$\frac{1}{(xq)_\infty} = \sum_{i,j \geq 0} \frac{x^i q^{\binom{2j}{2}+i}}{(-q;q)_i (q)_{2j} (q)_{i-2j}} + \sum_{i,j \geq 0} \frac{x^i q^{\binom{2j+1}{2}+2i}}{(-q;q)_i (q)_{2j+1} (q)_{i-2j-1}}$$

$$= \sum_{i,j \geq 0} \frac{x^i q^{\binom{2j}{2}+2i}}{(-q;q)_i (q)_{2j} (q)_{i-2j}} + \sum_{i,j \geq 0} \frac{x^i q^{\binom{2j+1}{2}+i}}{(-q;q)_i (q)_{2j+1} (q)_{i-2j-1}}, \quad (10.19)$$

which appear to be new.

We note that, as in Sect. 9, similar finite results can be obtained by setting $y = 0$ and $y = 1$ in Theorem 12. The $y = 0$ case reduces to instances of the classical q-binomial series [7, p. 36, eq. (3.3.7)]. The $y = 1$ case again appears to be new.

11 Part size and parity indices

The previous two sections were devoted to considerations of partitions enumerated according to parity indices. Surprisingly, when we connect part size to a parity index, we come upon a new interpretation of the Rogers-Ramanujan identities.

Definition We say that a partition λ has even (resp. odd) *ample* part size if each part λ_i is larger than $I_{UE}(\lambda)$ (resp. $I_{UO}(\lambda)$).

Let $R_e(N, r, m, n)$ denote the number of partitions of n into m distinct parts each $\leq 2N$; in addition, we require even ample part size with upper even parity index $= r$.
Next

$$\rho_e(N, y, x, q) = \sum_{r,m,n \geq 0} R_e(N, r, m, n) y^r x^m q^n. \tag{11.1}$$

Theorem 14

$$\rho_e(N, y, x, q) = \sum_{j=0}^{N} \begin{bmatrix} N \\ j \end{bmatrix}_{q^2} q^{j^2} x^i (-yq)_j. \tag{11.2}$$

Proof We note by the Largest Part Decomposition Principle, that

$$\rho_e(N, y, x, q) = (1 + xq^{2N-1}) \rho_e(N-1, y, x, q) + xyq^{2N} \rho_e(N-1, y, xq, q). \tag{11.3}$$

The last term in the above by accounts for those partitions in which $2N$ appears; the xq in $\rho_e(N-1, y, xq, q)$ both increases the part size by 1 (which is necessary because the xyq^{2N} out front accounts for an increase in the upper even parity index) and changes the upper even index to the upper odd index to which $2N$ is attached to recover the upper even index.

In addition, we see that

$$\rho_e(0, y, x, q) = 1. \tag{11.4}$$

We observe that (11.3) and (11.4) completely determine $\rho_e(N, y, x, q)$.
Now we consider

$$\overline{\rho}(N, y, x, q) = \sum_{j=0}^{N} \begin{bmatrix} N \\ j \end{bmatrix}_{q^2} q^{j^2} z^j (-yq)_j, \tag{11.5}$$

$$\overline{\rho}(N, y, x, q) - \overline{\rho}(N-1, y, x, q)$$

$$= \sum_{j \geq 0} \left(\begin{bmatrix} N \\ j \end{bmatrix}_{q^2} - \begin{bmatrix} N-1 \\ j \end{bmatrix}_{q^2} \right) q^{j^2} x^j (-yq)_j$$

$$= \sum_{j \geq 0} q^{2(N-j)} \begin{bmatrix} N-1 \\ j-1 \end{bmatrix}_{q^2} q^{j^2} x^j (-yq)_j \quad \text{(by (2.12))}$$

$$= xq^{2N-1} \sum_{j \geq 0} \begin{bmatrix} N-1 \\ j \end{bmatrix}_{q^2} q^{j^2} x^j (-yq)_j (1 + yq^{j+1})$$

$$= xq^{2N-1} \overline{\rho}(N-1, y, x, q) + xyq^{2N} \sum_{j \geq 0} \begin{bmatrix} N-1 \\ j \end{bmatrix}_{q^2} q^{j^2} (xq)^j (-yq)_j$$

$$= xq^{2N-1} \overline{\rho}(N-1, y, x, q) + xyq^{2N} \overline{\rho}(N-1, y, xq, q). \tag{11.6}$$

Thus in light of the fact that $\overline{\rho}(0, y, x, q) = \rho_e(0, y, x, q)$ and that (11.3) and (11.6) are identical recurrences, we see by mathematical induction that for each $N \geq 0$

$$\rho_e(N, y, x, q) = \overline{\rho}(N, y, x, q),$$

and this proves our theorem.   □

It is now a straightforward matter to connect partitions into distinct parts of ample size with the Rogers-Ramanujan identities [7, p. 113].

Corollary 15 *The number of partitions of n into distinct parts of even ample size equals the number of partitions of n into parts $\equiv \pm 1$ (mod 5).*

Proof Setting $x = y = 1$ in (11.2) and letting $N \to \infty$, we see that the generating function for partitions into distinct parts with even ample part size is

$$\sum_{j \geq 0} \frac{q^{j^2}}{(q^2; q^2)_j} (-q)_j = \sum_{j \geq 0} \frac{q^{j^2}}{(q)_j} = \frac{1}{(q; q^5)_\infty (q^4; q^5)_\infty}$$

(by [4, p. 113, Cor. 7.9])

and the infinite product expression is just the generating function for partitions with parts $\equiv \pm 1$ (mod 5).   □

It is also easy to obtain a similar interpretation for the second Rogers-Ramanujan identity.

Corollary 16 *The number of partitions of n into distinct parts of odd ample size equals the number of partitions of n into parts $\equiv \pm 2$ (mod 5).*

Proof If we replace x by xq in (11.1), we see that the exponent on y now measures the odd upper index. Thus setting $x = q$, $y = 1$ in (11.2) and letting $N \to \infty$, we see that the generating function for partitions into distinct parts with odd ample part size is

$$\sum_{j \geq 0} \frac{q^{j^2+j}(-q)_j}{(q^2; q^2)_j} = \sum_{j \geq 0} \frac{q^{j^2+j}}{(q)_j} = \frac{1}{(q^2; q^5)_\infty (q^3; q^5)_\infty} \qquad \text{(by [7, p. 113, Cor. 7.10])}$$

and the infinite product expression is just the generating function for partitions with parts $\equiv \pm 2$ (mod 5).   □

12 Mock theta functions and little q-Jacobi polynomials

In the last section we saw that

$$\rho_e(\infty, y, x, q) = \sum_{N \geq 0} \frac{q^{n^2} x^n (-yq; q)_n}{(q^2; q^2)_n} \qquad (12.1)$$

is the generating function for partitions into distinct parts with even ample part size. Setting $y = 1$ yielded the connection with the Rogers-Ramanujan identities. Setting $x = 1$, $y = -1$ yields

$$f_o(q) = \sum_{n \geq 0} \frac{q^{n^2}}{(-q;q)_n} = 1 + q - q^2 + q^3 - q^6 + q^7 + q^9 - 2q^{10} + \cdots \quad (12.2)$$

one of Ramanujan's fifth order mock theta functions (cf. [9]). Thus $f_o(q)$ is the generating function for the excess of the number of partitions of n into distinct parts of even ample part size and an *even* upper parity index over those with an odd upper parity index. This is Theorem 17 as stated in the introduction.

For example, at $n = 10$, the six partitions in question are 10, $9 + 1$, $8 + 2$, $7 + 3$, $6 + 4$, $5 + 3 + 2$. Of these $9 + 1$ and $7 + 3$ have even upper parity index 0 while the other four have index 1; thus the coefficient of q^{10} in $f_o(q)$ must be $2 - 4 = -2$.

In this section, we shall relate $\rho_e(\infty, y, a, q)$ to an identity involving the little q-Jacobi polynomials [9, p. 27].

Theorem 18

$$\rho_e(\infty, y, a, q)$$

$$= \sum_{n \geq 0} \frac{q^{n^2} a^n (-yq;q)_n}{(q^2;q^2)_n}$$

$$= \frac{1}{(aq;q)_\infty} \sum_{n \geq 0} \frac{(-1)^n a^n q^{2n^2} (a^2;q^2)_n (1 - aq^{2n})}{(q^2;q^2)_n (1 - a)} p_n\left(y; -\frac{a}{q}, -1 : q\right), \quad (12.3)$$

where the little q-Jacobi polynomial is given by [12, p. 27]

$$p_n(y; A, B : q) = {}_2\phi_1\left(\begin{matrix} q^{-n}, ABq^{n+1}; q, qy \\ Aq \end{matrix}\right). \quad (12.4)$$

Proof We recall the weak Bailey lemma [9, p. 116, eq. (2.5) with $n, \rho_1, \rho_2 \to \infty$] (cf. [9, p. 116, eq. (3.1) corrected])

$$\sum_{n=0}^{\infty} q^{n^2} a^n \beta_n = \frac{1}{(aq;q)_\infty} \sum_{n=0}^{\infty} q^{n^2} a^n \alpha_n, \quad (12.5)$$

where

$$\beta_n = \sum_{r=0}^{n} \frac{\alpha_r}{(q;q)_{n-r}(aq;q)_{n+r}}.$$

So if we take $\beta_n = (-yq;q)_n/(q^2;q^2)_n$, then by [9, p. 115, eq. (2.7)]

$$\alpha_n = (1 - aq^{2n}) \sum_{j=0}^{n} \frac{(aq;q)_{n+j-1}(-1)^{n-j} q^{\binom{n-j}{2}}(-yq;q)_j}{(q;q)_{n-j}(q^2;q^2)_j}$$

$$= \frac{(-1)^n q^{\binom{n}{2}}(1-aq^{2n})(a;q)_n}{(1-a)(q;q)_n} {}_3\phi_2\left(\begin{matrix} q^{-n}, aq^n, -yq; q, q \\ -q, 0 \end{matrix}\right)$$

(by [12, p. 241, eq. (III.7)])

$$= \frac{(-1)^n q^{n^2}(a^2; q^2)_n(1-aq^{2n})}{(q^2; q^2)_n(1-a)} {}_2\phi_1\left(\begin{matrix} q^{-n}, aq^n; q, yq \\ -a \end{matrix}\right)$$

$$= (-1)^n q^{n^2}(a^2; q^2)_n(1-aq^{2n}) p_n\left(y; -\frac{a}{q}, -1:q\right),$$

and substituting these values for α_n and β_n into (12.5), we obtain the theorem. $\square$

Our first corollary is the classical identity of Rogers and Ramanujan [12, pp. 36–37, eq. (2.7.5) with $n \to \infty$] from which one deduces both Rogers-Ramanujan identities (the cases $a = 1$ and $a = q$).

Corollary 19

$$\sum_{n=0}^{\infty} \frac{q^{n^2} a^n}{(q;q)_n} = \frac{1}{(aq;q)_\infty} \sum_{n=0}^{\infty} \frac{(a;q)_n(-1)^n a^n q^{n(5n-1)/2}(1-aq^{2n})}{(q;q)_n(1-a)}. \tag{12.6}$$

Proof Setting $y = 1$, and noting by [12, p. 11, eq. (1.5.3)] that

$$p_n(1; A, B : q) = \frac{(ABq^{n+1})^n(B^{-1}q^{-n}; q)_n}{(Aq; q)_n} = \frac{A^n q^{\binom{n+1}{2}}(Bq; q)_n}{(Aq; q)_n},$$

we find that (12.3) simplifies to (12.6). $\square$

In order to obtain the formula [9, p. 114, eq. (1.4)] for $f_o(q)$, we need some lemmas concerning the little q-Jacobi polynomials.

Lemma 20

$$p_n(-1; -1, -1 : q) = (-1)^n q^{\binom{n+1}{2}} \sum_{j=-n}^{n} (-1)^j q^{-j^2}.$$

Proof This result follows directly by mathematical induction provided we can show that

$$(-1)^n q^{-\binom{n+1}{2}} p_n(-1; -1, -1 : q) - (-1)^{n-1} q^{-\binom{n}{2}} p_{n-1}(-1; -1, -1 : q)$$
$$= 2(-1)^n q^{-n^2},$$

or equivalently

$${}_2\phi_1\left(\begin{matrix} q^{-n}, q^{n+1}; q, -q \\ -q \end{matrix}\right) + q^n {}_2\phi_1\left(\begin{matrix} q^{-n+1}, q^n; q, -q \\ -q \end{matrix}\right) = 2q^{-\binom{n}{2}}.$$

But

$$
{}_2\phi_1\left(\begin{matrix} q^{-n}, q^{n+1}; q, -q \\ -q \end{matrix}\right) + q^n {}_2\phi_1\left(\begin{matrix} q^{-n+1}, q^n; q, -q \\ -q \end{matrix}\right)
$$

$$
= \sum_{j=0}^{n} \frac{(q^{-n+1}; q)_{j-1}(q^{n+1}; q)_{j-1}(-q)^j}{(q; q)_j(-q; q)_j}
$$

$$
\times \left((1 - q^{-n})(1 - q^{n+j}) + q^n(1 - q^{-n+j})(1 - q^n)\right)
$$

$$
= (1 + q^n) {}_2\phi_1\left(\begin{matrix} q^{-n}, q^n; q, -q \\ -q \end{matrix}\right)
$$

$$
= 2q^{-\binom{n}{2}}
$$

by [12, p. 11, eq. (1.5.2)].   □

Lemma 21

$$
2p_n\left(-1; -\frac{1}{q}, -1 : q\right) - 2p_n\left(-1; -\frac{1}{q}, -\frac{1}{q} : q\right)
$$

$$
= (1 - q^n)p_{n-1}(-1; -1, -1 : q).
$$

Proof

$$
2p_n\left(-1; -\frac{1}{q}, -1 : q\right) - 2p_n\left(-1; -\frac{1}{q}, -\frac{1}{q} : q\right)
$$

$$
= 2{}_2\phi_1\left(\begin{matrix} q^{-n}, q^n; q, -q \\ -1 \end{matrix}\right) - 2{}_2\phi_1\left(\begin{matrix} q^{-n}, q^{n-1}; q, -q \\ -1 \end{matrix}\right)
$$

$$
= 2\sum_{j=0}^{n} \frac{(q^{-n}, q)_j(-q)^j}{(q; q)_j(-1; q)_j}\left((1 - q^{n+j-1}) - (1 - q^{n-1})\right)
$$

$$
= 2\sum_{j=1}^{n} \frac{(q^{-n}, q)_j(-q)^j(q^n)_{j-1}}{(q; q)_{j-1}(-1; q)_j}
$$

$$
= (1 - q^n){}_2\phi_1\left(\begin{matrix} q^{-n+1}, q^n; q, -q \\ -q \end{matrix}\right)
$$

$$
= (1 - q^n)p_{n-1}(-1; -1, -1 : q).   □
$$

Lemma 22

$$
2p_n\left(-1; -\frac{1}{q}, -1 : q\right)
$$

$$
= (-1)^n q^{\binom{n+1}{2}} \sum_{j=-n}^{n} (-1)^j q^{-j^2} - (-1)^n q^{\binom{n}{2}} \sum_{j=-n+1}^{n-1} (-1)^i q^{-j^2}.
$$

Proof Noting that by [12, p. 11, eq. (1.5.2)]

$$p_n\left(-1; -\frac{1}{q}, -\frac{1}{q} : q\right) = {}_2\phi_1\left(\begin{matrix} q^{-n}, q^{n-1}; q, -q \\ -1 \end{matrix}\right) = \frac{(-q^{1-n}; q)_n}{(-1; q)_n} = q^{-\binom{n}{2}},$$

we combine Lemmas 20 and 21 to see that

$$2p_n\left(-1; -\frac{1}{q}, -1 : q\right)$$

$$= 2q^{-\binom{n}{2}} + (1 - q^n)(-1)^{n-1}q^{\binom{n}{2}} \sum_{j=-n+1}^{n-1} (-1)^j q^{-j^2}$$

$$= (-1)^n q^{\binom{n+1}{2}} \sum_{j=-n}^{n} (-1)^j q^{-j^2} - (-1)^n q^{\binom{n}{2}} \sum_{j=-n+1}^{n-1} (-1)^j q^{-j^2},$$

as desired.   □

Corollary 23 ([9, p. 114, eq. (1.4)])

$$f_0(q) = \frac{1}{(q; q)_\infty} \sum_{n=0}^{\infty} \sum_{|j| \leq n} (-1)^n q^{n(5n+1)/2 - j^2}(1 - q^{4n+2}).$$

Proof Setting $y = -1$, $a = 1$ in (12.3) and invoking Lemma 22, we see that

$$f_0(q) = \sum_{n=0}^{\infty} \frac{q^{n^2}}{(-q; q)_n}$$

$$= \frac{1}{(q; q)_\infty} \sum_{n=0}^{\infty} q^{n^2}(-1)^n \left((-1)^n q^{\binom{n+1}{2}} \sum_{j=-n}^{n} (-1)^j q^{-j^2}\right.$$

$$\left. - (-1)^n q^{\binom{n}{2}} \sum_{j=-n+1}^{n-1} (-1)^j q^{-j^2}\right)$$

$$= \frac{1}{(q; q)_\infty} \sum_{\substack{n=0 \\ |j| \leq n}}^{\infty} (-1)^j q^{n(5n+1)/2 - j^2}(1 - q^{4n+2}),$$

where we have replaced n by $n + 1$ in the second double sum.   □

13 Conclusion and open problems

As was mentioned in Sect. 1 and again in Sect. 5, K. Kursungoz has found combinatorial proofs of many of the results in this paper. In particular, he has sieve-theoretic proofs of equations (7.4), (7.5), (8.4), (8.5) and (8.12); i.e. all the theorems in Sects. 7 and 8. He has bijective proofs of (9.3), (9.4), (10.3), (10.4) and (11.2); i.e. all the theorems in Sects. 9–11. Additionally, as was noted in Sect. 5 he has a combinatorial, bijective proof of Theorem 4 for all k, not just $k = 2$ or 3, and has beautifully generalized the result to cover all specializations to (1.6) with $1 \leq a \leq k$. On top of this he has bijective explanations of each of the multiple series generating functions given in Sects. 3 and 4.

Furthermore, he has found a reasonably natural interpretation for

$$\sum_{n_1,\ldots,n_{k-1} \geq 0} \frac{q^{N_1^2+\cdots+N_{k-1}^2} x^{N_1+\cdots+N_{k-1}} (-yq;q)_{n_1} \cdots (-yq;q)_{n_{k-1}}}{(q^2;q^2)_{n_1} (q^2;q^2)_{n_2} \cdots (q^2;q^2)_{n_{k-1}}}.$$

The following are further problems that suggest themselves either from comparison of this work with results in the literature or from empirical studies.

1. Show bijectively that $W_{3,3}(n)$ is equal to the number of partitions of n into parts that differ by at least two and by more than 2 if the parts are even.
2. Show bijectively that $W_{3,1}(n)$ is equal to the number of partitions of n into parts (each > 2) that differ by at least two and by more than 2 if the parts are even.

We know from Theorem 1 and a comparison of it with the Göllnitz-Gordon identities that in each case the sets of partitions involved are equinumerous. Thus the important word in each problem is "bijectively." More generally

3. Prove bijectively that

$$W_{2k-1,2a-1}(n) = D_{k,a}(n),$$

where $D_{k,a}(n)$ denotes the number of partitions of n of the form $n = \sum_{i \geq 1} f_i \cdot i$ (here we use the frequency notation f_i for the appearances of i in the partition) with $f_1 + f_2 \leq a - 1$ and for all $i \geq 1$,

$$f_{2i-1} \leq 1 \quad \text{and} \quad f_{2i} + f_{2i+1} + f_{2i+2} \leq k - 1.$$

The assertion is true by comparing Theorem 1 here in the odd case with Theorem 1 in [2]. This problem and this observation leads naturally to a more general topic.

4. The overriding theme of this paper is parity. Can one generalize these theorems to moduli other than 2? In light of the fact that the above $D_{k,a}(n)$ is the special case $\lambda + 1 = 2$ of the general partition theorem from [4] and [5], it would seem that theorems comparable to those in Sects. 3–5 might be found by attempting to generalize the observation in Problem 3 to these more general results.
5. It follows from an old formula of Rogers [19, p. 333, eq. (6)] that

$$P_o(-1, 1; q) = \sum_{n=0}^{\infty} q^{n(3n+1)/2}(1 - q^{2n+1}).$$

Prove combinatorially that

$$\sum_{r,m\geq 0} p_o(r,m,N)(-1)^r = \begin{cases} 1 & \text{if } N = n(3n+1)/2, \\ -1 & \text{if } N = n(3n+5)/2+1, \\ 0 & \text{otherwise.} \end{cases}$$

6. Prove combinatorially that $U_o(-1,-1;q)$ is equal to the third order mock theta function

$$f(q) = 1 + \sum_{n>0} \frac{q^{n^2}}{((-q;q)_n)^2}.$$

In Problems 7 and 8, let $uc(n)$ denote the coefficient of q^n in the series for $(q,q)_\infty U_o(-1,-1;q)$.

7. Prove that $uc(n) = 8$ if n is a prime congruent to 7 or 11 mod 12, $-uc(2) = uc(3) = uc(5) = 4$, and for all other primes $uc(n) = 0$.
8. Prove that $uc(n^2) = 0$ if n is an odd prime power and that $uc(2^n) = -4$.
9. Prove that if $y = 1$ and $x = -1$ in the second sum in (9.15) the result is

$$\sum_{n=1}^{\infty} (-1)^n q^{n^2}.$$

10. Prove that if $y = 1$ and $x = -1$ in the first sum in (9.16) the result is

$$\sum_{n=1}^{\infty} (-1)^n q^{n^2}.$$

11. Extend the parity indices to overpartitions in a manner that will provide natural generalizations of the work of Corteel and Lovejoy [11, 18].
12. Provide a combinatorial proof of Lemma 20.
13. Noting that the theorems in Sects. 9 and 10 are devoted to bi-basic generalizations of the q-binomial theorem and q-binomial series, find a full bi-basic generalization to bases other than q and q^2.
14. An appealing, long-term project would be a full analytic proof of Theorem 4 and its extension to a more general q-hypergeometric series. We note that (5.1) can be rewritten as

$$\sum_{\substack{n_1,\ldots,n_{k-1}\geq 0 \\ j_1,\ldots,j_{k-1}\geq 0}} q^{(N_1+J_1)^2+(N_2+J_2)^2+\cdots+(N_{k-1}+J_{k-1})^2}$$

$$\times \frac{x^{N_1+\cdots+N_{k-1}+J_1+\cdots+J_{k-1}} y^{j_1+\cdots+j_{k-1}}}{(q^2;q^2)_{n_1}\cdots(q^2;q^2)_{n_{k-1}}(q^2;q^2)_{j_1}\cdots(q^2;q^2)_{j_{k-1}}}.$$

where $N_i = n_i + n_{i+1} + \cdots + n_{k-1}$ and $J_i = j_i + j_{i+1} + \cdots + j_{k-1}$. This latter formulation seems likely to be a limiting case of series analogous to those considered in [3].

15. Find a combinatorial interpretation for

$$(-xq^2; q^2)_\infty Q_{\frac{k}{2}, \frac{q}{2}}(x^2; q^2)$$

with k even and a odd. This is the "missing case" in Sects. 3 and 4.

References

1. Andrews, G.E.: An analytic proof of the Rogers-Ramanujan-Gordon identities. Am. J. Math. **88**, 844–846 (1966)
2. Andrews, G.E.: A generalization of the Göllnitz-Gordon partition theorems. Proc. Am. Math. Soc. **8**, 945–952 (1967)
3. Andrews, G.E.: On q-difference equations for certain well-poised basic hypergeometric series. Quart. J. Math. Oxford Ser. **19**(2), 433–447 (1968)
4. Andrews, G.E.: A generalization of the classical partition theorems. Trans. Am. Math. Soc. **145**, 205–221 (1968)
5. Andrews, G.E.: On the general Rogers-Ramanujan theorem. Mem. Am. Math. Soc. **152**, 86 (1974)
6. Andrews, G.E.: An analytic generalization of the Rogers-Ramanujan identities for odd moduli. Proc. Nat. Acad. Sci. USA **71**, 4082–4085 (1974)
7. Andrews, G.E.: The Theory of Partitions. Addison-Wesley, Reading (1976) reissued: Cambridge University Press, Cambridge (1998)
8. Andrews, G.E.: Ramanujan's "lost" notebook. IV: stacks and alternating parity in partitions. Adv. Math. **53**, 55–74 (1984)
9. Andrews, G.E.: The fifth and seventh order mock theta functions. Trans. Am. Math. Soc. **293**, 113–134 (1986)
10. Andrews, G.E., Berndt, B.C.: Ramanujan's Lost Notebook, Part II. Springer, New York (2008)
11. Corteel, S., Lovejoy, J.: Overpartitions. Trans. Am. Math. Soc. **356**, 1623–1635 (2004)
12. Gasper, G., Rahman, M.: Basic Hypergeometric Series. Cambridge University Press, Cambridge (1990)
13. Göllnitz, H.: Einfache Partitionen. Diplomarbeit W.S. Göttingen, 65 pp. (1960)
14. Göllnitz, H.: Partitionen mit Differenzenbedingungen. J. Reine Angew. Math. **225**, 154–190 (1967)
15. Gordon, B.: Some Ramanujan-like continued fractions. Abstracts of Short Communications. Int. Congr. of Math., Stockholm, pp. 29–30 (1962)
16. Gordon, B.: A combinatorial generalization of the Rogers-Ramanujan identities. Am. J. Math. **83**, 393–399 (1961)
17. Gordon, B.: Some continued fractions of the Rogers-Ramanujan type. Duke Math. J. **31**, 741–748 (1965)
18. Lovejoy, J.: Overpartition theorems of the Rogers-Ramanujan type. J. Lond. Math. Soc. **69**, 562–574 (2004)
19. Rogers, L.J.: On two theorems of combinatorial analysis and some allied identities. Proc. Lond. Math. Soc. **16**(2), 315–336 (1917)

Ramanujan J (2010) 23: 91–105
DOI 10.1007/s11139-009-9185-x

Some combinatorial properties of hook lengths, contents, and parts of partitions

Richard P. Stanley

Dedicated to George Andrews for his 70th birthday

Received: 15 December 2008 / Accepted: 10 June 2009 / Published online: 15 October 2009
© Springer Science+Business Media, LLC 2009

Abstract The main result of this paper is a generalization of a conjecture of Guoniu Han, originally inspired by an identity of Nekrasov and Okounkov. Our result states that if F is any symmetric function (say over $\mathbb{Q}$) and if

$$\Phi_n(F) = \frac{1}{n!} \sum_{\lambda \vdash n} f_\lambda^2 F(h_u^2 : u \in \lambda),$$

where h_u denotes the hook length of the square u of the partition λ of n and f_λ is the number of standard Young tableaux of shape λ, then $\Phi_n(F)$ is a polynomial function of n. A similar result is obtained when $F(h_u^2 : u \in \lambda)$ is replaced with a function that is symmetric separately in the contents c_u of λ and the shifted parts $\lambda_i + n - i$ of λ.

Keywords Partition · Hook length · Content · Shifted part · Standard Young tableau

Mathematics Subject Classification (2000) Primary 05E10 · Secondary 05E05

1 Introduction

We assume basic knowledge of symmetric functions such as given in [13, Chap. 7].
Let f_λ denote the number of standard Young tableaux (SYT) of shape $\lambda \vdash n$. Recall

---

This material is based upon work supported by the National Science Foundation under Grant No. 0604423. Any opinions, findings and conclusions or recommendations expressed in this material are those of the author and do not necessarily reflect those of the National Science Foundation.

R.P. Stanley (✉)
Department of Mathematics, Massachusetts Institute of Technology, Cambridge, MA 02139, USA
e-mail: rstan@math.mit.edu

the hook length formula of Frame, Robinson, and Thrall [3], [13, Cor. 7.21.6]:

$$f_\lambda = \frac{n!}{\prod_{u \in \lambda} h_u},$$

where u ranges over all squares in the (Young) diagram of λ, and h_u denotes the hook length at u. A basic property of the numbers f_λ is the formula

$$\sum_{\lambda \vdash n} f_\lambda^2 = n!,$$

which has an elegant bijective proof (the RSK algorithm). We will be interested in generalizing this formula by weighting the sum on the left by various functions of λ. Our primary interest is the sum

$$\Phi_n(F) = \frac{1}{n!} \sum_{\lambda \vdash n} f_\lambda^2 F(h_u^2 : u \in \lambda),$$

where $F = F(x_1, x_2, \dots)$ is a symmetric function, say over $\mathbb{Q}$ (denoted $F \in \Lambda_\mathbb{Q}$). The notation $F(h_u^2 : u \in \lambda)$ means that we are substituting for n of the variables in F the quantities h_u^2 for $u \in \lambda$, and setting all other variables equal to 0. For instance, if $F = p_k := \sum x_i^k$, then

$$F(h_u^2 : u \in \lambda) = \sum_{u \in \lambda} h_u^{2k}.$$

This paper is motivated by the conjecture [7, Conj. 3.1] of Guoniu Han that for all $k \in \mathbb{P} = \{1, 2, \dots\}$, we have that $\Phi_n(p_k) \in \mathbb{Q}[n]$, i.e.,

$$\frac{1}{n!} \sum_{\lambda \vdash n} f_\lambda^2 \sum_{u \in \lambda} h_u^{2k}$$

is a polynomial function of n. This conjecture in turn was inspired by the remarkable identity of Nekrasov and Okounkov [10] (later given a more elementary proof by Han [6])

$$\sum_{n \geq 0} \left(\sum_{\lambda \vdash n} f_\lambda^2 \prod_{u \in \lambda} (t + h_u^2) \right) \frac{x^n}{n!^2} = \prod_{i \geq 1} (1 - x^i)^{-1-t}. \tag{1}$$

(We have stated this identity in a slightly different form than given in [6, 10].) Our main result (Theorem 4.3) states that $\Phi_n(F) \in \mathbb{Q}[n]$ for any $F \in \Lambda_\mathbb{Q}$, i.e., for fixed F, $\Phi_n(F)$ is a polynomial function of n. In the course of the proof we also show that

$$\frac{1}{n!} \sum_{\lambda \vdash n} f_\lambda^2 G(\{c_u : u \in \lambda\}; \{\lambda_i + n - i : 1 \leq i \leq n\}) \in \mathbb{Q}[n].$$

Here $G = G(x; y)$ is any formal power series of bounded degree over $\mathbb{Q}$ that is symmetric in the x and y variables separately. Moreover, c_u denotes the content of $u \in \lambda$ [13, p. 373]; and we write $\lambda = (\lambda_1, \dots, \lambda_n)$, adding 0's at the end so that there are exactly n parts.

2 Contents

In the next section we will obtain a stronger result than the main result of this section (Theorem 2.1). Since Theorem 2.1 may be of independent interest and may be helpful for understanding the next section, we treat it separately.

If $t \in \mathbb{P}$ and F is a symmetric function in the variables $x_1, x_2, \ldots$, then we write $F(1^t)$ for the result of setting $x_1 = x_2 = \cdots = x_t = 1$ and all other $x_j = 0$ in F. For instance, $p_\lambda(1^t) = t^{\ell(\lambda)}$, where $\ell(\lambda)$ is the number of (positive) parts of λ. The *hook-content* formula for the case $q = 1$ [13, Cor. 7.21.4] asserts that

$$s_\lambda(1^t) = \frac{\prod_{u \in \lambda}(t + c_u)}{H_\lambda},\tag{2}$$

where s_λ is a Schur function and

$$H_\lambda = \prod_{u \in \lambda} h_u,$$

the product of the hook lengths of λ (so $f_\lambda = n!/H_\lambda$).

Theorem 2.1 *For any $F \in \Lambda_\mathbb{Q}$ we have*

$$\frac{1}{n!} \sum_{\lambda \vdash n} f_\lambda^2 F(c_u : u \in \lambda) \in \mathbb{Q}[n].$$

Proof By linearity it suffices to take $F = e_\mu$, the elementary symmetric function indexed by μ. Let $k \in \mathbb{P}$, and for $1 \le i \le k$ let $x^{(i)}$ denote the set of variables $x_1^{(i)}, x_2^{(i)}, \ldots$. Let $\mathfrak{S}_n$ denote the symmetric group of all permutations of $\{1, \ldots, n\}$. For $w \in \mathfrak{S}_n$ write $\rho(w)$ for the cycle type of w, i.e., $\rho(w)$ is the partition of n whose parts are the cycle lengths of w. We use the identity [5, Prop. 2.2], [13, Exer. 7.70]

$$\sum_{\lambda \vdash n} H_\lambda^{k-2} s_\lambda(x^{(1)}) \cdots s_\lambda(x^{(k)})$$

$$= \frac{1}{n!} \sum_{\substack{w_1 w_2 \cdots w_k = 1 \\ \text{in } \mathfrak{S}_n}} p_{\rho(w_1)}(x^{(1)}) \cdots p_{\rho(w_k)}(x^{(k)}).\tag{3}$$

Make the substitution $x^{(i)} = 1^{t_i}$ as explained above. Letting $c(w)$ denote the number of cycles of $w \in \mathfrak{S}_n$, we obtain

$$\sum_{\lambda \vdash n} H_\lambda^{-2} \prod_{u \in \lambda} \prod_{i=1}^k (t_i + c_u) = \frac{1}{n!} \sum_{\substack{w_1 w_2 \cdots w_k = 1 \\ \text{in } \mathfrak{S}_n}} t_1^{c(w_1)} \cdots t_k^{c(w_k)}.\tag{4}$$

For any $n \geq \mu_1$ let $\mu = (\mu_1, \ldots, \mu_k)$ be a partition with k parts, and take the coefficient of $t_1^{n-\mu_1} \cdots t_k^{n-\mu_k}$ on both sides of (4). Using $f_\lambda = n!/H_\lambda$, we obtain

$$\frac{1}{n!} \sum_{\lambda \vdash n} f_\lambda^2 e_\mu(c_u : u \in \lambda)$$

$$= \#\{(w_1, \ldots, w_k) \in \mathfrak{S}_n^k : w_1 \cdots w_k = 1, \ c(w_i) = n - \mu_i\}. \tag{5}$$

We therefore need to show that the right-hand side of (5) is a polynomial function of n.

Suppose that $c(w_i) = n - \mu_i$ and that the union F of the non-fixed points of all the w_i's has r elements. Then

$$1 + \mu_1 \leq r \leq 2 \sum \mu_i. \tag{6}$$

We can choose the set F in $\binom{n}{r}$ ways. Once we make this choice there is a certain number of ways (depending on r but independent of n) that we can have $w_1 \cdots w_k = 1$. (In more algebraic terms, $\mathfrak{S}_n$ acts on S_μ by conjugation, where S_μ is the set on the right-hand side of (5), and the number of orbits of this action is independent of n.) Hence for $n \geq 1 + \mu_1$, $\#S_\mu$ is a finite linear combination (over $\mathbb{N} = \{0, 1, 2, \ldots\}$) of polynomials $\binom{n}{r}$, and is thus a polynomial $N_\mu(n)$ as desired.

If $n < 1 + \mu_1$, then it is clear from the previous paragraph that the polynomial N_μ satisfies $N_\mu(n) = 0$. On the other hand, if $\lambda \vdash n$ then we also have $e_\mu(c_u : u \in \lambda) = 0$. Hence the two sides of (5) agree for $0 \leq n < 1 + \max \mu_i$, and the proof is complete. $\square$

Note that the proof of Theorem 2.1 shows that $N_\mu(n)$ is a *nonnegative* integer linear combination of the polynomials $\binom{n}{r}$. It can be shown that either $N_\mu = 0$ or $\deg N_\mu = \sum \mu_i$. Moreover $N_\mu \neq 0$ if and only if $\sum \mu_i$ is even, say $2r$, and $\mu_1 \leq r$. The nonzero polynomials $N_\mu(n)$ for $|\mu| \leq 6$ are given by

$$N_{1,1}(n) = \frac{n(n-1)}{2},$$

$$N_{2,2}(n) = \frac{n(n-1)(n-2)(3n-1)}{24},$$

$$N_{2,1,1}(n) = \frac{n(n-1)(n-2)(n+1)}{4},$$

$$N_{1,1,1,1}(n) = \frac{n(n-1)(3n^2+n-12)}{4},$$

$$N_{3,3}(n) = \frac{n^2(n-1)^2(n-2)(n-3)}{48},$$

$$N_{3,2,1}(n) = \frac{n(n-1)(n-2)(n-3)(3n^2+5n+4)}{48},$$

$$N_{3,1,1,1}(n) = \frac{n(n-1)(n-2)(n-3)(n^2+3n+4)}{8},$$

$$N_{2,2,2}(n) = \frac{n(n-1)(n-2)(3n^3 - 9n - 46)}{24},$$

$$N_{2,2,1,1}(n) = \frac{n(n-1)(n-2)(15n^3 + 20n^2 - 59n - 312)}{48},$$

$$N_{2,1,1,1,1}(n) = \frac{n(n-1)(n-2)(3n^3 + 8n^2 - 7n - 96)}{4},$$

$$N_{1,1,1,1,1,1}(n) = \frac{n(n-1)(15n^4 + 30n^3 - 105n^2 - 700n + 1344)}{8}.$$

A slight modification of the proof of a special case of Theorem 2.1 leads to a "content Nekrasov-Okounkov formula."

Theorem 2.2 *We have*

$$\sum_{n \geq 0} \left(\sum_{\lambda \vdash n} f_\lambda^2 \prod_{u \in \lambda} (t + c_u^2) \right) \frac{x^n}{n!^2} = (1-x)^{-t}.$$

Proof By the "dual Cauchy identity" [13, Thm. 7.14.3] we have

$$\sum_{\lambda \vdash n} s_\lambda(x) s_{\lambda'}(y) = \frac{1}{n!} \sum_{w \in \mathfrak{S}_n} \varepsilon_w p_{\rho(w)}(x) p_{\rho(w)}(y),$$

where $\varepsilon(w)$ is given by (15), and where λ' denotes the conjugate partition to λ. Substitute $x = 1^t$ and $y = 1^t$. Since the contents of λ' are the negative of those of λ, we obtain

$$\sum_{\lambda \vdash n} H_\lambda^{-2} \prod_{u \in \lambda} (t^2 - c_u^2) = \frac{1}{n!} \sum_{w \in \mathfrak{S}_n} \varepsilon_w t^{2c(w)}.$$

It is a well-known and basic fact that the sum on the right is $\binom{t^2}{n}$. Put $-t$ for t^2, multiply by $(-x)^n$ and sum on $n \geq 0$ to get the stated formula. □

A simple variant of Theorem 2.2 follows from considering the usual Cauchy identity (the case $k = 2$ of (3)) instead of the dual one:

$$\sum_{n \geq 0} \left(\sum_{\lambda \vdash n} f_\lambda^2 \prod_{u \in \lambda} (t + c_u)(v + c_u) \right) \frac{x^n}{n!^2} = (1-x)^{-tv}.$$

A related identity is due to Fujii *et al.* [4, Appendix], namely, for any $r \geq 0$ we have

$$\frac{1}{n!} \sum_{\lambda \vdash n} (f^\lambda)^2 \sum_{u \in \lambda} \prod_{i=0}^{r-1} (c_u^2 - i^2) = \frac{(2r)!}{(r+1)!^2} \langle n \rangle_{r+1}, \tag{7}$$

where $\langle n \rangle_{r+1} = n(n-1)\cdots(n-r)$. It follows from this formula that

$$\frac{1}{n!}\sum_{\lambda \vdash n}(f^\lambda)^2 \sum_{u \in \lambda}c_u^{2k} = \sum_{j=1}^{k}T(k,j)\frac{(2j)!}{(j+1)!^2}\langle n \rangle_{j+1}, \tag{8}$$

where $T(k,j)$ is a *central factorial number* [13, Exer. 5.8]. One of several equivalent definitions of $T(k,j)$ is the explicit formula

$$T(k,j) = 2\sum_{i=1}^{j}\frac{(-1)^{j-i}i^{2k}}{(j-i)!(j+i)!}.$$

Another definition is the generating function

$$\sum_{k\geq 0}T(k,j)x^k = \frac{x^j}{(1-1^2x)(1-2^2x)\cdots(1-j^2x)}. \tag{9}$$

The equivalence of (7) and (8) is a simple consequence of (9). For "hook length analogues" of (7) and (8), see the Note at the end of Sect. 4.

3 Shifted parts

In this section we write partitions λ of n as $(\lambda_1,\ldots,\lambda_n)$, placing as many 0's at the end as necessary. Thus for instance the three partitions of 3 are $(3,0,0)$, $(2,1,0)$, and $(1,1,1)$. Let $G(x;y)$ be a formal power series over $\mathbb{Q}$ of bounded degree that is symmetric in the variables $x = (x_1,x_2,\ldots)$ and $y = (y_1,y_2,\ldots)$ separately; in symbols, $G \in \Lambda_{\mathbb{Q}}[x] \otimes \Lambda_{\mathbb{Q}}[y]$. We are interested in the quantity

$$\Psi_n(G) = \frac{1}{n!}\sum_{\lambda \vdash n}f_\lambda^2 G(\{c_u : u \in \lambda\}; \{\lambda_i + n - i : 1 \leq i \leq n\}). \tag{10}$$

The case $y_i = 0$ for all i reduces to what was considered in the previous section. We will show that $\Psi_n(G)$ is a polynomial in n by an argument similar to the proof of Theorem 2.1. In addition to the substitution $x^{(i)} = 1^{t_i}$ we use a certain linear transformation φ which we now define.

Let $x^{(1)},\ldots,x^{(j)}$ and $y^{(1)},\ldots,y^{(k)}$ be disjoint sets of variables. We will work in the ring R of all bounded formal power series over $\mathbb{Q}$ that are symmetric in each set of variables separately. Define a map $\varphi: R \to \mathbb{Q}[v_1,\ldots,v_k]$ by the conditions:

- The map φ is linear over $\Lambda_{\mathbb{Q}}[x^{(1)}] \otimes \cdots \otimes \Lambda_{\mathbb{Q}}[x^{(j)}]$, i.e., the $x^{(i)}$-variables are treated as scalars.
- We have

$$\varphi\big(s_\lambda(y^{(h)})\big) = \frac{\prod_{i=1}^{n}(v_h + \lambda_i + n - i)}{H_\lambda},$$

where $\lambda \vdash n$.

- We have

$$\varphi\big(G_1(y^{(1)})\cdots G_k(y^{(k)})\big)=\varphi\big(G_1(y^{(1)})\big)\cdots\varphi\big(G_k(y^{(k)})\big),$$

where $G_h \in \Lambda_{\mathbb{Q}}[x^{(1)},\ldots,x^{(j)},y^{(h)}]$.

More algebraically, let $\Psi = \Lambda_{\mathbb{Q}}[x^{(1)}]\otimes\cdots\otimes\Lambda_{\mathbb{Q}}[x^{(j)}]$, and let $\varphi_h\colon \Psi[y^{(h)}]\to\mathbb{Q}[v_h]$ be the Ψ-linear transformation defined by

$$\varphi_h\big(s_\lambda(y^{(h)})\big)=H_\lambda^{-1}\prod_{i=1}^{n}(v_h+\lambda_i+n-i).$$

Then $\varphi=\varphi_1\otimes\cdots\otimes\varphi_k$ (tensor product over Ψ).

Write for simplicity f for $f(y^{(1)})$ and v for v_1. We would like to evaluate $\varphi(p_\mu)$, where p_μ is a power-sum symmetric function. We first need the following lemma. Define

$$A_\lambda(v)=H_\lambda^{-1}(v+\lambda_1+n-1)(v+\lambda_2+n-2)\cdots(v+\lambda_n).$$

Lemma 3.1 *For all $n\geq 0$ we have*

$$\sum_{i=0}^{n}\binom{v+i-1}{i}p_1^i e_{n-i}=\sum_{\lambda\vdash n}A_\lambda(v)s_\lambda. \tag{11}$$

Equivalently, we have

$$(1-p_1)^{-v}\sum_{n\geq 0}e_n=\sum_{n\geq 0}\sum_{\lambda\vdash n}A_\lambda(v)s_\lambda.$$

First proof (sketch) I am grateful to Guoniu Han for providing the following proof. Complete details may be found in his paper [8]. Denote the left-hand side of (11) by $L_n(v)$ and the right-hand side by $R_n(v)$. It is easy to see that $L_n(v)=L_n(v-1)+p_1 L_{n-1}(v)$, $L_n(0)=R_n(0)$, and $L_0(v)=R_0(v)$. Hence we need to show that

$$R_n(v)=R_n(v-1)+p_1 R_{n-1}(v). \tag{12}$$

Now for $\lambda\vdash n$ let

$$E_\lambda(v)=A_\lambda(v+n+1)-A_\lambda(v+n)-\sum_{\mu\in\lambda\setminus 1}A_\mu(v+n+1),$$

where $\lambda\setminus 1$ denotes the set of all partitions μ obtained from λ by removing one corner. Clearly $E_\lambda(v)$ is a polynomial in v of degree at most n, and it is not difficult to check that the degree in fact is at most $n-2$. The core of the proof (which we omit) is to show that $E_\lambda(i-\lambda_i)=0$ for $i=1,2,\ldots,n-1$. Since $E_\lambda(v)$ has degree at most $n-2$ and vanishes at $n-1$ distinct integers, we conclude that $E_\lambda(v)=0$. It is now straightforward to verify that (12) holds. $\qquad\square$

Second proof I am grateful to Tewodros Amdeberhan for helpful discussions. A formula of Andrews, Goulden, and Jackson [2] asserts that

$$\sum_{\lambda} s_\lambda(y_1, \ldots, y_n) s_\lambda(z_1, \ldots, z_m) \prod_{i=1}^{n} (v - \lambda_i - n + i)$$

$$= \prod_{j=1}^{n} \prod_{k=1}^{m} \frac{1}{1 - y_j z_k} \cdot [t_1 \cdots t_n](1 + t_1 + \cdots + t_n)^v \prod_{k=1}^{m} \left(1 - \sum_{j=1}^{n} \frac{t_j y_j z_k}{1 - y_j z_k}\right),$$

where the sum is over all partitions λ satisfying $\ell(\lambda) \leq n$, and where $[t_1 \cdots t_n]X$ denotes the coefficient of $t_1 \cdots t_n$ in X. Change v to $-v$ and multiply by $(-1)^n$ to get

$$\sum_{\lambda} s_\lambda(y_1, \ldots, y_n) s_\lambda(z_1, \ldots, z_m) \prod_{i=1}^{n} (v + \lambda_i + n - i)$$

$$= (-1)^n \prod_{j=1}^{n} \prod_{k=1}^{m} \frac{1}{1 - y_j z_k}$$

$$\times [t_1 \cdots t_n](1 + t_1 + \cdots + t_n)^{-v} \prod_{k=1}^{m} \left(1 - \sum_{j=1}^{n} \frac{t_j y_j z_k}{1 - y_j z_k}\right).$$

Let $m = n$, and take the coefficient of $z_1 \cdots z_n$ on both sides. The left-hand side becomes

$$\sum_{\lambda \vdash n} f_\lambda s_\lambda(y) \prod_{i=1}^{n} (v + \lambda_i + n - i).$$

Consider the coefficient of $z_1 \cdots z_n$ on the right-hand side. A term from this coefficient is obtained as follows. Pick a subset S of $[n] = \{1, 2, \ldots, n\}$, say $\#S = r$. Choose the coefficient of $\prod_{i \in S} z_i$ from $\prod_{j=1}^{n} \prod_{k=1}^{n} (1 - y_j z_k)^{-1}$. This coefficient is $p_1(y)^r$, and there are $\binom{n}{r}$ choices for S. We now must choose the coefficient $\prod_{i \in [n] - S} z_i$ from $\prod_{k=1}^{n} (1 - \sum_{j=1}^{n} \frac{t_j y_j z_k}{1 - y_j z_k})$. This coefficient is $(-1)^{n-r}(t_1 y_1 + \cdots + t_n y_n)^{n-r}$. Hence

$$\sum_{\lambda \vdash n} f_\lambda s_\lambda(y) \prod_{i=1}^{n} (t + \lambda_i + n - i)$$

$$= (-1)^n \sum_{r=0}^{n} \binom{n}{r} p_1(y)^r [t_1 \cdots t_n] \frac{(-1)^{n-r}(t_1 y_1 + \cdots + t_n y_n)^{n-r}}{(1 + t_1 + \cdots + t_n)^{-v}}.$$

Let $\{i_1, \ldots, i_{n-r}\}$ be an $(n - r)$-element subset of $[n]$, and let $\{j_1, \ldots, j_r\}$ be its complement. Then

$$[t_{i_1} \cdots t_{i_{n-r}}](t_1 y_1 + \cdots + t_n y_n)^{n-r} = (n - r)! y_{i_1} \cdots y_{i_{n-r}},$$

$$[t_{j_1} \cdots t_{j_r}](1 + t_1 + \cdots + t_n)^{-v} = \binom{-v}{r} r!$$

Hence

$$\sum_{\lambda \vdash n} f_\lambda s_\lambda(y) \prod_{i=1}^{n} (t + \lambda_i + n - i)$$

$$= \sum_{r=0}^{n} r!(n-r)! \binom{n}{r} p_1(y)^r (-1)^r \binom{-v}{r} e_{n-r}(y). \tag{13}$$

Write $(-1)^r \binom{-v}{r} = \binom{v+r-1}{r}$ and divide both sides of (13) by $n!$ to complete the proof. $\qquad \square$

Note (a) Amdeberhan [1] has simplified the second proof of Lemma 3.1; in particular, he avoids the use of the Andrews-Goulden-Jackson formula.

(b) Since the left-hand side of (11) is an *integral* linear combination of Schur functions when $v \in \mathbb{Z}$ (e.g., by Pieri's rule), it follows that for every $v \in \mathbb{Z}$ we have $A_\lambda(v) \in \mathbb{Z}$. By expanding the left-hand side of (11) in terms of Schur functions, we in fact obtain the following combinatorial expression for $A_\lambda(v)$:

$$A_\lambda(v) = \sum_{i=0}^{n} \binom{v+i-1}{i} f_{\lambda/1^{n-i}},$$

where $f_{\lambda/1^{n-i}}$ denotes the number of SYT of the skew shape $\lambda/1^{n-i}$.

We now turn to the evaluation of $\varphi(p_\mu)$.

Lemma 3.2 *For any partition $\mu \vdash n$ with $\ell = \ell(\mu)$ nonzero parts, we have*

$$\varphi(p_\mu) = (-1)^{n-\ell} \sum_{i=0}^{m} \binom{m}{i} (v)_i,$$

where $m = m_1(\mu)$, the number of parts of μ equal to 1, and $(v)_i = v(v+1) \cdots (v+i-1)$.

Proof We will work with two sets of variables $x = (x_1, x_2, \dots)$ and $y = (y_1, y_2, \dots)$. Recall that φ acts on symmetric functions in y only, regarding symmetric function in x as scalars. Thus using Lemma 3.1 we have

$$\varphi \sum_{\lambda \vdash n} s_\lambda(x) s_\lambda(y) = \sum_{\lambda \vdash n} A_\lambda(v) s_\lambda(x)$$

$$= \sum_{i=0}^{n} \binom{v+i-1}{i} p_1^i e_{n-i}. \tag{14}$$

A standard symmetric function identity [13, (7.23)] states that

$$e_{n-i} = \sum_{\rho \vdash n-i} \varepsilon_\rho z_\rho^{-1} p_\rho,$$

where

$$\varepsilon_\rho = (-1)^{|\rho|-\ell(\rho)}, \tag{15}$$

and if ρ has m_i parts equal to i then $z_\rho = 1^{m_1}m_1!2^{m_2}m_2!\cdots$. Let ν be the partition obtained from μ by removing all parts equal to 1. Write $(\nu, 1^j)$ for the partition obtained from ν by adjoining j 1's, so $\mu = (\nu, 1^m)$. Note that

$$\varepsilon_{(\nu,1^{m-i})} = (-1)^{|\nu|+m-i-\ell(\nu)-(m-i)} = (-1)^{|\nu|-\ell(\nu)} = (-1)^{n-\ell(\mu)}.$$

Note also that

$$z_{(\nu,1^{m-i})} = \frac{(m-i)!}{m!}z_\mu.$$

Hence if we expand the right-hand side of (14) in terms of power sum symmetric functions, then the coefficient of p_μ is

$$\sum_{i=0}^{m}\binom{\nu+i-1}{i}\varepsilon_{(\nu,1^{m-i})}z_{(\nu,1^{m-i})}^{-1}$$

$$= (-1)^{n-\ell}\sum_{i=0}^{m}\binom{m}{i}(\nu)_i z_\mu^{-1}. \tag{16}$$

It follows from the Cauchy identity [13, Thm. 7.12.1] (and is also the special case $k = 2$ of (3)) that

$$\sum_{\lambda\vdash n}s_\lambda(x)s_\lambda(y) = \sum_{\mu\vdash n}z_\mu^{-1}p_\mu(x)p_\mu(y). \tag{17}$$

Thus when we apply φ (acting on the y variables) to (17) and use (16), then we obtain

$$\sum_{\mu\vdash n}\varphi(p_\mu(y))p_\mu(x)$$

$$= \sum_{\mu\vdash n}\left((-1)^{n-\ell(\mu)}\sum_{i=0}^{m}\binom{m}{i}(\nu)_i\right)z_\mu^{-1}p_\mu(x).$$

Since the p_μ's are linearly independent, the proof follows.   □

Theorem 3.3 *For any $G \in \Lambda_{\mathbb{Q}}[x] \otimes \Lambda_{\mathbb{Q}}[y]$ we have*

$$\Psi_n(G) \in \mathbb{Q}[n],$$

where $\Psi_n(G)$ is given by (10).

Proof By linearity it suffices to take $G = e_\mu(x)e_\nu(y)$. Apply φ to the identity (3) in the variables $x^{(1)}, \ldots, x^{(j)}, y^{(1)}, \ldots, y^{(k)}$. Then make the substitution $x^{(h)} = 1^{th}$ and multiply by $n!$. By (2) and Lemma 3.2 we obtain

$$\frac{1}{n!} \sum_{\lambda \vdash n} f_\lambda^2 \prod_{h=1}^{j} \prod_{u \in \lambda} (t_h + c_u) \cdot \prod_{h=1}^{k} \prod_{i=1}^{n} (v_h + \lambda_i + n - i)$$

$$= \sum_{\substack{w_1 \cdots w_j w_1' \cdots w_k' = 1 \\ \text{in } \mathfrak{S}_n}} \prod_{h=1}^{j} t_h^{c(w_h)}$$

$$\times \prod_{h=1}^{k} \left((-1)^{n-\ell(\rho(w_h'))} \sum_{i=0}^{m_1(\rho(w_h'))} \binom{m_1(\rho(w_h'))}{i} (v_h)_i \right). \tag{18}$$

The remainder of the proof is a straightforward generalization of that of Theorem 2.1. Take the coefficient of $t_1^{n-\mu_1} \cdots t_j^{n-\mu_j} v_1^{n-\nu_1} \cdots v_k^{n-\nu_k}$. The left-hand side becomes $\Psi_n(e_\mu(x)e_\nu(y))$, so we need to show that the coefficient of $t_1^{n-\mu_1} \cdots t_j^{n-\mu_j} v_1^{n-\nu_1} \cdots$ $v_k^{n-\nu_k}$ on the right-hand side of (18) is a polynomial in n. Suppose that $n \geq \mu_1$ and $n \geq \nu_1$. The coefficient of $v_h^{n-\nu_h}$ in $v_h(v_h + 1) \cdots (v_h + n - i - 1)$ is the sign-less Stirling number $c(n - i, n - \nu_h)$. The coefficient of $v_h^{n-\nu_h}$ in (18) is 0 unless $n - m_1(\rho(w_h')) \leq i \leq \nu_h$. For each choice of $0 \leq i_h \leq i$ $(1 \leq h \leq k)$, there are only finitely many orbits of the action of $\mathfrak{S}_n$ by (coordinatewise) conjugation on the set of $(w_1, \ldots, w_j, w_1', \ldots, w_k') \in \mathfrak{S}_n^{j+k}$ for which $w_1 \cdots w_j w_1' \cdots w_k' = 1$, w_h has $n - \mu_h$ cycles, and w_h' has $n - i_h$ fixed points. The size of each of these orbits is a polynomial in n, as in the proof of Theorem 2.1. Moreover, the Stirling number $c(n - i, n - \nu_h)$ is a polynomial in n for fixed i and ν_h, and similarly for the binomial coefficient $\binom{n-i_h}{n-i}$, so $\Psi_n(e_\mu(x)e_\nu(y))$ is a polynomial $N_{\mu,\nu}(n)$ for $n \geq \max\{\mu_1, \nu_1\}$. If $0 \leq n < \max\{\mu_1, \nu_1\}$, then both $N_{\mu,\nu}(n)$ and $\Psi_n(e_\mu(x)e_\nu(y))$ are equal to 0 (as in the proof of Theorem 2.1), so the proof is complete. $\qquad\square$

Note Since n is a polynomial in n, it is easy to see that Theorem 3.3 still holds if we replace $\Psi_n(G)$ with

$$\frac{1}{n!} \sum_{\lambda \vdash n} f_\lambda^2 G(\{c_u : u \in \lambda\}; \{\lambda_i - i : 1 \leq i \leq n\}).$$

On the other hand, Theorem 3.3 becomes *false* if we replace $\Psi_n(G)$ with

$$\frac{1}{n!} \sum_{\lambda \vdash n} f_\lambda^2 G(\{c_u : u \in \lambda\}; \{\lambda_i : 1 \leq i \leq n\}).$$

For instance,

$$\frac{1}{n!} \sum_{\lambda \vdash n} f_\lambda^2 (\lambda_1^2 + \lambda_2^2 + \cdots + \lambda_n^2)$$

is not a polynomial function of n, nor is it integer valued.

4 Hook lengths squared

The connection between contents, hook lengths, and the shifted parts $\lambda_i + n - i$ is given by the following result, an immediate consequence [13, Lemma 7.21.1].

Lemma 4.1 *Let* $\lambda = (\lambda_1, \ldots, \lambda_n) \vdash n$. *Then we have the multiset equality*

$$\{h_u : u \in \lambda\} \cup \{\lambda_i - \lambda_j - i + j : 1 \le i < j \le n\}$$
$$= \{n + c_u : u \in \lambda\} \cup \{1^{n-1}, 2^{n-2}, \ldots, n - 1\}.$$

For example, when $\lambda = (3, 1)$ Lemma 4.1 asserts that

$$\{4, 2, 1, 1\} \cup \{3, 5, 6, 2, 3, 1\} = \{3, 4, 5, 6\} \cup \{1, 1, 1, 1, 2, 2, 3\}$$

as multisets.

Lemma 4.2 *For any* $F \in \Lambda_{\mathbb{Q}}$, *we have*

$$F(1^{n-1}, 2^{n-2}, \ldots, n - 1) \in \mathbb{Q}[n],$$

where the exponents denote multiplicity.

Proof It suffices to take $F = p_j$ since the polynomials in n form a ring. Thus we want to show that

$$\sum_{i=1}^{n-1} (n - i) i^j \in \mathbb{Q}[n],$$

which is routine.   □

We come to the main result of this paper. Recall the definition

$$\Phi_n(F) = \frac{1}{n!} \sum_{\lambda \vdash n} f_\lambda^2 F(h_u^2 : u \in \lambda).$$

Theorem 4.3 *For any symmetric function* $F \in \Lambda_{\mathbb{Q}}$ *we have* $\Phi_n(F) \in \mathbb{Q}[n]$.

Proof As usual it suffices to take $F = e_\mu$, where $\mu = (\mu_1, \ldots, \mu_k)$. Define the multisets (or *alphabets*)

$$A_\lambda = \{h_u^2 : u \in \lambda\},$$
$$B_\lambda = \{(\lambda_i - \lambda_j - i + j)^2 : 1 \le i < j \le n\},$$
$$C_\lambda = \{(n + c_u)^2 : u \in \lambda\},$$
$$D_n = \{b_1^{n-1}, b_2^{n-2}, \ldots, b_{n-1}\},$$

where $b_i = i^2 \in \mathbb{Z}$ (so for instance $D_4 = \{1, 1, 1, 4, 4, 9\}$). Write $\Omega(a, b, c) = (-1)^c e_a(C_\lambda) e_b(D_n) h_c(B_\lambda)$. Using standard λ-ring notation and manipulations (see e.g. Lascoux [9, Chap. 2]), we have from Lemma 4.1 that

$$\Phi_n(e_\mu) = \frac{1}{n!} \sum_{\lambda \vdash n} f_\lambda^2 e_\mu(A_\lambda)$$

$$= \frac{1}{n!} \sum_{\lambda \vdash n} f_\lambda^2 e_\mu(C_\lambda + D_n - B_\lambda)$$

$$= \frac{1}{n!} \sum_{\lambda \vdash n} f_\lambda^2 \prod_{i=1}^{k} \left(\sum_{\substack{a,b,c \geq 0 \\ a+b+c=\mu_i}} \Omega(a, b, c) \right)$$

$$= \sum_{\substack{a_1,b_1,c_1 \geq 0 \\ a_1+b_1+c_1=\mu_1}} \cdots \sum_{\substack{a_k,b_k,c_k \geq 0 \\ a_k+b_k+c_k=\mu_k}} \frac{1}{n!} \sum_{\lambda \vdash n} f_\lambda^2 \prod_{r=1}^{k} \Omega(a, b, c).$$

Consider the inner sum over λ, together with the factor $1/n!$. By Lemma 4.2 each $e_{b_r}(D_n)$ is a polynomial in n which we can factor out of the sum. Note that $h_{c_r}(B_\lambda)$ is a symmetric function of the numbers $\rho_i = \lambda_i + n - i$ since $(\rho_i - \rho_j)^2$ is symmetric in i and j. (This is the one point in the proof that requires the use of the alphabet $\{h_u^2 : u \in \lambda\}$ rather than the more general $\{h_u : u \in \lambda\}$.) What remains after factoring out each $e_{b_r}(D_n)$ is therefore a polynomial in n by Theorem 3.3, and the proof follows. $\square$

Note (a) The λ-ring computations in the proof of Theorem 4.3 can easily be replaced with more "naive" techniques such as generating functions. The λ-ring approach, however, makes the computation more routine.

(b) An interesting feature of the proofs of Theorems 2.1, 3.3, and 4.3 is that they don't involve just "formal" properties of symmetric functions; use of representation theory is required. This is because the only known proof of the crucial equation (3) involves representation theory, viz., the determination of the primitive orthogonal idempotents in the center of the group algebra of $\mathfrak{S}_n$. Is there a proof of (3) or of Theorems 2.1, 3.3, and 4.3 that doesn't involve representation theory?

Here is a small table of the polynomials $\Phi_n(e_\mu)$:

$$\Phi_n(e_1) = \frac{1}{2}n(3n - 1),$$

$$\Phi_n(e_2) = \frac{1}{24}n(n - 1)(27n^2 - 67n + 74),$$

$$\Phi_n(e_1^2) = \frac{1}{12}n(27n^3 - 14n^2 - 9n + 8),$$

$$\Phi_n(e_3) = \frac{1}{48}n(n - 1)(n - 2)(27n^3 - 174n^2 + 511n - 552),$$

$$\Phi_n(e_2 e_1) = \frac{1}{48} n(n-1)(81n^4 - 204n^3 + 137n^2 + 390n - 512),$$

$$\Phi_n(e_1^3) = \frac{1}{24} n(81n^5 - 45n^4 - 69n^3 - 31n^2 + 216n - 128).$$

Note Soichi Okada has conjectured [11] the following "hook analogue" of (7):

$$\frac{1}{n!} \sum_{\lambda \vdash n} f_\lambda^2 \sum_{u \in \lambda} \prod_{i=1}^{r} (h_u^2 - i^2) = \frac{1}{2(r+1)^2} \binom{2r}{r} \binom{2r+2}{r+1} \langle n \rangle_{r+1}. \qquad (19)$$

This conjecture has been proved by Greta Panova [12] using Theorem 4.3. From this result we get the following analogue of (8):

$$\frac{1}{n!} \sum_{\lambda \vdash n} f_\lambda^2 \sum_{u \in \lambda} h_u^{2k} = \sum_{j=1}^{k+1} T(k+1, j) \frac{1}{2j^2} \binom{2(j-1)}{j-1} \binom{2j}{j} \langle n \rangle_j.$$

Note Using Theorem 3.3 and the method of the proof of Theorem 4.3 to reduce hook lengths squared to contents and shifted parts, it is clear that we have the following "master theorem" subsuming both Theorems 3.3 and 4.3.

Theorem 4.4 *For any $K \in \Lambda_{\mathbb{Q}}[x] \otimes \Lambda_{\mathbb{Q}}[y] \otimes \Lambda_q[z]$, we have*

$$\frac{1}{n!} \sum_{\lambda \vdash n} f_\lambda^2 K_\lambda \in \mathbb{Q}[n],$$

where

$$K_\lambda = K(\{c_u : u \in \lambda\}; \{\lambda_i + n - i : 1 \le i \le n\}; \{h_u^2 : u \in \lambda\}).$$

5 Some questions

1. Can the Nekrasov-Okounkov formula (1) be proved using the techniques we have used to prove Theorem 4.3?
2. Can the Nekrasov-Okounkov formula (1) be generalized with the left-hand side replaced with the following expression (or some simple modification thereof)?

$$\sum_{n \ge 0} \left(\sum_{\lambda \vdash n} f_\lambda^{2k} \prod_{i=1}^{k} \prod_{u \in \lambda} (t_i + h_u^2) \right) \frac{x^n}{n!^{2k}}$$

Note that if we put each $t_i = 0$ then we obtain the partition generating function $\prod_{i \ge 1} (1 - x^i)^{-1}$. The same question can be asked with h_u^2 replaced with c_u^2 or c_u.
3. Define a linear transformation $\psi : \Lambda_{\mathbb{Q}} \to \mathbb{Q}[t]$ by

$$\psi(s_\lambda) = H_\lambda^{-1} \prod_{u \in \lambda} (t + h_u^2).$$

Is there a nice description of $\psi(p_\mu)$?

Acknowledgements   I am grateful to Soichi Okada for calling my attention to reference [4] and for providing conjecture (19). I also am grateful to an anonymous referee for many helpful suggestions.

References

1. Amdeberhan, T.: Differential operators, shifted parts, and hook lengths. Preprint arXiv:0807.2473
2. Andrews, G., Goulden, I., Jackson, D.M.: Generalizations of Cauchy's summation formula for Schur functions. Trans. Am. Math. Soc. **310**, 805–820 (1988)
3. Frame, J.S., de Robinson, G.B., Thrall, R.M.: The hook graphs of S_n. Can. J. Math. **6**, 316–324 (1954)
4. Fujii, S., Kanno, H., Moriyama, S., Okada, S.: Instanton calculus and chiral one-point functions in supersymmetric gauge theories. Adv. Theor. Math. Phys. **12**, 1401–1428 (2008)
5. Hanlon, P.J., Stanley, R., Stembridge, J.R.: Some combinatorial aspects of the spectra of normally distributed random matrices. Contemp. Math. **158**, 151–174 (1992)
6. Han, G.-N.: The Nekrasov-Okounkov hook length formula: refinement, elementary proof, extension, and applications. Preprint arXiv:0805.1398
7. Han, G.-N.: Some conjectures and open problems on partition hook lengths. Preprint available at www-irma.u-strasbg.fr/~guoniu/hook
8. Han, G.-N.: Hook lengths and shifted parts of partitions. Preprint arXiv:0807.1801
9. Lascoux, A.: Symmetric Functions and Combinatorial Operators on Polynomials. CBMS Regional Conference Series in Mathematics, vol. 99. American Mathematical Society, Providence (2003)
10. Nekrasov, N.A., Okounkov, A.: Seiberg-Witten theory and random partitions, in the unity of mathematics. In: Progress in Mathematics, vol. 244, pp. 525–596. Birkhäuser Boston, Boston (2006)
11. Okada, S.: Private communication. Dated 7 July (2008)
12. Panova, G.: Proof of a conjecture of Okada. Preprint arXiv:0811.3463
13. Stanley, R.: Enumerative Combinatorics, vol. 2. Cambridge University Press, New York/Cambridge (1999)

Ramanujan J (2010) 23: 107–126
DOI 10.1007/s11139-009-9194-9

The doubloon polynomial triangle

Dominique Foata · Guo-Niu Han

Dedicated to George Andrews, on the occasion of his seventieth birthday

Received: 17 November 2008 / Accepted: 12 August 2009 / Published online: 1 May 2010
© Springer Science+Business Media, LLC 2010

Abstract The doubloon polynomials are generating functions for a class of combinatorial objects called normalized doubloons by the compressed major index. They provide a refinement of the q-tangent numbers and also involve two major specializations: the Poupard triangle and the Catalan triangle.

Keywords Doubloon polynomials · Doubloon polynomial triangle · Poupard triangle · Catalan triangle · Reduced tangent numbers

Mathematics Subject Classification (2000) Primary 05A15 · 05A30 · 33B10

1 Introduction

The *doubloon*[1] *polynomials* $d_{n,j}(q)$ ($n \geq 1$, $2 \leq j \leq 2n$) introduced in this paper serve to globalize the *Poupard triangle* [15] and the classical *Catalan triangle* [16]. They also provide a refinement of the q-*tangent numbers*, fully studied in our previous paper [6]. Finally, as generating polynomials for the doubloon model, they constitute a common combinatorial set-up for the above integer triangles. They may be defined by the following recurrence:

---

[1] Although the word "doubloon" originally refers to a Spanish gold coin, it is here used to designate a permutation written as a two-row matrix.

D. Foata
Institut Lothaire, 1, rue Murner, 67000 Strasbourg, France
e-mail: foata@math.u-strasbg.fr

G.-N. Han (✉)
I.R.M.A. UMR 7501, Université Louis Pasteur et CNRS, 7, rue René-Descartes, 67084 Strasbourg, France
e-mail: guoniu@math.u-strasbg.fr

Fig. 1 The doubloon
polynomial triangle

$$d_{1,2}(q)$$
$$d_{2,2}(q) \; d_{2,3}(q) \; d_{2,4}(q)$$
$$d_{3,2}(q) \; d_{3,3}(q) \; d_{3,4}(q) \; d_{3,5}(q) \; d_{3,6}(q)$$
$$d_{4,2}(q) \; d_{4,3}(q) \; d_{4,4}(q) \; d_{4,5}(q) \; d_{4,6}(q) \; d_{4,7}(q) \; d_{4,8}(q)$$

$d_{1,2}(q) = 1; \qquad d_{2,2}(q) = q; \qquad d_{2,3}(q) = q + 1; \qquad d_{2,4}(q) = 1;$

$d_{3,2}(q) = 2q^3 + 2q^2; \qquad d_{3,3}(q) = 2q^3 + 4q^2 + 2q;$

$d_{3,4}(q) = q^3 + 4q^2 + 4q + 1; \qquad d_{3,5}(q) = 2q^2 + 4q + 2;$

$d_{3,6}(q) = 2q + 2; \qquad d_{4,2}(q) = 5q^6 + 12q^5 + 12q^4 + 5q^3;$

$d_{4,3}(q) = 5q^6 + 17q^5 + 24q^4 + 17q^3 + 5q^2;$

$d_{4,4}(q) = 3q^6 + 15q^5 + 29q^4 + 29q^3 + 15q^2 + 3q;$

$d_{4,5}(q) = q^6 + 9q^5 + 25q^4 + 34q^3 + 25q^2 + 9q + 1;$

$d_{4,6}(q) = 3q^5 + 15q^4 + 29q^3 + 29q^2 + 15q + 3;$

$d_{4,7}(q) = 5q^4 + 17q^3 + 24q^2 + 17q + 5; \qquad d_{4,8}(q) = 5q^3 + 12q^2 + 12q + 5.$

Fig. 1′ The first doubloon polynomials

(D1) $d_{0,j}(q) = \delta_{1,j}$ (Kronecker symbol).

(D2) $d_{n,j}(q) = 0$ for $n \geq 1$ and $j \leq 1$ or $j \geq 2n + 1$.

(D3) $d_{n,2}(q) = \sum_j q^{j-1} d_{n-1,j}(q)$ for $n \geq 1$.

(D4)

$$d_{n,j}(q) - 2 d_{n,j-1}(q) + d_{n,j-2}(q)$$

$$= -(1 - q) \sum_{i=1}^{j-3} q^{n+i+1-j} d_{n-1,i}(q)$$

$$- \left(1 + q^{n-1}\right) d_{n-1,j-2}(q) + (1 - q) \sum_{i=j-1}^{2n-1} q^{i-j+1} d_{n-1,i}(q)$$

for $n \geq 2$ and $3 \leq j \leq 2n$.

The polynomials $d_{n,j}(q)$ ($n \geq 1$, $2 \leq j \leq 2n$) are easily evaluated using (D1)–(D4) and form the *doubloon polynomial triangle*, as shown in Figs. 1 and 1′.

Notice the different symmetries of the coefficients of the polynomials $d_{n,j}(q)$, which will be fully exploited in Sect. 4 (Corollaries 4.3, 4.7, 4.8). Various specializations are displayed in Fig. 2 below, where $C_n = \frac{1}{2n+1} \binom{2n}{n}$ stands for the celebrated Catalan number, and t_n for the *reduced tangent number* occurring in the Taylor ex-

Fig. 2 The specializations of $d_{n,j}(q)$

$$d_{n,j}(0) \xleftarrow{q=0} d_{n,j}(q) \xrightarrow{q=1} d_{n,j}(1)$$

$$\downarrow \Sigma \qquad\qquad \downarrow \Sigma \qquad\qquad \downarrow \Sigma$$

$$C_n \xleftarrow{q=0} d_n(q) \xrightarrow{q=1} t_n$$

Fig. 3 The Poupard triangle $(d_{n,j}(1))$

$$
\begin{array}{c}
1 \\
1 \quad 2 \quad 1 \\
4 \quad 8 \quad 10 \quad 8 \quad 4 \\
34 \quad 68 \quad 94 \quad 104 \quad 94 \quad 68 \quad 34
\end{array}
$$

pansion

$$\sqrt{2}\tan(u/\sqrt{2}) = \sum_{n \geq 0} \frac{u^{2n+1}}{(2n+1)!} t_n$$

$$= \frac{u}{1!}1 + \frac{u^3}{3!}1 + \frac{u^5}{5!}4 + \frac{u^7}{7!}34 + \frac{u^9}{9!}496 + \frac{u^{11}}{11!}11056 + \cdots. \qquad (1.1)$$

The symbol Σ attached to each vertical arrow has the meaning "make the summation over j," and $d_n(q)$ is the polynomial further defined in (1.3).

When $q = 1$, the $(D1)$–$(D4)$ recurrence becomes

$(P1)$ $d_{0,j}(1) = \delta_{0,j}$ (Kronecker symbol).
$(P2)$ $d_{n,j}(1) = 0$ for $n \geq 1$ and $j \leq 1$ or $j \geq 2n+1$.
$(P3)$ $d_{n,2}(1) = \sum_j d_{n-1,j}(1)$ for $n \geq 1$.
$(P4)$ $d_{n,j}(1) - 2d_{n,j-1}(1) + d_{n,j-2}(1) = -2d_{n-1,j-2}(1)$
    for $n \geq 2$ and $3 \leq j \leq 2n$.

which is exactly the recurrence introduced by Christiane Poupard [15].

The integers $d_{n,j}(1)$ are easily evaluated using $(P1)$–$(P4)$ and form the *Poupard triangle*, as shown in Fig. 3.

When $q = 0$, relation $(D4)$ becomes

$$d_{n,j}(0) - 2d_{n,j-1}(0) + d_{n,j-2}(0) = -d_{n-1,j-2}(0) + d_{n-1,j-1}(0),$$

which can be rewritten as

$$d_{n,j}(0) - d_{n,j-1}(0) - d_{n-1,j-1}(0) = d_{n,j-1}(0) - d_{n,j-2}(0) - d_{n-1,j-2}(0),$$

so that by induction

$$d_{n,j}(0) - d_{n,j-1}(0) - d_{n-1,j-1}(0) = d_{n,2}(0) = \sum_j q^{j-1} d_{n-1,j} \mid_{q=0} = 0$$

using $(D2)$ and $(D3)$ when $n \geq 2$. Consequently, the integers $d_{n,j}(0)$ satisfy the recurrence relation

Fig. 4 The Catalan triangle
$(d_{n,j}(0))$

$$1$$
$$0\ 1\ 1$$
$$0\ 0\ 1\ 2\ 2$$
$$0\ 0\ 0\ 1\ 3\ 5\ 5$$
$$0\ 0\ 0\ 0\ 1\ 4\ 9\ 14\ 14$$
$$0\ 0\ 0\ 0\ 0\ 1\ 5\ 14\ 28\ 42\ 42$$

(C1)

$$d_{n,j}(0) = d_{n,j-1}(0) + d_{n-1,j-1}(0)$$

for $n \geq 2$ and $3 \leq j \leq 2n$ with the initial conditions
(C2)

$$d_{n,n+1}(0) = 1 \quad (n \geq 1);$$

$$d_{n,j}(0) = 0 \quad (n \geq 1 \text{ and } j \leq 1 \text{ or } j \geq 2n+1);$$

$$d_{n,2}(0) = d_{n,3}(0) = \cdots = d_{n,n}(0) = 0 \quad (n \geq 2).$$

In view of (C1) and (C2), the integers $d_{n,j}(0)$ $(n \geq 1,\ n+1 \leq j \leq 2n)$ obey the rules of the classical *Catalan triangle* that has been studied by many authors (see the sequence A00976 in 16 and its abundant bibliography). They form the *Catalan triangle* displayed in Fig. 4.

For each $n \geq 0$, let $A_n(t,q)$ be the Carlitz [2, 3] q-analog of the Eulerian polynomial defined by the identity

$$\frac{A_n(t,q)}{(t;q)_{n+1}} = \sum_{j \geq 0} t^j \left([j+1]_q\right)^n, \tag{1.2}$$

where $(t;q)_{n+1} = (1-t)(1-tq) \cdots (1-tq^n)$ and $[j+1]_q = 1 + q + q^2 + \cdots + q^n$ are the traditional q-ascending factorials and q-analogs of the positive integers.

The polynomial $d_n(q)$ under study was introduced in [6]. It is defined by

$$d_n(q) = \frac{(-1)^n q^{\binom{n}{2}} A_{2n+1}(-q^{-n}, q)}{(1+q)(1+q^2)\cdots(1+q^n)}. \tag{1.3}$$

It was shown to be a *polynomial* of degree $\binom{n-1}{2}$, with *positive integral* coefficients, having the following two properties:

$$d_n(1) = t_n; \tag{1.4}$$

$$d_n(0) = C_n \quad (n \geq 0) \tag{1.5}$$

(see the bottom row in the diagram of Fig. 2).

The first values of the polynomials $d_n(q)$ are: $d_0(q) = d_1(q) = 1$; $d_2(q) = 2 + 2q$; $d_3(q) = 5 + 12q + 12q^2 + 5q^3$; $d_4(q) = 14 + 56q + 110q^2 + 136q^3 + 110q^4 + 56q^5 + 14q^6$.

The main purpose of this paper is to prove the following theorem.

Theorem 1.1 *Let $(d_{n,j}(q))$ be the set of polynomials defined by (D1)–(D4), and $d_n(q)$ be defined by (1.3). Then, the following identity holds:*

$$\sum_j d_{n,j}(q) = d_n(q). \tag{1.6}$$

In other words, the diagram in Fig. 2 is commutative.

Even for $q = 1$, identity (1.6), which then reads $\sum_j d_{n,j}(1) = t_n$, is not at all straightforward. It was elegantly proved by Christiane Poupard [15] by means of the bivariable generating function

$$Z(u, v) = 1 + \sum_{n \geq 1} \sum_{1 \leq l \leq 2n-1} \frac{u^{2n-l}}{(2n-l)!} \frac{v^l}{l!} d_{n,l+1}(1). \tag{1.7}$$

She even obtained the following stronger result:

$$Z(u, v) = \frac{\cos((u-v)/\sqrt{2})}{\cos((u+v)/\sqrt{2})}, \tag{1.8}$$

so that

$$\frac{\partial}{\partial u} Z(u, v)\Big|_{\{v=0\}} = \sum_{n \geq 1} \frac{u^{2n-1}}{(2n-1)!} d_{n,2}(1) = \sqrt{2}\tan(u/\sqrt{2}), \tag{1.9}$$

which proves $\sum_j d_{n,j}(1) = t_n$ by appealing to (P3).

Finally, she obtains a combinatorial interpretation for the polynomial

$$d_n(s, 1) = \sum_j d_{n,j}(1) s^j \tag{1.10}$$

in terms of strictly ordered binary trees, or in an equivalent manner, of André permutations, which are also alternating (see, e.g., 7–9).

For $q = 0$, identity (1.6) reads $\sum_j d_{n,j}(0) = C_n = \frac{1}{n+1}\binom{2n}{n}$. This is a consequence of the identities

$$d_{n,2n}(0) = \sum_j d_{n-1,j}(0); \tag{1.11}$$

$$d_{n,j}(0) = \binom{j-2}{n-1} - \binom{j-2}{n} = \frac{2n-j+1}{n}\binom{j-2}{n-1}; \tag{1.12}$$

so that, in particular,

$$d_{n,2n}(0) = \frac{1}{n}\binom{2n-2}{n-1} = C_{n-1}, \tag{1.13}$$

easily obtained from $(C1)$ and $(C2)$ by induction and iteration, as well as an expression for the generating function

$$\sum_{n\geq 1} u^n \sum_j d_{n,j}(0)v^j = \frac{1}{1-v-uv}\left(uv^2 - \frac{v}{2}(1-\sqrt{1-4uv^2})\right). \qquad (1.14)$$

Again, the reader is referred to the excellent commented bibliography about the sequence A009766 in Sloane's On-Line Encyclopedia of Integer Sequences [16], in particular the contributions made by David Callan [1] and Emeric Deutsch [4], where identities (1.11)–(1.14) are actually derived with other initial conditions.

The proof of Theorem 1.1 will be of *combinatorial nature*. In our previous paper [6] we proved that each polynomial $d_n(q)$ was a polynomial with positive integral coefficients by showing that it was the generating function for the class $\mathcal{N}_{2n+1}^0$ of permutations called *normalized doubloons*, by an integral-valued statistic "cmaj" called the *compressed major index*:

$$d_n(q) = \sum_{\delta \in \mathcal{N}_{2n+1}^0} q^{\mathrm{cmaj}\,\delta}. \qquad (1.15)$$

In particular, $d_n(1) = \#\mathcal{N}_{2n+1}^0 = t_n$ by (1.4). Normalized doubloon and "cmaj" will be fully described in Sect. 2. Here we just recall that each normalized doubloon $\delta \in \mathcal{N}_{2n+1}^0$ is a two-row matrix $\delta = \binom{a_0\,a_1\,\cdots\,a_n}{b_0\,b_1\,\cdots\,b_n}$, where the word $\rho(\delta) = a_0a_1\cdots a_nb_nb_{n-1}\cdots b_0$ is a permutation of $012\cdots(2n+1)$ having further properties that will be given shortly. The set of all normalized doubloons $\delta = \binom{a_0\,a_1\,\cdots\,a_n}{b_0\,b_1\,\cdots\,b_n}$ such that $b_0 = j$ is denoted by $\mathcal{N}_{2n+1,j}^0$. It will be shown that $\mathcal{N}_{2n+1,j}^0$ is an empty set for $j = 0, 1$, and $2n+1$, so that the subsets $\mathcal{N}_{2n+1,j}^0$ $(j = 2, 3, \ldots, 2n)$ form a *partition* of $\mathcal{N}_{2n+1}^0$. Consequently, the following identity holds:

$$d_n(q) = \sum_{j=2}^{2n} \sum_{\delta \in \mathcal{N}_{2n+1,j}^0} q^{\mathrm{cmaj}\,\delta}. \qquad (1.16)$$

Accordingly, Theorem 1.1 is a simple corollary of the next theorem.

Theorem 1.2 *Let $(d_{n,j}(q))$ be the set of polynomials in one variable q defined by* $(D1)$–$(D4)$. *Then $d_{n,j}(q)$ is the generating polynomial for $\mathcal{N}_{2n+1,j}^0$ by the compressed major index. In other words,*

$$d_{n,j}(q) = \sum_{\delta \in \mathcal{N}_{2n+1,j}^0} q^{\mathrm{cmaj}\,\delta}. \qquad (1.17)$$

To prove that the polynomial $d_{n,j}(q)$, as expressed in (1.17), satisfies $(D4)$, symmetry properties must be derived (see Corollaries 4.3, 4.7, and 4.8). This requires a careful geometric study of those doubloons and how the statistic "cmaj" evolves. All this is developed in Sects. 3 and 4. The proof of the recurrence is completed in Sect. 5, and in the final section concluding remarks are made.

2 Doubloons

A *doubloon* of order $2n + 1$ is a $2 \times (n + 1)$-matrix $\delta = \begin{pmatrix} a_0\,a_1\,\cdots\,a_n \\ b_0\,b_1\,\cdots\,b_n \end{pmatrix}$ such that the word $\rho(\delta) = a_0 a_1 \cdots a_n b_n b_{n-1} \cdots b_0$, called the *reading* of δ, is a permutation of $012\cdots(2n+1)$. Let $\mathcal{D}_{2n+1}$ (resp. $\mathcal{D}^0_{2n+1}$) denote the set of all doubloons $\delta = \begin{pmatrix} a_0\,a_1\,\cdots\,a_n \\ b_0\,b_1\,\cdots\,b_n \end{pmatrix}$ of order $(2n + 1)$ (resp. the subset of all doubloons such that $a_0 = 0$).

Let $1 \le k \le n$; each doubloon $\delta = \begin{pmatrix} a_0\,\cdots\,a_n \\ b_0\,b_1\,\cdots\,b_n \end{pmatrix}$ is said to be *normalized at k* if the following two conditions are satisfied:

(N1) *Exactly one of the two integers a_k, b_k lies between a_{k-1} and b_{k-1} (we also say that δ is *interlaced at k*).

(N2) Either $a_{k-1} > a_k$ and $b_{k-1} > b_k$, or $a_{k-1} < b_k$ and $b_{k-1} < a_k$.

In an equivalent manner, δ is *normalized at k* if one of the four following orderings holds:

$$a_{k-1} < b_k < b_{k-1} < a_k;$$
$$b_k < b_{k-1} < a_k < a_{k-1};$$
$$b_{k-1} < a_k < a_{k-1} < b_k; \tag{2.1}$$
$$a_k < a_{k-1} < b_k < b_{k-1}.$$

For each pair of distinct integers (i, j), the symbol $\mathcal{N}^i_{2n+1,j}$ will denote the set of all doubloons $\delta = \begin{pmatrix} a_0\,a_1\,\cdots\,a_n \\ b_0\,b_1\,\cdots\,b_n \end{pmatrix}$, normalized at *every* $k = 1, 2, \ldots, n$, such that $a_0 = i$ and $b_0 = j$. The doubloons belonging to $\mathcal{N}^0_{2n+1,j}$ for some j are simply called *normalized*. Also $\mathcal{N}^0_{2n+1}$ designates the union of the sets $\mathcal{N}^0_{2n+1,j}$. For further results on permutations studied as two-row matrices, see [5, 12, 14].

Let $\delta = \begin{pmatrix} a_0\,a_1\,\cdots\,a_n \\ b_0\,b_1\,\cdots\,b_n \end{pmatrix}$ be a doubloon of order $(2n + 1)$, and h an integer. For each $k = 0, 1, \ldots, n$, let $a'_k = a_k + h$, $b'_k = b_k + h$ be expressed as *residues* $\mathrm{mod}(2n + 2)$. The two-row matrix $\begin{pmatrix} a'_0\,a'_1\,\cdots\,a'_n \\ b'_0\,b'_1\,\cdots\,b'_n \end{pmatrix}$, denoted by $\delta + h$, is still a doubloon. Let $T_h : \delta \mapsto \delta + h$.

Proposition 2.1 *The map T_h, restricted to $\mathcal{N}^i_{2n+1,j}$, is a bijection onto $\mathcal{N}^{i+h}_{2n+1,j+h}$ (superscript and subscript being taken $\mathrm{mod}(2n + 2)$).*

Proof The four orderings in (2.1) are *cyclic* rearrangements of each other, so that if δ is normalized at each i, the doubloon $T_h\delta = \delta + h$ has the same property. □

The *number of descents*, des δ, (resp. the *major index*, maj δ) of a doubloon $\delta = \begin{pmatrix} a_0\,a_1\,\cdots\,a_n \\ b_0\,b_1\,\cdots\,b_n \end{pmatrix}$ is defined to be the number of descents (resp. the major index) of the permutation $\rho(\delta) = a_0 a_1 \cdots a_n b_n b_{n-1} \cdots b_0$, so that if the descents ($a_i > a_{i+1}$, or $a_n > b_n$, or still $b_{i-1} > b_i$) occur at positions $l_1, l_2, \ldots, l_r$ in $\rho(\delta)$, then des $\delta = r$ and maj $\delta = l_1 + l_2 + \cdots + l_r$. The *compressed major index*, cmaj δ, of δ is defined by

$$\mathrm{cmaj}\,\delta = \mathrm{maj}\,\delta - (n + 1)\,\mathrm{des}\,\delta + \binom{n}{2}. \tag{2.2}$$

For example, there is one normalized doubloon of order 3 ($n = 1$): $\delta = \binom{0\,3}{2\,1}$ and cmaj $\delta = \text{maj}(0\,3\,1\,2) - (1+1)\,\text{des}(0\,3\,1\,2) + \binom{1}{2} = 0$, so that $d_{1,2}(q) = 1$. There are four normalized doubloons of order 5 ($n = 2$) and the partition of $\mathcal{N}_5^0$ reads:

$$\mathcal{N}_{5,2}^0 = \left\{ \binom{0\,4\,3}{2\,1\,5} \right\}, \quad \mathcal{N}_{5,3}^0 = \left\{ \binom{0\,5\,4}{3\,2\,1}, \binom{0\,4\,2}{3\,1\,5} \right\}, \quad \mathcal{N}_{5,4}^0 = \left\{ \binom{0\,5\,3}{4\,2\,1} \right\}.$$

We have cmaj$(043512) = \text{maj}(043512) - 3\,\text{des}(043512) + \binom{2}{2} = (2+4) - 3 \times 2 + 1 = 1$, so that $d_{2,2}(q) = q$. Furthermore, $d_{2,3}(q) = 1 + q$ and $d_{2,4}(q) = 1$, as expected (see Fig. 2).

3 Operations on doubloons

In this section we study the actions of several operators on the statistic "cmaj." First, we recall the action of the dihedral group on the traditional statistics "des" and "maj," in particular characterize the images of each permutation $\sigma = \sigma(1)\sigma(2)\cdots\sigma(n)$ under the *reversal* $\mathbf{r}$ (resp. *complement* $\mathbf{c}$) that maps σ onto its mirror image $\mathbf{r}\,\sigma = \sigma(n)\cdots\sigma(2)\sigma(1)$ (resp. onto its complement $\mathbf{c}\,\sigma = (n+1-\sigma(1))(n+1-\sigma(2))\cdots(n+1-\sigma(n))$). The next properties are well known (see, e.g., 10): if σ belongs to $\mathfrak{S}_n$, then

$$\text{des}\,\mathbf{r}\,\sigma = n - 1 - \text{des}\,\sigma; \qquad \text{maj}\,\mathbf{r}\,\sigma = \text{maj}\,\sigma - n\,\text{des}\,\sigma + \binom{n}{2}; \qquad (3.1)$$

$$\text{des}\,\mathbf{c}\,\sigma = n - 1 - \text{des}\,\sigma; \qquad \text{maj}\,\mathbf{c}\,\sigma = \binom{n}{2} - \text{maj}\,\sigma; \qquad (3.2)$$

$$\text{des}\,\mathbf{r}\,\mathbf{c}\,\sigma = \text{des}\,\sigma; \qquad \text{maj}\,\mathbf{r}\,\mathbf{c}\,\sigma = n\,\text{des}\,\sigma - \text{maj}\,\sigma. \qquad (3.3)$$

Now let i, j be the two integers defined by $\sigma(1) = n - j$ and $\sigma(n) = n - i$, and let σ' be the permutation mapping k onto

$$\sigma'(k) = \begin{cases} \sigma(k) + i & \text{if } \sigma(k) + i \leq n; \\ \sigma(k) + i - n & \text{if } \sigma(k) + i > n. \end{cases}$$

The operation $\sigma \mapsto \sigma'$ and property (3.5) below already appear in [13] for the study of the Z-statistic.

Lemma 3.1 *We have:*

$$\text{des}\,\sigma - \text{des}\,\sigma' = \begin{cases} 0 & \text{if } i < j; \\ 1 & \text{if } i > j; \end{cases} \qquad (3.4)$$

$$\text{maj}\,\sigma - \text{maj}\,\sigma' = i. \qquad (3.5)$$

Proof As $\sigma(n) = n - i$, we can factorize the word $\sigma = \sigma(1)\sigma(2)\cdots\sigma(n)$ as a product $p_0 q_1 p_1 \cdots q_r p_r$ ($r \geq 1$), where the letters of all p_k (resp. all q_k) are smaller than or equal to (resp. greater than) $n - i$. Also, let $p_0' q_1' p_1' \cdots q_r' p_r'$ be the corresponding factorization of σ' such that $\lambda p_0' = \lambda p_0$, $\lambda q_1' = \lambda q_1$, ... ($\lambda$ being the word length). In particular, p_r (and p_r') is never empty since $\sigma(n) = n - i$ (and $\sigma'(n) = n$), but p_0 (resp. and p_0') is nonempty if and only if $\sigma(1) + i = n - j + i \leq n$, that is, if $i < j$.

As the rightmost letter of each factor q_k is larger than the leftmost letter of p_k, we have

$$\text{des } \sigma = \text{des } p_0 + \text{des } q_1 + 1 + \text{des } p_1 + \cdots + \text{des } q_r + 1 + \text{des } p_r.$$

In σ' the rightmost letter of each p_k' is greater than the leftmost letter of q_{k+1}', so that, if $i < j$,

$$\text{des } \sigma' = \text{des } p_0' + 1 + \text{des } q_1' + \cdots + \text{des } p_{r-1}' + 1 + \text{des } q_r' + \text{des } p_r'$$

$$= \text{des } p_0 + \text{des } q_1 + 1 + \cdots + \text{des } p_{r-1} + \text{des } q_r + 1 + \text{des } p_r$$

$$= \text{des } \sigma,$$

while, if $i > j$, the factor p_0' is empty and

$$\text{des } \sigma' = \text{des } q_1' + \cdots + \text{des } p_{r-1}' + 1 + \text{des } q_r' + \text{des } p_r'$$

$$= \text{des } \sigma - 1.$$

However, in both cases we have

$$\text{maj } \sigma - \text{maj } \sigma' = \lambda(p_0 q_1) + \lambda(p_0 q_1 p_1 q_2) + \cdots + \lambda(p_0 q_1 \cdots p_{r-1} q_r)$$

$$- (\lambda(p_0') + \lambda(p_0' q_1' p_1') + \cdots + \lambda(p_0' q_1' \cdots p_{r-1}'))$$

$$= \lambda(q_1) + \lambda(q_2) + \cdots + \lambda(q_r)$$

$$= \#\{k : \sigma(k) > n - i\} = i. \qquad \square$$

Lemma 3.2 Let $\delta = \begin{pmatrix} a_0 \, a_1 \, \cdots \, a_n \\ b_0 \, b_1 \, \cdots \, b_n \end{pmatrix}$ be a doubloon such that $a_0 = i$ and $b_0 = j$. Also, let $\delta' = T_{-i}(\delta) = \delta - i = \begin{pmatrix} 0 \, a_1 - i \, \cdots \, a_n - i \\ j - i \, b_1 - i \, \cdots \, b_n - i \end{pmatrix}$. Then

$$\text{cmaj } \delta - \text{cmaj } \delta' = \begin{cases} -i & \text{if } i < j; \\ n + 1 - i & \text{if } i > j. \end{cases}$$

Proof Add 1 to each letter of the reading $\rho(\delta) = a_0 a_1 \cdots a_n b_n b_{n-1} \cdots b_0$ of δ and do the same for the reading $\rho(\delta')$ of δ'. We obtain two permutations σ, σ' of $12 \cdots (2n + 2)$ of the form

$$\sigma = (i + 1) \, \sigma(2) \, \cdots \, \sigma(2n + 1) \, (j + 1);$$

$$\sigma' = 1 \, \sigma'(2) \, \cdots \, \sigma'(2n + 1) \, (j - i + 1).$$

When the transformation $\mathbf{r}\,\mathbf{c}$ is applied to each of them, we get

$$\mathbf{r}\,\mathbf{c}\,\sigma = (2n+2-j) \cdots (2n+2-i);$$

$$\mathbf{r}\,\mathbf{c}\,\sigma' = (2n+2-j+i) \cdots (2n+2).$$

It follows from Lemma 3.1 (with $(2n+2)$ replacing n, $\mathbf{r}\,\mathbf{c}\,\sigma$ instead of σ, and $\mathbf{r}\,\mathbf{c}\,\sigma'$ instead of σ') that

$$\mathrm{des}\,\mathbf{r}\,\mathbf{c}\,\sigma - \mathrm{des}\,\mathbf{r}\,\mathbf{c}\,\sigma' = \begin{cases} 0 & \text{if } i < j; \\ 1 & \text{if } i > j; \end{cases}$$

$$\mathrm{maj}\,\mathbf{r}\,\mathbf{c}\,\sigma - \mathrm{maj}\,\mathbf{r}\,\mathbf{c}\,\sigma' = i;$$

so that by (3.3)

$$\mathrm{des}\,\sigma - \mathrm{des}\,\sigma' = \begin{cases} 0 & \text{if } i < j; \\ 1 & \text{if } i > j; \end{cases}$$

$$\mathrm{maj}\,\sigma - \mathrm{maj}\,\sigma' = (2n+2)(\mathrm{des}\,\sigma - \mathrm{des}\,\sigma') - (\mathrm{maj}\,\mathbf{r}\,\mathbf{c}\,\sigma - \mathrm{maj}\,\mathbf{r}\,\mathbf{c}\,\sigma')$$

$$= (2n+2)(\mathrm{des}\,\sigma - \mathrm{des}\,\sigma') - i.$$

As $\mathrm{cmaj}\,\delta = \mathrm{maj}\,\sigma - (n+1)\,\mathrm{des}\,\sigma + \binom{n}{2}$, we get

$$\mathrm{cmaj}\,\delta - \mathrm{cmaj}\,\delta' = (\mathrm{maj}\,\sigma - \mathrm{maj}\,\sigma') - (n+1)(\mathrm{des}\,\sigma - \mathrm{des}\,\sigma')$$

$$= (n+1)(\mathrm{des}\,\sigma - \mathrm{des}\,\sigma') - i$$

$$= \begin{cases} -i & \text{if } i < j; \\ n+1-i & \text{if } i > j. \end{cases} \qquad \square$$

Lemma 3.3 *Let* $\delta = \begin{pmatrix} 0\ a_1\ \cdots\ a_n \\ b_0\ b_1\ \cdots\ b_n \end{pmatrix} \mapsto \delta' = \begin{pmatrix} a_0'\ a_1'\ \cdots\ a_n' \\ b_0'\ b_1'\ \cdots\ b_n' \end{pmatrix}$ *be the transformation mapping* $\delta \in \mathcal{D}^0_{2n+1}$ *onto the doubloon* δ' *obtained from* δ *by replacing each entry* a_k *(resp.* b_k*) by* $a_k' = (2n+1) - a_k$ *(resp.* $b_k' = (2n+1) - b_k$*). Then* $\mathrm{cmaj}\,\delta = n(n-1) - \mathrm{cmaj}\,\delta'$.

Proof Again, add 1 to each element of $\rho(\delta)$ and of $\rho(\delta')$ to obtain two permutations σ and σ' of $12 \cdots (2n+2)$. We have $\sigma' = \mathbf{c}\,\sigma$. By (3.2) $\mathrm{des}\,\sigma + \mathrm{des}\,\sigma' = 2n+1$ and $\mathrm{maj}\,\sigma + \mathrm{maj}\,\sigma' = \binom{2n+2}{2} = (n+1)(2n+1)$. Hence, $\mathrm{cmaj}\,\delta + \mathrm{cmaj}\,\delta' = \mathrm{maj}\,\sigma + \mathrm{maj}\,\sigma' - (n+1)(\mathrm{des}\,\sigma + \mathrm{des}\,\sigma') + n(n-1) = (n+1)(2n+1) - (n+1)(2n+1) + n(n-1) = n(n-1)$. $\square$

Let $\Gamma : \begin{pmatrix} 0\ a_1\ \cdots\ a_n \\ b_0\ b_1\ \cdots\ b_n \end{pmatrix} \mapsto \begin{pmatrix} 0 & 2n+2-a_1\ \cdots\ 2n+2-a_n \\ 2n+2-b_0\ 2n+2-b_1\ \cdots\ 2n+2-b_n \end{pmatrix}$ $[\Gamma : \delta \mapsto \delta''$ in short] be the transformation mapping each doubloon $\delta = \begin{pmatrix} 0\ a_1\ \cdots\ a_n \\ b_0\ b_1\ \cdots\ b_n \end{pmatrix} \in \mathcal{D}^0_{2n+1}$ onto the doubloon $\delta'' = \begin{pmatrix} 0\ a_1''\ \cdots\ a_n'' \\ b_0''\ b_1''\ \cdots\ b_n'' \end{pmatrix}$ obtained from δ by replacing each entry a_k (resp. b_k) by the residue $a_k'' = (2n+2) - a_k$ (resp. $b_k'' = (2n+2) - b_k$).

Remark If δ is interlaced at each k (condition ($N1$) holds at each k), the same property holds for $\Gamma(\delta)$. However, if δ is normalized at each k, the property is *not* preserved under the transformation Γ.

Lemma 3.4 *For each* $\delta \in \mathcal{D}^0_{2n+1}$, *we have* $\operatorname{cmaj} \delta + \operatorname{cmaj} \Gamma(\delta) = n^2$.

Proof First, transform δ into the doubloon δ' defined in Lemma 3.3, so that $\delta' = \left(\begin{smallmatrix} 2n+1 & a'_1 \cdots a'_n \\ 2n+1-j & b'_1 \cdots b'_n \end{smallmatrix}\right)$ and $\operatorname{cmaj} \delta = n(n-1) - \operatorname{cmaj} \delta'$. Next, apply Lemma 3.2 to δ' with $i = 2n + 1$, so that the new doubloon is of the form $\delta'' = \left(\begin{smallmatrix} 0 & a''_1 \cdots a''_n \\ 2n+2-j & b''_1 \cdots b''_n \end{smallmatrix}\right)$. Hence, $\operatorname{cmaj} \delta' - \operatorname{cmaj} \delta'' = (n+1) - (2n+1) = -n$ and $\operatorname{cmaj} \delta = n(n-1) - \operatorname{cmaj} \delta'' + n = n^2 - \operatorname{cmaj} \delta''$. $\qquad\square$

4 Further operations on doubloons

In our previous paper [6] we also introduced a class of transformations ϕ_i ($0 \le i \le n$) on $\mathcal{D}_{2n+1}$, called *micro flips*, which permute the entries in a given column. By definition,

$$\phi_i : \begin{pmatrix} a_0 \cdots a_{i-1} \, a_i \, a_{i+1} \cdots a_n \\ b_0 \cdots b_{i-1} \, b_i \, b_{i+1} \cdots b_n \end{pmatrix} \mapsto \begin{pmatrix} a_0 \cdots a_{i-1} \, b_i \, a_{i+1} \cdots a_n \\ b_0 \cdots b_{i-1} \, a_i \, b_{i+1} \cdots b_n \end{pmatrix} \quad (0 \le i \le n).$$

Next, the *macro flips* Φ_i are defined by $\Phi_i = \phi_i \phi_{i+1} \cdots \phi_n$ ($1 \le i \le n$). Note that both ϕ_i and Φ_i are *involutions* of $\mathcal{D}_{2n+1}$. In particular,

$$\Phi_1 \Phi_2 \cdots \Phi_i = \prod_{\substack{j \text{ odd,} \\ j \le i}} \phi_j. \tag{4.1}$$

By means of the transformation Γ (see Lemma 3.4 above) and the involutions Φ_i we now construct a bijection of $\mathcal{N}^0_{2n+1,j}$ onto $\mathcal{N}^0_{2n+1,2n+2-j}$.

Lemma 4.1 *Let* $\delta = \left(\begin{smallmatrix} 0 & a_1 \cdots a_n \\ b_0 & b_1 \cdots b_n \end{smallmatrix}\right)$ *be a normalized doubloon, and let* $\Gamma(\delta) = \delta'' = \left(\begin{smallmatrix} 0 & a''_1 \cdots a''_n \\ b''_0 & b''_1 \cdots b''_n \end{smallmatrix}\right)$. *Then, for each* $i = 1, 2, \ldots, n$, *the doubloon* $\Phi_1 \Phi_2 \cdots \Phi_i(\delta'')$ *is normalized at each integer* $1, 2, \ldots, i$. *In particular,* $\Phi_1 \Phi_2 \cdots \Phi_n(\delta'')$ *is normalized.*

Proof Using (4.1), we have

$$\delta^* = \Phi_1 \Phi_2 \cdots \Phi_i(\delta'') = \begin{cases} \left(\begin{smallmatrix} 0 & b''_1 \, a''_2 \cdots a''_{i-1} \, b''_i \\ b''_0 \, a''_1 \, b''_2 \cdots b''_{i-1} \, a''_i \end{smallmatrix}\right) & \text{if } i \text{ odd;} \\[2mm] \left(\begin{smallmatrix} 0 & b''_1 \, a''_2 \cdots b''_{i-1} \, a''_i \\ b''_0 \, a''_1 \, b''_2 \cdots a''_{i-1} \, b''_i \end{smallmatrix}\right) & \text{if } i \text{ even.} \end{cases}$$

The doubloon δ'' is not normalized (see the remark before Lemma 3.4), but is interlaced at each i, so that the relations $a''_{j-1} < a''_j$, $b''_{j-1} < b''_j$ or $a''_{j-1} > a''_j$, $b''_{j-1} > b''_j$ hold for all j. This shows that δ^* is normalized whenever j is odd or even. $\qquad\square$

For the next proposition, we need the following property proved in our previous paper [6, Theorem 3.5]:

Let $\delta = \begin{pmatrix} 0\,a_1\cdots a_n \\ b_0\,b_1\cdots b_n \end{pmatrix}$ be an interlaced doubloon, normalized at i.

Then $\operatorname{cmaj} \Phi_i(\delta) - \operatorname{cmaj} \delta = n - i + 1 \quad (1 \le i \le n).$ $\qquad(4.2)$

Proposition 4.2 *The transformation* $\Phi_1\Phi_2\cdots\Phi_n\Gamma : \delta \mapsto \delta^*$ *is a bijection of* $\mathcal{N}^0_{2n+1,j}$ *onto* $\mathcal{N}^0_{2n+1,2n+2-j}$ *having the property*

$$\operatorname{cmaj} \delta + \operatorname{cmaj} \delta^* = \binom{n}{2}. \qquad(4.3)$$

Proof Let $\delta'' = \Gamma(\delta)$. As $\Phi_1\Phi_2\cdots\Phi_i(\delta'')$ is normalized at i, we have

$$\operatorname{cmaj} \Phi_i\,\Phi_1\Phi_2\cdots\Phi_i(\delta'') - \operatorname{cmaj}\Phi_1\Phi_2\cdots\Phi_i(\delta'') = n - i + 1$$

by (4.2). Summing over all $i = 1, 2, \ldots, n$, we get

$$\sum_{i=1}^{n}\left(\operatorname{cmaj}\Phi_1\Phi_2\cdots\Phi_{i-1}(\delta'') - \operatorname{cmaj}\Phi_1\Phi_2\cdots\Phi_i(\delta'')\right) = \sum_{i=1}^{n}(n-i+1),$$

$$\operatorname{cmaj} \delta'' - \operatorname{cmaj}\Phi_1\Phi_2\cdots\Phi_n(\delta'') = n^2 - \binom{n}{2}.$$

Using Lemma 3.4, we conclude that $(n^2 - \operatorname{cmaj}\delta) - \operatorname{cmaj}\delta^* = n^2 - \binom{n}{2}$ and then $\operatorname{cmaj}\delta + \operatorname{cmaj}\delta^* = \binom{n}{2}$. $\qquad\qquad\square$

Example For $n = 2$, we have $\Phi_1\Phi_2 = \phi_1$. The four normalized doubloons of order 5 ($n = 2$) displayed at the end of Sect. 2 are mapped under $\Phi_1\Phi_2\Gamma = \phi_1\Gamma$ as shown in Fig. 5. Next to each doubloon appears the value of its "cmaj."

Let $d_{n,j}(q) = \sum_{\delta \in \mathcal{N}^0_{2n+1,j}} q^{\operatorname{cmaj}\delta}$. The final purpose is to show that $d_{n,j}(q)$ satisfies relations (D1)–(D4). This will be done in Sect. 5. In the rest of this section we state some symmetry properties of the $d_{n,j}(q)$'s.

Fig. 5 The transformation $\Phi_1\Phi_2\cdots\Phi_n\Gamma$

$$\begin{pmatrix} 0\,4\,3 \\ 2\,1\,5 \end{pmatrix},1 \qquad \begin{pmatrix} 0\,5\,4 \\ 3\,2\,1 \end{pmatrix},0 \qquad \begin{pmatrix} 0\,4\,2 \\ 3\,1\,5 \end{pmatrix},1 \qquad \begin{pmatrix} 0\,5\,3 \\ 4\,2\,1 \end{pmatrix},0$$

$$\downarrow\Gamma \qquad\qquad \downarrow\Gamma \qquad\qquad \downarrow\Gamma \qquad\qquad \downarrow\Gamma$$

$$\begin{pmatrix} 0\,2\,3 \\ 4\,5\,1 \end{pmatrix},3 \qquad \begin{pmatrix} 0\,1\,2 \\ 3\,4\,5 \end{pmatrix},4 \qquad \begin{pmatrix} 0\,2\,4 \\ 3\,5\,1 \end{pmatrix},3 \qquad \begin{pmatrix} 0\,1\,3 \\ 2\,4\,5 \end{pmatrix},4$$

$$\downarrow\phi_1 \qquad\qquad \downarrow\phi_1 \qquad\qquad \downarrow\phi_1 \qquad\qquad \downarrow\phi_1$$

$$\begin{pmatrix} 0\,5\,3 \\ 4\,2\,1 \end{pmatrix},0 \qquad \begin{pmatrix} 0\,4\,2 \\ 3\,1\,5 \end{pmatrix},1 \qquad \begin{pmatrix} 0\,5\,4 \\ 3\,2\,1 \end{pmatrix},0 \qquad \begin{pmatrix} 0\,4\,3 \\ 2\,1\,5 \end{pmatrix},1$$

Corollary 4.3 *For $2 \leq j \leq 2n$, we have*

$$d_{n,j}(q) = q^{\binom{n}{2}} d_{n,2n+2-j}(q^{-1}). \tag{4.4}$$

Proof This follows from the previous proposition:

$$d_{n,j}(q) = \sum_{\delta \in \mathcal{N}^0_{2n+1,j}} q^{\mathrm{cmaj}\,\delta} = \sum_{\delta^* \in \mathcal{N}^0_{2n+1,2n+1-j}} q^{\binom{n}{2}-\mathrm{cmaj}\,\delta^*} = q^{\binom{n}{2}} d_{n,2n+2-j}(q^{-1}). \qquad \square$$

In the next lemma we study the action of the sole transposition ϕ_0 that permutes the leftmost entries of each doubloon.

Lemma 4.4 *For each normalized doubloon δ,*

$$\mathrm{cmaj}\,\delta - \mathrm{cmaj}\,\phi_0(\delta) = -n. \tag{4.5}$$

Proof Let $\rho(\delta) = 0a_1 \cdots a_n b_n \cdots b_1 b_0$ be the reading of δ, so that $\rho(\phi_0(\delta)) = b_0 a_1 \cdots a_n b_n \cdots b_1 0$. As δ is normalized (and, in particular, interlaced), we have $0 < b_1 < b_0 < a_1$. Thus, $\rho(\delta)$ starts and ends with a rise ($0 < a_1$ and $b_1 < a_0$), while $\rho(\phi_0(\delta))$ starts with a rise $b_0 < a_1$ and ends with a descent $b_1 > 0$. Hence, $\mathrm{des}\,\rho(\delta) = \mathrm{des}\,\rho(\phi_0(\delta)) - 1$ and $\mathrm{maj}\,\rho(\delta) = \mathrm{maj}\,\rho(\phi_0(\delta)) - (2n+1)$ and $\mathrm{cmaj}\,\delta - \mathrm{cmaj}\,\rho(\phi_0(\delta)) = -(2n+1) - (n+1) - 1 = -n$ $\qquad \square$

The involution $\phi_0 \Phi_1 = \Phi_1 \phi_0 = \phi_0 \phi_1 \cdots \phi_n$ transposes the two rows of each doubloon $\delta \in \mathcal{D}_{2n+1}$.

Lemma 4.5 *For each normalized doubloon δ,*

$$\mathrm{cmaj}\,\delta = \mathrm{cmaj}\,\phi_0 \Phi_1(\delta). \tag{4.6}$$

Proof First, $\mathrm{cmaj}\,\delta = \mathrm{cmaj}\,\phi_0(\delta) - n$ by the previous lemma. Furthermore, as $\phi_0 \Phi_1(\delta)$ remains normalized at 1, relation (4.2) implies that $\mathrm{cmaj}\,\Phi_1(\phi_0 \Phi_1(\delta)) - \mathrm{cmaj}\,\phi_0 \Phi_1(\delta) = n - 1 + 1$. Thus,

$$\mathrm{cmaj}\,\phi_0 \Phi_1(\delta) = \mathrm{cmaj}\,\phi_0(\delta) - n$$
$$= \mathrm{cmaj}\,\delta + n - n = \mathrm{cmaj}\,\delta. \qquad \square$$

We next study the joint action of the operators T_{-j} (introduced in Lemma 3.2), ϕ_0, and Φ_1.

Proposition 4.6 *The transformation $T_{-j}\phi_0\Phi_1 : \delta \mapsto \delta'$ is a bijection of $\mathcal{N}^0_{2n+1,j}$ onto $\mathcal{N}^0_{2n+1,2n+2-j}$ having the property that*

$$\mathrm{cmaj}\,\delta - \mathrm{cmaj}\,\delta' = n + 1 - j. \tag{4.7}$$

Fig. 6 The transformation
$T_{-j}\phi_0\Phi_1$

$$\begin{pmatrix}043\\215\end{pmatrix},1 \qquad \begin{pmatrix}054\\321\end{pmatrix},0 \qquad \begin{pmatrix}042\\315\end{pmatrix},1 \qquad \begin{pmatrix}053\\421\end{pmatrix},0$$

$$\Big\downarrow \phi_0\Phi_1 \qquad\quad \Big\downarrow \phi_0\Phi_1 \qquad\quad \Big\downarrow \phi_0\Phi_1 \qquad\quad \Big\downarrow \phi_0\Phi_1$$

$$\begin{pmatrix}215\\043\end{pmatrix},1 \qquad \begin{pmatrix}321\\054\end{pmatrix},0 \qquad \begin{pmatrix}315\\042\end{pmatrix},1 \qquad \begin{pmatrix}421\\053\end{pmatrix},0$$

$$\Big\downarrow T_{-2} \qquad\quad \Big\downarrow T_{-3} \qquad\quad \Big\downarrow T_{-3} \qquad\quad \Big\downarrow T_{-4}$$

$$\begin{pmatrix}053\\421\end{pmatrix},0 \qquad \begin{pmatrix}054\\321\end{pmatrix},0 \qquad \begin{pmatrix}042\\315\end{pmatrix},1 \qquad \begin{pmatrix}043\\215\end{pmatrix},1$$

Proof Let $\delta \in \mathcal{N}^0_{2n+1,j}$. Then, $\phi_0\Phi_1(\delta)$ is normalized at each $i = 1, 2, \ldots, n$ and also $T_{-j}\phi_0\Phi_1(\delta)$ by the remark made before Lemma 3.2. Thus, $\delta' \in \mathcal{N}^0_{2n+1,2n+2-j}$, and the map $\delta \mapsto \delta'$ is bijective. Furthermore, $\mathrm{cmaj}\,\delta = \mathrm{cmaj}\,\phi_0\Phi_1(\delta)$ by Lemma 4.5 and $\mathrm{cmaj}\,\phi_0\Phi_1(\delta) = \mathrm{cmaj}\,\delta' + n + 1 - j$ by Lemma 3.2. $\qquad\square$

Example Again consider the four normalized doubloons of order 5. We get the display of Fig. 6 with the value of "cmaj" next to each doubloon.

Corollary 4.7 *For* $2 \le j \le 2n$, *we have*

$$d_{n,j}(q) = q^{n+1-j}\, d_{n,2n+2-j}(q). \tag{4.8}$$

The corollary immediately follows from Proposition 4.6. In its turn the next corollary is a consequence of Corollaries 4.3 and 4.7.

Corollary 4.8 *For* $2 \le j \le 2n$, *we have*

$$d_{n,j}(q) = q^{\binom{n+1}{2}+1-j}\, d_{n,j}(q^{-1}). \tag{4.9}$$

As could be seen in Fig. 2 for the first values, the polynomial $d_{n,j}(q)$ is a multiple of $d_{n,2n+2-j}(q)$ (Corollary 4.7), and Corollary 4.8 indicates a symmetry between its coefficients.

5 The recurrence itself

In this section we prove Theorem 1.2, that is, we show that relations $(D1)$–$(D4)$ hold for $d_{n,j}(q) = \sum_{\delta \in \mathcal{N}^0_{2n+1,j}} q^{\mathrm{cmaj}\,\delta}$. The first two relations $(D1)$–$(D2)$ are evidently true. For proving that relations $(D3)$–$(D4)$ hold for such a $d_{n,j}(q)$, we start with a doubloon $\delta \in \mathcal{N}^0_{2n+1}$, drop its leftmost column, and compare the compressed major indices of δ and of the doubloon obtained after deletion. In so doing we get the following result.

Lemma 5.1 *To each doubloon* $\delta \in \mathcal{N}^0_{2n+1,j}$, *there corresponds a unique triplet* (k, l, δ') *such that* $k \ge j - 1$, $l \le j - 2$, *and* $\delta' \in \mathcal{N}^k_{2n-1,l}$ *having the property that*

$$\mathrm{cmaj}\,\delta = \mathrm{cmaj}\,\delta' + (n - 1). \tag{5.1}$$

Proof Let $\delta = \begin{pmatrix} 0 \, a_1 \, \cdots \, a_n \\ j \, b_1 \, \cdots \, b_n \end{pmatrix} \in \mathcal{N}^0_{2n+1,j}$ and define

$$a'_i = \begin{cases} a_i - 1 & \text{if } a_i < j; \\ a_i - 2 & \text{if } j < a_i \leq 2n + 1; \end{cases}$$

with an analogous definition for the b'_i, the b's replacing the a's. Then, $\delta' = \begin{pmatrix} a'_1 \, \cdots \, a'_n \\ b'_1 \, \cdots \, b'_n \end{pmatrix}$ is a normalized doubloon of order $(2n+1)$, but its left-top corner a'_1 is not necessarily 0. Call it the *reduction* of δ. As $a_1 > j > b_1$, we see that $b'_1 = b_1 - 1 < j - 1$ and $a'_1 = a_1 - 2 > j - 2$, that is, $b'_1 \leq j - 2 < j - 1 \leq a'_1$. Thus, $\delta' \in \mathcal{N}^k_{2n-1,l}$ with $k \geq j - 1$ and $l \leq j - 2$. Conversely, given the triplet (δ', k, l) with the above properties, we can uniquely reconstruct the normalized doubloon δ.

As δ is normalized at 1, the inequalities $0 < b_1 < j < a_1$ hold, so that $\rho(\delta)$ starts and ends with a rise: $0 < a_1, b_1 < j$. In particular, $\operatorname{des} \rho(\delta) = \operatorname{des} \rho(\delta')$. However, $\operatorname{maj} \rho(\delta) = \operatorname{maj} \rho(\delta') + \operatorname{des} \delta$, as the first letter 0 is dropped when going from $\rho(\delta)$ to $\rho(\delta')$. Hence, $\operatorname{cmaj} \delta - \operatorname{cmaj} \delta' = \operatorname{des} \delta - ((n+1) - n) \operatorname{des} \delta + \binom{n}{2} - \binom{n-1}{2} = n - 1$. $\square$

The previous lemma yields the first passage from the generating polynomial $d_{n,j}(q)$ for the normalized doubloons $\delta \in \mathcal{N}^0_{2n+1,j}$ by "cmaj" to the generating polynomial for doubloons of order $(2n - 1)$, as expressed in the next corollary.

Corollary 5.2 *For $2 \leq j \leq 2n$, we have*

$$d_{n,j}(q) = q^{n-1} \sum_{k \geq j-1, l \leq j-2} q^{n-k} d_{n-1,2n+l-k}(q). \tag{5.2}$$

Proof By the previous lemma we can write

$$d_{n,j}(q) = \sum_{\delta \in \mathcal{N}^0_{2n+1,j}} q^{\operatorname{cmaj} \delta} = q^{n-1} \sum_{k \geq j-1, l \leq j-2} \sum_{\delta' \in \mathcal{N}^k_{2n-1,l}} q^{\operatorname{cmaj} \delta'}.$$

With each $\delta' \in \mathcal{N}^k_{2n-1,l}$, associate $\delta'' = T_{-k}\delta' = \delta' - k$. By Lemma 3.2 we get $\delta'' \in \mathcal{N}^0_{2n-1,2n+l-k}$ (note that the residue $\operatorname{mod}(2n+2)$ must be considered in the subscript for $l < k$) and $\operatorname{cmaj} \delta' - \operatorname{cmaj} \delta'' = n - k$. Hence,

$$\sum_{\delta' \in \mathcal{N}^k_{2n-1,l}} q^{\operatorname{cmaj} \delta'} = q^{n-k} \sum_{\delta'' \in \mathcal{N}^0_{2n-1,2n+l-k}} q^{\operatorname{cmaj} \delta''} = q^{n-k} d_{n-1,2n+l-k}(q).$$

$\square$

When $j = 2$ in (5.2), we get

$$d_{n,2}(q) = \sum_{k \geq 2} q^{2n-k-1} d_{n-1,2n-k}(q)$$

$$= \sum_{k \geq 2} q^{i-1} d_{n-1,i}(q) \quad [\text{by the change of variables } i = 2n - k]$$

and also

$$d_{n,2}(q) = q^{n-1} \sum_{i \geq 2} d_{n-1,i}(q) \quad \left[\text{by using (4.7)} \right].$$

Consequently, relation $(D3)$ holds for the polynomial $d_{n,j}(q) = \sum_{\delta \in \mathcal{N}_{2n+1,j}^0} q^{\mathrm{cmaj}\,\delta}$.

Using (5.2), it is also easy to derive the identity

$$q\,d_{n,3}(q) = (q+1)\,d_{n,2}(q).$$

Proposition 5.3 *For $2 \leq j \leq 2n$, we have*

$$d_{n,j}(q) = \sum_i \frac{q^{\max(0,i+1-j)} - q^{\min(i,2n+1-j)}}{1-q}\,d_{n-1,i}(q). \tag{5.3}$$

Proof Let $i = 2n + l - k$ with $0 \leq l \leq j - 2$ and $j - 1 \leq k \leq 2n - 1$. This implies $0 \leq i + k - 2n \leq j - 2$, or still $2n - i \leq k \leq 2n + j - i - 2$. Taking the two relations keeping k within bounds into account, we get the double inequality

$$\max(2n - i, j - 1) \leq k \leq \min(2n + j - i - 2, 2n - 1). \tag{5.4}$$

Identity (5.2) may then be rewritten as

$$d_{n,j}(q) = q^{2n-1} \sum_i d_{2n-1,i}(q) \sum_k q^{-k} \tag{5.5}$$

with k ranging over the interval defined in (5.4). The geometric sum over k is equal to

$$\frac{q^{-\max(2n-i,j-1)} - q^{-\min(2n+j-i-2,2n-1)-1}}{1-q^{-1}}.$$

Now, $2n - \max(2n - i, j - 1) = 2n + \min(-2n + i, -j + 1) = \min(i, 2n - j + 1)$ and $2n - 1 - \min(2n + j - i - 2, 2n - 1) = \max(0, -j + i + 1)$. Accordingly,

$$q^{2n-1} \sum_{k \text{ subject to (5.4)}} q^{-k} = \frac{q^{\max(0,i+1-j)} - q^{\min(i,2n+1-j)}}{1-q}.$$

This proves (5.3) when reporting the latter expression into (5.5). $\qquad\square$

For getting rid of "max" and "min" from identity (5.3), we decompose the sum into four subsums, assuming $j \geq 3$ (when $j = 2$, identity (5.3) gives back $(D3)$). We obtain

$$(1-q)d_{n,j}(q) = \sum_{i=0}^{j-1} d_{n-1,i}(q) + \sum_{i=j}^{2n+1} q^{i+1-j} d_{n-1,i}(q)$$

$$- \sum_{i=0}^{2n-j} q^i d_{n-1,i}(q) - \sum_{i=2n-j+1}^{2n+1} q^{2n+1-j} d_{n-1,i}(q). \tag{5.6}$$

By means of (5.6) we calculate $d_{n,j}(q) - d_{n,j-1}(q)$ and then $-d_{n,j-1}(q) + d_{n,j-2}(q)$, whose sum is the left-hand side of (D4). In the computation of $(1-q)(d_{n,j}(q) - d_{n,j-1}(q))$ the contribution of the first subsum is simply $d_{n-1,j-1}(q)$. Next, $q^{2n-j+1}d_{n-1,2n-j+1}(q)$ is the contribution of the second subsum. For the third subsum, we get

$$-q\, d_{n-1,j-1}(q) + \sum_{i=j}^{2n+1} q^{i-j+1}(1-q)d_{n-1,i}(q)$$

and, for the fourth one,

$$-q^{2n+1-j}d_{n-1,2n+1-j}(q) + \sum_{i=2n-j+2}^{2n+1} q^{2n+1-j}(q-1)d_{n-1,i}(q).$$

Altogether,

$$d_{n,j}(q) - d_{n,j-1}(q)$$
$$= d_{n-1,j-1}(q) + \sum_{i=j}^{2n+1} q^{i-j+1}d_{n-1,i}(q) - \sum_{i=2n-j+2}^{2n+1} q^{2n+1-j}d_{n-1,i}(q); \qquad (5.7)$$
$$-d_{n,j-1}(q) + d_{n,j-2}(q)$$
$$= -d_{n-1,j-2}(q) - \sum_{i=j-1}^{2n+1} q^{i-j+2}d_{n-1,i}(q) + \sum_{i=2n-j+3}^{2n+1} q^{2n+2-j}d_{n-1,i}(q). \quad (5.8)$$

Summing (5.7) and (5.8), we get

$$d_{n,j}(q) - 2d_{n,j-1}(q) + d_{n-1,j-2}(q)$$
$$= d_{n-1,j-1}(q) - d_{n-1,j-2}(q)$$
$$\quad - q\, d_{n-1,j-1}(q) + (1-q)\sum_{i=j}^{2n+1} q^{i-j+1}d_{n-1,i}(q)$$
$$\quad - q^{2n+1-j}\, d_{n-1,2n-j+2}(q) - (1-q)\sum_{i=2n-j+3}^{2n+1} q^{2n+1-j}d_{n-1,i}(q).$$

Now, $-q^{2n+1-j}\, d_{n-1,2n-j+2}(q) = -q^{n-1}d_{n-1,j-2}(q)$ by Corollary 4.7. Also,

$$\sum_{i=2n-j+3}^{2n+1} q^{2n+1-j}d_{n-1,i}(q) = \sum_{i=2n-j+3}^{2n-1} q^{2n+1-j}d_{n-1,i}(q)$$
$$= \sum_{k=1}^{j-3} q^{2n+1-j}d_{n-1,2n-k}(q) \quad [k=2n-i]$$
$$= \sum_{i=1}^{j-3} q^{n+i+1-j}d_{n-1,i}(q) \quad [\text{by Corollary 4.7}].$$

Hence,

$$d_{n,j}(q) - 2d_{n,j-1}(q) + d_{n,j-2}(q)$$

$$= -(1-q) \sum_{i=1}^{j-3} q^{n+i+1-j} d_{n-1,i}(q)$$

$$- \left(1 + q^{n-1}\right) d_{n-1,j-2}(q) + (1-q) \sum_{i=j-1}^{2n-1} q^{i-j+1} d_{n-1,i}(q)$$

for $n \geq 2$ and $3 \leq j \leq 2n$, which is precisely relation $(D4)$.

6 Concluding remarks

As mentioned in the Introduction, there has been a great number of papers deal-ing with the Catalan Triangle (the numbers $d_{n,j}(0)$). The recurrence for the num-bers $d_{n,j}(1)$, namely the set of conditions $(P1)$–$(P4)$, is definitely due to Christiane Poupard [15] and was recently rediscovered by Graham and Zang [11]. More exactly, the latter authors introduced the notion of *split-pair arrangement*. To show that the number of such arrangements of order n was equal to the reduced tangent number t_n, they set up an algebra for the coefficients $d_{n,j}(1)$ and again produced the recurrence $(P1)$–$(P4)$, but their proof of the identity $\sum_j d_{n,j}(1) = t_n$ was more elaborate than the original one made by Christiane Poupard [15], shortly sketched in the Introduc-tion.

Referring to Sect. 4, we say that two doubloons δ, $\delta' \in \mathcal{D}_{2n+1}$ are *equivalent* if there is a sequence ϕ_{i_1}, ϕ_{i_2}, ..., ϕ_{i_k} of micro flips such that $\delta' = \phi_{i_1} \phi_{i_2} \cdots \phi_{i_k}(\delta)$. A doubloon $\delta = \binom{a_0 \, a_1 \, \cdots \, a_n}{b_0 \, b_1 \, \cdots \, b_n}$ is said to be *minimal* if $a_k < b_k$ holds for every k. Clearly, every equivalence class of doubloons contains one and only one minimal doubloon. Also, as proved in [6], each doubloon $\delta \in \mathcal{D}^0_{2n+1}$ is equivalent to one and only one normalized doubloon.

The link between *split-pair arrangements* and *equivalence classes* of *interlaced doubloons* is the following: start with a *interlaced* and *minimal* doubloon $\delta = \binom{a_0 \, a_1 \, \cdots \, a_n}{b_0 \, b_1 \, \cdots \, b_n}$ and define the word $w = x_1 x_2 \cdots x_{2n}$ by $x_{l+1} = l$ if and only if a_k or b_k is equal to l. For instance,

$$\delta = \frac{0 \, 6 \, 3 \, 1 \, 9}{4 \, 2 \, 8 \, 7 \, 5} \mapsto w = 1 \, 4 \, 2 \, 3 \, 1 \, 5 \, 2 \, 4 \, 3 \, 5.$$

As δ is interlaced (condition $(N1)$), exactly one of the integers a_k, b_k lies between a_{k-1} and b_{k-1}. If $a_k = l < l' = b_k$, then $x_{l+1} = x_{l'+1} = k$. If $a_{k-1} = m < a_k = l < b_{k-1} = m' < b_k = l'$, then $x_{m+1} = x_{m'+1} = k - 1$, and there is exactly one letter equal to k, namely x_{l+1} between the two occurrences of $(k - 1)$, namely x_{m+1} and $x_{m'+1}$. The same conclusion if $a_{k-1} < a_k < b_{k-1} < b_k$. Those words w of length $2n$ having the property that exactly one letter equal to k lies between two occurrences of $(k - 1)$ for each $k = 2, 3, \ldots, n$ was called *split-pair arrangements* by Graham and

Zang [11]. The mapping $\delta \mapsto w$ provides a bijection between equivalence classes of interlaced doubloons and those arrangements.

A permutation $\sigma = \sigma(1)\sigma(2) \cdots \sigma(2n+1)$ of $12 \cdots (2n+1)$ is said to be *alternating* if $\sigma(2i) < \sigma(2i-1)$, $\sigma(2i+1)$ for each $i = 1, 2, \ldots, n$. For each even integer $2i$ ($1 \leq i \leq n$), let w_i' (resp. w_i'') be the longest right factor of $\sigma(1) \cdots \sigma(2i-1)$ (resp. longest left factor of $\sigma(2i+1) \cdots \sigma(2n+1)$) all letters of which are greater than $\sigma(2i)$. Let $\min w_i'$ (resp. $\min w_i''$) denote the minimum letter of w_i' (resp. of w_i''). Then σ is called an *alternating André* permutation if it is alternating and satisfies $\min w_i' > \min w_i''$ for every $i = 1, 2, \ldots, n$. The number of alternating André permutations of order $(2n+1)$ is equal to t_n (see, e.g., 7, Property 2.6 and (5.4), or [15]). The alternating André permutations of order 5 are the following: $53412, 51423, 41523, 31524$. Let $A_{2n+1,j}$ denote the set of alternating André permutations of order $(2n+1)$ *ending with j*.

Thanks to the identity

$$d_{n,j}(1) = \sum_{i \geq 0} \binom{2n+1-j}{2i+1} t_i \sum_{k=0}^{j-1} d_{n-i-1,k}(1), \tag{6.1}$$

valid for $n \geq 1$, $2 \leq j \leq 2n-1$ ($d_{0,j}(1) = \delta_{0,j}$), derived from the bivariable generating function (1.8), Christiane Poupard proved that

$$\# A_{2n+1,j} = d_{n,j}(1). \tag{6.2}$$

Three questions arise:

(1) Construct a natural (?) bijection of $\mathcal{N}_{2n+1,j}^0$ onto $A_{2n+1,j}$.
(2) Find an adequate statistic "stat" on the alternating André permutation such that the following identity holds:

$$d_{n,j}(q) = \sum_{\sigma \in A_{2n+1,j}} q^{\text{stat}\,\sigma}.$$

(3) As mentioned in the Introduction, we have an expression for the generating function for $d_{n,j}(1)$ (formula (1.8)) and for $d_{n,j}(0)$ (formula (1.14)). Find the exponential (or factorial) generating function for $d_{n,j}(q)$.

References

1. Callan, D.: A reference given in On-Line Encyclopedia of Integer Sequences for the sequence A009766
2. Carlitz, L.: q-Bernoulli and Eulerian numbers. Trans. Am. Math. Soc. **76**, 332–350 (1954)
3. Carlitz, L.: A combinatorial property of q-Eulerian numbers. Am. Math. Mon. **82**, 51–54 (1975)
4. Deutsch, E.: A reference given in On-Line Encyclopedia of Integer Sequences for the sequence A009766
5. Foata, D., Han, G.-N.: Word straightening and q-Eulerian calculus. In: Ismail, M.E.H., Stanton, D.W. (eds.) Contemporary Mathematics. q-Series from a Contemporary Perspective, vol. 254, pp. 141–156. AMS, Providence (2000)
6. Foata, D., Han, G.-N.: Doubloons and new q-tangent numbers. Preprint, 17 p. (2008)

7. Foata, D., Schützenberger, M.-P.: Nombres d'Euler et permutations alternantes. Manuscript (unabridged version), 71 p., University of Florida, Gainesville (1971). http://igd.univ-lyon1.fr/~slc/books/index.html

8. Foata, D., Schützenberger, M.-P.: Nombres d'Euler et permutations alternantes. In: Srivastava, J.N., et al. (eds.) A Survey of Combinatorial Theory, pp. 173–187. North-Holland, Amsterdam (1973)

9. Foata, D., Strehl, V.: Euler numbers and variations of permutations. In: Colloquio Internazionale sulle Teorie Combinatorie, 1973, Tome I. Atti dei Convegni Lincei, vol. 17, pp. 119–131. Accademia Nazionale dei Lincei, Rome (1976)

10. Foata, D., Schützenberger, M.-P.: Major index and inversion number of permutations. Math. Nachr. **83**, 143–159 (1978)

11. Graham, R., Zang, N.: Enumerating split-pair arrangements. J. Comb. Theory, Ser. A **115**, 293–303 (2008)

12. Han, G.-N.: Calcul Denertien. Thèse de Doctorat, Publ. l'I.R.M.A., Strasbourg, 476/TS-29 (1991), 119 p. http://www-irma.u-strasbg.fr/~guoniu/papers/p05these.pdf

13. Han, G.-N.: Une courte démonstration d'un résultat sur la Z-statistique. C. R. Acad. Sci. Paris Sér. I **314**, 969–971 (1992)

14. Han, G.-N.: Une transformation fondamentale sur les réarrangements de mots. Adv. Math. **105**(1), 26–41 (1994)

15. Poupard, C.: Deux propriétés des arbres binaires ordonnés stricts. Eur. J. Comb. **10**, 369–374 (1989)

16. Sloane, N.J.A.: On-line Encyclopedia of Integer Sequences

Ramanujan J (2010) 23: 127–135
DOI 10.1007/s11139-009-9170-4

Hook lengths and shifted parts of partitions

Guo-Niu Han

Dedicated to George Andrews on the occasion of his seventieth birthday

Received: 17 November 2008 / Accepted: 13 April 2009 / Published online: 9 December 2009
© Springer Science+Business Media, LLC 2009

Abstract Some conjectures on partition hook lengths, recently stated by the author, have been proved and generalized by Stanley, who also needed a formula by Andrews, Goulden and Jackson on symmetric functions to complete his derivation. Another identity on symmetric functions can be used instead. The purpose of this note is to prove it.

Keywords Partitions · Hook lengths · Hook formulas · Symmetric functions

Mathematics Subject Classification (2000) 05A15 · 05A17 · 05A19 · 05E05 · 11P81

1 Introduction

The hook lengths of partitions are widely studied in the Theory of Partitions, in Algebraic Combinatorics and Group Representation Theory. The basic notions needed here can be found in [7, p. 287], [5, p. 1]. A *partition* λ is a sequence of positive integers $\lambda = (\lambda_1, \lambda_2, \ldots, \lambda_\ell)$ such that $\lambda_1 \geq \lambda_2 \geq \cdots \geq \lambda_\ell > 0$. The integers $(\lambda_i)_{i=1,2,\ldots,\ell}$ are called the *parts* of λ, the number ℓ of parts being the *length* of λ denoted by $\ell(\lambda)$. The sum of its parts $\lambda_1 + \lambda_2 + \cdots + \lambda_\ell$ is denoted by $|\lambda|$. Let n be an integer, a partition λ is said to be a partition of n if $|\lambda| = n$. We write $\lambda \vdash n$. Each partition can be represented by its Ferrers diagram. For each box v in the Ferrers diagram of a partition λ, or for each box v in λ, for short, define the *hook length* of v, denoted by $h_v(\lambda)$ or h_v, to be the number of boxes u such that $u = v$, or u lies in the same

G.-N. Han (✉)
I.R.M.A. UMR 7501, Université de Strasbourg et CNRS, 7, rue René-Descartes, 67084 Strasbourg,
France
e-mail: guoniu@math.u-strasbg.fr

column as v and above v, or in the same row as v and to the right of v. The product of all hook lengths of λ is denoted by H_λ.

The hook length plays an important role in Algebraic Combinatorics thanks to the famous hook formula due to Frame, Robinson and Thrall [2]

$$f_\lambda = \frac{n!}{H_\lambda}, \tag{1.1}$$

where f_λ is the number of standard Young tableaux of shape λ.

For each partition λ let $\lambda \setminus 1$ be the set of all partitions μ obtained from λ by erasing one *corner* of λ. By the very construction of the standard Young tableaux and (1.1) we have

$$f_\lambda = \sum_{\mu \in \lambda \setminus 1} f_\mu \tag{1.2}$$

and then

$$\frac{n}{H_\lambda} = \sum_{\mu \in \lambda \setminus 1} \frac{1}{H_\mu}. \tag{1.3}$$

In this note we establish the following perturbation of formula (1.3). Define the *g-function* of a partition λ of n to be

$$g_\lambda(x) = \prod_{i=1}^{n} (x + \lambda_i - i), \tag{1.4}$$

where $\lambda_i = 0$ for $i \geq \ell(\lambda) + 1$.

Theorem 1.1 *Let x be a formal parameter. For each partition λ we have*

$$\frac{g_\lambda(x+1) - g_\lambda(x)}{H_\lambda} = \sum_{\mu \in \lambda \setminus 1} \frac{g_\mu(x)}{H_\mu}. \tag{1.5}$$

Theorem 1.1 is proved in Sect. 2. Some equivalent forms of Theorem 1.1 and remarks are given in Sect. 4. As an application we prove (see Sect. 3) the following result due to Stanley [8].

Theorem 1.2 *Let p, e and s be the usual symmetric functions [6, Chap. I]. Then*

$$\sum_{k=0}^{n} \binom{x+k-1}{k} p_1^k e_{n-k} = \sum_{\lambda \vdash n} H_\lambda^{-1} g_\lambda(x+n) s_\lambda. \tag{1.6}$$

Recently, the author stated some conjectures on partition hook lengths [3], which were suggested by hook length expansion techniques (see [4]). Later, Conjecture 3.1 in [3] was proved by Stanley [8]. One step of his proof is formula (1.6), based on a result by Andrews, Goulden and Jackson [1]. In this paper we provide a simple and direct proof of formula (1.6).

Remark Let D be the difference operator defined by

$$D(f(x)) = f(x+1) - f(x).$$

By iterating formula (1.5) we obtain

$$D^n \frac{g_\lambda(x)}{H_\lambda} = f_\lambda,$$

which is precisely the hook length formula (1.1).

2 Proof of Theorem 1.1

Let

$$\epsilon(x) = \frac{g_\lambda(x+1) - g_\lambda(x)}{H_\lambda} - \sum_{\mu \in \lambda \setminus 1} \frac{g_\mu(x)}{H_\mu}. \tag{2.1}$$

We see that $\epsilon(x)$ is a polynomial in x whose degree is less than or equal to n. Moreover

$$[x^n]\epsilon(x) = [x^n]\frac{g_\lambda(x+1) - g_\lambda(x)}{H_\lambda} = 0.$$

Furthermore,

$$[x^{n-1}]g_\lambda(x+1) = \sum_{i=1}^{n}(\lambda_i - i + 1) = n + \sum_{i=1}^{n}(\lambda_i - i) = n + [x^{n-1}]g_\lambda(x)$$

and

$$[x^{n-1}]\epsilon(x) = [x^{n-1}]\frac{g_\lambda(x+1) - g_\lambda(x)}{H_\lambda} - \sum_{\mu \in \lambda \setminus 1} \frac{1}{H_\mu} = \frac{n}{H_\lambda} - \sum_{\mu \in \lambda \setminus 1} \frac{1}{H_\mu} = 0.$$

The last equality is guaranteed by (1.3), so that $\epsilon(x)$ is a polynomial in x whose degree is less than and equal to $n - 2$. To prove that $\epsilon(x)$ is actually zero, it suffices to find $n - 1$ distinct values for x such that $\epsilon(x) = 0$. In the following we prove that $\epsilon(i - \lambda_i) = 0$ for $i - \lambda_i$ for $i = 1, 2, \ldots, n - 1$.

If $\lambda_i = \lambda_{i+1}$, or if the i-th row has no corner, the factor $x + \lambda_i - i$ lies in $g_\lambda(x)$ and also in $g_\mu(x)$ for all $\mu \in \lambda \setminus 1$. The factor $(x + 1) + \lambda_{i+1} - (i+1) = x + \lambda_i - i$ is furthermore in $g_\lambda(x + 1)$, so that $\epsilon(i - \lambda_i) = 0$.

Next, if $\lambda_i \geq \lambda_{i+1} + 1$, or if the i-th row has a corner, the factor $x + \lambda_i - i$ lies in $g_\lambda(x)$ and $g_\mu(x)$ for all $\mu \in \lambda \setminus 1$, except for $\mu = \lambda'$, which is the partition obtained from λ by erasing the corner from the i-th row. In this case equality (2.1) becomes

$$\epsilon(i - \lambda_i) = \frac{g_\lambda(i - \lambda_i + 1)}{H_\lambda} - \frac{g_{\lambda'}(i - \lambda_i)}{H_{\lambda'}}.$$

Fig. 1 Hook lengths of λ

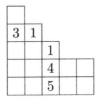

Fig. 2 Hook lengths of λ'

For proving Theorem 1.1, it remains to prove $\epsilon(i - \lambda_i) = 0$ or

$$\frac{H_\lambda}{H_{\lambda'}} = \frac{g_\lambda(i - \lambda_i + 1)}{g_{\lambda'}(i - \lambda_i)}. \tag{2.2}$$

Consider the following product

$$\frac{g_\lambda(x + 1)}{g_{\lambda'}(x)} = \frac{\prod_{j=1}^{n}(x + \lambda_j - j + 1)}{\prod_{j=1}^{n-1}(x + \lambda'_j - j)}. \tag{2.3}$$

The set of all $1 \le j \le n - 1$ such that $\lambda_j > \lambda_{j+1}$ is denoted by $\mathcal{T}$. For $1 \le j \le n - 1$ and $j \notin \mathcal{T}$ (which implies that $j \ne i$ and $\lambda'_j = \lambda_j = \lambda_{j+1}$), the numerator contains $x + \lambda_{j+1} - (j + 1) + 1 = x + \lambda_j - j$ and the denominator also contains $x + \lambda'_j - j = x + \lambda_j - j$. After cancellation of those common factors, (2.3) becomes

$$\frac{\frac{g_\lambda(x+1)}{g_{\lambda'}(x)} = \prod_{j \in \mathcal{B}}(x + \lambda_j - j + 1)}{\prod_{j \in \mathcal{T}}(x + \lambda'_j - j)} \tag{2.4}$$

where $\mathcal{B} = \{1\} \cup \{i + 1 \mid i \in \mathcal{T}\}$. Letting $x = i - \lambda_i$ in (2.4) yields

$$\frac{g_\lambda(i - \lambda_i + 1)}{g_{\lambda'}(i - \lambda_i)} = \frac{\prod_{j \in \mathcal{B}}(i - \lambda_i + \lambda_j - j + 1)}{\prod_{j \in \mathcal{T}}(i - \lambda_i + \lambda'_j - j)}. \tag{2.5}$$

We distinguish the factors in the right-hand side of (2.5) as follows.

(C1) For $j \in \mathcal{B}$ and $j > i$, $i - \lambda_i + \lambda_j - j + 1 = -(\lambda_i - \lambda_j + j - i - 1) = -h_v(\lambda)$, where v is the box $(i, \lambda_i + 1)$ in λ.
(C2) For $j \in \mathcal{B}$ and $j \le i$, $i - \lambda_i + \lambda_j - j + 1 = h_v(\lambda)$, where v is the box (j, λ_i) in λ.
(C3) For $j \in \mathcal{T}$ and $j > i$, $i - \lambda_i + \lambda_j - j = -(\lambda_i - \lambda_j + j - i) = -h_u(\lambda')$, where u is the box (i, λ_j) in λ'.
(C4) For $j \in \mathcal{T}$ and $j < i$, $i - \lambda_i + \lambda_j - j = h_u(\lambda')$, where u is the box (j, λ_i) in λ'.

Fig. 3 The boxes v in λ

Fig. 4 The boxes u in λ'

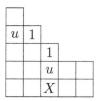

(C5) For $j \in \mathcal{T}$ and $j = i$, $i - \lambda_i + \lambda'_j - j = i - \lambda_i + \lambda'_i - i = -1$. See Figs. 3 and 4 for an example.

Since each $j \in \mathcal{B}$ such that $j > i$ is associated with $j - 1 \in \mathcal{T}$ and $j - 1 \geq i$, the right-hand side of (2.5) is positive and can be re-written

$$\frac{g_\lambda(i - \lambda_i + 1)}{g_{\lambda'}(i - \lambda_i)} = \frac{\prod_v h_v(\lambda)}{\prod_u h_u(\lambda')}, \qquad (2.6)$$

where v, u range over the boxes described in (C1)–(C4). Finally $H_\lambda / H_{\lambda'}$ is equal to the right-hand side of (2.6), since the hook lengths of all other boxes cancel. We have completed the proof of (2.2). □

For example, consider the partition $\lambda = 55331$ and $i = 4$. We have $\lambda' = 55321$ and

$$\frac{H_\lambda}{H_{\lambda'}} = \frac{4 \cdot 2 \cdot 1 \cdot 2 \cdot 5 \cdot 6}{3 \cdot 1 \cdot 1 \cdot 4 \cdot 5} = \frac{4 \cdot 2 \cdot 2 \cdot 6}{3 \cdot 4}.$$

On the other hand, $\mathcal{T} = \{2, 4, 5\}$, $\mathcal{B} = \{1, 3, 5, 6\}$ and

$$\frac{g_\lambda(x + 1)}{g_{\lambda'}(x)} = \frac{(x + 5)(x + 1)(x - 3)(x - 5)}{(x + 3)(x - 2)(x - 4)}.$$

Letting $x = i - \lambda_i = 4 - 3 = 1$ yields

$$\frac{g_\lambda(2)}{g_{\lambda'}(1)} = \frac{(6)(2)(-2)(-4)}{(4)(-1)(-3)} = \frac{6 \cdot 2 \cdot 2 \cdot 4}{4 \cdot 3}.$$

3 Proof of Theorem 1.2

Let $R_n(x)$ be the right-hand side of (1.6). By Theorem 1.1

$$R_n(x) = \sum_{\lambda \vdash n} \left(\frac{g_\lambda(x + n - 1)}{H_\lambda} + \sum_{\mu \in \lambda \backslash 1} \frac{g_\mu(x + n - 1)}{H_\mu} \right) s_\lambda$$

$$= R_n(x-1) + \sum_{\lambda \vdash n} \sum_{\mu \in \lambda \backslash 1} \frac{g_\mu(x+n-1)}{H_\mu} s_\lambda$$

$$= R_n(x-1) + \sum_{\mu \vdash n-1} \sum_{\lambda : \mu \in \lambda \backslash 1} \frac{g_\mu(x+n-1)}{H_\mu} s_\lambda$$

$$= R_n(x-1) + \sum_{\mu \vdash n-1} \frac{g_\mu(x+n-1)}{H_\mu} p_1 s_\mu,$$

where the next to last equality is

$$\sum_{\lambda : \mu \in \lambda \backslash 1} s_\lambda = p_1 s_\mu$$

by using Pieri's rule [6, p. 73]. We obtain the following recurrence for $R_n(x)$.

$$R_n(x) = R_n(x-1) + p_1 R_{n-1}(x). \tag{3.1}$$

Let $L_n(x)$ be the left-hand side of (1.6). Using elementary properties of binomial coefficients

$$L_n(x) = \sum_{k=0}^{n} \binom{x+k-1}{k} p_1^k e_{n-k}$$

$$= e_n + \sum_{k=1}^{n} \left(\binom{x+k-2}{k} + \binom{x+k-2}{k-1} \right) p_1^k e_{n-k}$$

$$= L_n(x-1) + p_1 \sum_{k=1}^{n} \binom{x+k-2}{k-1} p_1^{k-1} e_{n-k}$$

$$= L_n(x-1) + p_1 L_{n-1}(x). \tag{3.2}$$

We verify that $L_1(x) = R_1(x)$ and $L_n(0) = R_n(0)$, so that $L_n(x) = R_n(x)$ by 3.1 and (3.2). $\qquad\square$

4 Equivalent forms and further remarks

Let $\lambda = \lambda_1 \lambda_2 \cdots \lambda_\ell$ be a partition of n. The set of all $1 \le j \le n$ such that $\lambda_j > \lambda_{j+1}$ is denoted by $\mathcal{T}$ and let $\mathcal{B} = \{1\} \cup \{i+1 \mid i \in \mathcal{T}\}$. Those two sets can be viewed as the *in-corner* and *out-corner* index sets, respectively. Notice that $\#\mathcal{B} = \#\mathcal{T} + 1$. For each $i \in \mathcal{T}$ we define λ^{i-} to be the partition of $n-1$ obtained form λ by erasing the right-most box from the i-th row. Hence

$$\lambda \backslash 1 = \{\lambda^{i-} \mid i \in \mathcal{T}\}. \tag{4.1}$$

We verify that

$$g_{\lambda^{i-}}(x) = \frac{g_\lambda(x)(x + \lambda_i - i - 1)}{(x + \lambda_i - i)(x - n)}. \tag{4.2}$$

Fig. 5 In-corner

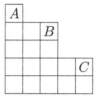

Fig. 6 Out-corner

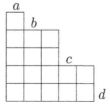

From Theorem 1.1

$$\frac{g_\lambda(x+1) - g_\lambda(x)}{H_\lambda} = \sum_{i \in \mathcal{T}} \frac{g_\lambda(x)(x + \lambda_i - i - 1)}{(x + \lambda_i - i)(x - n)} \frac{1}{H_\lambda^{i-}}$$

or

$$\sum_{i \in \mathcal{T}} \frac{H_\lambda}{H_{\lambda^{i-}}} \times \left(1 - \frac{1}{x + \lambda_i - i} \right) = n - x + \frac{(x - n)g_\lambda(x+1)}{g_\lambda(x)}. \tag{4.3}$$

Let us re-write (1.3)

$$\sum_{\mu \in \lambda \backslash 1} \frac{H_\lambda}{H_\mu} = n. \tag{4.4}$$

By subtracting (4.3) from (4.4) we obtain the following equivalent form of Theorem 1.1.

Theorem 4.1 *We have*

$$\sum_{i \in \mathcal{T}} \frac{H_\lambda}{H_{\lambda^{i-}}} \times \frac{1}{x + \lambda_i - i} = x - \frac{(x - n)g_\lambda(x+1)}{g_\lambda(x)}. \tag{4.5}$$

By the definitions of $\mathcal{T}$ and $\mathcal{B}$ we have

$$\frac{(x - n)g_\lambda(x+1)}{g_\lambda(x)} = \frac{\prod_{i \in \mathcal{B}}(x + \lambda_i - i + 1)}{\prod_{i \in \mathcal{T}}(x + \lambda_i - i)}, \tag{4.6}$$

so that Theorem 1.1 is also equivalent to the following result.

Theorem 4.2 *We have*

$$\sum_{i \in \mathcal{T}} \frac{H_\lambda}{H_{\lambda^{i-}}} \times \frac{1}{x + \lambda_i - i} = x - \frac{\prod_{i \in \mathcal{B}}(x + \lambda_i - i + 1)}{\prod_{i \in \mathcal{T}}(x + \lambda_i - i)}. \tag{4.7}$$

For example, take $\lambda = 55331$. Then $\mathcal{T} = 2, 4, 5$ and $\mathcal{B} = 1, 3, 5, 6 = \{1, 2+1, 4+1, 5+1\}$.

Hence $\lambda^{2-} = 54331$, $\lambda^{4-} = 55321$ and $\lambda^{5-} = 55330$. Equality (4.7) becomes

$$\frac{H_\lambda}{H_{\lambda^{5-}}} \times \frac{1}{x-4} + \frac{H_\lambda}{H_{\lambda^{4-}}} \times \frac{1}{x-1} + \frac{H_\lambda}{H_{\lambda^{2-}}} \times \frac{1}{x+3}$$
$$= x - \frac{(x-5)(x-3)(x+1)(x+5)}{(x-4)(x-1)(x+3)}.$$
$$= \frac{17x^2 - 38x - 75}{(x-4)(x-1)(x+3)}.$$

Theorems 4.1 and 4.2 can be proved directly using the method used in the proof of Theorem 1.1. First, we must verify that the numerator in the right-hand side of (4.5) is a polynomial in x whose degree is less than ($\leq$) #$\mathcal{T}$ − 1. By the partial fraction expansion technique it suffices to verify that (4.7) is true for all $x = i - \lambda_i$ $(i \in \mathcal{T})$. This direct proof contains the main part of the proof of Theorem 1.1. However it does not make use of the fundamental relation (1.3) or (4.4). Thus, the following corollary of Theorem 4.2 makes sense.

Corollary 4.3 *We have*

$$\sum_{\mu \in \lambda \backslash 1} \frac{H_\lambda}{H_\mu} = n. \tag{4.8}$$

Proof Let #$\mathcal{T} = k$. The right-hand side of (4.7) has the following form

$$\frac{Cx^{k-1} + \cdots}{x^k + \cdots}.$$

We now evaluate the coefficient C. By (4.6) we can write $C = A - B$ with

$$A = [x^{n-1}]x \prod_{i=1}^n (x + \lambda_i - i) = \sum_{1 \leq i < j \leq n} (\lambda_i - i)(\lambda_j - j)$$

and

$$B = [x^{n-1}](x - n) \prod_{i=1}^n (x + \lambda_i - i + 1)$$
$$= \sum_{1 \leq i < j \leq n} (\lambda_i - i + 1)(\lambda_j - j + 1) - n \sum_{1 \leq i \leq n} (\lambda_i - i + 1).$$
$$= B_1 - n \sum_{1 \leq i \leq n} (\lambda_i - i + 1),$$

where

$$B_1 = \sum_{1 \le i < j \le n} (\lambda_i - i + 1)(\lambda_j - j + 1)$$

$$= \sum_{1 \le i < j \le n} \left((\lambda_i - i)(\lambda_j - j) + (\lambda_i - i) + (\lambda_j - j) + 1 \right)$$

$$= A + \sum_{1 \le i < j \le n} (\lambda_i - i) + \sum_{1 \le i < j \le n} (\lambda_j - j) + \binom{n}{2}$$

$$= A + \sum_{1 \le i \le n} (n - i)(\lambda_i - i) + \sum_{1 \le j \le n} (j - 1)(\lambda_j - j) + \binom{n}{2}$$

$$= A + \sum_{1 \le i \le n} (n - 1)(\lambda_i - i) + \binom{n}{2}.$$

Finally

$$C = A - B$$

$$= -\sum_{1 \le i \le n} (n - 1)(\lambda_i - i) - \binom{n}{2} + n \sum_{1 \le i \le n} (\lambda_i - i + 1)$$

$$= -\sum_{1 \le i \le n} n(\lambda_i - i) + \sum_{1 \le i \le n} (\lambda_i - i) - \binom{n}{2} + n \sum_{1 \le i \le n} (\lambda_i - i) + n^2$$

$$= \sum_{1 \le i \le n} (\lambda_i - i) - \binom{n}{2} + n^2$$

$$= n - \binom{n + 1}{2} - \binom{n}{2} + n^2 = n. \qquad \square$$

References

1. Andrews, G., Goulden, I., Jackson, D.M.: Generalizations of Cauchy's summation formula for Schur functions. Trans. Am. Math. Soc. **310**, 805–820 (1988)
2. Frame, J., Robinson, S., de Beauregard, G., Thrall, R.M.: The hook graphs of the symmetric groups. Can. J. Math. **6**, 316–324 (1954)
3. Han, G.-N.: Some conjectures and open problems on partition hook lengths. Exp. Math. **18**, 97–106 (2009)
4. Han, G.-N.: Discovering hook length formulas by an expansion technique. Elec. J. Combin. **15**(1), R133 (2008) 41 pp.
5. Lascoux, A.: Symmetric Functions and Combinatorial Operators on Polynomials. CBMS Regional Conference Series in Mathematics, vol. 99 (2001)
6. Macdonald, I.G.: Symmetric Functions and Hall Polynomials, 2nd edn. Clarendon Press, Oxford (1995)
7. Stanley, R.P.: Enumerative Combinatorics, vol. 2. Cambridge University Press, Cambridge (1999)
8. Stanley, R.P.: Some combinatorial properties of hook lengths, contents, and parts of partitions. arXiv:0807.0383 [math.CO] (2008), 18 pp.

Ramanujan J (2010) 23: 137–149
DOI 10.1007/s11139-008-9156-7

A unification of two refinements of Euler's partition theorem

William Y.C. Chen · Henry Y. Gao · Kathy Q. Ji · Martin Y.X. Li

Dedicated to Professor George Andrews for his seventieth birthday

Received: 14 December 2008 / Accepted: 19 December 2008 / Published online: 6 January 2010
© Springer Science+Business Media, LLC 2009

Abstract We obtain a unification of two refinements of Euler's partition theorem respectively due to Bessenrodt and Glaisher. A specialization of Bessenrodt's insertion algorithm for a generalization of the Andrews-Olsson partition identity is used in our combinatorial construction.

Keywords Partition · Euler's partition theorem · Refinement · Bijection · Bessenrodt's bijection · Andrews-Olsson's theorem

Mathematics Subject Classification (2000) 05A17 · 11P81

1 Introduction

There are several bijective proofs and refinements of the classical partition theorem of Euler. This paper will be concerned with two remarkable bijections obtained by Sylvester [19] and Glaisher [16], see also, [6, pp. 8–9]. Glaisher's bijection implies a refinement of Euler's theorem involving the number of odd parts in a partition with

This work was supported by the 973 Project, the PCSIRT Project of the Ministry of Education, the Ministry of Science and Technology, and the National Science Foundation of China.

W.Y.C. Chen (✉) · H.Y. Gao · K.Q. Ji · M.Y.X. Li
Center for Combinatorics, LPMC-TJKLC, Nankai University, Tianjin 300071,
People's Republic of China
e-mail: chen@nankai.edu.cn

H.Y. Gao
e-mail: gaoyong@cfc.nankai.edu.cn

K.Q. Ji
e-mail: ji@nankai.edu.cn

M.Y.X. Li
e-mail: lyz6988@yahoo.com.cn

distinct parts and the number of parts repeated odd times in a partition with odd parts. On the other hand, as observed by Bessenrodt [9], Sylvester's bijection also leads to a refinement of Euler's theorem. The main result of this paper is a unification of these two refinements that does not directly follow from Sylvester's bijection and Glaisher's bijection.

Let us give an overview of the background and terminology. We will adopt the common notation on partitions used in Andrews [3, Chap. 1]. *A partition* λ of a positive integer n is a finite nonincreasing sequence of positive integers

$$\lambda = (\lambda_1, \lambda_2, \ldots, \lambda_r)$$

such that $\sum_{i=1}^{r} \lambda_i = n$. The entries λ_i are called the parts of λ, and λ_1 is the largest part. The number of parts of λ is called the length of λ, denoted by $l(\lambda)$. The weight of λ is the sum of its parts, denoted by $|\lambda|$. A partition λ can also be represented in the following form

$$\lambda = (1^{m_1}, 2^{m_2}, 3^{m_3}, \ldots),$$

where m_i is the multiplicity of the part i in λ. The conjugate partition of λ is defined by $\lambda' = (\lambda_1', \lambda_2', \ldots, \lambda_t')$, where λ_i' is the number of parts of λ that are greater than or equal to i.

Euler's partition theorem reads as follows.

Theorem 1.1 (Euler) *The number of partitions of n with distinct parts is equal to the number of partitions of n with odd parts.*

Let $\mathcal{D}$ denote the set of partitions with distinct parts, and let $\mathcal{D}(n)$ denote the set of partitions of n in $\mathcal{D}$. Similarly, let $\mathcal{O}$ denote the set of partitions with odd parts, and let $\mathcal{O}(n)$ denote the set of partitions of n in $\mathcal{O}$. Sylvester's fish-hook bijection [19], also referred to as Sylvester's bijection, and Glaisher's bijection [6, pp. 8–9] have established direct correspondences between $\mathcal{D}(n)$ and $\mathcal{O}(n)$. These two bijections imply refinements of Euler's theorem. There are also several other refinements of Euler's partition theorem, see, for example, [1, 2, 4, 9, 17, 18, 21], [14, pp. 51–52], [15, pp. 46–47].

Sylvester's refinement [3, p. 24] is stated as follows. Recall that a chain in a partition with distinct parts is a maximal sequence of parts consisting of consecutive integers. The number of chains in a partition λ is denoted by $n_c(\lambda)$. The number of different parts in a partition μ is denoted by $n_d(\mu)$. For example, the partition $(8, 7, 5, 3, 2, 1)$ has three chains, and the partition $(8, 6, 6, 5, 4, 4, 2, 1)$ has six different parts.

Theorem 1.2 (Sylvester) *The number of partitions of n into distinct parts with exactly k chains is equal to the number of partitions of n into odd parts (repetitions allowed) with exactly k different parts. In the notation of generating functions, we have*

$$\sum_{\lambda \in \mathcal{D}} z^{n_c(\lambda)} q^{|\lambda|} = \sum_{\mu \in \mathcal{O}} z^{n_d(\mu)} q^{|\mu|}. \tag{1.1}$$

Fine [15, pp. 46–47] has derived a refinement of Euler's theorem.

Theorem 1.3 (Fine) *The number of partitions of n into distinct parts with largest part k is equal to the number of partitions of n into odd parts such that the largest part plus twice the number of parts equals $2k + 1$. In the notation of generating functions, we have*

$$\sum_{\lambda \in \mathcal{D}} x^{\lambda_1} q^{|\lambda|} = \sum_{\mu \in \mathcal{O}} x^{(\mu_1 - 1)/2 + l(\mu)} q^{|\mu|}. \tag{1.2}$$

Bessenrodt [9] has shown that Sylvester's bijection implies the following refinement, which is a limiting case of the lecture hall theorem due to Bousquet-Mélou and Erikssonin [12, 13]. Let $l_a(\lambda)$ denote the alternating sum of λ, namely,

$$l_a(\lambda) = \lambda_1 - \lambda_2 + \lambda_3 - \lambda_4 + \cdots.$$

Theorem 1.4 (Bessenrodt) *The number of partitions of n into distinct parts with alternating sum l is equal to the number of partitions of n with l odd parts. In terms of generating functions, we have*

$$\sum_{\lambda \in \mathcal{D}} y^{l_a(\lambda)} q^{|\lambda|} = \sum_{\mu \in \mathcal{O}} y^{l(\mu)} q^{|\mu|}. \tag{1.3}$$

It has also been shown by Bessenrodt [9] that Sylvester's bijection maps the parameter $n_c(\lambda)$ to the parameter $n_d(\mu)$. Combining the above Theorems 1.2 and 1.3, we arrive at the following equidistribution result.

Theorem 1.5 (Sylvester-Bessenrodt) *The number of partitions of n into distinct parts with largest part k, alternating sum l and m chains is equal to the number of partitions of n into l odd parts with exactly m different parts such that the largest part plus twice the number of parts equals $2k + 1$. In terms of generating functions, we have*

$$\sum_{\lambda \in \mathcal{D}} x^{\lambda_1} y^{l_a(\lambda)} z^{n_c(\lambda)} q^{|\lambda|} = \sum_{\mu \in \mathcal{O}} x^{(\mu_1 - 1)/2 + l(\mu)} y^{l(\mu)} z^{n_d(\mu)} q^{|\mu|}. \tag{1.4}$$

Recently, Zeng [21] has found a generating function proof of the above three-parameter refinement (1.4).

From a different angle, Glaisher [16], see also [6, pp. 8–9], has given a refinement of Euler's partition theorem. Let $l_o(\lambda)$ denote the number of odd parts in λ, and let $n_o(\mu)$ denote the number of different parts in μ with odd multiplicities.

Theorem 1.6 (Glaisher) *The number of partitions of n into distinct parts with k odd parts is equal to the number of partitions of n with odd parts such that there are exactly k different parts repeated odd times. In terms of generating functions, we have*

$$\sum_{\lambda \in \mathcal{D}} x^{l_o(\lambda)} q^{|\lambda|} = \sum_{\mu \in \mathcal{O}} x^{n_o(\mu)} q^{|\mu|}. \tag{1.5}$$

Given the two bijections of Sylvester and Glaisher, it is natural to ask the question whether the joint distribution of the statistics $(l_o(\lambda), l_a(\lambda))$ of partitions of n with distinct parts coincides with the joint distribution of the statistics $(n_o(\mu), l(\mu))$ of partitions with odd parts. It turns out that this is indeed the case. However, neither Sylvester's bijection nor Glaisher's bijection implies this result. To give a combinatorial proof of this result, we need Bessenrodt's insertion algorithm.

It should be noted that the equidistribution of $(l_o(\lambda), l_a(\lambda))$ and $(n_o(\mu), l(\mu))$ can also be deduced from a recent result of Boulet [11] by the manipulation of generating functions.

This paper is organized as follows. In Sect. 2, we present the main result and some lemmas. Section 3 is devoted to a brief review of Bessenrodt's insertion algorithm. In Sect. 4, we utilize Boulet's formula to give a generating function proof of the two-parameter refinement of Euler's theorem. In Sect. 5, we give a combinatorial proof of the unification of the refinements of Bessenrodt (1.3) and Glaisher (1.5).

2 The main result

The main result of this paper is the following unification of the refinements of Bessenrodt and Glaisher.

Theorem 2.1 *The number of partitions of n into distinct parts with l odd parts and alternating sum m is equal to the number of partitions of n into exactly m odd parts and l parts repeated odd times. In terms of generating functions, we have*

$$\sum_{\lambda \in \mathcal{D}} x^{l_o(\lambda)} y^{l_a(\lambda)} q^{|\lambda|} = \sum_{\mu \in \mathcal{O}} x^{n_o(\mu)} y^{l(\mu)} q^{|\mu|}. \tag{2.1}$$

For example, Table 1 illustrates the case of $n = 7$.

It is clear that the above theorem reduces to Bessenrodt's refinement (1.3) when $x = 1$ and to Glaisher's refinement (1.5) when $y = 1$.

To prove Theorem 2.1, we proceed to construct a bijection Δ between $\mathcal{D}(n)$ and $\mathcal{O}(n)$ such that for $\lambda \in \mathcal{D}(n)$ and $\mu = \Delta(\lambda) \in \mathcal{O}(n)$, we have

$$l_o(\lambda) = n_o(\mu), \qquad l_a(\lambda) = l(\mu).$$

Let $\mathcal{A}_1(n)$ denote the set of partitions of n subject to the following conditions:

Table 1 The case of $n = 7$ for Theorem 2.1	$\lambda \in \mathcal{D}(7)$	$l_o(\lambda)$	$l_a(\lambda)$	$\mu \in \mathcal{O}(7)$	$n_o(\mu)$	$l(\mu)$
	(7)	1	7	(1^7)	1	7
	$(1, 6)$	1	5	$(1^4, 3)$	1	5
	$(2, 5)$	1	3	$(1, 3^2)$	1	3
	$(3, 4)$	1	1	(7)	1	1
	$(1, 2, 4)$	1	3	$(1^2, 5)$	1	3

1. Only parts divisible by 2 may be repeated.
2. The difference between successive parts is at most 4 and strictly less than 4 if either part is divisible by 2.
3. The smallest part is less than 4.

By considering the conjugate of the 2-modular representation of a partition, it is easy to establish a bijection between $\mathcal{D}(n)$ and $\mathcal{A}_1(n)$.

Lemma 2.2 *There is a bijection φ between $\mathcal{D}(n)$ and $\mathcal{A}_1(n)$. Furthermore, for $\lambda \in \mathcal{D}(n)$ and $\alpha = \varphi(\lambda) \in \mathcal{A}_1(n)$, we have*

$$l_o(\lambda) = l_o(\alpha), \qquad l_a(\lambda) = 2r_2(\alpha) + l_o(\alpha), \tag{2.2}$$

where $r_2(\alpha)$ denotes the number of parts congruent to 2 modulo 4 in α.

Let $\mathcal{A}_2(n)$ denote the set of partitions of n subject to the following conditions:

1. No part divisible by 4.
2. Only parts divisible by 2 may be repeated.

We then establish a bijection between $\mathcal{O}(n)$ and $\mathcal{A}_2(n)$ in the spirit of Glaisher's bijection.

Lemma 2.3 *There is a bijection ψ between $\mathcal{O}(n)$ and $\mathcal{A}_2(n)$. Furthermore, for $\mu \in \mathcal{O}(n)$ and $\beta = \psi(\mu) \in \mathcal{A}_2(n)$, we have*

$$n_o(\mu) = l_o(\beta), \qquad l(\mu) = 2r_2(\beta) + l_o(\beta). \tag{2.3}$$

In view of the above two lemmas, we see that Theorem 2.1 can be deduced from the following theorem.

Theorem 2.4 *There is a bijection ϕ between $\mathcal{A}_1(n)$ and $\mathcal{A}_2(n)$. Furthermore, for $\alpha \in \mathcal{A}_1(n)$ and $\beta = \phi(\alpha) \in \mathcal{A}_2(n)$, we have*

$$l_o(\alpha) = l_o(\beta), \qquad r_2(\alpha) = r_2(\beta). \tag{2.4}$$

We find that Theorem 2.4 can be deduced from Bessenrodt's insertion algorithm which was devised as a combinatorial proof of a generalization of Andrews-Olsson's theorem [10]. Combining the bijection φ for Lemma 2.2, ψ for Lemma 2.3 and ϕ for Theorem 2.4, we are led to a new bijection Δ for Euler's partition theorem which implies the equidistribution of the statistics $(l_o(\lambda), l_a(\lambda))$ of partitions with distinct parts and the statistics $(n_o(\mu), l(\mu))$ of partitions with odd parts.

3 Bessenrodt's insertion algorithm

To provide a purely combinatorial proof of Andrews-Olsson's theorem [7], Bessenrodt [8] constructs an explicit bijection on the sets of partitions in Andrews-Olsson's theorem, which we call Bessenrodt's insertion algorithm. The original insertion algorithm does not imply the bijection in Theorem 2.4, but we find that the generalized insertion algorithm given by Bessenrodt [10] in 1995 can be used to establish the bijection required by Theorem 2.4.

We give an overview of Bessenrodt's insertion algorithm. Let N be an integer, and let $\mathbb{A}_N = \{a_1, a_2, \ldots, a_r\}$ with $1 \leq a_1 < a_2 < \cdots < a_r < N$. Andrews-Olsson's theorem involves two sets $\mathcal{AO}_1(\mathbb{A}_N; n, N)$ and $\mathcal{AO}_2(\mathbb{A}_N; n, N)$ defined below.

Definition 3.1 Let $\mathcal{AO}_1(\mathbb{A}_N; n, N)$ denote the set of partitions of n satisfying the following conditions:

1. Each part is congruent to 0 or some a_i modulo N;
2. Only the multiples of N can be repeated;
3. The difference between two successive parts is at most N and strictly less than N if either part is divisible by N;
4. The smallest part is less than N.

Definition 3.2 Let $\mathcal{AO}_2(\mathbb{A}_N; n, N)$ denote the set of partitions of n satisfying the following conditions:

1. Each part is congruent to some a_i modulo N;
2. No part can be repeated.

The cardinalities of $\mathcal{AO}_1(\mathbb{A}_N; n, N)$ and $\mathcal{AO}_2(\mathbb{A}_N; n, N)$ are denoted by $p_1(\mathbb{A}_N; n, N)$ and $p_2(\mathbb{A}_N; n, N)$ respectively. Andrews-Olsson's theorem is stated as follows.

Theorem 3.3 (Andrews-Olsson) *For any $n \in \mathbb{N}$, we have*

$$p_1(\mathbb{A}_N; n, N) = p_2(\mathbb{A}_N; n, N).$$

By examining the two sets $\mathcal{A}_1(n)$ and $\mathcal{A}_2(n)$ in Theorem 2.4, we find they are somehow analogous to the two sets $\mathcal{AO}_1(\mathbb{A}_N; n, N)$ and $\mathcal{AO}_2(\mathbb{A}_N; n, N)$ in Andrews-Olsson's theorem. Moreover, we could also apply Bessenrodt's insertion algorithm to establish a bijection between $\mathcal{A}_1(n)$ and $\mathcal{A}_2(n)$. Here we present a more general bijection Φ between the two sets $\mathcal{C}_1(\mathbb{A}_{2N}; n, 2N)$ and $\mathcal{C}_2(\mathbb{A}_{2N}; n, 2N)$, and we can restrict the bijection Φ to $\mathcal{A}_1(n)$ and $\mathcal{A}_2(n)$ by setting $N = 2$ and $\mathbb{A}_4 = \{1, 2, 3\}$.

Definition 3.4 Let $\mathcal{C}_1(\mathbb{A}_{2N}; n, 2N)$ denote the set of partitions of n satisfying the following conditions:

1. Each part is congruent to 0 or some a_i modulo $2N$;
2. Only the multiples of N can be repeated;
3. The difference between two successive parts is at most $2N$ and strictly less than $2N$ if either part is divisible by N;
4. The smallest part is less than $2N$.

Definition 3.5 Let $\mathcal{C}_2(\mathbb{A}_{2N}; n, 2N)$ denote the set of partitions of n satisfying the following conditions:

1. Each part is congruent to some a_i modulo $2N$;
2. Only multiples of N may be repeated.

The cardinalities of $\mathcal{C}_1(\mathbb{A}_{2N}; n, 2N)$ and $\mathcal{C}_2(\mathbb{A}_{2N}; n, 2N)$ are denoted by $c_1(\mathbb{A}_{2N}; n, 2N)$ and $c_2(\mathbb{A}_{2N}; n, 2N)$ respectively. Then we have the following theorem which will be needed to prove Theorem 2.4.

Theorem 3.6 *For any $n \in \mathbb{N}$, we have*

$$c_1(\mathbb{A}_{2N}; n, 2N) = c_2(\mathbb{A}_{2N}; n, 2N).$$

Theorem 3.6 can be proved either by a variant of Bessenrodt's insertion algorithm obtained in 1991, or by specializing a generalization of Bessenrodt's algorithm obtained in 1995.

We outline the first approach by constructing a bijection Φ between $\mathcal{C}_1(\mathbb{A}_{2N}; n, 2N)$ and $\mathcal{C}_2(\mathbb{A}_{2N}; n, 2N)$ based on a variant of Bessenrodt's insertion algorithm [8].

For $\lambda \in \mathcal{C}_1(\mathbb{A}_{2N}; n, 2N)$, we first extract some parts from λ to form a pair of partitions (α, β), where $\alpha \in \mathcal{C}_1(\mathbb{A}_{2N}; n, 2N) \cap \mathcal{C}_2(\mathbb{A}_{2N}; n, 2N)$ and β is a partition with parts divisible by $2N$. Then we insert β into α to get a partition $\gamma \in \mathcal{C}_2(\mathbb{A}_{2N}; n, 2N)$.

The bijection Φ consists of the following two steps.

Step 1: Extract certain parts from $\lambda = (\lambda_1, \lambda_2, \ldots, \lambda_{l(\lambda)}) \in \mathcal{C}_1(\mathbb{A}_{2N}; n, 2N)$.

We now construct a pair of partitions (α, β) based on the partition λ. Let λ_j be a part divisible by $2N$, and λ_t be the smallest part bigger than λ_j. We remove λ_j if λ_t does not exist or the difference between λ_t and λ_{j+1} satisfies the difference condition for $\mathcal{C}_1(\mathbb{A}_{2N}; n, 2N)$. After removing these parts λ_j, we obtain a partition α^1 in $\mathcal{C}_1(\mathbb{A}_{2N}; n, 2N)$, and we can rearrange these parts that have been removed to form partition β^1.

Assume that there are l parts divisible by $2N$ in α^1. Let $t = 1$. We may iterate the following procedure until we get a pair of partitions $(\alpha^{l+1}, \beta^{l+1})$.

- Let α_i^t be the largest part divisible by $2N$ in α^t.
- Subtract $2N$ from $\alpha_1^t, \alpha_2^t, \ldots, \alpha_{i-1}^t$ and remove α_i^t from α^t.
- Rearrange the remaining parts to give a new partition α^{t+1} and add one part $(i - 1) \cdot 2N + \alpha_i^t$ to β^t to get β^{t+1}.

Then let $\alpha = \alpha^{l+1}$, $\beta = \beta^{l+1}$. It can be seen that $\alpha \in \mathcal{C}_1(\mathbb{A}_{2N}; n, 2N) \cap \mathcal{C}_2(\mathbb{A}_{2N}; n, 2N)$ and $\beta_1 \leq 2N \cdot l(\alpha)$.

Step 2: Insert β into α to generate a partition $\gamma \in \mathcal{C}_2(\mathbb{A}_{2N}; n, 2N)$.

For each β_i, we add $2N$ to the first $\beta_i/2N$ parts of α: $\alpha_1, \alpha_2, \ldots, \alpha_{\beta_i/2N}$, then denote the resulted partition by γ. It can be shown that $\gamma \in \mathcal{C}_2(\mathbb{A}_{2N}; n, 2N)$ for $\beta_1 \leq 2N \cdot l(\alpha)$. For the details of the proof, see [8, 20].

The inverse map Φ^{-1} can be described as follows. For $\gamma \in \mathcal{C}_2(\mathbb{A}_{2N}; n, 2N)$, we extract certain parts from γ to get a pair of partitions (α, β), where $\alpha \in \mathcal{C}_1(\mathbb{A}_{2N}; n, 2N) \cap \mathcal{C}_2(\mathbb{A}_{2N}; n, 2N)$ and β is a partition with parts divisible by $2N$. Then we insert β into α to form a partition $\lambda \in \mathcal{C}_1(\mathbb{A}_{2N}; n, 2N)$.

Formally speaking, the inverse map Φ^{-1} consists of the following two steps.

Step 1: Extraction of parts from γ.

Suppose $\gamma \in C_2(\mathbb{A}_{2N}; n, 2N)$. Let $\alpha = \gamma$, $\beta = \emptyset$ and $t = l(\gamma)$. We can obtain a pair of partitions (α, β) by the following procedure:

- If α_t is divisible by N, then there exists an integer i such that $\alpha_t - \alpha_{t+1} = i \cdot 2N + r_t$, where $0 \leq r_t < 2N$;
- If α_t is not divisible by N, then there exists an integer i such that $\alpha_t - \alpha_{t+1} = i \cdot 2N + r_t$, where $0 < r_t \leq 2N$;
- Subtract $i \cdot 2N$ from the parts $\alpha_1, \alpha_2, \ldots, \alpha_t$; Rearrange these parts to generate a new partition α and add i parts of size $t \cdot 2N$ to β.
- If $t \geq 2$, then replace t by $t - 1$ and repeat the above procedure. If $t = 1$, we get a pair of partitions (α, β).

Step 2: Insert β into α.

Assume that (α, β) is a pair of partitions such that $\alpha \in C_1(\mathbb{A}_{2N}; n, 2N) \cap C_2(\mathbb{A}_{2N}; n, 2N)$ and β is a partition with parts divisible by $2N$. We can construct a partition $\lambda \in C_1(\mathbb{A}_{2N}; n, 2N)$. If $\beta_1 \leq \alpha_1 + 2N - 1$, we insert all the parts of β into α to generate a new partition λ.

If $\beta_1 > \alpha_1 + 2N - 1$, we set $t = 1$ initially and iterate the following procedure until $\beta_t \leq \alpha_1 + 2N - 1$:

- Let i be the largest positive integer such that $\beta_t - i \cdot 2N \geq \alpha_i$, namely for $j > i$ we have $\beta_t - j \cdot 2N < \alpha_j$.
- Add $2N$ to the first i parts $\alpha_1, \alpha_2, \ldots, \alpha_i$, then insert $\beta_t - i \cdot 2N$ into α in the position before the part α_{i+1}.
- Rearrange the resulted parts to form a new partition α and replace t by $t + 1$.

Finally, we arrive at the condition $\beta_t \leq \alpha_1 + 2N - 1$. Then we insert all the remaining parts of β into α to generate a new partition λ. It can be shown that $\lambda \in C_1(\mathbb{A}_{2N}; n, 2N)$. For the details of the proof, see [8, 20]. □

We now turn to the generalization of Bessenrodt's insertion algorithm and we will show how one can derive Theorem 3.6 from this generalized algorithm. Let $\mathbb{A}_N = \mathbb{A}'_N \cup \mathbb{A}''_N$ with $\mathbb{A}'_N \cap \mathbb{A}''_N = \emptyset$.

Definition 3.7 Let $\mathcal{B}_1(\mathbb{A}'_N, \mathbb{A}''_N; n, N)$ denote the set of partitions of n satisfying the following conditions:

1. Each part is congruent to 0 or some a_i modulo N;
2. Only the part congruent to 0 or some a_i belonging to $\mathbb{A}'_N$ modulo N can be repeated;
3. The difference between two successive parts is at most N and strictly less than N if either part congruent to 0 or some a_i belonging to $\mathbb{A}'_N$ modulo N;
4. The smallest part is less than N.

Definition 3.8 Let $\mathcal{B}_2(\mathbb{A}'_N, \mathbb{A}''_N; n, N)$ denote the set of partitions of n satisfying the following conditions:

1. Each part is congruent to some a_i modulo N;
2. Only part congruent to some a_i belonging to $\mathbb{A}'_N$ modulo N can be repeated.

The cardinalities of $\mathcal{B}_1(\mathbb{A}'_N, \mathbb{A}''_N; n)$ and $\mathcal{B}_2(\mathbb{A}'_N, \mathbb{A}''_N; n)$ are denoted by $b_1(\mathbb{A}'_N, \mathbb{A}''_N; n)$ and $b_2(\mathbb{A}'_N, \mathbb{A}''_N; n, N)$ respectively.

Bessenrodt's generalization of the Andrews-Olsson theorem is stated as follows.

Theorem 3.9 (Bessenrodt) *For any $n \in \mathbb{N}$, we have*

$$b_1(\mathbb{A}'_N, \mathbb{A}''_N; n, N) = b_2(\mathbb{A}'_N, \mathbb{A}''_N; n, N).$$

Clearly, Andrews-Olsson's Theorem 3.3 can be viewed as the special case $\mathbb{A}'_N = \emptyset$ of Theorem 3.9. Theorem 3.6 is the special case for $2N$ and $\mathbb{A}'_{2N} = \{N\}$. As noted by Bessenrodt [10], the special case $N = 2$, $\mathbb{A}'_2 = \{1\}$ and $\mathbb{A}''_2 = \emptyset$, reduces to Euler's partition theorem, and Bessenrodt's insertion algorithm for this case coincides with Sylvester's bijection.

4 Connection to Boulet's formula

In this section, we show that our two-parameter refinement (2.1) can be derived from a formula of Boulet. The following four-parameter weight was introduced by Boulet [11] as a generalization of the weight defined by Andrews [5]. Let a, b, c and d be commuting indeterminants. Define the following weight function $\omega(\lambda)$ on the set of partitions:

$$\omega(\lambda) = a^{\sum_{i \geq 1} \lceil \lambda_{2i-1}/2 \rceil} b^{\sum_{i \geq 1} \lfloor \lambda_{2i-1}/2 \rfloor} c^{\sum_{i \geq 1} \lceil \lambda_{2i}/2 \rceil} d^{\sum_{i \geq 1} \lfloor \lambda_{2i}/2 \rfloor},$$

where $\lceil x \rceil$ (resp. $\lfloor x \rfloor$) stands for the smallest (resp. largest) integer greater (resp. less) than or equal to x for a given real number x. Boulet obtained the following formula:

$$\sum_{\lambda \in P} \omega(\lambda) = \prod_{j=1}^{\infty} \frac{(1 + a^j b^{j-1} c^{j-1} d^{j-1})(1 + a^j b^j c^j d^{j-1})}{(1 - a^j b^j c^j d^j)(1 - a^j b^j c^{j-1} d^{j-1})(1 - a^j b^{j-1} c^j d^{j-1})}, \quad (4.1)$$

where P denotes the set of integer partitions. It can be easily checked that the generating function of partitions in which every part appears an even number of times is

$$\prod_{j=1}^{\infty} \frac{1}{(1 - a^j b^j c^j d^j)(1 - a^j b^{j-1} c^j d^{j-1})}.$$

From (4.1), Boulet deduced the generating function for the weight function $\omega(\lambda)$ when λ runs over partitions with distinct parts ([11, Corollary 2]):

$$\sum_{\lambda \in \mathcal{D}} \omega(\lambda) = \prod_{j=1}^{\infty} \frac{(1 + a^j b^{j-1} c^{j-1} d^{j-1})(1 + a^j b^j c^j d^{j-1})}{(1 - a^j b^j c^{j-1} d^{j-1})}. \quad (4.2)$$

Making the substitutions $a \mapsto xyq, b \mapsto x^{-1}yq, c \mapsto xy^{-1}q, d \mapsto x^{-1}y^{-1}q$ in (4.1), Boulet derived the following identity due to Andrews [5].

Theorem 4.1 (Andrews) *We have*

$$\sum_{\lambda \in P} x^{l_o(\lambda)} y^{l_o(\lambda')} q^{|\lambda|} = \prod_{j=1}^{\infty} \frac{(1 + xyq^{2j-1})}{(1 - q^{4j})(1 - x^2 q^{4j-2})(1 - y^2 q^{4j-2})}.$$

Using the same substitution in (4.2), we find obtain the following formula for partitions with distinct parts.

Theorem 4.2 *We have*

$$\sum_{\lambda \in D} x^{l_o(\lambda)} y^{l_o(\lambda')} q^{|\lambda|} = \prod_{j=1}^{\infty} \frac{1 + xyq^{2j-1}}{1 - y^2 q^{4j-2}}. \tag{4.3}$$

On the other hand, it is easy to derive the following generating function.

Theorem 4.3 *We have*

$$\sum_{\lambda \in \mathcal{O}} x^{n_o(\lambda)} y^{l(\lambda)} q^{|\lambda|} = \prod_{j=1}^{\infty} \frac{1 + xyq^{2j-1}}{1 - y^2 q^{4j-2}}. \tag{4.4}$$

Proof We have

$$\prod_{j=1}^{\infty} \frac{1 + xyq^{2j-1}}{1 - y^2 q^{4j-2}} = \prod_{j=1}^{\infty} \left((1 + xyq^{2j-1}) \sum_{i=0}^{\infty} y^{2i} q^{(2i)\cdot(2j-1)} \right)$$

$$= \prod_{j=1}^{\infty} \sum_{i=0}^{\infty} \left(y^{2i} q^{(2i)\cdot(2j-1)} + xy^{(2i+1)} q^{(2i+1)\cdot(2j-1)} \right)$$

$$= \prod_{j=1}^{\infty} (1 + xyq^{(2j-1)} + y^2 q^{2\cdot(2j-1)} + xy^3 q^{3\cdot(2j-1)} + \cdots)$$

$$= \sum_{\lambda \in \mathcal{O}} x^{n_o(\lambda)} y^{l(\lambda)} q^{|\lambda|},$$

as desired. □

Since $l_a(\lambda) = l_o(\lambda')$ for any partition λ, combining Theorems 4.2 and 4.3 yields Theorem 2.1.

5 A combinatorial proof of the main result

In this section, we give a combinatorial proof of Theorem 2.1. We will use a restricted version of the variant of Bessenrodt's insertion algorithm given in Sect. 3. We now proceed to give the proofs of Lemma 2.2, Lemma 2.3 and Theorem 2.4.

Proof of Lemma 2.2 For $\lambda \in \mathcal{D}(n)$, define $\alpha = \varphi(\lambda)$ to be the 2-modular diagram conjugate of λ. It is necessary to show that $\alpha \in \mathcal{A}_1(n)$. On the one hand, it is easy to see that there is no odd part in α that can be repeated, since there is at most one "1" in each row of the 2-modular diagram of λ. Moreover, the condition that λ is a partition with distinct parts implies that the difference between successive parts in α is at most 4 and strictly less than 4 if either part is divisible by 2, and that the smallest part of α is less than 4.

The reverse map φ^{-1} can be easily constructed. For $\alpha \in \mathcal{A}_1(n)$, we note that its 2-modular diagram conjugate is a partition with distinct parts, namely, $\lambda = \varphi^{-1}(\alpha) \in \mathcal{D}(n)$. Thus φ is a bijection. Furthermore, it is not difficult to check $l_o(\lambda) = l_o(\alpha)$ and $l_a(\lambda) = 2r_2(\alpha) + l_o(\alpha)$. This completes the proof. $\qquad\square$

Proof of Lemma 2.3 Let $\mu = (1^{m_1}, 3^{m_3}, 5^{m_5}, \ldots, (2t-1)^{m_{2t-1}}) \in \mathcal{O}(n)$. For every multiplicity m_i, we write $m_i = 2h_i + s_i$ ($s_i = 0, 1$). Then we define $\beta = \psi(\mu) = (1^{m'_1}, 2^{m'_2}, 3^{m'_3}, \ldots, k^{m'_k})$, where $m'_{2i+1} = s_{2i+1}$ and $m'_{2i} = h_i$. Clearly, $m'_{4i} = h_{2i} = 0$ and $m'_{2i+1} \leq 1$, and so $\beta \in \mathcal{A}_2(n)$. For example, let $\mu = (1, 3, 7^2, 9, 15)$. Then we have $\beta = (1, 3, 9, 14, 15)$ whose 2-modular diagram is illustrated in Fig. 1.

The inverse map ψ^{-1} can be easily described. Let $\beta = (1^{m_1}, 2^{m_2}, 3^{m_3}, \ldots, t^{m_t}) \in \mathcal{A}_2(n)$. Then we have $m_{4i} = 0$ and $m_{2i-1} = 0$ or 1 for $i \geq 1$. Let

$$\mu = \psi^{-1}(\beta) = (1^{m'_1}, 2^{m'_2}, 3^{m'_3}, \ldots, k^{m'_k}),$$

where $m'_{2i-1} = 2m_{4i-2} + m_{2i-1}$ and $m'_{2i} = 0$ for $i = 1, 2, \ldots$. Obviously, $\psi^{-1}(\beta) \in \mathcal{O}(n)$. It follows that $n_o(\mu) = l_o(\beta)$ and $l(\mu) = 2r_2(\beta) + l_o(\beta)$. This completes the proof. $\qquad\square$

It is easy to see that Theorem 2.4 follows from Theorem 3.6 by setting $N = 2$ and $\mathbb{A}_4 = \{1, 2, 3\}$, that is, $\mathcal{C}_1(\{1, 2, 3\}; n, 4) = \mathcal{A}_1(n)$ and $\mathcal{C}_2(\{1, 2, 3\}; n, 4) = \mathcal{A}_2(n)$.

Figures 2 and 3 illustrate the procedure in Theorem 2.4.

To conclude, we combine the maps ψ, ϕ and φ to construct the desired map Δ from the set of partitions with distinct parts to the set of partitions with odd parts: $\Delta = \psi^{-1} \circ \phi \circ \varphi$. The properties of Δ lead to a proof of Theorem 2.1.

Fig. 1 The diagram representation of $\beta = (1, 3, 7^2, 9, 15)$

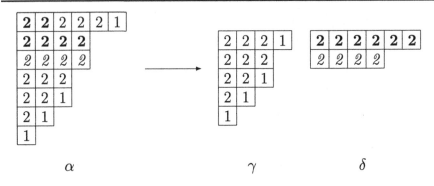

Fig. 2 Extraction of parts from α

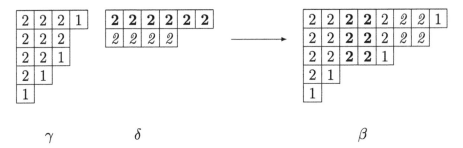

Fig. 3 Insertion all parts of δ into γ

For example, let $\lambda = (17, 16, 14, 10, 7, 4, 2, 1) \in \mathcal{D}(71)$. Then we have

$$\alpha = \varphi(\lambda) = (15, 12, 10, 9, 8, 6, 6, 4, 1),$$

$$\beta = \phi(\alpha) = (19, 18, 13, 10, 6, 5),$$

$$\mu = \psi^{-1}(\beta) = (3^2, 5^3, 9^2, 13, 19) \in \mathcal{O}(71).$$

Moreover, $l_o(\lambda) = 3$, $l_a(\lambda) = 9$, $n_o(\mu) = 3$ and $l(\mu) = 9$.

References

1. Alder, H.L.: Partition identities-from Euler to the present. Am. Math. Mon. **76**, 733–746 (1969)
2. Andrews, G.E.: On generalizations of Euler's partition theorem. Mich. Math. J. **1**, 491–498 (1966)
3. Andrews, G.E.: The Theory of Partitions. Addison–Wesley, Reading (1976)
4. Andrews, G.E.: On a partition theorem of N.J. Fine. J. Natl. Acad. Math. India **1**, 105–107 (1983)
5. Andrews, G.E.: On a partition function of Richard Stanley. Electron. J. Comb. **11**(2), #R1 (2004)
6. Andrews, G.E., Eriksson, K.: Integer Partitions. Cambridge Univ. Press, Cambridge (2004)
7. Andrews, G.E., Olsson, J.B.: Partition identities with an application to group representation theory. J. Reine Angew. Math. **413**, 198–212 (1991)
8. Bessenrodt, C.: A combinatorial proof of a refinement of the Andrews-Olsson partition identity. Eur. J. Comb. **12**, 271–276 (1991)
9. Bessenrodt, C.: A bijection for Lebesgue's partition identity in the spirit of Sylvester. Discrete Math. **132**, 1–10 (1994)

10. Bessenrodt, C.: Generalizations of the Andrews-Olsson partition identity. Discrete Math. **141**, 11–22 (1995)
11. Boulet, C.: A four parameter partition identity. Ramanujan J. **12**(3), 315–320 (2006)
12. Bousquet-Mélou, M., Eriksson, K.: Lecture hall partitions. Ramanujan J. **1**, 101–111 (1997)
13. Bousquet-Mélou, M., Eriksson, K.: Lecture hall partitions II. Ramanujan J. **1**, 165–185 (1997)
14. Bressoud, D.: Proofs and Confirmations: The Story of the Alternating Sign Matrix Conjecture. Cambridge Univ. Press, Cambridge (1999)
15. Fine, N.J.: Basic Hypergeometric Series and Applications. Math. Surveys, vol. 27. AMS, Providence (1988)
16. Glaisher, J.W.L.: A theorem in partitions. Messenger Math. **12**, 158–170 (1883)
17. Kim, D., Yee, A.J.: A note on partitions into distinct parts and odd parts. Ramanujan J. **3**, 227–231 (1999)
18. Pak, I.: On Fine's partition theorems, Dyson, Andrews, and missed opportunities. Math. Intell. **25**, 10–16 (2003)
19. Sylvester, J.: A constructive theory of partitions, arranged in three acts, an interact and an exodion. Am. J. Math. **5**, 251–330 (1882)
20. Yee, A.J.: A combinatorial proof of Andrews' partition functions related to Schur's partition theorem. Proc. Am. Math. Soc. **130**, 2229–2235 (2002)
21. Zeng, J.: The q-variations of Sylvester's bijection between odd and strict partitions. Ramanujan J. **9**(3), 289–303 (2005)

Ramanujan J (2010) 23: 151–157
DOI 10.1007/s11139-009-9200-2

Identities and congruences for Ramanujan's $\omega(q)$

Jan H. Bruinier · Ken Ono

For George Andrews on his 70-th birthday

Received: 16 December 2008 / Accepted: 24 August 2009 / Published online: 30 January 2010
© Springer Science+Business Media, LLC 2010

Abstract Recently, the authors constructed *generalized Borcherds products* where modular forms are given as infinite products arising from weight $1/2$ harmonic Maass forms. Here we illustrate the utility of these results in the special case of Ramanujan's mock theta function $\omega(q)$. We obtain identities and congruences modulo 512 involving the coefficients of $\omega(q)$.

Keywords Mock theta functions · Harmonic Maass forms

Mathematics Subject Classification (2000) Primary 11F37 · 33D15

1 Introduction and statement of results

In a recent paper, the authors [3] obtained results concerning *generalized Borcherds products*. Loosely speaking, these are modular forms which are infinite products whose exponents are coefficients of weight 1/2 harmonic Maass forms (see [6] for a survey on harmonic Maass forms in number theory). The authors then employed these results to study the vanishing of derivatives of modular L-functions.

The second author thanks the support of the NSF, the Hilldale Foundation and the Manasse family.

J.H. Bruinier
Fachbereich Mathematik, Technische Universität Darmstadt, Schlossgartenstrasse 7,
64289 Darmstadt, Germany
e-mail: bruinier@mathematik.tu-darmstadt.de

K. Ono (✉)
Department of Mathematics, University of Wisconsin, Madison, WI 53706, USA
e-mail: ono@math.wisc.edu

Here we illustrate the implications of these results for partitions and q-series. We consider the special case of Ramanujan's mock theta function $\omega(q)$

$$\omega(q) = \sum_{n=0}^{\infty} a_\omega(n)q^n := \sum_{n=0}^{\infty} \frac{q^{2n(n+1)}}{(q;q^2)_{n+1}^2} = 1 + 2q + 3q^2 + 4q^3 + 6q^4 + 8q^5 + \cdots .$$

(1.1)

As usual, we use the customary notation

$$(a;q)_n := (1-a)(1-aq)(1-aq^2)\cdots(1-aq^{n-1}).$$

Thanks to Fine's identity (see (26.84) of [4])[1]

$$q\omega(q) = \sum_{n=0}^{\infty} \frac{q^{n+1}}{(q;q^2)_{n+1}} = \sum_{n=1}^{\infty} \frac{q^n}{(1-q^{1+0})(1-q^{2+1})\cdots(1-q^{n+(n-1)})},$$

(1.2)

we find that $q\omega(q)$ is a generating function for an elegant partition function. The coefficient $a_\omega(n)$ denotes the number of partitions of $n-1$ whose summands, apart from one of maximal size, form pairs of consecutive non-negative integers.

Example Here are the partitions of 6:

$$6,\ 5+1,\ 4+2,\ 4+1+1,\ 3+3,\ 3+2+1,\ 3+1+1+1,$$
$$2+2+2,\ 2+2+1+1,\ 2+1+1+1+1,\ 1+1+1+1+1+1.$$

Eight of these partitions correspond to partitions whose summands, apart from one of the largest summands, occur in pairs of consecutive non-negative integers:

$$6,\ 5+(1+0),\ 4+(1+0)+(1+0),\ 3+(2+1),\ 3+(1+0)+(1+0)+(1+0),$$
$$2+(2+1)+(1+0),\ 2+(1+0)+(1+0)+(1+0)+(1+0),$$
$$1+(1+0)+(1+0)+(1+0)+(1+0)+(1+0).$$

This corresponds to our observation that $a_\omega(5) = 8$.

Here we investigate the arithmetic properties of the partition function $a_\omega(n)$. We shall relate this function to the classical divisor functions

$$\sigma_\nu(n) := \sum_{1 \le d|n} d^\nu$$

(1.3)

which play central roles in the theory of modular forms. To this end, we define a "strange" divisor function using the coefficients $a_\omega(n)$, the Legendre symbol $(\frac{\bullet}{3})$,

---

[1] The reader is also encouraged to see Andrews's recent paper [1] for more on the combinatorial interpretation of $\omega(q)$.

and the classical Jacobi-symbol character $\chi(m) := (\frac{-8}{m})$. We define $\widehat{\sigma}_\omega(n)$ by

$$\widehat{\sigma}_\omega(n) := \sum_{1 \leq d \mid n} \left(\frac{d}{3}\right) \chi(n/d) d \cdot a_\omega\left(\frac{2d^2 - 2}{3}\right), \tag{1.4}$$

and we consider the following two generating functions:

$$L_\omega(q) := \sum_{n \geq 1} \widehat{\sigma}_\omega(n) q^n = q - 6q^2 + q^3 + 116q^4 - 506q^5 - 6q^6 + \cdots, \tag{1.5}$$

$$\widetilde{L}_\omega(q) := \sum_{\substack{n \geq 1 \\ \gcd(n,6)=1}} \widehat{\sigma}_\omega(n) q^n = q - 506q^5 + 9736q^7 - 3638260q^{11} + \cdots. \tag{1.6}$$

We prove the following curious theorem.

Theorem 1.1 *The q-series $L_\omega(q)$ (resp., $\widetilde{L}_\omega(q)$) is the Fourier expansion of a weight 2 meromorphic modular form on $\Gamma_0(6)$ (resp., $\Gamma_0(216)$), where $q := e^{2\pi i z}$.*

An explicit form of this result (see Sect. 2) gives the following congruences.

Theorem 1.2 *The following are true:*

1. *We have that*

$$L_\omega(q) \equiv \sum_{n=0}^{\infty} \left(q^{(2n+1)^2} + q^{3(2n+1)^2}\right) \pmod{2}.$$

2. *We have that*

$$\widetilde{L}_\omega(q) \equiv \sum_{\substack{n \geq 1 \\ \gcd(n,6)=1}} \sigma_1(n) q^n \pmod{512}.$$

In particular, if $p \geq 5$ is prime, then

$$a_\omega\left(\frac{2p^2 - 2}{3}\right) \equiv \begin{cases} (\frac{p}{3}) \pmod{512} & \text{if } p \equiv 1, 3 \pmod 8, \\ (\frac{p}{3})(1 + 2p^{255}) \pmod{512} & \text{if } p \equiv 5, 7 \pmod 8. \end{cases}$$

Example If $p = 7$, then Theorem 1.2(2) implies that

$$a_\omega(32) = 1391 \equiv 367 \equiv \left(\frac{7}{3}\right)(1 + 2 \cdot 7^{255}) \pmod{512}.$$

Three Remarks

(1) It is natural to ask whether there is a combinatorial explanation for the fact that $a_\omega(\frac{2p^2-2}{3}) \equiv \pm 1 \pmod{512}$ for the "half" of the primes which satisfy the congruence $p \equiv 1, 3 \pmod 8$.

(2) The results presented here are examples of a general theory in the case of a single generalized Borcherds product for $\omega(q)$. There are infinitely many such Borcherds products for $\omega(q)$. For any given product, one may obtain congruences modulo arbitrary powers of infinitely many primes (for example, see [2]). For $L_\omega(q)$, these are the primes p for which $(\frac{-2}{p}) \in \{0, -1\}$. For these p, we have that $L_\omega(q)$ is a p-adic modular form in the sense of Serre [7] (for example, see [2]). In the present paper, we are content with $p = 2$ and the p-power modulus $2^9 = 512$.

(3) More generally, one may construct such generalized Borcherds products for all of Ramanujan's mock theta functions using Theorems 6.1 and 6.2 of [3]. These modular forms will have twisted Heegner divisors, as well as logarithmic derivatives which are meromorphic weight 2 modular forms, which for certain primes p will turn out to be p-adic modular forms.

In Sect. 2, we prove Theorems 1.1 and 1.2 using the results of [3] combined with various standard arguments from the theory of modular forms.

2 Proofs

Our results follow from a generalized Borcherds product obtained in [3]. Using the coefficients of $\omega(q)$, we define the formal power series

$$
B_\omega(z) := \prod_{m=1}^{\infty} \left(\frac{1 + \sqrt{-2}q^m - q^{2m}}{1 - \sqrt{-2}q^m - q^{2m}} \right)^{-4(\frac{m}{3})a_\omega(\frac{2m^2-2}{3})}
$$

$$
= \left(\frac{1 + \sqrt{-2}q - q^2}{1 - \sqrt{-2}q - q^2} \right)^{-4} \cdot \left(\frac{1 + \sqrt{-2}q^2 - q^4}{1 - \sqrt{-2}q^2 - q^4} \right)^{12} \cdots
$$

$$
= 1 - 8\sqrt{-2}q - \left(64 - 24\sqrt{-2}\right)q^2 + \left(384 + 168\sqrt{-2}\right)q^3 \cdots . \quad (2.1)
$$

This formal power series, where $q := e^{2\pi i z}$ for $z \in \mathbb{H}$, is discussed in Example 8.2 of [3]. Thanks to work of Zwegers [8], Theorems 6.1 and 6.2 of [3] imply the following theorem.

Theorem 2.1 *The q-series $B_\omega(z)$ is the Fourier expansion of a weight 0 modular form on the congruence subgroup $\Gamma_0(6)$.*

Proof of Theorem 1.1 That $\widetilde{L}_\omega(q)$ is a meromorphic modular form on $\Gamma_0(216)$ will follow from the assertion that $L_\omega(q)$ is the Fourier expansion of a weight 2 meromorphic modular form on $\Gamma_0(6)$. One simply uses the standard U and V operators (for example, see Sect. 2.4 of [5]).

Let $\Theta := q \cdot \frac{d}{dq} = \frac{1}{2\pi i} \cdot \frac{d}{dz}$. If $f(z)$ is a meromorphic modular form (for example, see Sect. 2.3 of [5]), it is a standard fact that $\Theta(f)/f$ is a weight 2 meromorphic modular form. Therefore, it follows that

$$\frac{\Theta(B_\omega(z))}{B_\omega(z)} = -8\sqrt{-2}q + 48\sqrt{-2}q^2 - 8\sqrt{-2}q^3 - 928\sqrt{-2}q^4 + 4048\sqrt{-2}q^5 + \cdots$$

$$= -8\sqrt{-2}\left(q - 6q^2 + q^3 + 116q^4 - 506q^5 - 6q^6 + 9736q^7 - \cdots\right)$$

$$= -8\sqrt{-2} \cdot G_\omega(q)$$

is a weight 2 meromorphic modular form on $\Gamma_0(6)$. It suffices to prove that $G_\omega(q) = L_\omega(q)$.

To prove this assertion, we let

$$P(X) := \frac{1 + \sqrt{-2}X - X^2}{1 - \sqrt{-2}X - X^2}.$$

If m is a positive integer, then a straightforward calculation reveals that

$$\frac{\Theta(P(q^m))}{P(q^m)} = 2m\sqrt{-2}\sum_{n=1}^{\infty}\chi(n)q^{mn}.$$

Using this result, it follows that

$$\frac{\Theta(B_\omega(z))}{B_\omega(z)} = -8\sqrt{-2}\sum_{m=1}^{\infty}m\left(\frac{m}{3}\right)a_\omega\left(\frac{2m^2 - 2}{3}\right)\sum_{n=1}^{\infty}\chi(n)q^{mn}.$$

That $G_\omega(q) = L_\omega(q)$ now follows immediately, and so $L_\omega(q)$ is a meromorphic modular form on $\Gamma_0(6)$. $\qquad\square$

Now we turn to the proof of Theorem 1.2.

Proof of Theorem 1.2 We recall the explicit description of the meromorphic modular form $B_\omega(z)$ given in Example 8.2 of [3]. Let $j_6^*(z)$ be the usual Hauptmodul for $\Gamma_0^*(6)$, the extension of $\Gamma_0(6)$ by all the Atkin–Lehner involutions. It is not difficult to verify that

$$j_6^*(z) := \left(\frac{\eta(z)\eta(2z)}{\eta(3z)\eta(6z)}\right)^4 + 4 + 3^4\left(\frac{\eta(3z)\eta(6z)}{\eta(z)\eta(2z)}\right)^4$$

$$= q^{-1} + 79q + 352q^2 + 1431q^3 + \cdots,$$

where $\eta(z) := q^{1/24}\prod_{n=1}^{\infty}(1 - q^n)$ is Dedekind's eta-function. Let α_1 and α_2 be the Heegner points

$$\alpha_1 := \frac{-2 + \sqrt{-2}}{6} \quad \text{and} \quad \alpha_2 := \frac{2 + \sqrt{-2}}{6}.$$

We have that $j_6^*(\alpha_1) = j_6^*(\alpha_2) = -10$. Therefore, it follows that $j_6^*(z) + 10$ is a rational modular function on $X_0(6)$ whose divisor consists of the 4 cusps with multiplicity -1 and the points α_1 and α_2 with multiplicity 2.

Let $E_4(z) = 1 + 240 \sum_{n=1}^{\infty} \sigma_3(n)q^n$ be the standard weight 4 Eisenstein series for $SL_2(\mathbb{Z})$, and let

$$\delta(z) := \eta(z)^2 \eta(2z)^2 \eta(3z)^2 \eta(6z)^2 = q - 2q^2 - 3q^3 + 4q^4 + \cdots.$$

Using $E_4(z)$ and $\delta(z)$, we define the weight 4 holomorphic $\Gamma_0(6)$-modular form $\phi(z)$ by

$$450\phi(z) := \left(3360 - 1920\sqrt{-2}\right)\delta(z) + \left(1 - 7\sqrt{-2}\right)E_4(z)$$
$$+ \left(4 - 28\sqrt{-2}\right)E_4(2z) + \left(89 + 7\sqrt{-2}\right)E_4(3z)$$
$$+ \left(356 + 28\sqrt{-2}\right)E_4(6z).$$

In terms of $\phi(z)$, $j_6^*(z)$ and $\delta(z)$, one easily checks that

$$B_\omega(z) = \frac{\phi(z)}{(j_6^*(z) + 10)\delta(z)}.$$

By Theorem 1 of [2], generalized to $\Gamma_0(6)$ and $B_\omega(z)$ in the obvious way, we have that $-8\sqrt{-2}L_\omega(q) = \Theta(B_\omega(z))/B_\omega(z)$ is a 2-adic modular form of weight 2. This then implies that $L_\omega(z)$ (mod 2^k), for every positive integer k, is the reduction of a holomorphic modular form.

To obtain Theorem 1.2, we now employ the identity

$$\mathcal{E}(z) := \frac{\eta(4z)^8}{\eta(2z)^4} = \sum_{n=0}^{\infty} \sigma_1(2n+1)q^{2n+1} = q + 4q^3 + 6q^5 + \cdots. \qquad (2.2)$$

Congruence (1) is equivalent to the assertion that

$$L_\omega(q) \equiv \mathcal{E}(z) + \mathcal{E}(3z) \pmod{2},$$

while (2) is equivalent to the assertion that

$$\widetilde{L}_\omega(q) \equiv \mathcal{E}(z) - \mathcal{E}(z)\big|U(3)\big|V(3) \pmod{512}.$$

These congruences are easily confirmed using the constructive proof of Theorem 1 of [2], combined with Sturm's Theorem (see Theorem 2.58 of [5]). That

$$a_\omega\left(\frac{2p^2 - 2}{3}\right) \equiv \begin{cases} \left(\frac{p}{3}\right) \pmod{512} & \text{if } p \equiv 1, 3 \pmod 8, \\ \left(\frac{p}{3}\right)(1 + 2p^{255}) \pmod{512} & \text{if } p \equiv 5, 7 \pmod 8 \end{cases}$$

follows easily from (2), namely that

$$\sigma_1(p) \equiv \widehat{\sigma}_\omega(p) \pmod{512},$$

and the definition of $\widehat{\sigma}_\omega(p)$. $\square$

References

1. Andrews, G.E.: Partitions, Durfee symbols, and the Atkin–Garvan moments of ranks. Invent. Math. **169**, 37–73 (2007)
2. Bruinier, J.H., Ono, K.: Arithmetic of Borcherds's exponents. Math. Ann. **327**, 293–303 (2003)
3. Bruinier, J.H., Ono, K.: Heegner divisors, L-functions, and Maass forms. Ann. Math. (accepted for publication)
4. Fine, N.J.: Basic Hypergeometric Series and Applications. Math. Surveys and Monographs, vol. 27. Amer. Math. Soc., Providence (1988)
5. Ono, K.: The Web of Modularity: Arithmetic of the Coefficients of Modular Forms and q-Series. CBMS Regional Conference Series in Mathematics, vol. 102. Amer. Math. Soc., Providence (2004)
6. Ono, K.: Unearthing the visions of a master: harmonic Maass forms and number theory. In: Proceedings of the 2008 Harvard–MIT Current Developments in Mathematics Conference (accepted for publication)
7. Serre, J.-P.: Formes modulaires et fonctions zêta p-adiques. In: Springer Lect. Notes. Math., vol. 350, pp. 191–268 (1973)
8. Zwegers, S.: Mock ϑ-functions and real analytic modular forms. In: Berndt, B.C., Ono, K. (eds.) q-Series with Applications to Combinatorics, Number Theory, and Physics. Contemp. Math., vol. 291, pp. 269–277. Amer. Math. Soc., Providence (2001)

Ramanujan J (2010) 23: 159–167
DOI 10.1007/s11139-009-9202-0

On the subpartitions of the ordinary partitions

Byungchan Kim

Dedicated to George E. Andrews for his seventieth birthday

Received: 24 December 2008 / Accepted: 27 August 2009 / Published online: 15 April 2010
© Springer Science+Business Media, LLC 2010

Abstract Let $a_1 \geq a_2 \geq \cdots \geq a_\ell$ be an ordinary partition. A subpartition with gap d of an ordinary partition is defined as the longest sequence satisfying $a_1 > a_2 > \cdots > a_s$ and $a_s > a_{s+1}$, where $a_i - a_j \geq d$ for all $i < j \leq s$. This is a generalization of the Rogers–Ramanujan subpartition which was introduced by L. Kolitsch. In this note, we will study various properties of subpartitions, and as an application we will give a combinatorial proof of two entries which are in Ramanujan's lost notebook.

Keywords Partition · Subpartition · Partial theta function

Mathematics Subject Classification (2000) 11P81 · 05A17

1 Introduction

Let $a_1 \geq a_2 \geq \cdots \geq a_\ell$ be an ordinary partition. In a recent paper [7], L. Kolitsch introduced the Rogers–Ramanujan subpartitions and established their connection to other partitions. The Rogers–Ramanujan subpartition is the longest sequence satisfying $a_1 > a_2 > \cdots > a_s$ and $a_s > a_{s+1}$, where $a_i - a_j \geq 2$ for all $i < j \leq s$. In this note, we will generalize his result with an arbitrary gap condition and will study their connections to other partitions. Let us fix a positive integer d. Then, for a given partition, a subpartition with gap d is defined as the longest sequence satisfying $a_1 > a_2 > \cdots > a_s$ and $a_s > a_{s+1}$, where $a_i - a_j \geq d$ for all $i < j \leq s$. Note that Kolitsch's Rogers–Ramanujan subpartition is the case $d = 2$. For convenience, we will define the subpartition of the empty partition as the empty partition. We define the length of the subpartition with gap d as the number of parts in the subpartition. When the gap d is clear in the context, we will say the subpartition instead of the

B. Kim (✉)
Department of Mathematics, University of Illinois, 1409 West Green Street, Urbana, IL 61801, USA
e-mail: bkim4@illinois.edu

subpartition with gap d. In Sect. 2, we will give a generating function of the ordinary partition that keeps track of the length of the subpartition with gap d. In Sect. 3, we will study their connection to the partial theta function, which is of the form

$$\sum_{n=0}^{\infty}(-1)^n q^{n(n-1)/2} x^n,$$

by attaching a proper weight to the generating function. In Sect. 4, we will focus on the subpartition with gap 1. By using the results from Sects. 2 and 3, we will give combinatorial proofs of the identities:

$$\frac{1}{(q)_{\infty}^2}\sum_{n=0}^{\infty}(-1)^n q^{(n^2+n)/2} = \sum_{n=0}^{\infty}\frac{q^n}{(q)_n^2}, \tag{1}$$

$$\frac{1}{(q)_{\infty}^2}\left(1+2\sum_{n=1}^{\infty}(-1)^n q^{(n^2+n)/2}\right) = \sum_{n=0}^{\infty}\frac{q^{2n}}{(q)_n^2}, \tag{2}$$

which are entries in Ramanujan's lost notebook [3, p. 19, Entries 1.4.10 and 1.4.11, 8]. Here and in the sequel, we use the customary notation for the q-series:

$$(a)_0 := (a; q)_0 := 1,$$

$$(a)_n := (a; q)_n := (1-a)(1-aq)\cdots\left(1-aq^{n-1}\right), \quad n \geq 1,$$

and

$$(a)_{\infty} := (a; q)_{\infty} := \lim_{n\to\infty}(a; q)_n, \quad |q| < 1.$$

We will conclude this note with some comments on possible future projects.

2 A generating function

For a given partition λ, we always write it in the form $a_1 \geq a_2 \geq \cdots \geq a_{\ell}$. Before finding a generating function, we need to introduce some notation. Let us fix a positive integer d and define, for each nonnegative integer k,

$$S_{d,k} = \begin{cases} 1 + (1+d) + (1+2d) + \cdots + (k-1)d + 1 = \frac{dk^2 - (d-2)k}{2}, & \text{if } k \neq 0, \\ 0, & \text{if } k = 0. \end{cases}$$

Then, for a given partition λ, there are three cases:

(I) There is no subpartition in λ.
(II) The subpartition of λ is λ. In this case, we will say the partition λ is a complete partition after Kolitsch.
(III) λ is not complete and it has a subpartition with length ℓ.

In Case (I), i.e., to have no subpartition in λ, we should have $a_1 = a_2$. By using a standard argument [1, Chap. 1], we can easily see that

$$\sum_{i=1}^{\infty} \frac{q^{2i}}{(q)_i} \tag{3}$$

generates such partitions.

In Case (II), i.e., when λ is a complete partition, the gaps between successive parts of λ should be at least d. Such partitions are generated by

$$\sum_{\ell=0}^{\infty} \frac{q^{S_{d,\ell}}}{(q)_\ell}. \tag{4}$$

Note that the length of the subpartition in the above is ℓ.

In Case (III), suppose that a given partition λ has the subpartition with length ℓ and $a_\ell = j$. Note that since λ is not a complete partition, there are at least $\ell + 1$ parts in λ and, by definition, $a_\ell > a_{\ell+1}$. Then, there are two possibilities:

(i) $a_\ell - a_{\ell+1}$ is less than d.
(ii) $a_\ell - a_{\ell+1} \geq d$, but $a_{\ell+1} = a_{\ell+2}$.

In Case (i), we have the generating function

$$\sum_{\ell=1}^{\infty} \frac{q^{S_{d,\ell}}}{(q)_{\ell-1}} \left(\sum_{j=2}^{\infty} \frac{q^{(\ell+1)(j-1)}}{(q)_{j-1}} + \cdots + \sum_{j=d}^{\infty} \frac{q^{\ell(j-1)+(j-d+1)}}{(q)_{j-d+1}} \right)$$

$$= \sum_{\ell=1}^{\infty} \frac{q^{S_{d,\ell}}}{(q)_{\ell-1}} \left(\sum_{n=1}^{\infty} \frac{q^{(\ell+1)n}}{(q)_n} + \cdots + q^{\ell(d-2)} \sum_{n=1}^{\infty} \frac{q^{(\ell+1)n}}{(q)_n} \right)$$

$$= \sum_{\ell=1}^{\infty} \frac{q^{S_{d,\ell}}(1 - q^{\ell(d-1)})}{(q)_\ell} \sum_{n=1}^{\infty} \frac{q^{(\ell+1)n}}{(q)_n}$$

$$= \sum_{\ell=1}^{\infty} \frac{q^{S_{d,\ell}}(1 - q^{\ell(d-1)})}{(q)_\ell} \left(\frac{1}{(q^{\ell+1})_\infty} - 1 \right)$$

$$= \sum_{\ell=1}^{\infty} \frac{q^{S_{d,\ell}}(1 - q^{\ell(d-1)})}{(q)_\infty} - \sum_{\ell=1}^{\infty} \frac{q^{S_{d,\ell}}(1 - q^{\ell(d-1)})}{(q)_\ell}, \tag{5}$$

where in the penultimate line we used the q-binomial theorem [5, p. 8].

In Case (ii), we have the generating function

$$\sum_{\ell=1}^{\infty} \frac{q^{S_{d,\ell}}}{(q)_{\ell-1}} \sum_{j=d+1}^{\infty} q^{\ell(j-1)} \sum_{i=1}^{j-d} \frac{q^{2i}}{(q)_i}$$

$$= \sum_{\ell=1}^{\infty} \frac{q^{S_{d,\ell}}}{(q)_{\ell-1}} \sum_{i=1}^{\infty} \sum_{j=d+i}^{\infty} q^{\ell(j-1)} \frac{q^{2i}}{(q)_i}$$

$$= \sum_{\ell=1}^{\infty} \frac{q^{S_{d,\ell}+\ell(d-1)}}{(q)_\ell} \sum_{i=1}^{\infty} \frac{q^{(\ell+2)i}}{(q)_i}$$

$$= \sum_{\ell=1}^{\infty} \frac{q^{S_{d,\ell}+\ell(d-1)}}{(q)_\ell} \left(\frac{1}{(q^{\ell+2})_\infty} - 1 \right), \tag{6}$$

by the q-binomial theorem.

Since all partitions fall into one of the above three cases, the sum of the above generating functions (3), (4), (5), and (6) should be $\frac{1}{(q)_\infty}$. Thus, we have

$$\frac{1}{(q)_\infty} = \sum_{i=1}^{\infty} \frac{q^{2i}}{(q)_i} + \sum_{\ell=0}^{\infty} \frac{q^{S_{d,\ell}}}{(q)_\ell}$$

$$+ \sum_{\ell=1}^{\infty} \frac{q^{S_{d,\ell}}(1-q^{\ell(d-1)})}{(q)_\infty} - \sum_{\ell=1}^{\infty} \frac{q^{S_{d,\ell}}(1-q^{\ell(d-1)})}{(q)_\ell}$$

$$+ \sum_{\ell=1}^{\infty} \frac{q^{S_{d,\ell}+\ell(d-1)}}{(q)_\ell} \left(\frac{1}{(q^{\ell+2})_\infty} - 1 \right)$$

$$= \sum_{\ell=0}^{\infty} \frac{q^{S_{d,\ell}+\ell(d-1)}}{(q)_\ell(q^{\ell+2})_\infty} + \sum_{\ell=1}^{\infty} \frac{q^{S_{d,\ell}}(1-q^{\ell(d-1)})}{(q)_\infty}$$

$$= \frac{1}{(q^2)_\infty} + \frac{1}{(q)_\infty} \sum_{\ell=1}^{\infty} (q^{S_{d,\ell}} - q^{S_{d,\ell}+\ell d+1})$$

$$= \frac{1}{(q^2)_\infty} + \frac{1}{(q)_\infty} \sum_{\ell=1}^{\infty} (q^{S_{d,\ell}} - q^{S_{d,\ell+1}}),$$

since

$$\sum_{i=1}^{\infty} \frac{q^{2i}}{(q)_i} = \frac{1}{(q^2)_\infty} - 1,$$

by the q-binomial theorem. Thus, we have proved our first theorem.

Theorem 1 *Let ℓ be the length of the subpartition with gap d, then we have*

$$\frac{1}{(q)_\infty} = \frac{1}{(q^2)_\infty} + \frac{1}{(q)_\infty} \sum_{\ell=1}^{\infty} (q^{S_{d,\ell}} - q^{S_{d,\ell+1}}). \tag{7}$$

Remark An analytic proof of Theorem 1 is very simple; thus we will omit it. Note that, by setting $d = 2$, we can recover Kolitsch's Theorem 1.

Define $p(n, \ell, d)$ to be the number of partitions of n having a subpartition of length ℓ with gap d. Then, by observing coefficients of q^n in (7), we can easily deduce that

Corollary 2 *For all nonnegative integers n and ℓ and a positive integer d, we have*

$$p(n, \ell, d) = p(n - S_{d,\ell}) - p(n - S_{d,\ell+1}).$$

3 Subpartitions with parity condition

Let us define $p_e(n, d)$ to be the sum $\sum_{\ell \text{ even}} p(n, \ell, d)$, i.e., the number of partitions of n that have subpartitions with even lengths, and similarly for $p_o(n, d)$. Then, we have

$$\sum_{n=0}^{\infty} (p_e(n, d) - p_o(n, d))q^n = \frac{1}{(q^2)_\infty} + \frac{1}{(q)_\infty} \sum_{\ell=1}^{\infty} (-1)^\ell \left(q^{S_{d,\ell}} - q^{S_{d,\ell+1}}\right)$$

$$= \frac{1}{(q)_\infty} \left(1 + 2 \sum_{\ell=1}^{\infty} (-1)^\ell q^{S_{d,\ell}}\right)$$

$$= \frac{1}{(q)_\infty} \left(2 \sum_{\ell=0}^{\infty} (-1)^\ell q^{S_{d,\ell}} - 1\right). \tag{8}$$

Note that, when $d = 2$, $1 + 2 \sum_{k=1}^{\infty} (-1)^k q^{S_{d,k}}$ becomes a theta function, which implies Kolitsch's Theorem 4. For other d's, $\sum_{k=0}^{\infty} (-1)^k q^{S_{d,k}}$ is a partial theta function of the form

$$\sum_{k=0}^{\infty} (-1)^k q^{\frac{dk^2 - (d-2)k}{2}}.$$

Since $p_e(n, d) + p_o(n, d) = p(n)$, we have

$$\sum_{n=0}^{\infty} p_e(n, d)q^n = \frac{1}{(q)_\infty} \sum_{k=0}^{\infty} (-1)^k q^{S_{d,k}}. \tag{9}$$

By replacing a and q with q and q^d, respectively, in the identity [4, (2.1b)]

$$\sum_{k=0}^{\infty} (-1)^k a^k q^{(k^2-k)/2} = (a)_\infty \sum_{n=0}^{\infty} \frac{a^n q^{n^2}}{(q)_n (a)_n},$$

we obtain

$$\sum_{n=0}^{\infty} p_e(n, d)q^n = \frac{(q; q^d)_\infty}{(q)_\infty} \sum_{m=0}^{\infty} \frac{q^{dm^2+m}}{(q^d; q^d)_m (q; q^d)_m}. \tag{10}$$

We have the following partition theoretic interpretation of (10).

Theorem 3 *In the case of $d \geq 2$, the number of partitions of n with an even length subpartition with gap d is the same as the number of partitions of n such that the parts which are congruent to 1 modulo d have the following property: Consider the*

*d-modular diagram of the partition, which consists of such parts. If it has the Dur-
fee square of a side k, then the largest part of the partition below the Durfee square
should be less than or equal to $d(k-1)+1$. In the case of $d=1$, the number of partitions of n with subpartitions of even length is the same as the number of partitions
of n that have the following property: If it has the Durfee square of a side k, then the
number of parts in the partition on the right side of the Durfee square is k.*

4 Subpartitions with gap 1

In this section, we will investigate the subpartitions with gap 1. By (8) and (9) in
Sect. 3, we have

$$\sum_{n=0}^{\infty}\left(p_e(n,1)-p_o(n,1)\right)q^n = \frac{1}{(q)_\infty}\left(1+2\sum_{k=1}^{\infty}(-1)^k q^{(k^2+k)/2}\right),$$

$$\sum_{n=0}^{\infty} p_e(n,1)q^n = \frac{1}{(q)_\infty}\sum_{k=0}^{\infty}(-1)^k q^{(k^2+k)/2}.$$

Thus, Entries 1.4.10 and 1.4.11 of [3, p. 19] are equivalent to the following identities:

$$\sum_{n\geq 0} p_e(n,1)q^n = \sum_{n=0}^{\infty}\frac{q^n}{(q)_n}\left(q^{n+1}\right)_\infty, \tag{11}$$

$$\sum_{n\geq 0}\left(p_e(n,1)-p_o(n,1)\right)q^n = \sum_{n=0}^{\infty}\frac{q^{2n}}{(q)_n}\left(q^{n+1}\right)_\infty. \tag{12}$$

Now we will give a combinatorial proof for the above identities. Throughout the
proofs, $t(\lambda)$ denotes the number of parts in λ.

Proof of (11) Note that

$$\frac{q^n}{(q)_n}\left(q^{n+1}\right)_\infty$$

generates partition pairs $(\pi(n), \sigma(n))$, where $\pi(n)$ is a partition with the largest
part n, and $\sigma(n)$ is a partition into distinct parts such that the smallest part is larger
than n, and the exponent of (-1) is $t(\sigma(n))$. For a given partition λ, suppose that
λ has the subpartition of length ℓ. Then, λ is of the form $\lambda_1 > \lambda_2 > \cdots > \lambda_\ell >
\lambda_{\ell+1} \geq \lambda_{\ell+2} \geq \cdots$. Thus, on the right hand side of (11), λ is generated $\ell+1$
times as $(\pi(\lambda_1), \emptyset), (\pi(\lambda_2), \sigma(\lambda_2)), \ldots,$ and $(\pi(\lambda_{\ell+1}), \sigma(\lambda_{\ell+1}))$. Note that, in fact,
$\lambda_{\ell+1} = \lambda_{\ell+2}$. If not, the length of the subpartition should be bigger than ℓ. Thus, λ is
not of the form $(\pi(\lambda_{\ell+2}), \sigma(\lambda_{\ell+2}))$. Note also that the exponent of (-1) in the previous partition pairs is $(-1)^0, (-1)^1, \ldots, (-1)^\ell$, respectively. Thus, their sum is 1 if
ℓ is even and is 0 if ℓ is odd. Thus, on the right side of (11), after cancelation we are
left with the partitions that have subpartitions with even length. $\qquad\square$

Proof of (12) Note that

$$\frac{q^{2n}}{(q)_n}(q^{n+1})_\infty$$

generates partition pairs $(\pi(n), \sigma(n))$, where $\pi(n)$ is a partition with $\pi_1(n) = \pi_2(n) = n$, $\sigma(n)$ is a partition into distinct parts such that the smallest part is larger than n, and the exponent of (-1) is $t(\sigma(n))$. For a given partition λ, suppose that λ has the subpartition with length ℓ. Then, as before, λ is of the form $\lambda_1 > \lambda_2 > \cdots > \lambda_\ell > \lambda_{\ell+1} \geq \lambda_{\ell+2} \geq \cdots$. Recall that $\lambda_{\ell+1} = \lambda_{\ell+2}$. Thus, on the right side, λ is generated as $(\pi(\lambda_{\ell+1}), \sigma(\lambda_{\ell+1}))$. Since the exponent of (-1) is ℓ in this partition pair, we are done. □

Note that the right side of (1) is a generating function of the number of stacks with summits. For the definition of the stack with a summit and the proof of the just mentioned fact, consult the paper of Andrews [2]. For other combinatorial proofs of Entry 1.4.10 and Entry 1.4.11, examine the work of Yee [9], or the previous work of the author [6].

Next, we will obtain another generating function for $p_e(n, 1) - p_o(n, 1)$ by using a simple Durfee square argument. For a given partition λ, let λ^r be the conjugate of the partition on the right side of the Durfee square and λ^b be the partition below the Durfee square. Let $s(\lambda)$ be the side of the Durfee square of λ. Then, the coefficient of q^n of

$$2\sum_{k=0}^{\infty}\frac{q^{k^2+k}}{(q)_k^2} - \frac{1}{(q)_\infty} = 2\sum_{k=0}^{\infty}\frac{q^{k^2+k}}{(q)_k^2} - \sum_{k=0}^{\infty}\frac{q^{k^2}}{(q)_k^2} \tag{13}$$

is (the number of partitions of n such that $\lambda_1^b = s(\lambda)$) plus (the number of partitions of n such that $\lambda_1^r = s(\lambda)$) minus $p(n)$, by symmetry. Since λ with $\lambda_1^b < s(\lambda)$ is not generated by $\sum_{k=0}^{\infty}\frac{q^{k^2+k}}{(q)_k^2}$ and we count λ twice if $\lambda_1^r = \lambda_1^b = s(\lambda)$, we have

$$2\sum_{k=0}^{\infty}\frac{q^{k^2+k}}{(q)_k^2} - \sum_{k=0}^{\infty}\frac{q^{k^2}}{(q)_k^2} = 1 + \sum_{k=1}^{\infty}\frac{q^{k^2+2k}}{(q)_k^2} - \sum_{k=1}^{\infty}\frac{q^{k^2}}{(q)_{k-1}^2}.$$

In summary, we have

$$\sum_{n\geq 0}(p_e(n, 1) - p_o(n, 1))q^n = \frac{1}{(q)_\infty}\left(1 + 2\sum_{k=1}^{\infty}(-1)^k q^{(k^2+k)/2}\right)$$

$$= 1 + \sum_{k=1}^{\infty}\frac{q^{k^2+2k}}{(q)_k^2} - \sum_{k=1}^{\infty}\frac{q^{k^2}}{(q)_{k-1}^2}$$

$$= (1 - q)\sum_{k=0}^{\infty}\frac{q^{k^2+2k}}{(q)_k^2}.$$

Therefore, we have proved the following theorem.

Theorem 4 *The difference between the number of partitions of n with subpartition (with gap 1) of even length and those of odd length is the number of partitions λ of n satisfying $\lambda_1^r = \lambda_1^b = s(\lambda)$ and 1 is not a part of λ^b.*

As an immediate corollary, we have

$$p_e(n, 1) \geq p_o(n, 1), \quad \text{for all } n \geq 2. \tag{14}$$

Note that equality holds if and only if $n = 2$.

Next, we will investigate the parity of $p_e(n, 1)$. We see that

$$\sum_{n \geq 0} p_e(n, 1) q^n \equiv \frac{1}{(q)_\infty} \sum_{n \geq 0} q^{(n^2 + n)/2} \pmod{2}$$

$$\equiv \frac{(q)_\infty^3}{(q)_\infty} \pmod{2}$$

$$\equiv \left(q^2; q^2\right)_\infty \pmod{2}$$

$$\equiv \sum_{m=-\infty}^{\infty} q^{m(3m-1)} \pmod{2},$$

where for the second congruence we used Jacobi's identity [5, p. 14], and for the last congruence we used the pentagonal number theorem [5, p. 12]. Thus, we can conclude that $p_e(n, 1)$ is almost always even. Hence, we have proved the following theorem.

Theorem 5 *For all nonnegative integers n, we have*

$$p_e(n, 1) \equiv \begin{cases} 1 \pmod{2}, & \text{if n is of the form } m(3m \pm 1) \\ & \text{for some nonnegative integer m,} \\ 0 \pmod{2}, & \text{otherwise.} \end{cases}$$

5 Concluding remarks

By Corollary 2, we have

$$p(n, \ell, 1) = p\left(n - \frac{(\ell^2 + \ell)}{2}\right) - p\left(n - \frac{(\ell + 1)(\ell + 2)}{2}\right).$$

From the famous Ramanujan congruences, we can deduce the following trivial congruences for $p(n, \ell, 1)$:

$$p(5n + 4, 5k + 4, 1) \equiv 0 \pmod{5},$$

$$p(7n + 5, 7k + 6, 1) \equiv 0 \pmod{7},$$

$$p(11n + 6, 11k + 10, 1) \equiv 0 \pmod{11},$$

for all nonnegative integers n and k. It would be very interesting to find a congruence for $p(n, \ell, d)$ besides the congruences that are inherited by the congruences of the ordinary partition function.

Secondly, are there interesting properties of the subpartitions of other partitions? For example, let us define the subpartition of a partition into distinct parts as the longest sequence where the gap between successive parts is 1. Then, by a similar argument that we used to prove Theorem 1, we can easily prove that

$$(-q)_\infty = \sum_{\ell=0}^{\infty} q^{(\ell^2+\ell)/2} + \sum_{\ell=1}^{\infty} q^{(\ell^2+\ell)/2} \sum_{i=1}^{\infty} q^{i\ell}(-q)_{i-1} \tag{15}$$

and

$$(-q)_\infty = 1 + \sum_{\ell=1}^{\infty} \frac{q^{(\ell^2+\ell)/2}}{1-q^\ell} + \sum_{\ell=1}^{\infty} q^{(\ell^2+\ell)/2} \sum_{i=1}^{\infty} q^{i\ell}\left((-q)_{i-1} - 1\right), \tag{16}$$

where ℓ is the length of the subpartition. It appears that (15) and (16) do not appear in the literature of the q-series. Can we prove (15) or (16) analytically?

Acknowledgements The author would like to thank Bruce Berndt for his encouragement and help. The author also thanks the referee for the valuable comments on an earlier version of this paper.

References

1. Andrews, G.E.: The Theory of Partitions. Addison-Wesley, Reading (1976). Reissued by Cambridge University Press, Cambridge (1984)
2. Andrews, G.E.: Ramanujan's "lost" notebook. IV. Stacks and alternating parity in partitions. Adv. Math. **53**, 55–74 (1984)
3. Andrews, G.E., Berndt, B.C.: Ramanujan's Lost Notebook, Part II. Springer, New York (2009)
4. Andrews, G.E., Warnaar, S.O.: The product of partial theta functions. Adv. Appl. Math. **39**, 116–120 (2007)
5. Berndt, B.C.: Number Theory in the Spirit of Ramanujan. Am. Math. Soc., Providence (2006)
6. Kim, B.: Combinatorial proofs of certain identities involving partial theta functions. Int. J. Number Theory (to appear)
7. Kolitsch, L.W.: Rogers–Ramanujan subpartitions and their connections to other partition. Ramanujan J. **16**, 163–167 (2008)
8. Ramanujan, S.: The Lost Notebook and Other Unpublished Papers. Narosa, New Delhi (1988)
9. Yee, A.J.: Bijective proofs of a theorem of Fine and related partition identities. Int. J. Number Theory **5**, 209–218 (2009)

Ramanujan J (2010) 23: 169–181
DOI 10.1007/s11139-009-9158-0

Arithmetic properties of partitions with even parts distinct

**George E. Andrews · Michael D. Hirschhorn ·
James A. Sellers**

Received: 30 January 2009 / Accepted: 6 February 2009 / Published online: 16 September 2010
© Springer Science+Business Media, LLC 2010

Abstract In this work, we consider the function $ped(n)$, the number of partitions of an integer n wherein the even parts are distinct (and the odd parts are unrestricted). Our goal is to consider this function from an arithmetical point of view in the spirit of Ramanujan's congruences for the unrestricted partition function $p(n)$. We prove a number of results for $ped(n)$ including the following: For all $n \geq 0$,

$$ped(9n + 4) \equiv 0 \pmod 4$$

and

$$ped(9n + 7) \equiv 0 \pmod{12}.$$

Indeed, we compute appropriate generating functions from which we deduce these congruences and find, in particular, the surprising result that

$$\sum_{n \geq 0} ped(9n + 7)q^n = 12 \frac{(q^2; q^2)_\infty^4 (q^3; q^3)_\infty^6 (q^4; q^4)_\infty}{(q; q)_\infty^{11}}.$$

We also show that $ped(n)$ is divisible by 6 at least $1/6$ of the time.

Keywords Congruence · Partition · Distinct even parts · Generating function · Lebesgue identity

Research of the first author supported in part by NSF Grant DMS-0801184.

G.E. Andrews · J.A. Sellers (✉)
Department of Mathematics, The Pennsylvania State University, University Park, PA 16802, USA
e-mail: sellersj@math.psu.edu

G.E. Andrews
e-mail: andrews@math.psu.edu

M.D. Hirschhorn
School of Mathematics and Statistics, UNSW, Sydney 2052, Australia
e-mail: m.hirschhorn@unsw.edu.au

Mathematics Subject Classification (2000) 05A17 · 11P83

1 Introduction

In recent years, the function which enumerates those integer partitions wherein even parts are distinct (and odd parts are unrestricted) has arisen quite naturally. For example, the generating function for these partitions appears in the following classic identity of Lebesgue [8]:

$$\sum_{n\geq 0}\left(\prod_{i=1}^{n}\frac{1+q^i}{1-q^i}\right)q^{\binom{n+1}{2}} = \frac{(-q^2;q^2)_\infty}{(q;q^2)_\infty} = \frac{(q^4;q^4)_\infty}{(q;q)_\infty}$$

where $(a;q)_m = \prod_{n=0}^{m-1}(1-aq^n)$ and $(a;q)_\infty = \lim_{m\to\infty}(a;q)_m$.

As the above equation shows, the number of partitions of n wherein even parts are distinct equals the number of partitions of n with no parts divisible by 4. These are often referred to as 4-regular partitions and much has been written about arithmetic properties of such partitions. The reader interested in work involving regular partitions may wish to see Alladi [1], Andrews [2, Theorem 9], [3], Dandurand and Penniston [5], Granville and Ono [6], Gordon and Ono [7], Patkowski [9], and Penniston [10].

Our goal in this work is to consider $ped(n)$, the number of partitions of n wherein the even parts are distinct, from an arithmetic point of view in the spirit of Ramanujan's congruences for the unrestricted partition function $p(n)$. In this vein, we will prove various congruence properties satisfied by $ped(n)$ as well as a number of explicit results on generating function dissections. In particular, we will prove that

$$\sum_{n\geq 0}ped(9n+1)q^n = \frac{(q^2;q^2)_\infty^2(q^3;q^3)_\infty^4(q^4;q^4)_\infty}{(q;q)_\infty^5(q^6;q^6)_\infty^2}$$

$$+ 24q\frac{(q^2;q^2)_\infty^3(q^3;q^3)_\infty^3(q^4;q^4)_\infty(q^6;q^6)_\infty^3}{(q;q)_\infty^{10}},$$

$$\sum_{n\geq 0}ped(9n+4)q^n = 4\frac{(q^2;q^2)_\infty(q^3;q^3)_\infty(q^4;q^4)_\infty(q^6;q^6)_\infty}{(q;q)_\infty^4}$$

$$+ 48q\frac{(q^2;q^2)_\infty^2(q^4;q^4)_\infty(q^6;q^6)_\infty^6}{(q;q)_\infty^9},$$

and

$$\sum_{n\geq 0}ped(9n+7)q^n = 12\frac{(q^2;q^2)_\infty^4(q^3;q^3)_\infty^6(q^4;q^4)_\infty}{(q;q)_\infty^{11}}$$

from which it is immediate that for $n \geq 0$,

$$ped(9n+4) \equiv 0 \pmod 4$$

and

$$ped(9n + 7) \equiv 0 \pmod{12}.$$

We also deduce that for $\alpha \geq 1$ and all $n \geq 0$,

$$ped\left(3^{2\alpha+1}n + \frac{17 \times 3^{2\alpha} - 1}{8}\right) \equiv 0 \pmod 6,$$

$$ped\left(3^{2\alpha+2}n + \frac{19 \times 3^{2\alpha+1} - 1}{8}\right) \equiv 0 \pmod 6$$

and that $ped(n)$ is divisible by 6 at least $1/6$ of the time.

In the proofs below, we use nothing deeper than Ramanujan's $_1\psi_1$ summation formula, which, as noted in [4, Theorem 10.5.1], is given as follows: For $|q| < 1$ and $|b/a| < |x| < 1$,

$$\sum_{n=-\infty}^{\infty} \frac{(a; q)_n}{(b; q)_n} x^n = \frac{(ax; q)_\infty (q/ax; q)_\infty (q; q)_\infty (b/a; q)_\infty}{(x; q)_\infty (b/ax; q)_\infty (b; q)_\infty (q/a; q)_\infty}. \tag{1}$$

2 Preliminaries

We shall require several properties of the functions denoted $\phi(q)$ and $\psi(q)$ by Ramanujan, namely

$$\phi(q) = \sum_{n=-\infty}^{\infty} q^{n^2}$$

and

$$\psi(q) = \sum_{n=0}^{\infty} q^{(n^2+n)/2} = \sum_{n=-\infty}^{\infty} q^{2n^2-n}.$$

The necessary properties are given in the following lemmas. We include proofs for the sake of completeness.

Lemma 2.1

$$\phi(-q) = \frac{(q; q)_\infty^2}{(q^2; q^2)_\infty}.$$

Proof

$$\phi(-q) = \sum_{n=-\infty}^{\infty} (-1)^n q^{n^2}$$

$$= (q; q^2)_\infty^2 (q^2; q^2)_\infty \quad \text{by Jacobi's Triple Product Identity}$$

$$= (q; q)_\infty (q; q^2)_\infty$$

$$= \frac{(q; q)_\infty^2}{(q^2; q^2)_\infty}. \qquad \square$$

Lemma 2.2

$$\psi(-q) = \frac{(q^2; q^2)_\infty}{(-q; q^2)_\infty}.$$

Proof We have

$$\psi(-q) = \sum_{n=-\infty}^{\infty} (-1)^n q^{2n^2-n}$$

$$= (q; q^4)_\infty (q^3; q^4)_\infty (q^4; q^4)_\infty \quad \text{by Jacobi's Triple Product Identity}$$

$$= \frac{(q; q)_\infty}{(q^2; q^4)_\infty}$$

$$= \frac{(q^2; q^2)_\infty (q; q^2)_\infty}{(q^2; q^4)_\infty}$$

$$= \frac{(q^2; q^2)_\infty}{(-q; q^2)_\infty}. \qquad \qquad \square$$

Lemma 2.3

$$\phi(-q)\psi(-q) = \frac{(q; q)_\infty (q^2; q^2)_\infty}{(-q; q)_\infty (-q; q^2)_\infty}.$$

Proof We have

$$\phi(-q)\psi(-q) = \frac{(q; q)_\infty^2}{(q^2; q^2)_\infty} \frac{(q^2; q^2)_\infty}{(-q; q^2)_\infty} \quad \text{by Lemmas 2.1 and 2.2}$$

$$= \frac{(q; q)_\infty^2}{(-q; q^2)_\infty}$$

$$= \frac{(q; q)_\infty (q^2; q^2)_\infty}{(-q; q)_\infty (-q; q^2)_\infty}. \qquad \qquad \square$$

Lemma 2.4

$$\frac{\phi(-q)}{\psi(-q)} = \frac{(q; q^2)_\infty}{(-q^2; q^2)_\infty}.$$

Proof We have

$$\frac{\phi(-q)}{\psi(-q)} = \frac{(q; q)_\infty^2}{(q^2; q^2)_\infty} \frac{(-q; q^2)_\infty}{(q^2; q^2)_\infty} \quad \text{by Lemmas 2.1 and 2.2}$$

$$= \frac{(-q; q^2)_\infty}{(-q; q)_\infty^2}$$

$$= \frac{1}{(-q; q)_\infty (-q^2; q^2)_\infty}$$

$$= \frac{(q;q)_\infty}{(q^2;q^2)_\infty(-q^2;q^2)_\infty}$$

$$= \frac{(q;q^2)_\infty}{(-q^2;q^2)_\infty}. \qquad \square$$

We shall also need the following results.

Lemma 2.5

$$\frac{(q;q)_\infty^2(ax;q)_\infty(q/ax;q)_\infty}{(a;q)_\infty(q/a;q)_\infty(x;q)_\infty(q/x;q)_\infty} = \sum_{n=-\infty}^{\infty} \frac{x^n}{1-aq^n}.$$

Proof In Ramanujan's $_1\psi_1$ summation formula (1) set $b = aq$, then divide by $1 - a$. $\qquad \square$

Lemma 2.6

$$\phi(-q) = a(q^3) - 2qb(q^3)$$

where $a(q) = \phi(-q^3)$ and $b(q) = \frac{(q;q)_\infty(q^6;q^6)_\infty^2}{(q^2;q^2)_\infty(q^3;q^3)_\infty}$.

Proof

$$\phi(-q) = \sum_{n=-\infty}^{\infty} (-1)^n q^{n^2}$$

$$= \sum_{n=-\infty}^{\infty} (-1)^{3n} q^{(3n)^2} + \sum_{n=-\infty}^{\infty} (-1)^{3n-1} q^{(3n-1)^2} + \sum_{n=-\infty}^{\infty} (-1)^{3n+1} q^{(3n+1)^2}$$

$$= \sum_{n=-\infty}^{\infty} (-1)^n q^{9n^2} - 2q \sum_{n=-\infty}^{\infty} (-1)^n q^{9n^2-6n}$$

$$= a(q^3) - 2qb(q^3)$$

where

$$a(q) = \sum_{n=-\infty}^{\infty} (-1)^n q^{3n^2} = \phi(-q^3)$$

and

$$b(q) = \sum_{n=-\infty}^{\infty} (-1)^n q^{3n^2-2n}$$

$$= (q;q^6)_\infty(q^5;q^6)_\infty(q^6;q^6)_\infty \quad \text{by Jacobi's Triple Product Identity}$$

$$= \frac{(q;q)_\infty(q^6;q^6)_\infty^2}{(q^2;q^2)_\infty(q^3;q^3)_\infty}. \qquad \square$$

Lemma 2.7

$$a(q)^3 - 8qb(q)^3 = \frac{\phi(-q)^4}{\phi(-q^3)}.$$

Proof First,

$$(q;q)_\infty(\omega q;\omega q)_\infty(\omega^2 q;\omega^2 q)_\infty = \prod_{n=1}^{\infty}(1-q^n)(1-\omega^n q^n)(1-\omega^{2n}q^n)$$

$$= \prod_{3\mid n}(1-q^n)^3 \prod_{3\nmid n}(1-q^{3n})$$

$$= \frac{\prod_{n\geq 1}(1-q^{3n})^4}{\prod_{3\mid n}(1-q^{3n})}$$

$$= \frac{\prod_{n\geq 1}(1-q^{3n})^4}{\prod_{n\geq 1}(1-q^{9n})}$$

$$= \frac{(q^3;q^3)_\infty^4}{(q^9;q^9)_\infty}.$$

Next,

$$\phi(-q)\phi(-\omega q)\phi(-\omega^2 q) = \frac{(q;q)_\infty^2(\omega q;\omega q)_\infty^2(\omega^2 q;\omega^2 q)_\infty^2}{(q^2;q^2)_\infty(\omega^2 q^2;\omega^2 q^2)_\infty(\omega q^2;\omega q^2)_\infty} \qquad \text{by Lemma 2.1}$$

$$= \left(\frac{(q^3;q^3)_\infty^4}{(q^9;q^9)_\infty}\right)^2 \frac{(q^{18};q^{18})_\infty}{(q^6;q^6)_\infty^4}$$

$$= \left(\frac{(q^3;q^3)_\infty^2}{(q^6;q^6)_\infty}\right)^4 \frac{(q^{18};q^{18})_\infty}{(q^9;q^9)_\infty^2}$$

$$= \frac{\phi(-q^3)^4}{\phi(-q^9)}.$$

Finally,

$$a(q^3)^3 - 8q^3 b(q^3)^3 = (a(q^3) - 2qb(q^3))(a(q^3) - 2\omega qb(q^3))(a(q^3) - 2\omega^2 qb(q^3))$$

$$= \phi(-q)\phi(-\omega q)\phi(-\omega^2 q)$$

$$= \frac{\phi(-q^3)^4}{\phi(-q^9)}.$$

If we replace q^3 by q, we obtain the required result.   □

Lemma 2.8

$$\frac{1}{(a-2qb)^2} = \frac{a^4 + 4qa^3 b + 12q^2 a^2 b^2 + 16q^3 ab^3 + 16q^4 b^4}{(a^3 - 8q^3 b^3)^2}.$$

Remark 2.9 Note that this is an elementary lemma—it holds for all a and b provided $a - 2qb \neq 0$.

Proof

$$\frac{1}{(a-2qb)^2} = \left(\frac{a^2 + 2qab + 4q^2b^2}{a^3 - 8q^3b^3}\right)^2$$

$$= \frac{a^4 + 4qa^3b + 12q^2a^2b^2 + 16q^3ab^3 + 16q^4b^4}{(a^3 - 8q^3b^3)^2}. \qquad \square$$

With the above lemmas proved, we can now move to the dissections of the generating function for $ped(n)$ with the goal of proving the desired congruences.

3 Main results

We shall begin by proving the following theorem which provides our first set of important dissections.

Theorem 3.1

$$\sum_{n\geq 0} ped(3n)q^n = \frac{(q^4;q^4)_\infty(q^6;q^6)_\infty^4}{(q;q)_\infty^3(q^{12};q^{12})_\infty^2},$$

$$\sum_{n\geq 0} ped(3n+1)q^n = \frac{\phi(-q^3)\psi(-q^3)}{\phi(-q)^2}, \quad and$$

$$\sum_{n\geq 0} ped(3n+2)q^n = 2\frac{(q^2;q^2)_\infty(q^6;q^6)_\infty(q^{12};q^{12})_\infty}{(q;q)_\infty^3}.$$

Proof

$$\sum_{n\geq 0} ped(n)q^n = \frac{(-q^2;q^2)_\infty}{(q;q^2)_\infty}$$

$$= \frac{(-q^2;q^6)_\infty(-q^4;q^6)_\infty(-q^6;q^6)_\infty}{(q;q^6)_\infty(q^3;q^6)_\infty(q^5;q^6)_\infty}$$

$$= \frac{(-q^6;q^6)_\infty(-q^3;q^6)_\infty^2}{(q^3;q^6)_\infty(q^6;q^6)_\infty^2}\left(\frac{(q^6;q^6)_\infty^2(-q^2;q^6)_\infty(-q^4;q^6)_\infty}{(-q^3;q^6)_\infty^2(q;q^6)_\infty(q^5;q^6)_\infty}\right)$$

$$= \frac{(-q^3;q^3)_\infty(-q^3;q^6)_\infty}{(q^3;q^3)_\infty(q^6;q^6)_\infty}\sum_{n=-\infty}^{\infty}\frac{q^n}{1+q^{6n+3}} \quad \text{by Lemma 2.5}$$

$$= \frac{1}{\phi(-q^3)\psi(-q^3)}\sum_{n=-\infty}^{\infty}\frac{q^n}{1+q^{6n+3}} \quad \text{by Lemma 2.3.}$$

It follows that

$$\sum_{n\geq 0} ped(3n)q^n = \frac{1}{\phi(-q)\psi(-q)} \sum_{n=-\infty}^{\infty} \frac{q^n}{1+q^{6n+1}}$$

$$= \frac{1}{\phi(-q)\psi(-q)} \left(\frac{(q^6;q^6)_\infty^2 (-q^2;q^6)_\infty (-q^4;q^6)_\infty}{(-q;q^6)_\infty (-q^5;q^6)_\infty (q;q^6)_\infty (q^5;q^6)_\infty} \right)$$

$$= \frac{(q^4;q^4)_\infty (q^6;q^6)_\infty^4}{(q;q)_\infty^3 (q^{12};q^{12})_\infty^2}$$

after simplification, using Lemmas 2.1, 2.2 and such devices as

$$(-q^2;q^6)_\infty(-q^4;q^6)_\infty = \frac{(-q^2;q^2)_\infty}{(-q^6;q^6)_\infty}, \qquad (-q;q)_\infty = \frac{(q^2;q^2)_\infty}{(q;q)_\infty},$$

$$(-q;q^2)_\infty = \frac{(q^2;q^4)_\infty}{(q;q^2)_\infty} \quad \text{and} \quad (q;q^2)_\infty = \frac{(q;q)_\infty}{(q^2;q^2)_\infty}.$$

Similarly,

$$\sum_{n\geq 0} ped(3n+2)q^n = \frac{1}{\phi(-q)\psi(-q)} \sum_{n=-\infty}^{\infty} \frac{q^n}{1+q^{6n+5}}$$

$$= \frac{1}{\phi(-q)\psi(-q)} \left(\frac{(q^6;q^6)_\infty^2 (-q^6;q^6)_\infty (-1;q^6)_\infty}{(-q;q^6)_\infty (-q^5;q^6)_\infty (q;q^6)_\infty (q^5;q^6)_\infty} \right)$$

$$= \frac{2}{\phi(-q)\psi(-q)} \left(\frac{(q^6;q^6)_\infty^2 (-q^6;q^6)_\infty (-q^6;q^6)_\infty}{(-q;q^6)_\infty (-q^5;q^6)_\infty (q;q^6)_\infty (q^5;q^6)_\infty} \right)$$

$$= 2 \frac{(q^2;q^2)_\infty (q^6;q^6)_\infty (q^{12};q^{12})_\infty}{(q;q)_\infty^3}$$

again after simplification. Lastly,

$$\sum_{n\geq 0} ped(3n+1)q^n = \frac{1}{\phi(-q)\psi(-q)} \sum_{n=-\infty}^{\infty} \frac{q^n}{1+q^{6n+3}}$$

$$= \frac{\phi(-q^3)\psi(-q^3)}{\phi(-q)\psi(-q)} \frac{(-q^2;q^2)_\infty}{(q;q^2)_\infty} \quad \text{from the above work}$$

$$= \frac{\phi(-q^3)\psi(-q^3)}{\phi(-q)^2} \quad \text{by Lemma 2.4.} \qquad \Box$$

With the above set of dissections complete, we now move to a second set of dissections which provide the proofs of the desired congruences. The main results are summarized in the following theorem:

Theorem 3.2

$$\sum_{n\geq 0}ped(9n+1)q^n = \frac{(q^2;q^2)_\infty^2(q^3;q^3)_\infty^4(q^4;q^4)_\infty}{(q;q)_\infty^5(q^6;q^6)_\infty^2}$$

$$+24q\frac{(q^2;q^2)_\infty^3(q^3;q^3)_\infty^3(q^4;q^4)_\infty(q^6;q^6)_\infty^3}{(q;q)_\infty^{10}},$$

$$\sum_{n\geq 0}ped(9n+4)q^n = 4\frac{(q^2;q^2)_\infty(q^3;q^3)_\infty(q^4;q^4)_\infty(q^6;q^6)_\infty}{(q;q)_\infty^4}$$

$$+48q\frac{(q^2;q^2)_\infty^2(q^4;q^4)_\infty(q^6;q^6)_\infty^6}{(q;q)_\infty^9},$$

$$\sum_{n\geq 0}ped(9n+7)q^n = 12\frac{(q^2;q^2)_\infty^4(q^3;q^3)_\infty^6(q^4;q^4)_\infty}{(q;q)_\infty^{11}}.$$

Proof By Lemmas 2.6 and 2.8, we have

$$\sum_{n\geq 0}ped(3n+1)q^n$$

$$=\frac{\phi(-q^3)\psi(-q^3)}{\phi(-q)^2}$$

$$=\frac{\phi(-q^3)\psi(-q^3)}{(a(q^3)-2qb(q^3))^2}$$

$$=\phi(-q^3)\psi(-q^3)$$

$$\times\frac{a(q^3)^4+4qa(q^3)^3b(q^3)+12q^2a(q^3)^2b(q^3)^2+16q^3a(q^3)b(q^3)^3+16q^4b(q^3)^4}{(a(q^3)^3-8q^3b(q^3)^3)^2}.$$

From Lemmas 2.6 and 2.7, it follows that

$$\sum_{n\geq 0}ped(9n+1)q^n = \phi(-q)\psi(-q)\frac{a(q)^4+16qa(q)b(q)^3}{(a(q)^3-8qb(q)^3)^2}$$

$$=\frac{\psi(-q)\phi(-q^3)^3}{\phi(-q)^7}\left(a(q)^3+16qb(q)^3\right)$$

$$=\frac{\psi(-q)\phi(-q^3)^3}{\phi(-q)^7}\left((a(q)^3-8qb(q)^3)+24qb(q)^3\right)$$

$$=\frac{\psi(-q)\phi(-q^3)^3}{\phi(-q)^7}\left(\frac{\phi(-q)^4}{\phi(-q^3)}+24q\left(\frac{(q;q)_\infty(q^6;q^6)_\infty^2}{(q^2;q^2)_\infty(q^3;q^3)_\infty}\right)^3\right)$$

$$= \frac{(q^2;q^2)_\infty^2 (q^3;q^3)_\infty^4 (q^4;q^4)_\infty}{(q;q)_\infty^5 (q^6;q^6)_\infty^2}$$
$$+ 24q \frac{(q^2;q^2)_\infty^3 (q^3;q^3)_\infty^3 (q^4;q^4)_\infty (q^6;q^6)_\infty^3}{(q;q)_\infty^{10}}$$

after simplification. Similarly,

$$\sum_{n\geq 0} ped(9n+4)q^n$$

$$= \phi(-q)\psi(-q) \frac{4a(q)^3 b(q) + 16q b(q)^4}{(a(q)^3 - 8q b(q)^3)^2}$$

$$= 4 \frac{\psi(-q)\phi(-q^3)^2}{\phi(-q)^7} \left(\frac{(q;q)_\infty (q^6;q^6)_\infty^2}{(q^2;q^2)_\infty (q^3;q^3)_\infty} \right) \left(a(q)^3 + 4q b(q)^3 \right)$$

$$= 4 \frac{\psi(-q)\phi(-q^3)^2}{\phi(-q)^7} \left(\frac{(q;q)_\infty (q^6;q^6)_\infty^2}{(q^2;q^2)_\infty (q^3;q^3)_\infty} \right) \left((a(q)^3 - 8q b(q)^3) + 12q b(q)^3 \right)$$

$$= 4 \frac{\psi(-q)\phi(-q^3)^2}{\phi(-q)^7} \left(\frac{(q;q)_\infty (q^6;q^6)_\infty^2}{(q^2;q^2)_\infty (q^3;q^3)_\infty} \right)$$
$$\times \left(\frac{\phi(-q)^4}{\phi(-q^3)} + 12q \left(\frac{(q;q)_\infty (q^6;q^6)_\infty^2}{(q^2;q^2)_\infty (q^3;q^3)_\infty} \right)^3 \right)$$

$$= 4 \frac{(q^2;q^2)_\infty (q^3;q^3)_\infty (q^4;q^4)_\infty (q^6;q^6)_\infty}{(q;q)_\infty^4}$$
$$+ 48q \frac{(q^2;q^2)_\infty^2 (q^4;q^4)_\infty (q^6;q^6)_\infty^6}{(q;q)_\infty^9}$$

after simplification. Lastly,

$$\sum_{n\geq 0} ped(9n+7)q^n = \phi(-q)\psi(-q) \frac{12a(q)^2 b(q)^2}{(a(q)^3 - 8q b(q)^3)^2}$$

$$= 12 \frac{\psi(-q)\phi(-q^3)^4}{\phi(-q)^7} \left(\frac{(q;q)_\infty (q^6;q^6)_\infty^2}{(q^2;q^2)_\infty (q^3;q^3)_\infty} \right)^2$$

$$= 12 \frac{(q^2;q^2)_\infty^4 (q^3;q^3)_\infty^6 (q^4;q^4)_\infty}{(q;q)_\infty^{11}}. \qquad \square$$

Corollary 3.3 *For all* $n \geq 0$,

$$ped(9n+4) \equiv 0 \pmod 4, \quad and$$
$$ped(9n+7) \equiv 0 \pmod{12}.$$

Proof These congruences are immediate from Theorem 3.2. $\qquad \square$

With these results in hand, we now wish to prove two infinite families of Ramanujan-like congruences modulo 6 satisfied by $ped(n)$. We begin by proving the following:

Theorem 3.4 *For $\alpha \geq 1$,*

$$\sum_{n \geq 0} ped\left(3^{2\alpha}n + \frac{3^{2\alpha} - 1}{8}\right)q^n \equiv \psi(q) \quad (\text{mod } 2),$$

$$\sum_{n \geq 0} ped\left(3^{2\alpha+1}n + \frac{3^{2\alpha+2} - 1}{8}\right)q^n \equiv \psi(q^3) \quad (\text{mod } 2),$$

$$\sum_{n \geq 0} ped\left(3^{2\alpha}n + \frac{3^{2\alpha} - 1}{8}\right)q^n \equiv (-1)^{\alpha-1}\psi(-q)\phi(-q^3) \quad (\text{mod } 3), \quad and$$

$$\sum_{n \geq 0} ped\left(3^{2\alpha+1}n + \frac{3^{2\alpha+2} - 1}{8}\right)q^n \equiv (-1)^{\alpha}\phi(-q)\psi(-q^3) \quad (\text{mod } 3).$$

Proof From the work above, we have

$$\sum_{n \geq 0} ped(9n + 1)q^n = \frac{\psi(-q)\phi(-q^3)^3}{\phi(-q)^7}\left(\frac{\phi(-q)^4}{\phi(-q^3)} + 24q\left(\frac{(q;q)_\infty(q^6;q^6)_\infty^2}{(q^2;q^2)_\infty(q^3;q^3)_\infty}\right)^3\right).$$

It follows that, modulo 2,

$$\sum_{n \geq 0} ped(9n + 1)q^n \equiv \frac{\psi(-q)\phi(-q^3)^2}{\phi(-q)^3} \equiv \psi(-q) \equiv \psi(q) = f(q^3, q^6) + q\psi(q^9)$$

where $f(a, b)$ is Ramanujan's theta-function,

$$f(a, b) = \sum_{n=-\infty}^{\infty} a^{(n^2+n)/2}b^{(n^2-n)/2}.$$

The mod 2 result now follows by induction on α. Similarly, modulo 3,

$$\sum_{n \geq 0} ped(9n + 1)q^n \equiv \frac{\psi(-q)\phi(-q^3)^2}{\phi(-q)^3} \equiv \psi(-q)\phi(-q^3)$$

$$= \phi(-q^3)\left(f(-q^3, q^6) - q\psi(-q^9)\right).$$

The mod 3 result now follows by induction on α, using the fact that

$$\phi(-q) = \phi(-q^9) - 2qf(-q^3, -q^{15}). \qquad \square$$

Theorem 3.5 *For $\alpha \geq 1$ and all $n \geq 0$,*

$$ped\left(3^{2\alpha+1}n + \frac{17 \times 3^{2\alpha} - 1}{8}\right) \equiv 0 \quad (\text{mod } 2),$$

$$ped\left(3^{2\alpha+2}n + \frac{11 \times 3^{2\alpha+1} - 1}{8}\right) \equiv 0 \pmod{2},$$

$$ped\left(3^{2\alpha+2}n + \frac{19 \times 3^{2\alpha+1} - 1}{8}\right) \equiv 0 \pmod{2},$$

$$ped\left(3^{2\alpha+1}n + \frac{17 \times 3^{2\alpha} - 1}{8}\right) \equiv 0 \pmod{3}, \quad and$$

$$ped\left(3^{2\alpha+2}n + \frac{19 \times 3^{2\alpha+1} - 1}{8}\right) \equiv 0 \pmod{3}.$$

Proof The results follow directly from Theorem 3.4, once we observe that $\psi(q)$, $\psi(-q)$ and $\phi(q)$ contain no term in which the power of q is 2 modulo 3, while $\psi(q^3)$ contains no term in which the power of q is 1 or 2 modulo 3. □

Corollary 3.6 *For $\alpha \geq 1$ and all $n \geq 0$,*

$$ped\left(3^{2\alpha+1}n + \frac{17 \times 3^{2\alpha} - 1}{8}\right) \equiv 0 \pmod{6} \quad and$$

$$ped\left(3^{2\alpha+2}n + \frac{19 \times 3^{2\alpha+1} - 1}{8}\right) \equiv 0 \pmod{6}.$$

Proof These results are immediate from Theorem 3.5. □

We close with a significant result regarding the density of $ped(n)$ modulo 6.

Theorem 3.7 *The function ped(n) is divisible by 6 at least $1/6$ of the time.*

Proof The arithmetic sequences $9n + 7$, $3^{2\alpha+1}n + \frac{17 \times 3^{2\alpha} - 1}{8}$ and $3^{2\alpha+2}n + \frac{19 \times 3^{2\alpha+1} - 1}{8}$ (for $\alpha \geq 1$), on which $ped(\cdot)$ is 0 modulo 6, do not intersect. These sequences account for

$$\frac{1}{9} + \frac{1}{27} + \frac{1}{81} + \cdots = \frac{1}{6}$$

of all positive integers. □

References

1. Alladi, K.: Partition identities involving gaps and weights. Trans. Am. Math. Soc. **349**, 5001–5019 (1997)
2. Andrews, G.E.: Euler's "De Partitio Numerorum". Bull. Am. Math. Soc. **44**, 561–573 (2007)
3. Andrews, G.E.: Partitions with distinct evens, preprint
4. Andrews, G.E., Askey, R., Roy, R.: Special Functions. Encyclopedia of Mathematics and Its Applications, vol. 71. Cambridge University Press, Cambridge (1999)

5. Dandurand, B., Penniston, D.: ℓ-divisibility of ℓ-regular partition functions. Ramanujan J. (2010, to appear)
6. Granville, A., Ono, K.: Defective zero p-blocks for finite simple groups. Trans. Am. Math. Soc. **348**, 331–347 (1996)
7. Gordon, B., Ono, K.: Divisibility of certain partition functions by powers of primes. Ramanujan J. **1**, 25–34 (1997)
8. Lebesgue, V.A.: Sommation de quelques series. J. Math. Pures Appl. **5**, 42–71 (1840)
9. Patkowski, A.: On some partitions where even parts do not repeat (2010, to appear)
10. Penniston, D.: Arithmetic of ℓ-regular partition functions. Int. J. Number Theory **4**, 295–302 (2008)

Ramanujan J (2010) 23: 183–193
DOI 10.1007/s11139-009-9213-x

Modularity and the distinct rank function

Amanda Folsom

Dedicated to George Andrews on his 70th birthday

Received: 16 February 2009 / Accepted: 21 October 2009 / Published online: 30 January 2010
© Springer Science+Business Media, LLC 2010

Abstract If $R(\omega, q)$ denotes Dyson's partition rank generating function, due to work of Bringmann and Ono, it is known that for roots of unity $\omega \neq 1$, $R(\omega, q)$ is the "holomorphic part" of a harmonic weak Maass form. Dating back to Ramanujan, it is also known that $\widehat{R}(\omega, q) := R(\omega, q^{-1})$ is given by Eichler integrals and modular forms. In analogy to these results, more recently Monks and Ono have shown that modular forms arise in a natural way from $G(\omega, q)$, the generating function for ranks of partitions into distinct parts. Moreover, Monks and Ono pose the following problem: determine whether the function $\widehat{G}(\omega, q) := G(\omega, q^{-1})$ appears naturally in the theory of modular forms. Here we answer this question of Monks and Ono, and show that $\widehat{G}(\omega, q)$, when combined with $\widehat{G}(\omega^{-1}, q)$ and a twisted third-order mock theta of Ramanujan, form a weight 1 modular form. We provide a more general result on the modularity of certain expressions involving basic hypergeometric series and then show that our result on $\widehat{G}(\omega, q)$ may be deduced from this as a special case.

Keywords Modular forms · Integer partitions · Basic hypergeometric series

Mathematics Subject Classification (2000) Primary 11F11 · Secondary 05A17 · 33D15

1 Introduction and statement of results

Of great interest in number theory are identities that relate modular forms to arithmetic or combinatorial generating functions. One of the most celebrated examples of

A. Folsom (✉)
Department of Mathematics, University of Wisconsin, Madison, WI 53706, USA
e-mail: folsom@math.wisc.edu

such an identity is the Rogers–Ramanujan identity

$$\sum_{n\geq 0}\frac{q^{n^2}}{(q;q)_n} = \prod_{n\geq 0}\left(1 - q^{5n+1}\right)^{-1}\left(1 - q^{5n+4}\right)^{-1}, \tag{1.1}$$

where $(a;q)_n = \prod_{j=0}^{n-1}(1 - aq^j)$, $n \geq 1$, and $(a;q)_0 = 1$. The infinite product on the right-hand side of (1.1) is essentially a weight 0 modular form. Further, the infinite series on the left-hand side of (1.1) may be interpreted combinatorially as the generating function for the number of partitions of an integer n with minimal difference equal to 2. Another highly lauded result involves Dyson's rank generating function

$$R(\omega, q) := \sum_{n\geq 0} N(m, n)\omega^m q^n = \sum_{n\geq 0}\frac{q^{n^2}}{(\omega q;q)_n(\omega^{-1}q;q)_n}.$$

The rank of an integer partition is defined to be the largest part of the partition minus the number of parts, and $N(m, n)$ denotes the number of partitions of n with rank m. In [2], Bringmann and Ono prove for roots of unity ω, that $R(\omega, q)$ is the "holomorphic part" of a weight $1/2$ harmonic weak Maass form, a form that satisfies an appropriate modular transformation under a subgroup of the modular group $SL_2(\mathbb{Z})$, is annihilated by a Laplacian operator, and satisfies a certain growth condition in the cusps.

More classically understood, yet in analogy to the modularity associated to $R(\omega, q)$ proved by Bringmann–Ono in [2], is a result dating back to Ramanujan regarding the function $\widehat{R}(\omega, q) := R(\omega, q^{-1})$. Namely, one now understands how $\widehat{R}(\omega, q)$ fits into the theory of modular forms: it can be expressed using a weight 0 modular form and two false theta series. (See [1, 5], and (1.3).)

Also in analogy to Dyson's rank generating function $R(\omega, q)$, more recently in [6], Monks and Ono study the generating function $G(\omega, q)$ for ranks of partitions into distinct parts. That is,

$$G(\omega, q) := \sum_{n\geq 0}\sum_{m\in\mathbb{Z}} Q(m, n)\omega^m q^n = \sum_{n\geq 0}\frac{q^{\frac{n^2+n}{2}}}{(\omega q;q)_n},$$

and $Q(m, n)$ counts the number of partitions of n into distinct parts with rank m. As is true with $R(\omega, q)$, the authors show that modular forms arise from $G(\omega, q)$ in a natural way. Moreover, the authors pose a problem in analogy to the known modularity associated to $\widehat{R}(\omega, q)$ as described above. That is, if $\widehat{G}(\omega, q) := G(\omega, q^{-1})$, Monks and Ono naturally ask the following:

Problem 1 (Monks–Ono [6]) *Determine whether the series $\widehat{G}(\omega, q)$ appears naturally within the theory of automorphic forms.*

In this paper we solve this problem and show that $\widehat{G}(\omega, q)$, combined with $\widehat{G}(\omega^{-1}, q)$ and a twisted third-order mock theta function of Ramanujan, form a weight 1 modular form. To describe our results, we begin with the following expression for $\widehat{G}(\omega, q)$, revealed in [6, (1.8)]:

$$\widehat{G}(\omega, q) = \sum_{n \geq 0} \frac{(-\omega^{-1})^n}{(\omega^{-1}q; q)_n}.$$

As Monks and Ono point out, the series is not well defined, yet one of their main theorems [6, Theorem 1.2] shows that, for each integer r such that $0 \leq r < m$ and any mth root of unity $-\omega^{-1} \neq 1$, the series $\lim_{n \to \infty} \widehat{G}_{mn+r}(\omega, q)$ is well defined, where $\widehat{G}_t(\omega, q)$ denotes the tth partial sum of $\widehat{G}_t(\omega, q)$. Moreover, they show that this well-defined series $\lim_{n \to \infty} \widehat{G}_{mn+r}(\omega, q)$ differs from $\widehat{G}(\omega, q) = \lim_{n \to \infty} \widehat{G}_{mn}(\omega, q)$ by a constant multiple of the infinite product $q^{\frac{1}{48}} \eta_0(\omega^{-1}, q)^{-1}$, where

$$\eta_0(\omega, q) := q^{\frac{1}{48}} \prod_{n \geq 1} (1 - \omega q^n). \tag{1.2}$$

The products $\eta_0(\omega, q)$ may be used to define weight 0 modular forms

$$\eta(\omega, q) := q^{\frac{1}{24}} \eta_0(\omega, q) \eta_0(\omega^{-1}, q), \tag{1.3}$$

where $\omega \neq 1$. Note that when $\omega = 1$, we have

$$\eta(1, q) = \eta^2(q),$$

where $\eta(q) := q^{\frac{1}{24}} \prod_{n \geq 1}(1 - q^n)$ is the Dedekind η-function, a weight $\frac{1}{2}$ modular form. To regard these functions as modular forms, one lets $q = e^{2\pi i z}$, where $z \in \mathbb{H}$, the upper-half complex plane. (See, for example, [4].) Similar to what holds for the function $G(\omega, q)$ [6, Theorem 1.1], we will show that $\widehat{G}(\omega, q)$ and $\widehat{G}(\omega^{-1}, q)$ become "factors" of a weight 1 modular form, along with a twisted third-order mock theta function of Ramanujan

$$\psi(\omega, q) := \sum_{n \geq 0} \frac{q^{n^2} \omega^n}{(q; q^2)_n}.$$

(The function $\psi(1, q)$ is the original 3rd-order Ramanujan mock theta function.) As we will see in Theorem 1.1, this weight one modular form is described in terms of the weight 0 modular forms $\eta(\omega, q)$, and the weight 1/2 modular form $\eta(q)$. More precisely, we define

$$\widehat{D}(\omega, q) := (1 + \omega^{-1})\widehat{G}(\omega, q) + (1 - \omega^{-2})(\psi(-\omega^2, q) - 1).$$

Replacing q by $e^{2\pi i z}$, $\widehat{D}(\omega, q)$ becomes a function on $\mathbb{H}$. Our main theorem is as follows.

Theorem 1.1 *Let* $-\omega^{-1} \neq 1$ *be a primitive* mth *root of unity. Then* $q^{-\frac{1}{12}} \widehat{D}(\omega, q) \times \widehat{D}(\omega^{-1}, q)$ *is the weight 1 modular form*

$$q^{-\frac{1}{12}} \widehat{D}(\omega, q) \widehat{D}(\omega^{-1}, q) = \frac{\eta^4(q^2)\eta^2(\omega^2, q)}{\eta^2(q)\eta^3(\omega^2, q^2)}.$$

Example As an example of Theorem 1.1, we consider the case $\omega = i$. In this case, by (3.4), we see that

$$\widehat{G}(i, q) = (1 + i) F\left(q^{-1}, 0; -1, q^2\right),$$

$$\widehat{G}(-i, q) = (1 - i) F\left(q^{-1}, 0; -1, q^2\right),$$

which shows that $\widehat{D}(i, q) = \widehat{D}(-i, q)$. Thus, we find the weight $1/2$ modular form

$$q^{-\frac{1}{24}} \widehat{D}(i, q) = \frac{\eta^7(q^2)}{\eta^3(q)\eta^3(q^4)} = \frac{q^{\frac{1}{12}}}{4} \cdot \frac{\eta(q^2)}{\eta(q)\eta(q^4)} \cdot D^2(i, q), \qquad (1.4)$$

where $D(\omega, q) := (1 + w)G(\omega, q) + (1 - \omega)G(-\omega, q)$ is defined by Monks and Ono in [6], where it is shown that $q^{\frac{1}{24}} D(i, q)$ is a weight $1/2$ modular form given by a specific η-quotient (agreeing with (1.4)). (See the example following Theorem 1.1 in [6].)

The remainder of the paper is structured as follows. In Sect. 2, beginning with previously known results due to Fine [3], we deduce a more general statement (Proposition 2.2) regarding the modularity of a certain expression involving basic hypergeometric series. In Sect. 3, we prove Theorem 1.1. To do this, we perform initial manipulations on $\widehat{G}(\omega; q)$ and then explain how one can deduce Theorem 1.1 from the results in Sect. 2.

2 Modularity and basic hypergeometric series

In this section, beginning with results due to Fine [3], we deduce a more general statement regarding the modularity of a certain expression involving basic hypergeometric series. The basic hypergeometric series $F(a, b; t, q)$ are defined by

$$F(a, b; t, q) := \sum_{n \geq 0} \frac{(aq; q)_n}{(bq; q)_n} t^n$$

for specified parameters a, b and t. It is well known that the series $F(a, b; t, q)$ are well defined for any choice of parameters a, b, and t with the exception of certain instances when $b = q^{-n}$ or $t = q^{-n}$, $n \in \mathbb{N}$. For a more explicit description, see [3], Chap. 1, Sect. 3. Here, we will consider the following difference of basic hypergeometric series, $D_F(\alpha, \beta, q)$, defined by

$$D_F(\alpha, \beta, q) := F(\alpha, 0; \beta, q) - \alpha^{-1}\beta^{-1} F\left(\beta^{-1}, 0; \alpha^{-1}, q\right).$$

Before stating the main result of this section, we must first introduce some notation. First, we define a function on $\frac{1}{2}\mathbb{Z}$ as follows:

$$h(r) := \begin{cases} 0, & r \in \mathbb{Z}, \\ 1, & r \in \frac{1}{2} + \mathbb{Z}. \end{cases}$$

Next, we set the following notation.

Definition 2.1 For $r, s \in \frac{1}{2}\mathbb{Z}$ and parameters $\zeta, \xi, \alpha, \beta$ such that the expressions below are defined, we let

$$P(\alpha, \beta) := \frac{(1 - \alpha^{-1}\beta^{-1})(1 - \alpha\beta)}{(1 - \alpha)(1 - \alpha^{-1})(1 - \beta)(1 - \beta^{-1})}, \tag{2.1}$$

$$c_1^0(\xi, r, q) := (-\xi^{-1})^{r-1} q^{-r(r-1)/2} \cdot \frac{(1 - \xi^{-1})}{(1 - \xi q^r)}, \tag{2.2}$$

$$c_{-1}^0(\xi, r, q) := (-\xi)^{-r-1} q^{-r(r+1)/2} \cdot \frac{(1 - \xi)}{(1 - \xi^{-1} q^{-r})}, \tag{2.3}$$

$$c_1^1(\xi, r, q) := (-\xi^{-1})^{r - \frac{1}{2}} q^{-\frac{1}{16} - \frac{r(r-1)}{2}} \cdot \left(1 - \xi q^r\right)^{-1}, \tag{2.4}$$

$$c_{-1}^1(\xi, r, q) := (-\xi)^{-r - \frac{1}{2}} q^{-\frac{1}{16} - \frac{r(r+1)}{2}} \cdot \left(1 - \xi^{-1} q^{-r}\right)^{-1}, \tag{2.5}$$

$$c_0^0(\xi, r, q) := 1, \tag{2.6}$$

$$\kappa \begin{bmatrix} \delta_1 \\ \delta_2 \\ \delta_3 \end{bmatrix} \begin{pmatrix} \alpha, \beta \\ \zeta, \xi \ ; \ q \\ r, s \end{pmatrix} := P(\alpha, \beta) \cdot \frac{(c_{\delta_1}^{h(r+s)}(\zeta\xi, r+s, q))^2}{c_{\delta_2}^{h(r)}(\zeta, r, q) \cdot c_{\delta_3}^{h(s)}(\xi, s, q)}, \tag{2.7}$$

where $\delta_i \in \{1, -1, 0\}$, $1 \le i \le 3$.

Next, we define the functions $N_1(\zeta, \xi, r, s; q)$ and $N_2(\zeta, r, s; q)$ by

$$N_1(\zeta, \xi, r, s; q) := \begin{cases} \eta(\zeta, q)\eta(\xi, q), & r, s \in \mathbb{Z}, \\[4pt] \eta(\zeta, q) \frac{\eta(\xi, q^{\frac{1}{2}})}{\eta(\xi, q)}, & r \in \mathbb{Z}, s \in \frac{1}{2} + \mathbb{Z}, \\[4pt] \eta(\xi, q) \frac{\eta(\zeta, q^{\frac{1}{2}})}{\eta(\zeta, q)}, & r \in \frac{1}{2} + \mathbb{Z}, s \in \mathbb{Z}, \\[4pt] \frac{\eta(\zeta, q^{\frac{1}{2}})\eta(\xi, q^{\frac{1}{2}})}{\eta(\zeta, q)\eta(\xi, q)}, & r, s \in \frac{1}{2} + \mathbb{Z}, \end{cases} \tag{2.8}$$

$$N_2(\zeta, r, s; q) := \begin{cases} \eta^2(\zeta, q), & r + s \in \mathbb{Z}, \\[4pt] \frac{\eta^2(\zeta, q^{\frac{1}{2}})}{\eta^2(\zeta, q)}, & r + s \in \frac{1}{2} + \mathbb{Z}. \end{cases} \tag{2.9}$$

The functions N_1 and N_2 are modular forms for roots of unity ζ and ξ. We define one more function depending on $r, s \in \frac{1}{2}\mathbb{Z}$:

$$p(r, s) := \begin{cases} 0, & r, s \in \mathbb{Z}, \\[4pt] \frac{1}{16}, & r + s \in \frac{1}{2} + \mathbb{Z}, \\[4pt] -\frac{1}{8}, & r, s \in \frac{1}{2} + \mathbb{Z}. \end{cases}$$

The following result shows that up to a finite product, the product $D_F(\zeta_m q^S, \zeta_n q^T, q) \cdot D_F(\zeta_m^{-1} q^{-S}, \zeta_n^{-1} q^{-T}, q)$ is a modular form (with easily computable weight) under certain hypotheses on the parameters defining the arguments, $\zeta_m, \zeta_n, S,$ and T.

Proposition 2.2 *Let $m, n \in \mathbb{N}$, and let ζ_m be an mth root of unity, and ζ_n an nth root of unity. Next, we define*

$$K := K(S, T, \zeta_m, \zeta_n; q) = \kappa^{-1} \begin{bmatrix} \operatorname{sgn}(S+T) \\ \operatorname{sgn}(S) \\ \operatorname{sgn}(T) \end{bmatrix} \begin{pmatrix} \zeta_m q^S, \zeta_n q^T \\ \zeta_m, \zeta_n \quad ; q \\ S, T \end{pmatrix}. \qquad (2.10)$$

If ζ_m, ζ_n, S, T satisfy one of the following four conditions

(i) $\zeta_m \neq 1, \zeta_n \neq 1$, and $S, T \in \frac{1}{2}\mathbb{Z}$,
(ii) $\zeta_m \neq 1, \zeta_n = 1, S \in \frac{1}{2}\mathbb{Z}$, and either $T \in \mathbb{Z}^{>0}$ or $T \in \frac{1}{2} + \mathbb{Z}$,
(iii) $\zeta_m = 1, \zeta_n \neq 1, T \in \frac{1}{2}\mathbb{Z}$, and either $S \in \mathbb{Z}^{<0}$ or $S \in \frac{1}{2} + \mathbb{Z}$,
(iv) $\zeta_m = \zeta_n = 1$, and either $S \in \mathbb{Z}^{<0}, T \in \mathbb{Z}^{>0}$, or $S, T \in \frac{1}{2} + \mathbb{Z}$,

then the following is true:

$$K \cdot q^{\frac{1}{12} - p(S,T)} \cdot D_F\left(\zeta_m q^S, \zeta_n q^T, q\right) \cdot D_F\left(\zeta_m^{-1} q^{-S}, \zeta_n^{-1} q^{-T}, q\right)$$
$$= \frac{\eta^2(q) N_2(\zeta_m \zeta_n, S, T; q)}{N_1(\zeta_m, \zeta_n, S, T; q)}. \qquad (2.11)$$

Moreover, the expression in (2.11) is a modular form whose weight is determined using (2.8) and (2.9).

We recall the definition of the signum function appearing in Proposition 2.2,

$$\operatorname{sgn}(x) := \begin{cases} 1, & x > 0, \\ -1, & x < 0, \\ 0, & x = 0. \end{cases}$$

We also remark that conditions (i)–(iv) in Proposition 2.2 are to ensure the expressions appearing for D_F are well defined. To prove Proposition 2.2, we require some known results regarding basic hypergeometric series. By observing that the $F(a, b; t, q)$ satisfy difference equations of the form $f_n = L_n + M_n f_{n+1}, n \geq 0$, iterative methods allow one to solve for f_0. This procedure is described by Fine in [3], Chap. 1, Sects. 5–7. One result given by these methods that will be useful here is the following.

Lemma 2.3 *Assuming all series are well defined, one has the equality*

$$F(a, 0; t, q) \cdot \frac{(ba^{-1}; q)_\infty}{(bq; q)_\infty (ba^{-1} t^{-1}; q)_\infty}$$
$$= F(a, b; t, q) + \frac{ba^{-1} t^{-1}}{1 - ba^{-1} t^{-1}} \sum_{n \geq 0} \frac{(ba^{-1}; q)_n q^n}{(bq; q)_n (bq a^{-1} t^{-1}; q)_n}. \qquad (2.12)$$

Here and throughout the expression $(a; q)_\infty := \prod_{n \geq 0}(1 - aq^n)$. By the iterative procedures mentioned above, one may also deduce the following (see [3]).

Lemma 2.4 *Assuming the series below are well defined, the following identities hold:*

$$\frac{(atq;q)_\infty}{(t;q)_\infty} = F(a,1;t,q), \tag{2.13}$$

$$\sum_{n\geq 0}\frac{(t;q)_n q^n}{(bq;q)_n(q;q)_n} = F(bt^{-1},0;t,q)\frac{(t;q)_\infty}{(bq;q)_\infty(q;q)_\infty}. \tag{2.14}$$

Armed with Lemma 2.3 and Lemma 2.4, we first apply (2.14) to the series appearing in the right-hand side of (2.12), where we take $b=1$. We find

$$\sum_{n\geq 0}\frac{(a^{-1};q)_n q^n}{(q;q)_n(qa^{-1}t^{-1};q)_n} = F(t^{-1},0;a^{-1},q)\cdot\frac{(a^{-1};q)_\infty}{(a^{-1}t^{-1}q;q)_\infty(q;q)_\infty}. \tag{2.15}$$

Next we substitute (2.15) into (2.12) (with $b=1$) and replace the $F(a,1;t,q)$ appearing in (2.12) by the appropriate infinite product given by (2.13). Simplifying, we find

$$\begin{aligned}
D_F(a,t,q) &= \frac{(q;q)_\infty(a^{-1}t^{-1};q)_\infty(atq;q)_\infty}{(a^{-1};q)_\infty(t;q)_\infty}\\[2mm]
&= \frac{(1-a^{-1}t^{-1})}{(1-a^{-1})(1-t)}\cdot q^{-\frac{1}{12}}\frac{\eta(q)\eta(at,q)}{\eta_0(a^{-1},q)\eta_0(t,q)} \tag{2.16}
\end{aligned}$$

for parameters a,t,q such that this expression is well defined. We will need to prove one additional lemma.

Lemma 2.5 *Assuming notation as above, the following are true.*

(i) *Let $R\in\mathbb{Z}$. Then we have*

$$\eta_0(\zeta q^R,q)\eta_0(\zeta^{-1}q^{-R},q) = c^0_{\mathrm{sgn}(R)}(\zeta,R,q)\cdot\eta_0(\zeta,q)\eta_0(\zeta^{-1},q).$$

(ii) *Let $R\in\frac{1}{2}+\mathbb{Z}$. Then we have*

$$\eta_0(\zeta q^R,q)\eta_0(\zeta^{-1}q^{-R},q) = c^1_{\mathrm{sgn}(R)}(\zeta,R,q)\cdot\frac{\eta_0(\zeta,q^{\frac{1}{2}})\eta_0(\zeta^{-1},q^{\frac{1}{2}})}{\eta_0(\zeta,q)\eta_0(\zeta^{-1},q)}.$$

Proof To prove (i), we first note that the case $R=0$ follows trivially by the definition of $c^0_0(\zeta,0,q)$. Next, we note that one need only consider the case $R\in\mathbb{Z}^{\geq 1}$. The result for the case $R\in\mathbb{Z}^{\leq -1}$ follows from the result for $R\geq 1$ after replacing R by $-R$, and ζ by ζ^{-1}. Thus, assuming $R\in\mathbb{Z}^{\geq 1}$, we have

$$\begin{aligned}
&\eta_0(\zeta q^R,q)\eta_0(\zeta^{-1}q^{-R},q)\\[1mm]
&= q^{\frac{1}{24}}\prod_{k\geq 1}\left(1-\zeta q^{R+k}\right)\left(1-\zeta^{-1}q^{-R+k}\right)
\end{aligned}$$

$$
= \eta_0(\zeta,q)\eta_0\!\left(\zeta^{-1},q\right)\cdot \prod_{k=-R+1}^{0}\left(1-\zeta^{-1}q^k\right)\prod_{k=1}^{R}\left(1-\zeta q^k\right)^{-1}
$$

$$
= \eta_0(\zeta,q)\eta_0\!\left(\zeta^{-1},q\right)\cdot \prod_{k=0}^{R-1}\left(1-\zeta^{-1}q^{-k}\right)\prod_{k=1}^{R}\left(1-\zeta q^k\right)^{-1}
$$

$$
= \eta_0(\zeta,q)\eta_0\!\left(\zeta^{-1},q\right)\cdot \frac{(1-\zeta^{-1})}{(1-\zeta q^R)}\prod_{k=1}^{R-1}\frac{(1-\zeta^{-1}q^{-k})}{(1-\zeta q^k)}
$$

$$
= \eta_0(\zeta,q)\eta_0\!\left(\zeta^{-1},q\right)\cdot \frac{(1-\zeta^{-1})}{(1-\zeta q^R)}\cdot \left(-\zeta^{-1}\right)^{R-1}q^{-R(R-1)/2}
$$

$$
= c_1^0(\zeta,R,q)\cdot \eta_0(\zeta,q)\eta_0\!\left(\zeta^{-1},q\right),
$$

where we understand that any empty product appearing above (which occurs only in the case $R=1$) equals 1. The proof of (ii) follows similarly. Again we need only consider the case $R\in \frac{1}{2}+\mathbb{Z}^{\geq 0}$, as the case $R\in \frac{1}{2}+\mathbb{Z}^{<0}$ follows from the established result for $R\in \frac{1}{2}+\mathbb{Z}^{\geq 0}$ after replacing R by $-R$, and ζ by ζ^{-1}. For $R\in \frac{1}{2}+\mathbb{Z}^{\geq 0}$, we compute

$$
\eta_0\!\left(\zeta q^R,q\right)\eta_0\!\left(\zeta^{-1}q^{-R},q\right)
$$

$$
= q^{\frac{1}{24}}\prod_{k\geq 1}\left(1-\zeta q^{R+k}\right)\left(1-\zeta^{-1}q^{-R+k}\right)
$$

$$
= q^{\frac{1}{24}}\prod_{k\geq 1}\left(1-\zeta q^{\frac{1}{2}(2(R+k-\frac{1}{2})+1)}\right)\left(1-\zeta^{-1}q^{\frac{1}{2}(2(-R+k+\frac{1}{2})-1)}\right)
$$

$$
= q^{\frac{1}{24}}\prod_{k\geq R+\frac{1}{2}}\left(1-\zeta q^{\frac{1}{2}(2k+1)}\right)\prod_{k\geq -R+\frac{3}{2}}\left(1-\zeta^{-1}q^{\frac{1}{2}(2k-1)}\right)
$$

$$
= q^{\frac{1}{24}}\prod_{k=0}^{R-\frac{1}{2}}\left(1-\zeta q^{\frac{1}{2}(2k+1)}\right)^{-1}\prod_{k\geq 0}\left(1-\zeta q^{\frac{1}{2}(2k+1)}\right)
$$

$$
\times \prod_{-R+\frac{3}{2}}^{1}\left(1-\zeta^{-1}q^{\frac{1}{2}(2k-1)}\right)\prod_{k\geq 2}\left(1-\zeta^{-1}q^{\frac{1}{2}(2k-1)}\right)
$$

$$
= \frac{q^{\frac{1}{24}}}{(1-\zeta q^R)}\prod_{k\geq 0}\left(1-\zeta q^{\frac{1}{2}(2k+1)}\right)\left(1-\zeta^{-1}q^{\frac{1}{2}(2k+1)}\right)
$$

$$
\times \prod_{k=0}^{R-\frac{3}{2}}\frac{(1-\zeta^{-1}q^{\frac{1}{2}(-2k-1)})}{(1-\zeta q^{\frac{1}{2}(2k+1)})}, \tag{2.17}
$$

where we understand the empty product in (2.17) above (which only occurs when $R = \frac{1}{2}$) to equal 1. Continuing, we simplify and see that (2.17) becomes

$$\frac{q^{\frac{1}{24}}}{(1 - \zeta q^R)} \prod_{k \geq 1} \frac{(1 - \zeta q^{k/2})(1 - \zeta^{-1} q^{k/2})}{(1 - \zeta q^k)(1 - \zeta^{-1} q^k)} \left(-\zeta^{-1} q^{-\frac{1}{2}}\right)^{R - \frac{1}{2}} q^{-(R - \frac{3}{2})(R - \frac{1}{2})/2}$$

$$= c_1^1(\zeta, r, q) \cdot \frac{\eta_0(\zeta, q^{\frac{1}{2}}) \eta_0(\zeta^{-1}, q^{\frac{1}{2}})}{\eta_0(\zeta, q) \eta_0(\zeta^{-1}, q)}. \qquad \square$$

To prove Proposition 2.2, one uses (2.16) to rewrite the product $D_F(\zeta_m q^S, \zeta_n q^T, q) \cdot D_F(\zeta_m^{-1} q^{-S}, \zeta_n^{-1} q^{-T}, q)$ in terms of the functions $\eta(q)$, $\eta(\omega, q)$, and $P(\alpha, \beta)$ (defined in (2.1)). Next, one applies Lemma 2.5 to the expression obtained. Proposition 2.2 follows after using the definition of the finite product κ in (2.7), and the functions N_1 and N_2 in (2.8) and (2.9).

3 Proof of Theorem 1.1

Before we are able to apply Proposition 2.2 to the function $\widehat{G}(\omega, q)$, we must first obtain a new expression for $\widehat{G}(\omega, q)$ to which Proposition 2.2 applies. We do this by showing that the function

$$\widehat{G}(\omega, t; q) = \sum_{n \geq 0} \frac{t^n}{(\omega^{-1} q; q)_n} = \sum_{n=0}^{\infty} g_n t^n$$

satisfies a certain recurrence relation. Iterating the recurrence relation, and letting $t = -\omega^{-1}$, reveals the following.

Proposition 3.1 *The function $\widehat{G}(\omega, q)$ satisfies*

$$\widehat{G}(\omega, q) = \frac{1}{1 + \omega^{-1}} \sum_{n=0}^{\infty} \frac{(-1)^n \omega^{-2n} q^{n^2}}{(\omega^{-2} q^2; q^2)_n}.$$

Proof One can see by definition of g_n that $g_{n+1}(1 - \omega^{-1} q^{n+1}) = g_n$. If one multiplies this equation by t^{n+1} and sums on n, one finds that

$$\widehat{G}(\omega, t; q) = \frac{-1 + \omega^{-1}}{t - 1} - \frac{\omega^{-1}}{t - 1} \widehat{G}(\omega, tq, q). \tag{3.1}$$

Moreover, the fact that $(1 - a)(aq; q)_{n-1} = (a; q)_n$ implies

$$\widehat{G}(\omega, t; q) = 1 + \frac{t}{1 - \omega^{-1} q} \widehat{G}(\omega q^{-1}, t; q). \tag{3.2}$$

Combining (3.1) and (3.2) shows that

$$\widehat{G}(\omega, t; q) = \frac{1}{1 - t} + \frac{t q \omega^{-1}}{(1 - t)(1 - \omega^{-1} q)} \widehat{G}(\omega q^{-1}, tq; q). \tag{3.3}$$

Iterating (3.3) two more times, we see

$$\widehat{G}(\omega, t; q) = \frac{1}{1-t}\left(1 + \frac{tq\omega^{-1}}{(tq;q)_1(\omega^{-1}q;q)_1} + \frac{t^2q^4\omega^{-2}}{(tq;q)_1(\omega^{-1}q)_2}\widehat{G}(\omega q^{-2}, tq^2, q)\right),$$

$$\widehat{G}(\omega, t; q) = \frac{1}{1-t}\left(1 + \frac{tq\omega^{-1}}{(tq;q)_1(\omega^{-1}q;q)_1} + \frac{t^2q^4\omega^{-2}}{(tq;q)_2(\omega^{-1}q;q)_2}\right.$$

$$+ \left.\frac{t^3q^9\omega^{-3}}{(tq;q)_2(\omega^{-1}q;q)_3}\widehat{G}(\omega q^{-3}, tq^3, q)\right).$$

Continuing in this manner and iterating (3.3) an infinite number of times proves the proposition, after letting $t = -\omega^{-1}$.   □

Next, we make use of another expression for basic hypergeometric series of the shape appearing in Proposition 3.1.

Lemma 3.2 *For any parameters a, t, and q such that $F(a, 0; t, q)$ is well defined, we have*

$$F(a, 0; t, q) = \frac{1}{1-t}\sum_{n\geq 0}\frac{(-at)^n q^{(n^2+n)/2}}{(tq;q)_n}.$$

Proof See [3], Chap. 1, Sects. 4–6.   □

By Lemma 3.2 and Proposition 3.1, we see that

$$\widehat{G}(\omega, q) = (1 - \omega^{-1})F(q^{-1}, 0; \omega^{-2}, q^2),\tag{3.4}$$

$$(-1 + \psi(-\omega^2, q)) = -\omega^2 q F(\omega^2, 0; q, q^2).\tag{3.5}$$

Thus,

$$\widehat{D}(\omega, q)(1 - \omega^{-2})^{-1} = \widehat{G}(\omega, q) \cdot (1 - \omega^{-1})^{-1} + (-1 + \psi(-\omega^2, q))$$

$$= D_F(q^{-1}, \omega^{-2}, q^2).\tag{3.6}$$

After a short calculation using (2.16), one finds that

$$D_F(q^{-1}, \omega^2, q^2) = D_F(q, \omega^2, q^2) \cdot \frac{(1 - q^{-1})(1 - q\omega^2)}{(1 - q^{-1}\omega^{-2})},\tag{3.7}$$

or equivalently (after applying (3.6)) that

$$\widehat{D}(\omega, q)\widehat{D}(\omega^{-1}, q) = D_F(q^{-1}, \omega^{-2}, q^2)D_F(q, \omega^2, q^2)$$

$$\times \frac{(1 - q\omega^2)(1 - \omega^{-2})(1 - \omega^2)(1 - q^{-1})}{(1 - q^{-1}\omega^{-2})}.\tag{3.8}$$

We compute the relevant factor

$$\kappa := \kappa \begin{bmatrix} -1 \\ -1 \\ 0 \end{bmatrix} \begin{pmatrix} q^{-1}, \omega^{-2} \\ 1, \omega^{-2} \\ -\frac{1}{2}, 0 \end{pmatrix} = P(q^{-1}, \omega^{-2}) \cdot \frac{(c^1_{-1}(\omega^{-2}, -\frac{1}{2}, q^2))^2}{c^1_{-1}(1, -\frac{1}{2}, q^2)}$$

$$= q^{\frac{1}{8}} \frac{(1 - q^{-1}\omega^{-2})}{(1 - q^{-1})(1 - \omega^{-2})(1 - \omega^2)(1 - \omega^2 q)}.$$

$$(3.9)$$

Next, we apply Proposition 2.2 to the expression in (3.8) and find using (3.9) that

$$\widehat{D}(\omega, q)\widehat{D}(\omega^{-1}, q) = \kappa \cdot q^{-\frac{1}{6} + 2p(-\frac{1}{2}, 0)} \frac{(1 - \omega^2)(1 - \omega^{-2})(1 - q^{-1})(1 - q\omega^2)}{(1 - q^{-1}\omega^{-2})}$$

$$\times \eta^2(q^2) \frac{N_2(\omega^{-2}, -\frac{1}{2}, 0; q^2)}{N_1(1, \omega^{-2}, -\frac{1}{2}, 0; q^2)}$$

$$= q^{\frac{1}{12}} \frac{\eta^4(q^2)\eta^2(\omega^2, q)}{\eta^2(q)\eta^3(\omega^2, q^2)}.$$

This proves Theorem 1.1.

Acknowledgements The author wishes to thank the Ken Ono, George Andrews, and the referee for many helpful comments and suggestions while writing this paper.

References

1. Andrews, G.E.: An introduction to Ramanujan's "lost" notebook. In: Ramanujan: Essays and Surveys. Hist. Mat, vol. 22, pp. 165–184. Am. Math. Soc., Providence (2001)
2. Bringmann, K., Ono, K.: Dyson's ranks and Maass forms. Ann. Math. (to appear)
3. Fine, N.J.: Basic Hypergeometric Series and Applications. Math. Surveys and Monographs, vol. 27. Am. Math. Soc., Providence (1988)
4. Lang, S.: Elliptic Functions. Springer, New York (1987)
5. Lawrence, R., Zagier, D.: Modular forms and quantum invariants of 3-manifolds. Asian J. Math. **3**, 93–107 (1999)
6. Monks, M., Ono, K.: Modular forms arising from $Q(n)$ and Dyson's rank. Preprint (2009)

Ramanujan J (2010) 23: 195–213
DOI 10.1007/s11139-009-9209-6

Cluster parity indices of partitions

Kağan Kurşungöz

Dedicated to George Andrews for his 70th birthday

Received: 26 March 2009 / Accepted: 7 October 2009 / Published online: 30 January 2010
© Springer Science+Business Media, LLC 2010

Abstract Andrews recently made an extensive study of parity in partition identities. One of the open questions he listed was to describe the partitions enumerated by a series. The series resembled the series side of the Andrews–Gordon Identities with an extra parameter, and it seemed to have properties related to the parity indices, which are defined by Andrews. We define cluster parity indices, and settle the problem.

Keywords Integer partition · Cluster parity index

Mathematics Subject Classification (2000) Primary 05A17 · 05A15

1 Introduction

A partition λ of a positive integer n is a nonincreasing sequence of positive integers $\lambda_1 \geq \cdots \geq \lambda_k > 0$ such that $n = \lambda_1 + \cdots + \lambda_k$ [2, Chap. 1]. One may impose some constraints such as requiring distinct parts, parts that belong to certain residue classes modulo some positive integer, and so on.

In 1961, Gordon considered the partitions of n in which pairs of consecutive parts appear at most $k - 1$ times and 1 appears at most $a - 1$ times [4]. However, a generating function was not given for partitions subject to this constraint unless $k = 2$.

In 1974, Andrews, in his Proceedings of the National Academy of Sciences paper [1, (2.5)], discovered the generating function for $b_{k,a}(m, n)$, the number of partitions of n into m parts, where any consecutive pair of integers together occur at most $k - 1$ times, and 1 appears at most $a - 1$ times as

$$\sum_{m,n \geq 0} b_{k,a}(m, n) x^m q^n = \sum_{n_1,\ldots,n_{k-1} \geq 0} \frac{q^{N_1^2 + \cdots + N_{k-1}^2 + N_a + \cdots + N_{k-1}} x^{N_1 + \cdots + N_{k-1}}}{(q)_{n_1} \cdots (q)_{n_{k-1}}}. \quad (1.1)$$

K. Kurşungöz (✉)
The Pennsylvania State University, University Park, PA, USA
e-mail: kursun@math.psu.edu

Here,

$$(a)_n = (a; q)_n = (1 - a)(1 - aq) \cdots \left(1 - aq^{n-1}\right)$$

and

$$N_r = n_r + \cdots + n_{k-1}, \quad r = 1, \ldots, k - 1.$$

He used functional equations derived from recurrences satisfied by $b_{k,a}(m, n)$ and established the right hand side of (1.1) as a solution. Here, the exponent of q is the number being partitioned (n), and the exponent of x is the number of parts (m).

In a recent paper [3], Andrews revisited his generating function (1.1) and extended his results by considering some additional restrictions involving parities. He achieved those generalizations by using double recurrences satisfied by $b_{k,a}(m, n)$ where additional constraints are imposed. He discovered many new identities and accompanying classes of partitions, and gave a list of open problems.

One of the open problems Andrews listed was to investigate the function

$$\sum_{n_1, \ldots, n_{k-1} \geq 0} \frac{q^{N_1^2 + \cdots + N_{k-1}^2} x^{N_1 + \cdots + N_{k-1}} (-yq)_{n_1} \cdots (-yq)_{n_{k-1}}}{(q^2; q^2)_{n_1} \cdots (q^2; q^2)_{n_{k-1}}},$$

$$N_r = n_r + \cdots + n_{k-1}, r = 1, \ldots, k - 1, \tag{1.2}$$

which seems to have properties related to parity indices [3, 6], and which satisfy the conditions for $b_{k,k}(m, n)$ at the same time.

In Sect. 2, the *Gordon marking* of a partition is defined, and a set of attributes to the partition is described. Forward and backward moves are introduced, which are restrictions of adding one or subtracting one from some part in the partition. Several facts related to backward and forward moves follow. The proofs to these may be found in [5]. In Sect. 3, cluster parity indices are defined, hence the definition of parity indices is extended in a way compatible with Gordon marking. Then, an interpretation using ordinary partitions is given to (1.2), utilizing the methods in [5].

2 Background

Let $\lambda = \lambda_1 + \cdots + \lambda_m$ be a partition of n.

Definition 2.1 The *Gordon marking* of a partition λ is an assignment of positive integers (marks) to λ such that parts equal to any given integer a are assigned distinct marks from the set $\mathbb{Z}_{>0} \setminus \{r \mid \exists r\text{-marked } \lambda_j = a - 1\}$ such that the smallest possible marks are used first. Let $\lambda^{(r)}$ denote the sub-partition of λ that consists of all r-marked parts. Let N_r be the number of r-marked parts (i.e., the number of parts in $\lambda^{(r)}$), and let $n_r = N_r - N_{r+1}$ for any positive integer r.

For instance, if $\lambda = 18 + 17 + 16 + 15 + 15 + 13 + 13 + 11 + 9 + 7 + 6 + 6 + 5 + 4 + 3 + 2 + 2$, ($|\lambda| = 162$), then its Gordon marking would be

$$\lambda = 18_2 + 17_1 + 16_3 + 15_2 + 15_1 + 13_2 + 13_1 + 11_1 + 9_1 + 7_2 + 6_3 + 6_1 + 5_2$$

$$+ 4_1 + 3_3 + 2_2 + 2_1.$$

In fact, we can represent the Gordon marking by an array where the column indicates the value of a part, and the row (counted from bottom to top) indicates the mark, so the Gordon marking of λ above would be:

$$\lambda = \begin{Bmatrix} & 3 & & 6 & & & & & & 16 & & \\ 2 & & 5 & 7 & & & 13 & 15 & & & 18 \\ 2 & 4 & 6 & & 9 & 11 & 13 & 15 & & 17 & \end{Bmatrix};$$

we will use this representation throughout.

There are several things to note here. First of all, Gordon marking is unique.

$\lambda^{(r)}$s are sub-partitions with distinct non-consecutive parts because no consecutive parts are assigned the same mark by definition. Also, for any r-marked λ_j, $r > 1$, there is a unique $(r-1)$-marked $\lambda_{j_0} = \lambda_j$ or $\lambda_{j_0} = \lambda_j - 1$. This implies $N_1 \geq N_2 \geq \ldots$, and hence $n_1, n_2, \ldots \geq 0$.

Last, if λ is enumerated by $b_{k,a}(m, n)$, then there are no k or greater marked parts, since each consecutive pair of integers together occur at most $(k-1)$ times. In this case, we can restrict our attention on $N_1, \ldots, N_{k-1}$, and $n_1, \ldots, n_{k-1}$.

Definition 2.2 Let $\lambda = \lambda_1 + \cdots + \lambda_m$ be a Gordon marked partition. Let λ_j be an r-marked part such that

(a) There are no $(r + 1)$ or higher marked parts equal to λ_j or $\lambda_j + 1$,

and either

(b1) There is an r_0 marked part $\lambda_{j_0} = \lambda_j - 1$, $r_0 < r$ such that there are no r_0-marked parts equal to $\lambda_j + 1$, and $r_0 + 1$ or higher marked parts equal to $\lambda_j - 1$

or

(b2) There are $1, \ldots, r - 1$ marked parts equal to λ_j or $\lambda_j + 1$, and no r-marked parts equal to $\lambda_j + 2$.

A *forward move of the rth kind* is replacing r_0-marked λ_{j_0} with an r_0-marked $\lambda_{j_0} + 1$ if (a) and (b1) hold; and replacing r-marked λ_j with an r-marked $\lambda_j + 1$ if (a) and (b2) hold, and (b1) fails; hence $|\lambda| \leftarrow |\lambda| + 1$.

For example,

$$\lambda = \begin{Bmatrix} & 3 & & 6 & & & & & & 16 & & \\ 2 & & 5 & 7 & & & 13 & 15 & & & 18 \\ 2 & 4 & 6 & & 9 & 11 & 13 & 15 & & 17 & \end{Bmatrix}$$

$\downarrow$ a forward move of the 3rd kind on 3-marked 16

$$\lambda' = \begin{Bmatrix} & 3 & & 6 & & & & & & 16 & & \\ 2 & & 5 & 7 & & & 13 & & \mathbf{16} & & 18 \\ 2 & 4 & 6 & & 9 & 11 & 13 & 15 & & 17 & \end{Bmatrix}.$$

Observe that a forward move of the 2nd kind is not possible on 2-marked 13 in λ, since neither (b1) nor (b2) holds. Similarly, a forward move of the 2nd kind on 2-marked 2 is not possible, since (a) fails. To be more precise, we can replace that

2-marked 2 with a 2-marked 3 as some forward move, but that would be a forward move of the 3rd kind for 3-marked 3.

We remark here also that a forward move of the rth kind preserves the Gordon marking of unchanged parts.

Definition 2.3 Let $\lambda = \lambda_1 + \cdots + \lambda_m$ be a Gordon marked partition. Let $\lambda_j \neq 1$ be an r-marked part such that

(c) There are no $(r + 1)$ or greater marked parts that are equal to λ_j or $\lambda_j + 1$.

(d) There is an $r_0 \leq r$ such that there is an r_0-marked $\lambda_{j_0} = \lambda_j$, but no r_0-marked parts that are equal to $\lambda_j - 2$.

Choose the smallest r_0 described in (d), and a *backward move of the rth kind* on λ_j is replacing r_0-marked λ_{j_0} with an r_0-marked $\lambda_{j_0} - 1$, and hence $|\lambda| \leftarrow |\lambda| - 1$.

For instance,

$$\lambda = \left\{ \begin{array}{cccccccccc} & 3 & & 6 & & & & & 16 & \\ 2 & & 5 & 7 & & & 13 & 15 & & 18 \\ 2 & 4 & 6 & & 9 & 11 & 13 & 15 & 17 & \end{array} \right\}$$

$$\downarrow \quad \text{a backward move of the 3rd kind on 3-marked 6}$$

$$\lambda'' = \left\{ \begin{array}{cccccccccc} & 3 & 5 & & & & & & 16 & \\ 2 & & 5 & 7 & & & 13 & 15 & & 18 \\ 2 & 4 & 6 & & 9 & 11 & 13 & 15 & 17 & \end{array} \right\}.$$

Note that a backward move of the 2nd kind is not possible for 2-marked 2, since (c) does not hold. Similarly, a backward move of the 1st kind on 1-marked 11 is not possible, since (d) fails.

We note that a backward move of the rth kind preserves the marking of other unchanged parts.

Definition 2.4 A *double move* is two moves of the same kind made in succession on the same part, when conditions exist. Same part refers to λ_j if another strictly smaller marked part is altered, and to $\lambda_j \pm 1$ if λ_j itself is altered.

Proposition 2.5 *Let $\lambda = \lambda_1 + \cdots + \lambda_m$ be a Gordon marked partition. Let $\lambda_j \neq 1$ be an r-marked part. If conditions exist for a backward move of the rth kind on λ_j, then conditions will exist for a forward move of the rth kind on the same part after the backward move is performed. Conversely, if conditions exist for a forward move of the rth kind on λ_j, then conditions will exist for a backward move of the rth kind on the same part after the forward move is performed. Moreover, the moves made in given orders will fix λ.*

In other words, so many forward and that many backward moves, or vice versa, on the same part are inverse transformations on λ when conditions exist for the former sequence of moves.

Proposition 2.6 *Let* $\lambda = \lambda_1 + \cdots + \lambda_m$ *be a Gordon marked partition. Let* $\lambda_{j_1} < \lambda_{j_2}$ *be two r-marked parts such that there are no r or higher marked* λ_{j_3} *for which* $\lambda_{j_1} < \lambda_{j_3} < \lambda_{j_2}$.

(i) *If conditions exist for a backward move of the rth kind on* λ_{j_1}, *and (c) is satisfied for* λ_{j_2}, *then the move made on* λ_{j_1} *will enable a backward move of the rth kind on* λ_{j_2}.

(ii) *If conditions exist for a forward move of the rth kind on* λ_{j_2}, *(a) is satisfied for* λ_{j_1}, *and either there are no* $r + 1$ *or higher marked parts equal to* $\lambda_{j_1} - 1$, *or for all* $r_0 < r$ *there are* r_0 *marked parts equal to* λ_{j_1}, *then the move made on* λ_{j_2} *will enable a forward move of the rth kind on* λ_{j_1}.

As an example, a backward move of the 1st kind on 1-marked 9 in λ as defined above enables a backward move of the 1st kind for 1-marked 11.

We provide another example and a non-example for part (ii) of Proposition 2.6. Let

$$
\eta = \left\{
\begin{matrix}
 & 2 & & \\
 & 2 & & \\
1 & & 3 & 5 \\
1 & & 3 & 5 \\
1 & & 3 & 5
\end{matrix}
\right\}, \quad \text{and} \quad
\eta' = \left\{
\begin{matrix}
 & & & & 5 & & & \\
 & & & 4 & & 6 & & 8 \\
 & & 3 & & 5 & & 7 & \\
 & 2 & & 4 & & 6 & & 8 \\
1 & & 3 & & 5 & & 7 &
\end{matrix}
\right\}.
$$

For η, let η_{j_1} be the 3-marked 3, and η_{j_2} be the 3-marked 5, and for η', let η'_{j_1} be the 4-marked 6, and η'_{j_2} be the 4-marked 8. A forward move of the 3rd kind on η_{j_2} enables a forward move of the 3rd kind on η_{j_1}, in spite of the fact that there are 4- and 5-marked 2s. However, a forward move of the 4th kind on η'_{j_2} does not enable a forward move of the 4th kind on η'_{j_1}, since both (b1) and (b2) fail for the 4-marked 6. These last examples indicate that we do need a longer hypothesis for the second part in Proposition 2.6.

3 Cluster parity indices

In this section, we reconcile ideas from [5] and [6]. We begin with a definition and related constructions.

Let a partition $\lambda = \lambda_1 + \cdots + \lambda_m$ along with its Gordon marking be given.

Definition 3.1 *An r-cluster in* $\lambda = \lambda_1 + \cdots + \lambda_m$ *is a sub-partition* $\lambda_{i_1} \leq \cdots \leq \lambda_{i_r}$ *such that* λ_{i_j} *is j-marked for* $j = 1, \ldots, r$, $\lambda_{i_{j+1}} - \lambda_{i_j} \leq 1$ *for* $j = 1, \ldots, r - 1$, *and there are no* $(r + 1)$*-marked parts that are equal to* λ_{i_r} *or* $\lambda_{i_r} + 1$.

Note that the absence of $(r + 2)$-marked parts that are equal to λ_{i_r} or $\lambda_{i_r} + 1$ is not required.

Example Below is a partition where its r-clusters are indicated. There is a 5-cluster that consists of parts 4, 5, 5, 6, 6, a 3-cluster consisting of parts 6, 7, 8, and a 2-cluster consisting of parts 8, 9.

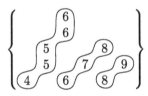

There is an obvious seemingly-alternative definition of an r-cluster. We would take an r-marked part λ_i where there are no $(r + 1)$-marked parts that are equal to λ_i or $\lambda_i + 1$. Then, we associate to that r-marked part λ_i an $(r - 1)$-marked part, an $(r - 2)$-marked part, …, a 1-marked part, all equal to λ_i or $\lambda_i - 1$. Observe that this definition is more intuitive as it reflects the idea of Gordon marking better. When we try to apply this definition to the above example, however, we would have

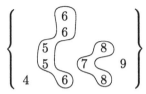

where the 2-marked 9 is dangling, or it defines a 2-cluster which overlaps with the 3-cluster that is defined by the 3-marked 8. This is not a desirable situation, and Definition 3.1 leads to much nicer consequences, as the subsequent results show.

Proposition 3.2 *Any partition* $\lambda = \lambda_1 + \cdots + \lambda_m$ *along with its Gordon marking has a unique decomposition into non-overlapping r-clusters.*

Proof It is evident that $\lambda = \lambda_1 + \cdots + \lambda_m$ is enumerated by $b_{k,k}(m, n)$ for some k. We will construct n_r r-clusters for $r = k - 1, k - 2, \ldots, 1$ in the given decreasing order. Here,

$$n_r = N_r - N_{r+1}$$

and N_rs are the number of r-marked parts in the Gordon marking of λ, as described in Definition 2.1.

Let

$$\lambda_1^{(r)} \leq \cdots \leq \lambda_{n_r}^{(r)}$$

be the r-marked parts that do not belong to any previously constructed r_1-cluster for $r_1 > r$. Then, there is no $(r + 1)$-marked $\lambda_j^{(r+1)}$ that is equal to $\lambda_j^{(r)}$ or $\lambda_j^{(r)} + 1$ for $j = 1, \ldots, n_r$, by the construction of the previous clusters. Also,

$$\lambda_j^{(r)} - \lambda_i^{(r)} \geq 2$$

for $1 \leq i < j \leq n_r$, by the Gordon marking of λ.

Now, there are exactly n_r $(r-1)$-marked

$$\lambda_1^{(r-1)} \leq \cdots \leq \lambda_{n_r}^{(r-1)}$$

such that

$$\lambda_j^{(r-1)} = \lambda_j^{(r)} \quad \text{or} \quad \lambda_j^{(r-1)} = \lambda_j^{(r)} - 1$$

for $j = 1, \ldots, n_r$, again by the Gordon marking of λ. These $\lambda_j^{(r-1)}$s are the $(r-1)$-marked parts in the r-clusters that contain $\lambda_j^{(r)}$s for $j = 1, \ldots, n_r$. Then, given indices i, j such that for $i, j, 1 \leq i < j \leq n_r$,

$$0 \leq \lambda_j^{(r)} - \lambda_j^{(r-1)} \leq 1 \quad \text{and} \quad 0 \leq \lambda_i^{(r)} - \lambda_i^{(r-1)} \leq 1.$$

It follows that

$$-1 \leq \left(\lambda_j^{(r)} - \lambda_i^{(r)}\right) - \left(\lambda_j^{(r-1)} - \lambda_i^{(r-1)}\right) \leq 1.$$

Since the difference in the first parenthesis is at least 2, the difference in the second parenthesis must be at least 1. So the $\lambda_j^{(r-1)}$s are distinct for $j = 1, \ldots, n_r$. Therefore,

$$\lambda_j^{(r-1)} - \lambda_i^{(r-1)} \geq 2,$$

by the Gordon marking of λ. Moreover, $\lambda_j^{(r-1)}$s do not belong to any r_1 cluster for $r_1 > r$, $j = 1, \ldots, n_r$. Because if they did, this would force some $\lambda_j^{(r)}$s also to be contained in those r_1-clusters, contradicting the choice of $\lambda_j^{(r)}$s, $j = 1, \ldots, n_r$.

We repeat the procedure to find the s-marked parts in the n_r r-clusters that are being constructed for $s = r - 2, r - 3, \ldots, 1$. At each step, by the above arguments, we have $\lambda_1^{(s)} \leq \cdots \leq \lambda_{n_r}^{(s)}$, which are s-marked parts that are pairwise at least two apart, and which do not belong to any r_1 cluster for $r_1 > r$. This gives us n_r r-clusters which do not overlap with the already constructed r_1-clusters, $r_1 > r$.

For each $r = k - 1, k - 2, \ldots, 1$, the construction makes unique choices of parts, given the uniqueness of the previously constructed clusters. For $r = k - 1$, there are only n_{k-1} $(k-1)$-marked parts in λ to begin with, so uniqueness of the whole decomposition follows.

Finally, we have used $r n_r$ parts for r-clusters, $r = k - 1, k - 2, \ldots, 1$, which are pairwise non-overlapping. Therefore, all $N_1 + \cdots + N_{k-1} = n_1 + 2n_2 + \cdots + (k-1)n_{k-1}$ parts eventually belong to an r-cluster for some $r = k-1, k-2, \ldots, 1$. $\square$

The partition λ from the previous example has a unique such decomposition.

We need a few key definitions and a related auxiliary result before the main theorem of this section (Theorem 3.7).

Definition 3.3 *Parity of an r-cluster* is the opposite parity of the number of even parts in that r-cluster.

Example We revisit

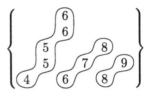

Here, the 5-cluster containing the 5-marked 6 is an even cluster, since there are an odd number of even parts in it, namely 4, 6, 6. The 3-cluster containing the 3-marked 8 is an odd cluster, and the 2-cluster containing the 2-marked 9 is an even cluster.

The reason for taking the opposite parity is that 1-clusters are just numbers, so we should keep the definition consistent with the parity of an integer. The reason why we look at the count of even parts instead of odd parts will become clear in the course of the proof of the main theorem of this section (Theorem 3.7).

Definition 3.4 *Lower even r-cluster parity index* of a partition λ is the number of times that the r-cluster parity changes from the r-cluster with the smallest r-marked part to the one with the largest r-marked part, beginning with an even r-cluster parity.

Definition 3.5 Given a partition $\lambda = \lambda_1 + \cdots + \lambda_m$ enumerated by $b_{k,k}(m, n)$, the *full lower even cluster parity index* of λ is the sum of all lower even 1-, 2-, ..., $(k - 1)$-cluster parity indices.

Example Below, only the 3-clusters of a Gordon marked partition are shown.

$$
\left\{
\begin{array}{ccccccccc}
 & & \boxed{7} & & \boxed{11} & & \boxed{15} & & \\
2 & & 7 & 9 & 11 & & 14 & 16 & 19 \\
2 & 4 & ⑥ & 8 & ⑩ & ⑬ & 15 & 17 & 19
\end{array}
\right\}
$$

The 3-cluster containing the 3-marked 7 is an even cluster, the one with the 3-marked 11 is also an even cluster, and so is the one with the 3-marked 15. Therefore, the 3-cluster parities written in increasing order of the 3-marked parts are

$$E E E.$$

Beginning with an even 3-cluster parity, there is no parity change, so the lower even 3-cluster parity index of this partition is 1.

Now, the 2-clusters in the same Gordon marked partition are shown.

$$
\left\{
\begin{array}{ccccccccc}
 & & 7 & & 11 & & 15 & & \\
\boxed{2} & & 7 & 9 & 11 & & 14 & 16 & 19 \\
2 & 4 & 6 & ⑧ & 10 & 13 & ⑮ & 17 & 19
\end{array}
\right\}
$$

With similar reasoning as above, the 2-cluster parities written in increasing order of 2-marked parts are

$$O E E O.$$

There is one parity change after an even 2-cluster parity to start with, so the lower even 2-cluster parity index of this partition is 2.

Next is the same Gordon marked partition where the 1-clusters are shown.

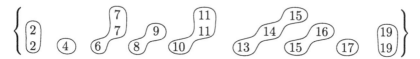

It is easy to see that the lower even 1-cluster parity index is 2.

Finally, all clusters of the Gordon marked partition in this example are shown.

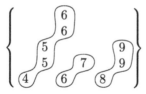

The full lower even cluster parity index of this partition is hence $5 = 2 + 2 + 1$.

Proposition 3.6 *Let $\lambda = \lambda_1 + \cdots + \lambda_m$ be a given partition along with its Gordon marking.*

(i) *If conditions exist for a single backward or forward move of the rth kind on an r-marked λ_i, then λ_i is the r-marked part in an r-cluster. Moreover, that move made will commute the r-cluster parity of the r-cluster containing λ_i (possibly altering the r-cluster parity index), and will keep the other parity indices fixed.*

(ii) *If conditions exist for a double backward or forward move of the rth kind on an r-marked λ_i, then that double move made will fix all cluster parity indices in λ.*

Before the proof is presented, we consider two examples.

Example Note that in

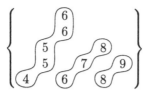

conditions exist for a forward move of the 3rd kind on the 3-marked 8. After the move is performed, the partition is transformed into

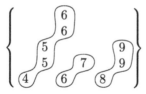

The 3-cluster is re-formed, and its parity changed from odd to even. The 5-cluster did not change at all; and although the 2-cluster is also re-formed, its parity is not altered.

Example In

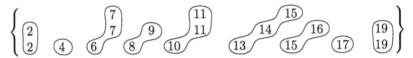

conditions exist for a double backward move of the 3rd kind on the 3-marked 11. After the move is made, the partition becomes

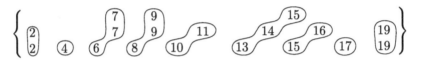

Observe that all cluster parities are preserved, as well as the respective orders of 1-, 2-, and 3-cluster parities. That is, all lower even cluster parity indices are fixed, and so is the full lower even cluster parity index.

Proof of Proposition 3.6 Throughout the proof, (a), (b1), and (b2) will refer to the conditions listed in Definition 2.2.

(i) We will prove this for a single forward move of the rth kind. The full result will follow by Proposition 2.5, since backward and forward moves of the same kind on the same part are inverses to each other.

When (a) holds, then there are no $(r + 1)$-marked parts that are equal to λ_i or $\lambda_i + 1$. Thus, λ_i does not belong to any r_1-cluster for $r_1 > r$, so it defines an r-cluster.

If (b1) held along with (a), then for some r_0, $r_0 < r$, there would be an r_0-marked λ_{i_0} which was equal to $\lambda_i - 1$, and there were $(r_0 + 1)$-, $(r_0 + 2)$-, $\ldots$, r-marked parts that were equal to λ_i. All of the mentioned $(r_0 + 1)$-, $(r_0 + 2)$-, $\ldots$, r-marked parts together with r_0-marked λ_{i_0} belong to the r-cluster that contains the r-marked λ_i. The forward move of the rth kind on the r-marked λ_i will replace the r_0-marked λ_{i_0} that equals $\lambda_i - 1$ by an r_0-marked $\lambda_{i_0} + 1$ that then will equal λ_i, and will still belong to the r-cluster that contains the r-marked λ_i along with the previously mentioned $(r_0 + 1)$-, $(r_0 + 2)$-, $\ldots$, r-marked parts that are equal to λ_i.

There are two cases to consider here:

Case (i) There is an $(r_0 - 1)$-marked λ_j that is equal to λ_{i_0}.

In this case, before the move $\lambda_{i_0} - \lambda_j = 0$, and after the move $(\lambda_{i_0} + 1) - \lambda_j = 1$. Thus, the $(r_0 - 1)$-marked λ_j is in the r-cluster containing the r-marked λ_i either before or after the move. Since no $(r_0 - 1)$ or lower marked part is altered by the forward move, the r-cluster parity of the r-cluster containing the r-marked λ_i is altered while all other cluster parities remain unchanged. The proof follows for this case.

Case (ii) There is an $(r_0 - 1)$-marked λ_j that is equal to $\lambda_{i_0} - 1$.

In this case, there is also an $(r_0 - 1)$-marked λ_{j_1} that is equal to $\lambda_{i_0} + 1$ (and hence to $\lambda_j + 2$) by the Gordon marking of λ. Again, by the Gordon marking, (a), and (b1), there are no r_0 or higher marked parts that are equal to $\lambda_{i_0} + 1$ or $\lambda_{i_0} + 2$. In particular, there are no r_0-marked parts that are equal to λ_{j_1} or $\lambda_{j_1} + 1$. Thus, λ_{j_1} is the $(r_0 - 1)$-marked part in an $(r_0 - 1)$-cluster before the forward move is performed. Please note that the forward move replaces λ_{i_0} by $\lambda_{i_0} + 1$. Therefore, after the move is performed, the $(r_0 - 1)$-cluster that contains λ_{j_1} will become part of the r-cluster containing the r-marked λ_i, and another $(r_0 - 1)$-cluster will emerge, this time containing the $(r_0 - 1)$-marked λ_j.

To conclude the proof in this case, we need to verify that the $(r_0 - 1)$-clusters containing λ_j and λ_{j_1} possess the same number of even and odd parts, once forgetting about the r_0- and higher marked parts. Call λ_j and λ_{j_1} as $\lambda_j^{(r_0-1)}$ and $\lambda_{j_1}^{(r_0-1)}$, respectively, where the superscript indicates the mark in the Gordon marking of λ. Call the s-marked parts in the $(r_0 - 1)$-cluster that contains the $(r_0 - 1)$-marked $\lambda_j^{(r_0-1)}$ as $\lambda_j^{(s)}$, and the s-marked parts in the $(r_0 - 1)$-cluster that contains the $(r_0 - 1)$-marked $\lambda_{j_1}^{(r_0-1)}$ as $\lambda_{j_1}^{(s)}$, $s < r_0 - 1$.

We inductively show that $\lambda_{j_1}^{(s)} = \lambda_j^{(s)} + 2$, $s \leq r_0 - 1$. We already know that $\lambda_{j_1} = \lambda_j + 2$, and that there is an r_0-marked part that is equal to $\lambda_j + 1$ in the original Gordon marking of λ, i.e., before the move is performed. This forms the base case of our induction on $s = r_0 - 1, r_0 - 2, \ldots, 1$.

For $s < r_0 - 1$, we assume that $\lambda_{j_1}^{(s+1)} = \lambda_j^{(s+1)} + 2$ and there is an $(s + 1)$-marked part that is equal to $\lambda_j^{(s+1)} + 1$. By the construction of clusters and by the Gordon marking of λ, $\lambda_{j_1}^{(s)} = \lambda_{j_1}^{(s+1)} - 1$, or $\lambda_{j_1}^{(s)} = \lambda_{j_1}^{(s+1)}$.

If $\lambda_{j_1}^{(s)} = \lambda_{j_1}^{(s+1)} - 1$, then it cannot be the case that $\lambda_j^{(s)} = \lambda_j^{(s+1)}$ since this would imply that $\lambda_{j_1}^{(s)} = \lambda_j^{(s)} + 1$, which violates the Gordon marking of λ. Thus, $\lambda_j^{(s)} = \lambda_j^{(s+1)} - 1$ also, and hence $\lambda_{j_1}^{(s)} = \lambda_j^{(s)} + 2$.

Else if $\lambda_{j_1}^{(s)} = \lambda_{j_1}^{(s+1)}$, then it cannot be the case that $\lambda_j^{(s)} = \lambda_j^{(s+1)} - 1$. This is because there is an $(s + 1)$-marked part that is equal to $\lambda_j^{(s+1)} + 1$, and when $\lambda_j^{(s)} = \lambda_j^{(s+1)} - 1$, the mark s is spared for a part that is equal to $\lambda_j^{(s+1)} + 1$. This violates the Gordon marking of λ. Thus, $\lambda_j^{(s)} = \lambda_j^{(s+1)}$, and consequently $\lambda_{j_1}^{(s)} = \lambda_j^{(s)} + 2$. The proof is complete in this case.

So far, we have assumed that (a) and (b1) hold. Else if (b1) fails, and (b2) holds along with (a), then the r-marked λ_i is replaced by r-marked $\lambda_i + 1$, and the

above argument carries over word by word with the only changes $r_0 \leftarrow r$ and
(b1) $\leftarrow$ (b2).

(ii) This is an iterated application of the previous part.   □

Now we are ready to prove the main result.

Theorem 3.7 *The function*

$$\sum_{n_1,\ldots,n_{k-1}\geq 0} \frac{q^{N_1^2+\cdots+N_{k-1}^2} x^{N_1+\cdots+N_{k-1}} (-yq)_{n_1} \cdots (-yq)_{n_{k-1}}}{(q^2; q^2)_{n_1} \cdots (q^2; q^2)_{n_{k-1}}},$$

$$N_r = n_r + \cdots + n_{k-1}, r = 1, \ldots, k-1, \tag{3.1}$$

*generates partitions enumerated by $b_{k,k}(m, n)$, where the exponent of q is the number
being partitioned (n), the exponent of x is the number of parts (m), and the exponent
of y is the full lower even cluster parity index.*

Proof (Cf. proof of Andrews' Conjecture in [5]) Let a partition $\lambda = \lambda_1 + \cdots + \lambda_m$ be
given along with its Gordon marking enumerated by $b_{k,k}(m, n)$, and a non-negative
integral power of y the exponent of which equals the full lower even cluster parity
index. We produce

- nonnegative integers $n_1, \ldots, n_{k-1}$,
- a base partition $\tilde{\lambda}$ such that

$$\tilde{\lambda}^{(r)} = 1, 3, \ldots, 2N_r - 1$$

  (contribution from the factor $q^{N_1^2+\cdots+N_{k-1}^2} x^{N_1+\cdots+N_{k-1}}$),
- a partition $\mu^{(r)}$ into at most n_r even parts (contribution from the factor $\frac{1}{(q^2;q^2)_{n_r}}$)
  for each $r = 1, \ldots, k-1$,
- and a partition $\nu^{(r)}$ into at most n_r distinct parts $\leq n_r$ (contribution from the factor
  $(-yq)_{n_r}$) for each $r = 1, \ldots, k-1$.

The number of parts in $\nu^{(r)}$ will give the lower even r-cluster parity index. This will
constitute the backward phase of the proof.

Conversely, given nonnegative integers $n_1, \ldots, n_{k-1}$, and for each $r = 1, \ldots, k-1$
a partition $\mu^{(r)}$ into at most n_r even parts, and a partition $\nu^{(r)}$ into at most n_r distinct
parts $\leq n_r$, we construct a partition $\lambda = \lambda_1 + \cdots + \lambda_m$ enumerated by $b_{k,k}(m, n)$
along with a nonnegative integral power of y, the exponent of which equals the full
lower even cluster parity index. The lower even r-cluster parity index of λ will be the
number of parts in $\nu^{(r)}$ for $r = 1, \ldots, k-1$. This will be the forward phase of the
proof.

Finally, we argue that the constructions are inverses to each other.

For the backward direction, let a partition $\lambda = \lambda_1 + \cdots + \lambda_m$ be given along with
its Gordon marking enumerated by $b_{k,k}(m, n)$, and a non-negative integral power of
y the exponent of which equals the full lower even cluster parity index.

Using Definition 2.1, we set $n_r = N_r - N_{r+1}$, where N_r is the number of r marked
parts in λ. $\tilde{\lambda}$ as described above is then constructed. Note that in $\tilde{\lambda}$, all clusters are odd

clusters (there are no even numbers at all), hence the full lower even cluster parity index is zero.

To transform λ into $\widetilde{\lambda}$, we consider each $r = k - 1, k - 2, \ldots, 1$ in the given decreasing order. At each step, we take the n_r-largest r-marked parts. By construction of $\widetilde{\lambda}$, all $(r + 1)$-marked parts are strictly less than the n_r largest r-marked parts by the time we consider any particular r. Consequently, by Definition 3.1, those n_r r-marked parts are also the r-marked parts in the r-clusters.

We perform so many double backward moves of the rth kind on the ith largest r-marked part for $i = n_r, n_r - 1, \ldots, 1$ in the given decreasing order until no more double backward moves are possible. We assign

$$\mu_i^{(r)} = 2 \times \begin{array}{l} \text{number of double backward moves} \\ \text{performed on the } i\text{th largest } r\text{-marked part.} \end{array}$$

Proposition 2.6(i) ensures

$$\mu_{n_r}^{(r)} \leq \cdots \leq \mu_1^{(r)}.$$

This will give us $\mu^{(r)}$, partitions into at most n_r even parts. The lower even r-cluster parity index remain fixed by Proposition 3.6(ii).

Let λ_i be the smallest r-marked part that belongs to an even r-cluster. It is the jth largest r-marked part for some $j \leq n_r$. By the construction so far, the parts smaller than the r-marked λ_i are all odd, and λ_i is even. To justify this claim, please note that by the hypothesis, we have obtained the $(r + 1), (r + 2), \ldots, (k - 1)$-clusters in $\widetilde{\lambda}$ in their respective places, which consist of all odd parts. Then, if there are r-clusters with smaller r-marked parts than λ_i, they are odd clusters. Hence, the number of even parts are even. Let λ_{i_0} be the smallest r-marked part $\lambda_{i_0} < \lambda_i$ such that there are at least two even parts in the r-cluster defined by λ_{i_0}. Then, a double backward move of the rth kind is possible on λ_{i_0}, contradicting the construction of $\mu^{(r)}$.

Now, since λ_i is even, the smaller r-marked part (if any) λ_{i_0} is odd, then (c) and (d) in Definition 2.3 are satisfied ((d) for $r_0 = r$), so a single backward move of the rth kind is possible on r-marked λ_i so as to move λ_i into its respective place, as the jth largest r-marked part in $\widetilde{\lambda}$. By Proposition 2.6(i), then single backward moves of the rth kind become possible for each of the $(j - 1)$th, $(j - 2)$th, $\ldots$, 1st largest r-marked parts. This gives us exactly j moves.

Note that λ_i is the smallest r-marked part to belong in an even r-cluster, and that even r-cluster becomes an odd one after the single backward move of the rth kind is performed. Moreover, all r-cluster parities of r-clusters with larger r-marked parts than λ_i are commuted, which means that the lower even r-cluster parity index is decreased by one, hence so is the full lower even cluster parity index. Therefore, we let the j single backward moves of the rth kind be accompanied by a factor of y.

We repeat the procedure of finding the smallest r-marked part that belong to an even r-cluster until there are none, and keep track of the single backward moves of the rth kind we perform to construct $\widetilde{\lambda}$. Eventually, the single moves that become possible make a partition into distinct parts $\leq n_r$ which is accompanied by a power of y whose exponent counts the lower even r-cluster parity index. That partition is $\nu^{(r)}$, as described above. This concludes the backward phase of the construction, and hence the first part of the proof.

Let us go over an example on the fly. Below is a Gordon marked partition from a preceding example in this section, where all 3-clusters are indicated. Note that the construction takes into account the r-clusters in decreasing order of $r = k - 1$, $k - 2, \ldots, 1$, and $k = a = 4$ here.

$$\left\{ \begin{array}{cccccccccc} & & & 7 & & 11 & & 15 & & \\ 2 & & & 7 & 9 & 11 & 14 & & 16 & 19 \\ 2 & 4 & 6 & 8 & & 10 & 13 & 15 & 17 & 19 \end{array} \right\}$$

We first perform 5 double backward moves on the 3-marked 7, 6 double backward moves on the 3-marked 11, and 8 double backward moves on the 3-marked 15. This will give us

$$\left\{ \begin{array}{cccccccc} & 2 & 4 & 6 & & & & \\ 1 & 3 & 5 & 8 & & 13 & 16 & 19 \\ 1 & 3 & 5 & 8 & 10 & 12 & 15 & 17 & 19 \end{array} \right\}$$

recovering $\mu^{(3)} = 16, 12, 10$, since the mentioned backward moves are double moves. Please note that no further double backward move of 3rd kind is possible on any of the 3-marked parts.

At this stage, we observe that a single backward move of 3rd kind is possible on the 3-marked 2, and hence two more single backward moves on the larger 3-marked parts. This will account for $\nu^{(3)} = 3$ along with a factor y, which contributed as the lower even 3-cluster parity index 1. We now have

$$\left\{ \begin{array}{cccccccc} 1 & 3 & 5 & & & & & \\ 1 & 3 & 5 & 8 & & 13 & 16 & 19 \\ 1 & 3 & 5 & 8 & 10 & 12 & 15 & 17 & 19 \end{array} \right\}$$

Next, we look at the 2-clusters

$$\left\{ \begin{array}{cccccccc} 1 & 3 & 5 & & & & & \\ 1 & 3 & 5 & 8 & & 13 & 16 & 19 \\ 1 & 3 & 5 & 8 & 10 & 12 & 15 & 17 & 19 \end{array} \right\}$$

1 double backward move of the 2nd kind on the 2-marked 8 is possible, after which 2 double backward moves on the 2-marked 13, then 3 double backward moves on the 2-marked 16, and lastly, 3 double backward moves on the 2-marked 19. This brings

$$\left\{ \begin{array}{cccccccc} 1 & 3 & 5 & & & & & \\ 1 & 3 & 5 & 7 & 10 & 12 & 14 & & \\ 1 & 3 & 5 & 7 & 9 & 11 & 14 & 16 & 19 \end{array} \right\}$$

No further double backward moves of the 2nd kind are possible, so $\mu^{(2)} = 6, 6, 4, 2$. The lower even 2-cluster parity index is still 2.

A single backward move of 2nd kind is now possible on the 2-marked 10, enabling two more single backward moves on 2-marked 12, and 14. This accounts for 3 single moves, along with a factor y.

$$
\left\{
\begin{array}{ccccccccc}
1 & 3 & 5 \\
1 & 3 & 5 & \boxed{7} & \boxed{9} & \boxed{11} & & \boxed{14} \\
1 & 3 & 5 & \boxed{7} & \boxed{9} & \boxed{11} & \boxed{13} & & 16 & & 19
\end{array}
\right\}
$$

The $r = 2$ stage is not over because we still need to take care of one more 2-cluster parity change. A single backward move of 2nd kind is now possible on the 2-marked 14 with no further single backward moves. This accounts for the second factor y for the lower even 2-cluster parity index 2, and yields $v^{(2)} = 3, 1$.

$$
\left\{
\begin{array}{ccccccc}
1 & 3 & 5 \\
1 & 3 & 5 & \boxed{7} & \boxed{9} & \boxed{11} & \boxed{13} \\
1 & 3 & 5 & \boxed{7} & \boxed{9} & \boxed{11} & \boxed{13} & 16 & & 19
\end{array}
\right\}
$$

Lastly, we consider $r = 1$

$$
\left\{
\begin{array}{ccccccccc}
1 & 3 & 5 \\
1 & 3 & 5 & 7 & 9 & 11 & 13 \\
1 & 3 & 5 & 7 & 9 & 11 & 13 & \boxed{16} & & \boxed{19}
\end{array}
\right\}
$$

No double backward moves of 1st kind is possible on any of the 2 largest 1-marked parts. In other words, we cannot subtract 2 from either. Therefore, $\mu^{(1)}$ is the empty partition.

It is easy to see that the lower even 1-cluster parity index is 2. A single backward move of 1st kind on the 1-marked 16 will enable a single backward move on the 1-marked 19. It accounts for a factor y as well.

$$
\left\{
\begin{array}{ccccccccc}
1 & 3 & 5 \\
1 & 3 & 5 & 7 & 9 & 11 & 13 \\
1 & 3 & 5 & 7 & 9 & 11 & 13 & \boxed{15} & & \boxed{18}
\end{array}
\right\}
$$

Then, another single backward move of 1st kind is possible on the 1-marked 18, along with another factor y. This yields the base partition $\widetilde{\lambda}$, and $v^{(1)} = 2, 1$.

$$
\widetilde{\lambda} =
\left\{
\begin{array}{ccccccccc}
1 & 3 & 5 \\
1 & 3 & 5 & 7 & 9 & 11 & 13 \\
1 & 3 & 5 & 7 & 9 & 11 & 13 & 15 & & 17
\end{array}
\right\}
$$

We have accounted for 5 y factors, since the full lower even cluster parity index of the original partition is 5, and $\widetilde{\lambda}$ has all odd clusters, therefore its full lower even cluster parity index is zero.

At this point, it becomes clear why the definition of cluster parity counts the even parts in a cluster. We want the base partition $\widetilde{\lambda}$ to have full lower even cluster parity index zero. To ensure this, clusters consisting of a single file of odd numbers must have odd cluster parity.

For the forward phase of the proof, given integers $n_1, \ldots, n_{k-1}$, we first set $N_r = n_r + n_{r+1} + \cdots + n_{k-1}$ for $r = 1, \ldots, k-1$. Then, we construct $\widetilde{\lambda}$ as described above. The r-marked parts in $\widetilde{\lambda}$ are

$$\widetilde{\lambda}^{(r)} = 1, 3, \ldots, 2N_r - 1$$

for $r = 1, \ldots, k-1$.

In the procedure that follows, we consider $r = 1, \ldots, k-1$ in the given increasing order.

For a particular r, we take the n_r largest r-marked parts, which are in their original places in $\widetilde{\lambda}$. Thus, they are the r-marked parts in the n_r odd r-clusters. The initial lower even r-cluster parity index is zero. For the largest r-marked part, (a) and (b1) in Definition 2.2 are satisfied, so we can always perform a single forward move of the rth kind on it.

Next, we realize $\nu^{(r)}$, accompanied with a power of y the exponent of which equals the number of parts in $\nu^{(r)}$. For each part j of $\nu^{(r)}$ from largest to smallest, we perform single forward moves of the rth kind on the j largest r-marked parts. The forward move on the largest one is possible as argued in the preceding paragraph. The subsequent singe forward moves are possible by Proposition 2.6(ii).

After all parts of $\nu^{(r)}$ are realized as sequences of single forward moves of the rth kind, note that we have exactly as many r-cluster parity changes counted from the r-cluster with the smallest r-marked part to the one with the largest as there are parts in $\nu^{(r)}$ by Proposition 3.6(i). The first parity change is to an even r-cluster, since all r-clusters were originally odd. Therefore, the exponent of the accompanying power of y now counts the lower even r-cluster parity index. Also, no double backward move of the rth kind is possible at this point, since parts of $\nu^{(r)}$ are distinct.

We realize $\mu^{(r)}$s as double forward moves of the rth kind, from the largest ($\mu_1^{(r)}$) to smallest ($\mu_{n_r}^{(r)}$, possibly zero), as

$$\text{number of double forward moves to be performed on the } i\text{th largest } r\text{-marked part} = \mu_i^{(r)}/2$$

(a) and (b2) in Definition 2.2 are always satisfied (possibly (b1)) for the largest r-marked part. The moves for the remaining r-marked parts are then possible by Proposition 2.6(ii) and by the fact that

$$\mu_1^{(r)} \geq \cdots \geq \mu_{n_r}^{(r)}.$$

The lower even r-cluster parity index remain fixed by Proposition 3.6(ii). This way, λ is constructed.

Finally, by Definition 3.5, the power of y accompanying the constructed λ has exponent equal to full lower even cluster parity index of λ. This establishes the forward phase of the construction, and hence the second part of the proof.

We see that the two constructions above consist of moves performed in the exact reverse order, thus they are inverses to each other, since backward and forward moves are, by Proposition 2.5. $\qquad\square$

We give another example here, and work in the forward direction as described in the preceding proof. Let $n_1 = 0$, $n_2 = 4$, $n_3 = 0$, and $n_4 = 3$. Let

$$\mu^{(2)} = 2, 2; \qquad \mu^{(4)} = 16, 10, 6;$$
$$\nu^{(2)} = 4, 3, 1; \qquad \nu^{(4)} = 2, 1.$$

$\mu^{(1)}$, $\mu^{(3)}$, $\nu^{(1)}$, and $\nu^{(3)}$ are essentially empty partitions. Since $\nu^{(2)}$ and $\nu^{(4)}$ have three and two parts, respectively, the resulting partition λ will have lower even 2-cluster parity index 3, and lower even 4-cluster parity index 2. The full lower even parity index of λ will therefore be $5 = 3 + 0 + 2 + 0$.

We first construct the base partition $\tilde{\lambda}$ as described in the proof

$$\tilde{\lambda} = \begin{Bmatrix} 1 & 3 & 5 & & & & \\ 1 & 3 & 5 & & & & \\ 1 & 3 & 5 & 7 & 9 & 11 & 13 \\ 1 & 3 & 5 & 7 & 9 & 11 & 13 \end{Bmatrix}.$$

$\tilde{\lambda}$ has all odd clusters, hence the full lower even cluster parity index is zero.

Since $n_1 = n_3 = 0$, we will skip the procedure for $r = 1$ and $r = 3$. We thus start with $r = 2$ and realize $\nu^{(2)} = 4, 3, 1$. This will introduce lower even 2-cluster parity index 3. We first apply a single forward move of the 2nd kind on the largest 2-marked part 13. Then, single forward moves on the 3 largest 2-marked parts, and single forward moves on the 4 largest 2-marked parts follow.

The 2-clusters are indicated, and their parities listed in increasing order of the 2-marked parts in them are $EOOE$. The lower even 2-cluster parity index is 3, as $\nu^{(2)}$ is accompanied with y^3. Then we realize $\mu^{(2)}$ as described in the above proof

Next, for $r = 4$, we first take into account $\nu^{(4)} = 2, 1$ as a single forward move of the 4th kind on the largest 4-marked part 5, followed by single forward moves on the 2 largest 4-marked parts. Indicating the 4-clusters, we have

Here, the lower even 4-cluster parity index is 2, since $v^{(4)}$ is accompanied by y^2. Then, we realize $\mu^{(4)}$, and indicate all clusters in the resulting partition λ

In this final picture, the full lower even cluster parity index of λ is 5, as expected.

It should be remarked that the proof of Theorem 3.7 just with appropriate nota-tional changes applies to prove a slightly more general result that

$$\sum_{n_1,\ldots,n_{k-1}\geq 0} \frac{q^{N_1^2+\cdots+N_{k-1}^2}x^{N_1+\cdots+N_{k-1}}(-y_1q)_{n_1}\cdots(-y_{k-1}q)_{n_{k-1}}}{(q^2;q^2)_{n_1}\cdots(q^2;q^2)_{n_{k-1}}} \qquad (3.2)$$

generates partitions enumerated by $b_{k,k}(m,n)$, where the exponent of q is the number being partitioned (n), the exponent of x is the number of parts (m), and the exponent of y_j is the lower even j-cluster parity index for $j=1,\ldots,k-1$. Upon setting $y_1=\cdots=y_{k-1}=y$, Theorem 3.7 follows.

Lastly, we recall another definition from [3, Sect. 11] for a comparison.

Definition 3.8 A partition λ is said to have *even ample part size* if each part λ_i is larger than the upper even parity index of λ.

For $k=2$, Theorem 3.7 reduces to the statement that

$$\sum_{n\geq 0} \frac{q^{n^2}x^n(-yq)_n}{(q^2;q^2)_n} \qquad (3.3)$$

generates partitions with no same or consecutive parts, where the exponent of q is the number being partitioned, the exponent of x is the number of parts, and the exponent of y is the lower even parity index.

The function (3.3) also generates partitions into distinct parts with even ample part size [3, Theorem 14], [6, Theorem 4.2].

The interpretation of Theorem 3.7 is different from even ample part size. It will be convenient to rework the example given in [6] for even ample part size in this context. We have $k=2$, $n=n_1=6$, $\mu=4+4+4+2$, and $v=v^{(1)}=4+2+1$. Then, the base partition will be

$$\tilde{\lambda}=\{1 \quad 3 \quad 5 \quad 7 \quad 9 \quad 11\}$$

which becomes

$$\{1 \quad 3 \quad 6 \quad 8 \quad 11 \quad 14\}$$

after the realization of v, and

$$\lambda=\{1 \quad 3 \quad 8 \quad 12 \quad 15 \quad 18\}$$

after the realization of μ as described in the proof of Theorem 3.7, with lower even parity index 3, but not having even ample part size; as opposed to the partition

$$15 + 13 + 12 + 8 + 5 + 4$$

which is not enumerated by $b_{2,2}(6, 57)$, but has even ample part size.

References

1. Andrews, G.E.: An analytic generalization of the Rogers–Ramanujan identities for odd moduli. Proc. Natl. Acad. Sci. USA **71**, 4082–4085 (1974)
2. Andrews, G.E.: The Theory of Partitions. Addison-Wesley, Reading (1976). Reissued: Cambridge University Press (1998)
3. Andrews, G.E.: Parity in partition identities. Ramanujan J. (accepted)
4. Gordon, B.: A combinatorial generalization of the Rogers–Ramanujan identities. Am. J. Math. **83**, 393–399 (1961)
5. Kurşungöz, K.: Parity considerations in Andrews–Gordon identities. Eur. J. Combin. (2009, in press). doi:10.1016/j.ejc.2009.06.002
6. Kurşungöz, K.: Parity indices of partitions (submitted)

Ramanujan J (2010) 23: 215–225
DOI 10.1007/s11139-009-9190-0

Ramanujan's partial theta series and parity in partitions

Ae Ja Yee

Dedicated to George Andrews for his 70th birthday

Received: 13 April 2009 / Accepted: 3 July 2009 / Published online: 11 June 2010
© Springer Science+Business Media, LLC 2010

Abstract A partial theta series identity from Ramanujan's lost notebook has a connection with some parity problems in partitions studied by Andrews in Ramanujan J., to appear where 15 open problems are listed. In this paper, the partial theta series identity of Ramanujan is revisited and answers to Questions 9 and 10 of Andrews are provided.

Keywords Integer partitions · Parity indices in partitions · Partial theta series · Ramanujan's lost notebook

Mathematics Subject Classification (2000) Primary 05A17 · Secondary 11P81

1 Introduction

In his recent paper [3], George Andrews investigated a variety of parity questions in partition identities. At the end of the paper, he then listed 15 open problems, two of which have a connection with the following partial theta series from Ramanujan's lost notebook.

Theorem 1.1 ([7, p. 28], [4, Entry 1.6.2]) *For any complex number a,*

$$1 + \sum_{n=1}^{\infty} \frac{(-q;q)_{n-1} a^n q^{n(n+1)/2}}{(-aq^2;q^2)_n} = \sum_{n=0}^{\infty} a^n q^{n^2}. \tag{1.1}$$

Partially supported by National Science Foundation Grant DMS-0801184.

A.J. Yee (✉)
Department of Mathematics, Penn State University, University Park, PA 16802, USA
e-mail: yee@math.psu.edu

As customary, here and in the sequel, we employ the standard notation

$$(a;q)_0 := (a)_0 := 1, \qquad (a;q)_n := (a)_n := (1-a)(1-aq)\cdots(1-aq^{n-1}), \quad n \geq 1,$$

and

$$(a;q)_\infty := (a)_\infty := \lim_{n\to\infty} (a;q)_n, \quad |q| < 1.$$

For a partition λ, we define upper odd parity index $I_{UO}(\lambda)$ (resp. upper even parity index $I_{UE}(\lambda)$) by the maximum length of weakly decreasing subsequences of the parts whose terms alternate in parity starting with an odd (resp. even) part.

Example 1 $\lambda = (7,7,5,4,4,3,2,1,1)$. Then

$$I_{UO}(\lambda) = 5, \qquad I_{UE}(\lambda) = 4.$$

Let $\delta_o(N,r,m,n)$ (resp. $\delta_e(N,r,m,n)$) denote the number of partitions of n into m distinct parts $\leq N$ with upper odd (resp. even) parity index equal to r. We define

$$D_o(N, y, x; q) := D_o(N) = \sum_{r,m,n\geq 0} \delta_o(N,r,m,n)y^r x^m q^n,$$

$$D_e(N, y, x; q) := D_e(N) = \sum_{r,m,n\geq 0} \delta_e(N,r,m,n)y^r x^m q^n.$$

Then we have the following theorem.

Theorem 1.2 (Andrews [3])

$$D_o(\infty) = \sum_{i,j\geq 0} \frac{x^i y^{2j} q^{(i-j)^2+j^2+i+j}}{(-q;q)_i (q)_{2j} (q)_{i-2j}} + \sum_{i,j\geq 0} \frac{x^i y^{2j-1} q^{(i-j)^2+j^2+i-j}}{(-q;q)_i (q)_{2j-1} (q)_{i-2j+1}}, \quad (1.2)$$

$$D_e(\infty) = \sum_{i,j\geq 0} \frac{x^i y^{2j} q^{(i-j)^2+j^2+j}}{(-q;q)_i (q)_{2j} (q)_{i-2j}} + \sum_{i,j\geq 0} \frac{x^i y^{2j+1} q^{(i-j)^2+j^2+3j+1}}{(-q;q)_i (q)_{2j+1} (q)_{i-2j-1}}. \quad (1.3)$$

The ninth and tenth questions of Andrews are as follows.

Question 9 Prove that if $x = -1$ and $y = 1$, then the second sum of (1.2) is

$$\sum_{n=1}^{\infty} (-1)^n q^{n^2}.$$

Question 10 Prove that if $x = -1$ and $y = 1$, then the first sum of (1.3) is

$$\sum_{n=0}^{\infty} (-1)^n q^{n^2}.$$

The primary purpose of this paper is to provide answers to these two questions of Andrews, which involve Ramanujan's partial theta series (1.1).

In Sect. 2, we give answers to Andrews' questions. Then, a combinatorial proof of Theorem 1.1 will be given in Sect. 3. Theorem 1.2 will be proved combinatorially in Sect. 4. In the last section, some remarks are made, in particular, a sketch of a solution to Question 5 of Andrews [3] is given.

2 Open Questions 9 and 10 of Andrews

In this section, we will provide answers to his two open questions. We first need the following lemma.

Lemma 2.1 *For any positive integer* n,

$$\sum_{k=0}^{n} q^{(n-k)^2+k^2+n-k} \begin{bmatrix} n \\ 2k-1 \end{bmatrix}_q = \sum_{k=0}^{n} q^{(n-k)^2+k^2+k} \begin{bmatrix} n \\ 2k \end{bmatrix}_q = (-q;q)_{n-1} q^{n(n+1)/2},$$

$$\tag{2.1}$$

where

$$\begin{bmatrix} a \\ b \end{bmatrix}_q = \begin{cases} \dfrac{(q;q)_a}{(q;q)_b (q;q)_{a-b}}, & \text{if } 0 \leq b \leq a, \\ 0, & \text{otherwise.} \end{cases}$$

Proof We will only show that

$$\sum_{k=0}^{n} q^{(n-k)^2+k^2+k} \begin{bmatrix} n \\ 2k \end{bmatrix}_q = (-q;q)_{n-1} q^{n(n+1)/2},$$

which is equivalent to

$$\sum_{k=0}^{n} q^{(n-2k)(n-2k-1)/2} \begin{bmatrix} n \\ 2k \end{bmatrix}_q = (-q;q)_{n-1}.$$

This follows from the recurrences of q-binomial coefficient, namely

$$\sum_{k=0}^{n} q^{(n-2k)(n-2k-1)/2} \begin{bmatrix} n \\ 2k \end{bmatrix}_q$$

$$= \sum_{k=0}^{n} q^{(n-2k)(n-2k-1)/2} \left(\begin{bmatrix} n-1 \\ n-2k-1 \end{bmatrix}_q + q^{n-2k} \begin{bmatrix} n-1 \\ n-2k \end{bmatrix}_q \right)$$

$$= \sum_{k=0}^{n-1} q^{k(k+1)/2} \begin{bmatrix} n-1 \\ k \end{bmatrix}_q$$

$$= (-q;q)_{n-1}. \qquad \square$$

We now give answers to Questions 9 and 10 in the following theorem.

Theorem 2.2 *We have*

$$\sum_{n,k\geq0}\frac{(-1)^nq^{(n-k)^2+k^2+n-k}}{(-q;q)_n(q)_{2k-1}(q)_{n-2k+1}}=\sum_{n=1}^{\infty}(-1)^nq^{n^2},$$

$$\sum_{n,k\geq0}\frac{(-1)^nq^{(n-k)^2+k^2+k}}{(-q;q)_n(q)_{2n}(q)_{n-2k}}=\sum_{n=0}^{\infty}(-1)^nq^{n^2}.$$

Proof Rewrite the left-hand side of the first identity as

$$\sum_{n,k\geq0}\frac{(-1)^nq^{(n-k)^2+k^2+n-k}}{(-q;q)_n(q)_{2k-1}(q)_{n-2k+1}}=\sum_{n,k\geq0}\frac{(-1)^nq^{(n-k)^2+k^2+n-k}}{(q^2;q^2)_n}\begin{bmatrix}n\\2k-1\end{bmatrix}_q$$

$$=\sum_{n\geq1}\frac{(-1)^n(-q;q)_{n-1}q^{n(n+1)/2}}{(q^2;q^2)_n},$$

where the second equality follows from Lemma 2.1. Setting $a=-1$ in (1.1), we arrive at

$$1+\sum_{n=1}^{\infty}\frac{(-1)^n(-q;q)_{n-1}q^{n(n+1)/2}}{(q^2;q^2)_n}=\sum_{n=0}^{\infty}(-1)^nq^{n^2}$$

which completes the proof.

Since the proof of the second identity is similar, we omit it. $\square$

3 Combinatorics of Ramanujan's partial theta series

In this section, we provide a combinatorial proof of the partial theta series identity (1.1) from Ramanujan's lost notebook.

We introduce some terminology needed in the rest of this paper. For a partition λ, $\ell(\lambda)$ denotes the number of positive parts, and we define $\lambda_i=0$ for $i>\ell(\lambda)$. We also denote by $\emptyset$ the empty partition of 0. For partitions λ and μ, we define the sum $\lambda+\mu$ of λ and μ to be the partition whose ith part is $\lambda_i+\mu_i$. We denote the conjugate of a partition λ by λ'.

Theorem 3.1 *For any positive integer n, the generating function of partitions λ into n distinct parts with $\lambda_i-\lambda_{i+1}\leq2$ and even upper even parity index is*

$$(-q;q)_{n-1}q^{n(n+1)/2}.$$

Note that the smallest part of the partitions λ stated in Theorem 3.1 must be 1 since $\lambda_{\ell(\lambda)}\leq2$ and their upper even parity index is even.

Proof Let $\tau = (n, n-1, \ldots, 1)$ and μ be a partition generated by $(-q; q)_{n-1}$. We add each part of μ to τ vertically from the largest part and denote the resulting partition λ. That is,

$$\lambda = \tau + \mu'.$$

Since the distinct parts of μ are added, the adjacent parts of λ differ by at most 2. This process is reversible. Let i be the smallest integer such that $\lambda_i - \lambda_{i+1} = 2$. We then subtract 1 from each of the largest i parts of λ. Repeating this until there is no such i, i.e., the resulting partition is τ.

We now show that the upper even parity index of the λ is $2\lfloor (n - \ell(\mu))/2 \rfloor$. We use induction on the number of parts of μ. If $\mu = \emptyset$, then $\lambda = \tau$, the upper even parity index of which is $2\lfloor n/2 \rfloor$. Suppose that μ has k parts, and let λ be the partition resulting from insertion of all parts but the smallest one of μ. Since the insertion process is performed vertically, we see that for $i < \mu_{k-1}$,

$$\lambda_i - \lambda_{i+1} = 1.$$

We also see that vertical insertion of μ_k changes the parity of only the μ_k largest parts of λ. Thus, the μ_kth part and the $(\mu_k + 1)$th part have the same parity, from which we see that the upper even parity index reduces by 1. Furthermore, if $n + k$ is even, the parity index increases by 1; while if $n + k$ is odd, the parity index decreases by 1. Hence, we see that the parity index is

$$\begin{cases} 2\lfloor (n - k + 1)/2 \rfloor, & \text{if } n + k = \text{even}, \\ 2\lfloor (n - k + 1)/2 \rfloor - 2, & \text{if } n + k = \text{odd}, \end{cases}$$

which is equivalent to

$$2\lfloor (n - k)/2 \rfloor.$$

Therefore, the generating function is

$$(-q; q)_{n-1} q^{n(n+1)/2}. \qquad \square$$

Note that a partition with even upper even parity index has odd upper odd parity index, and vice versa. Thus, a similar result on the generating function of partitions into distinct parts and odd upper odd parity index follows from Theorem 3.1. We state this in the following corollary.

Corollary 3.2 *For any positive integer n, the generating function of partitions λ into n distinct parts with $\lambda_i - \lambda_{i+1} \leq 2$, $\lambda_n = 1$, and odd upper odd parity index is*

$$(-q; q)_{n-1} q^{n(n+1)/2}.$$

It follows from Theorem 3.1 and Corollary 3.2 that the generating function of partitions into n distinct parts with even (resp. odd) upper even (resp. odd) parity index is

$$\frac{(-q;q)_{n-1}q^{n(n+1)/2}}{(q^2;q^2)_n}.$$

We now recall Theorem 1.1 with a replaced by $-a$ and give a combinatorial proof.

Theorem 3.3 ([7, p. 28], [4, Entry 1.6.2])

$$1 + \sum_{n=1}^{\infty} \frac{(-a)^n(-q;q)_{n-1}q^{n(n+1)/2}}{(aq^2;q^2)_n} = \sum_{n=0}^{\infty}(-a)^n q^{n^2}. \tag{3.1}$$

Proof Let $D_e(E)$ be the set of partitions into distinct parts with even upper even parity index. As noted after Corollary 3.2, the left-hand side of (3.1) generates partitions π in $D_e(E)$. We will prove the theorem by setting up a sign reversing involution.

Let λ be a partition into distinct parts with $\lambda_i - \lambda_{i+1} \leq 2$ and even upper even parity index, and let σ be a partition into even parts $\leq 2\ell(\lambda)$. We cut each part of σ into halves so that the resulting partition ν has an even number of parts $\leq \ell(\lambda)$. Then it follows from the remark made after Corollary 3.2 that

$$\lambda + \nu' \in D_e(E).$$

For a partition $\pi \in D_e(E)$, by taking out

$$2i\lfloor(\pi_i - \pi_{i+1})/2\rfloor$$

boxes in columns from right to left in the Ferrers graph of π, we can also decompose it into λ and σ, where λ is a partition into distinct parts with $\lambda_i - \lambda_{i+1} \leq 2$ and even upper even parity index, and σ is a partition into even parts $\leq 2\ell(\lambda)$. Note that it follows from the decomposition of π into λ and σ that π is counted with weight $(-1)^{\ell(\pi)}a^{w(\pi)}$, where

$$w(\pi) = \sum_{i=1}^{\ell(\pi)} \lceil(\pi_i - \pi_{i+1})/2\rceil = \ell(\lambda) + \ell(\sigma).$$

Let $\ell(\lambda) = n$. By Theorem 3.1, λ can be decomposed uniquely into $\tau = (n, n - 1, \ldots, 1)$ and a partition μ generated by $(-q;q)_{n-1}$, namely

$$\lambda = \tau + \mu'.$$

If $\mu \neq \emptyset$, then we now consider the sequence s_i,

$$s_i = n + i - 1 + \mu_i,$$

which is weakly decreasing since $\mu_j > \mu_{j+1}$. Also,

$$s_{\ell(\mu)} = n + \ell(\mu) - 1 + \mu_{\ell(\mu)} \geq n + \ell(\mu) = \lambda_1. \tag{3.2}$$

Let e be the least i such that s_i is even. For convenience, we define $s_e = 0$ if there is no such even s_e or $\mu = \emptyset$. Let λ_E be the largest even part of λ. We also define $\lambda_E = 0$ if there is no even part in λ. We now compare $m = \max(s_e, \lambda_E)$ and the largest part of σ, namely σ_1.

Case 1: If $\sigma \neq \emptyset$ and $\sigma_1 > m$, then we remove σ_1 from σ and add σ_1 boxes to λ as follows. If $\sigma_1 \leq \lambda_1 + 2$, then we just add σ_1 to λ as a part. Clearly, the resulting partition λ^* satisfies the part difference condition $\lambda_i^* - \lambda_{i+1}^* \leq 2$. If $\sigma_1 > \lambda_1 + 2$, then since $\lambda = \tau + \mu'$, we see that

$$\sigma_1 - n - \ell(\mu) = \sigma_1 - \lambda_1 > 2.$$

Since $\sigma_1 - n - i$ for $1 \leq i \leq \ell(\mu) + 1$ is a strictly decreasing sequence between 2 and $n - 1$ and $\mu_{\ell(\mu)+1} = 0$, there exists an i such that

$$\mu_i < \sigma_1 - n - i \leq \mu_{i-1},$$

where $\mu_0 = \infty$. Let c be the smallest such i. We now define

$$\mu^* = (\mu_1 + 1, \mu_2 + 1, \ldots, \mu_{c-1} + 1, \sigma_1 - n - c, \mu_c, \mu_{c+1}, \ldots),$$
$$\tau^* = (n + 1, n, \ldots, 1),$$
$$\lambda^* = \tau^* + (\mu^*)'. \tag{3.3}$$

Clearly, the parts of μ^* are distinct and less than $n + 1$. Thus the adjacent parts of λ^* differ by at most 2. The largest part of the resulting partition σ^* is less than or equal to $2n$.

Case 2: $m \neq 0$ and $\sigma_1 \leq m$. In this case, if $m = \lambda_E$, then we remove the part λ_E from λ and denote the resulting partition by λ^*. If $\lambda_1 = \lambda_E$, then any adjacent parts of λ^* still differ by 2. If λ_1 is odd, then by (3.2) if $s_e > 0$,

$$\lambda_E \geq s_e > \lambda_1,$$

which is a contradiction. So, $s_e = 0$. That is, $\mu = \emptyset$ or every s_i is odd. If $\mu = \emptyset$, then $\lambda = (n, n - 1, \ldots, 1)$, so clearly λ^* satisfies the part difference condition. We now show that if every s_i is odd, then

$$\lambda_i - \lambda_{i+1} = 2 \quad \text{iff} \quad \lambda_i \text{ is odd}.$$

Since $s_1 = n + \mu_1$ is odd, the least two adjacent parts differing by 2 in λ are both odd. Also, since s_2 is odd, the next least two adjacent parts differing by 2 are both odd, and so on. Thus, the part after λ_E has to be odd. Thus any two adjacent parts of λ^* still differ by at most 2. If $m = s_e \neq \lambda_E$, then we subtract s_e boxes from λ as follows. Let

$$\mu^* = (\mu_1 - 1, \mu_2 - 1, \ldots, \mu_{e-1} - 1, \mu_{e+1}, \mu_{e+2}, \ldots),$$
$$\tau^* = (n - 1, n - 2, \ldots, 1),$$
$$\lambda^* = \tau^* + (\mu^*)'. \tag{3.4}$$

Since the parts of μ are distinct and less than n, the parts of μ^* are distinct and less than $n - 1$. Thus the adjacent parts of λ^* differ by at most 2. In either case, we then add m to σ as a part and denote the resulting partition by σ^*. From the definition of m, we see that

$$m = \max(\lambda_E, s_e) \leq \max(\lambda_1, s_1) \leq 2n - 1,$$

so $m \leq 2n - 2$ since m is even. Thus the largest part of the resulting partition σ^* is less than or equal to $2n - 2$.

Let $\pi^* = \lambda^* + \sigma^{*'}$. Indeed,

$$w(\pi) = \ell(\lambda) + \ell(\sigma) = \ell(\lambda^*) + \ell(\sigma^*) = w(\pi^*),$$
$$|\ell(\pi) - \ell(\pi^*)| = 1.$$

Thus, it now suffices to show that this is an involution under which $\pi = (2n - 1, 2n - 3, \ldots, 3, 1)$ remains fixed. For any $n \geq 1$, if $\pi = (2n - 1, 2n - 3, \ldots, 3, 1)$, then $\sigma = \emptyset$ since the adjacent parts differ by exactly 2. Thus

$$\lambda = (2n - 1, 2n - 3, \ldots, 3, 1),$$
$$\mu = (n - 1, n - 2, \ldots, 1),$$
$$s_i = n + i - 1 + \mu_i = n + i - 1 + (n - i) = 2n - 1,$$

from which it follows that $m = \max(s_e, \lambda_E) = 0$. Therefore, when $\pi = (2n - 1, 2n - 3, \ldots, 3, 1)$, it belongs to neither Case 1 nor Case 2. This shows that such π are fixed under the map. The generating function of such partitions is the right-hand side of (3.1).

We now show that the map is a bijection on the set of the remaining partitions. If $\sigma_1 > m$ and $\sigma_1 > \lambda_1 + 2$, then $\sigma_1 > s_e$, and by (3.3) we get

$$s_i^* = n + 1 + i - 1 + \mu_i^* = n + 1 + i - 1 + \mu_i + 1 = s_i + 2 \quad \text{for } i < c,$$
$$s_c^* = n + 1 + c - 1 + \mu_c^* = n + 1 + c - 1 + \sigma_1 - n - c = \sigma_1,$$
$$s_i^* = s_{i-1} \quad \text{for } i > c.$$

If s_i^* is even for some $i < c$, then since $s_i^* = s_i + 2$ and $s_e < \sigma_1$, we see that $s_i + 2 \leq \sigma_1$. If $s_i + 2 = \sigma_1$, then

$$\mu_i < \sigma_1 - n - i = \mu_i + 1 \leq \mu_{i-1}.$$

This contradicts the minimality of c. Thus s_c^* is the largest even number in the sequence. On the other hand, by the definition,

$$\lambda_E^* \leq \lambda_1^* = n + 1 + \ell(\mu^*) = n + 1 + \ell(\mu) + 1 = \lambda_1 + 2 < \sigma_1.$$

So, $m^* = \max(s_e^*, \lambda_E^*) = \sigma_1$. Since $\sigma_1^* \leq \sigma_1$, we subtract σ_1 boxes back from λ^* adding to σ^* as defined in (3.4).

If $\sigma \neq \emptyset$ and $m < \sigma_1 \leq \lambda_1 + 2$, then we added σ_1 to λ as a part. In this case, since $\lambda_E < \sigma_1$ and the adjacent parts of λ differ by at most 2, there exists a unique j such that

$$\lambda_j = \sigma_1 + 1 = \lambda_{j+1} + 2.$$

Then $\lambda^* = (\lambda_1, \ldots, \lambda_j, \sigma_1, \lambda_{j+1}, \ldots, \lambda_n)$. By the definition,

$$(\mu^*)' = (\mu_1' - 1, \ldots, \mu_j' - 1, \mu_j' - 1, \mu_{j+1}', \ldots),$$

which is equivalent to

$$\mu_i^* = \begin{cases} \mu_i + 1, & \text{if } \mu_i > j, \\ \mu_{i+1}, & \text{if } \mu_i < j. \end{cases}$$

We compute

$$s_i^* = n + 1 + i - 1 + \mu_i^* = n + 1 + i - 1 + \mu_i + 1 = s_i + 2 \quad \text{if } \mu_i > j,$$
$$s_i^* = n + 1 + i - 1 + \mu_i^* = n + 1 + i - 1 + \mu_{i+1} = s_{i+1} \quad \text{if } \mu_i < j,$$

from which it follows that $s_e^* \leq s_e + 2 \leq \sigma_1$. Since $\lambda_E^* = \sigma_1$, we obtain $m^* = \max(s_e^*, \lambda_E^*) = \lambda_E^* = \sigma_1$. By subtracting σ_1 from λ^* and adding it to σ^*, we recover the original λ and σ.

Similarly, we can show that the map is an involution for partitions with $m \geq \sigma_1$, but we omit it. Therefore, the identity holds true. $\qquad\square$

4 Upper parity indices in partitions into distinct parts

In this section, we prove Theorem 1.2 combinatorially. We first rewrite the theorem as follows.

$$D_o(\infty) = \sum_{n=0}^{\infty} \sum_{k=0}^{n} \frac{x^n y^{2k} q^{(n-k)^2 + k^2 + n + k}}{(q^2; q^2)_n} \begin{bmatrix} n \\ 2k \end{bmatrix}_q$$
$$+ \sum_{n=0}^{\infty} \sum_{k=0}^{n} \frac{x^n y^{2k-1} q^{(n-k)^2 + k^2 + n - k}}{(q^2; q^2)_n} \begin{bmatrix} n \\ 2k - 1 \end{bmatrix}_q, \qquad (4.1)$$

$$D_e(\infty) = \sum_{n=0}^{\infty} \sum_{k=0}^{n} \frac{x^n y^{2k} q^{(n-k)^2 + k^2 + k}}{(q^2; q^2)_n} \begin{bmatrix} n \\ 2k \end{bmatrix}_q$$
$$+ \sum_{n=0}^{\infty} \sum_{k=0}^{n} \frac{x^n y^{2k+1} q^{(n-k)^2 + k^2 + 3k + 1}}{(q^2; q^2)_n} \begin{bmatrix} n \\ 2k + 1 \end{bmatrix}_q. \qquad (4.2)$$

In the following theorem, we first show that the inner summation of the first double summations in (4.2) is the generating function of partitions into n distinct parts with even upper even parity index.

Theorem 4.1 *The generating function of partitions into n distinct parts with even upper even parity index is*

$$\sum_{k=0}^{n} \frac{q^{(n-k)^2+k^2+k}}{(q^2;q^2)_n} \begin{bmatrix} n \\ 2k \end{bmatrix}_q.$$

Proof By Theorem 3.1, it suffices to show that

$$\sum_{k=0}^{n} q^{(n-k)^2+k^2+k} \begin{bmatrix} n \\ 2k \end{bmatrix}_q = (-q;q)_{n-1}q^{n(n+1)/2}, \tag{4.3}$$

which is equivalent to

$$\sum_{k=0}^{n} q^{(n-2k)(n-2k-1)/2} \begin{bmatrix} n \\ 2k \end{bmatrix}_q = (-q;q)_{n-1}.$$

This follows from Lemma 2.1. Indeed, one can fully show identity (4.3) in terms of partitions by rearranging the parts of partitions generated by one side to obtain the partitions generated by the other side. We omit the details. □

Similarly, we can show that the generating function of partitions into n distinct parts with odd upper odd parity index is

$$\sum_{k=0}^{n} \frac{q^{(n-k)^2+k^2+n-k}}{(q^2;q^2)_n} \begin{bmatrix} n \\ 2k-1 \end{bmatrix}_q.$$

For even upper odd parity index (resp. odd upper even parity index), we first take a partition λ with even upper even parity index (resp. odd upper odd parity index). Then the smallest part of λ must be odd. We now add 1 to each of the parts of λ. Then the resulting partition has even upper odd parity index (resp. odd upper even parity index). By Theorem 4.1, we can show that the generating function of partitions into n distinct parts with even upper odd parity index is

$$\sum_{k=0}^{n} \frac{q^{(n-k)^2+k^2+n+k}}{(q^2;q^2)_n} \begin{bmatrix} n \\ 2k \end{bmatrix}_q.$$

Similarly, we can show that the generating function of partitions into n distinct parts with odd upper even parity index is

$$\sum_{k=0}^{n} \frac{q^{(n-k)^2+k^2+3k+1}}{(q^2;q^2)_n} \begin{bmatrix} n \\ 2k+1 \end{bmatrix}_q.$$

5 Remarks

The proof of Theorem 1.2 given in Sect. 4 is essentially the same as that of K. Kursungoz [6].

Question 5 of Andrews follows from Franklin's involution for the Euler pentagonal number theorem [2, pp. 10–11]. We state the question and sketch a solution. For a partition λ, we denote by $I_{LO}(\lambda)$ (resp. I_{LE}) the maximum length of weakly increasing subsequences of the parts whose terms alternate in parity starting with an odd (resp. even) part. Let $p_o(r, m, n)$ be the number of partitions of n into m distinct parts with $I_{LO} = r$. We define

$$P_o(y, x; q) = \sum_{r,m,n \geq 0} p_o(r, m, n) y^r x^m q^n.$$

Then we have the following theorem.

Theorem 5.1 (Andrews [3])

$$P_o(y, x; q) = \sum_{n=0}^{\infty} \frac{x^n y^n q^{n(n+1)/2}(-q/y; q)_n}{(q^2; q^2)_n}.$$

Question 5. It follows from an old formula of Rogers that

$$P_o(-1, 1; q) = \sum_{n=0}^{\infty} q^{n(3n+1)/2}(1 - q^{2n+1}).$$

Prove combinatorially.

Note that $\lambda_1 \equiv I_{LO}(\lambda) \pmod 2$ for a partition λ. Thus,

$$P_o(-1, 1; q) = \sum_{n=0}^{\infty} \frac{(-1)^n q^{n(n+1)/2}}{(-q; q)_n} = \sum_{n=0}^{\infty} q^{n(3n+1)/2}(1 - q^{2n+1}),$$

where the second equality follows from Franklin's involution.

Another combinatorial proof of Theorem 1.1 is given by B. C. Berndt, B. Kim, and the author in [5]. K. Alladi [1] has devised a completely different proof of Theorem 1.1 and has also provided a number-theoretic interpretation of Theorem 1.1 as a weighted partition theorem.

Acknowledgements The author thanks the anonymous referee for helpful comments. The author also would like to thank William Y.C. Chen for pointing out an error in the original proof of Theorem 3.3.

References

1. Alladi, K.: A partial theta identity of Ramanujan and its number theoretic interpretation. Ramanujan J. **20**, 229–239 (2009)
2. Andrews, G.E.: The Theory of Partitions. Addison-Wesley, Reading (1976). Reissued: Cambridge University Press, Cambridge (1998)
3. Andrews, G.E.: Parity in partition identities. Ramanujan J. **23** (2010, to appear)
4. Andrews, G.E., Berndt, B.C.: Ramanujan's Lost Notebook, Part II. Springer, New York (2009)
5. Berndt, B.C., Kim, B., Yee, A.J.: Ramanujan's Lost Notebook: Combinatorial proofs of identities associated with Heine's transformation or partial theta functions. J. Comb. Theory Ser. A **117**, 957–973 (2010)
6. Kursungoz, K.: Parity considerations in Andrews-Gordon identities, and the k-marked Durfee symbols. Ph.D. Thesis, Penn. Sate University
7. Ramanujan, S.: The Lost Notebook and Other Unpublished Papers. Narosa, New Delhi (1988)

Ramanujan J (2010) 23: 227–241
DOI 10.1007/s11139-009-9188-7

A combinatorial study and comparison of partial theta identities of Andrews and Ramanujan

Krishnaswami Alladi

Dedicated to George Andrews on the occasion of his 70-th birthday

Received: 23 April 2009 / Accepted: 25 June 2009 / Published online: 12 August 2009
© Springer Science+Business Media, LLC 2009

Abstract We provide a simple proof of a partial theta identity of Andrews and study the underlying combinatorics. This yields a weighted partition theorem involving partitions into distinct parts with smallest part odd which turns out to be a companion to a weighted partition theorem involving the same partitions that we recently deduced from a partial theta identity in Ramanujan's Lost Notebook. We also establish some new partition identities from certain special cases of Andrews' partial theta identity.

Keywords Ramanujan's Lost Notebook · Partial theta identity · Heine's transformation · Partitions into distinct parts · Weighted partition identities

Mathematics Subject Classification (2000) Primary 05A17 · 05A19 · Secondary 05A17

1 Introduction

In a recent paper [3] we showed that the partial theta identity

$$1 + \sum_{n=1}^{\infty} \frac{(-a)^n q^{n(n+1)/2}(-q)_{n-1}}{(aq^2;q^2)_n} = \sum_{k=0}^{\infty} (-a)^k q^{k^2} \tag{1.1}$$

in Ramanujan's Lost Notebook [10; p. 38] is equivalent to:

Research supported in part by NSA Grants MSPF-06G-150 and MSPF-08G-154.

K. Alladi (✉)
Department of Mathematics, University of Florida, Gainesville, FL 32611, USA
e-mail: alladik@math.ufl.edu

Theorem 1 *Let π be a partition into distinct parts, $\pi : b_1 + b_2 + \cdots + b_\nu$ with b_ν odd. Consider the gaps $b_i - b_{i+1}$, with $b_{\nu+1}$ set equal to 0. Define the weight of the i-th gap to be $\omega_i = a^{\delta_i}$, where δ_i is the least integer $\geq (b_i - b_{i+1})/2$. Define the weight of the partition as*

$$\omega(\pi) = (-1)^\nu \prod_{i=1}^{\nu} a^{\delta_i}. \tag{1.2}$$

Then

$$\sum_{\sigma(\pi)=n,\pi\in P_{d,o}} \omega(\pi) = (-a)^k, \quad \text{if } n = k^2, \quad \text{and} \quad 0 \quad \text{otherwise.} \tag{1.3}$$

Here and throughout, $\sigma(\pi)$ is the sum of the parts of π and $P_{d,o}$ is the set of partitions into distinct parts with smallest part odd. Also in (1.1) we have used the standard notation

$$(a)_n = (a; q)_n = \prod_{j=0}^{n-1}(1 - aq^j) \tag{1.4}$$

for a non-negative integer n and complex number a. Using this we may define

$$(a)_\infty = \lim_{n \to \infty} (a)_n = \prod_{j=0}^{\infty}(1 - aq^j), \tag{1.5}$$

when $|q| < 1$. When the base is just q, we simply write $(a)_n$ for $(a; q)_n$ as in (1.1).

Recently Andrews [7] showed me another representation for the partial theta series in (1.1), namely

$$\sum_{n=0}^{\infty} q^{2n}(q^{2n+2}; q^2)_\infty (aq^{2n+1}; q^2)_\infty = \sum_{k=0}^{\infty}(-a)^k q^{k^2}, \tag{1.6}$$

and gave a proof using a special case of Heine's transformation.

It is not apparent from (1.6) that the real emphasis is on partitions belonging to $P_{d,o}$ because it appears that (1.6) deals with partitions with smallest part even. In Sect. 2 we will provide a simple proof of (1.6) and show that it is equivalent to the following weighted partition theorem:

Theorem 2 *Let $\pi \in P_{d,o}$, $\pi : b_1 + b_2 + \cdots + b_\nu$. Let $\nu_o(\pi)$ denote the number of odd parts of π. Define the weight $\omega_o(\pi)$ of the partition as*

$$\omega_o(\pi) = (-1)^\nu a^{\nu_o(\pi)}. \tag{1.7}$$

Then

$$\sum_{\sigma(\pi)=n,\pi\in P_{d,o}} \omega_o(\pi) = (-a)^k, \quad \text{if } n = k^2, \quad \text{and} \quad 0 \quad \text{otherwise.} \tag{1.8}$$

It is amazing that even though the weights in Theorems 1 and 2 on the same set of partitions are different, the sums of the weights over these partitions $\pi \in P_{d,o}$ of n are equal and non-zero precisely when n is a square. In the special case $a = 1$, even though the series on the left in (1.1) and (1.6) are different in form, we have $\omega(\pi) = \omega_o(\pi)$, and so Theorems 1 and 2 both yield the following elegant result:

Theorem 3 *Let $R_o(n)$ (resp. $R_e(n)$) denote the number of partitions of n into distinct parts such that the smallest part is odd, and the number of parts is odd (resp. even). Then*

$$R_e(n) - R_o(n) = (-1)^k, \quad \text{if } n = k^2, \quad \text{and} \quad 0, \quad \text{otherwise.} \tag{1.9}$$

Theorem 3 which can be compared to Euler's celebrated Pentagonal Number's Theorem was first observed (and proved) in [3], although Andrews [5] had previously noticed a result closely resembling Theorem 3 (see [3] for a comparison of Andrews' theorem and our Theorem 3).

In the next section we will provide two proofs of (1.6) after which we will study this identity and its consequences combinatorially.

2 Proofs of (1.6)

We will provide two proofs of (1.6).

Andrews' proof Observe that

$$\sum_{n=0}^{\infty} q^{2n}(q^{2n+2}; q^2)_\infty (aq^{2n+1}; q^2)_\infty$$

$$= (q^2; q^2)_\infty (aq; q^2)_\infty \left\{ \sum_{n=0}^{\infty} \frac{q^{2n}}{(q^2; q^2)_n (aq; q^2)_n} \right\}. \tag{2.1}$$

We can use Heine's transformation (see [6]) to evaluate the sum on the right of (2.1), and so it can be rewritten as

$$(q^2; q^2)_\infty (aq; q^2)_\infty \left\{ \frac{1}{(q^2; q^2)_\infty (aq; q^2)_\infty} \sum_{k=0}^{\infty} (-1)^k a^k q^{k^2} \right\} = \sum_{k=0}^{\infty} (-a)^k q^{k^2},$$

which proves (1.6). □

The proof of (1.1) given in [8] also utilizes Heine's transformation. In [9] a combinatorial proof of (1.1) was given by interpreting the left hand side in terms of vector partitions. In [3] we gave a very down to earth q-series proof of (1.1) without recourse to Heine's transformation. We shall now give a similar proof of (1.6).

Our proof of (1.6) By expanding $(aq^{2n+1}; q^2)_\infty$ as a series, rewrite the left hand side of (1.6) as

$$\sum_{n=0}^{\infty} q^{2n} (q^{2n+2}; q^2)_\infty \sum_{k=0}^{\infty} \frac{(-1)^k a^k q^{k^2+2kn}}{(q^2; q^2)_k}$$

$$= \sum_{k=0}^{\infty} \frac{(-1)^k a^k q^{k^2}}{(q^2; q^2)_k} \sum_{n=0}^{\infty} q^{2n(k+1)} (q^{2n+2}; q^2)_\infty. \tag{2.2}$$

By comparing coefficients of a^k on the right hand sides of (1.6) and (2.2), we see that in order to prove (1.6), we need to show that

$$\sum_{n=0}^{\infty} q^{2n(k+1)} (q^{2n+2}; q^2)_\infty = (q^2; q^2)_k. \tag{2.3}$$

Upon replacing q^2 by q, we see that (2.3) is equivalent to

$$\sum_{n=0}^{\infty} q^{n(k+1)} (q^{n+1})_\infty = (q)_k. \tag{2.4}$$

We divide both sides of (2.4) by $(q)_\infty$ to get

$$\sum_{n=0}^{\infty} \frac{q^{n(k+1)}}{(q)_n} = \frac{1}{(q^{k+1})_\infty}. \tag{2.5}$$

Now consider the well known identity for the generating function of unrestricted partitions, namely

$$\sum_{n=0}^{\infty} \frac{z^n q^n}{(q)_n} = \frac{1}{(zq)_\infty}. \tag{2.6}$$

The power of z on the left in (2.6) enumerates the number of parts, whereas the power of z on the right is the largest part, and the two are equal by conjugation of Ferrers graphs of partitions. In any case, (2.5) follows by setting $z = q^k$ in (2.6) and this proves (1.6). $\qquad\square$

3 Interpreting Andrews' identity as Theorem 2

Andrews' identity (1.6) appears to indicate that the series on the left deals with partitions into distinct parts with smallest part even. It turns out that the actual emphasis in (1.6) is on partitions into distinct parts with smallest part odd as we will show presently.

With the weight $\omega_o(\pi)$ defined as in (1.7) for partitions π belonging to the set P_d of ALL partitions into distinct parts, we see that

$$\sum_{\pi \in P_d, \sigma(\pi) \geq 0} \omega_o(\pi) q^{\sigma(\pi)} = (q^2; q^2)_\infty (aq; q^2)_\infty. \tag{3.1}$$

Observe that on the left in (1.6), the term corresponding to $n = 0$ is the product on the right in (3.1). Thus the first term in the series on the left in (1.6) is the generating function of ALL partitions into distinct parts counted with weight $\omega_o(\pi)$. Now let $P_{d,e}$ denote the set of partitions into distinct parts with smallest part *even*. Then the weighted generating function of partitions into distinct parts with smallest part even and positive is

$$\sum_{\pi \in P_{d,e}, \sigma(\pi) \geq 1} \omega_o(\pi) q^{\sigma(\pi)} = \sum_{n=1}^{\infty} -q^{2n}(q^{2n+2}; q^2)_{\infty}(aq; q^2)_{\infty}. \tag{3.2}$$

Note that the terms from $n = 1$ onwards on the left hand side of (1.6) yield the expression on the right in (3.2) but with the opposite sign! Thus from (3.1) and (3.2) we see that in the left hand side of (1.6), the generating function of the non-empty partitions in $P_{d,e}$ with weights $\omega_o(\pi)$ are actually subtracted off from the generating function of full set of partitions in P_d. Thus (1.6) is actually the identity

$$\sum_{\pi \in P_{d,o}, \sigma(\pi) \geq 0} \omega_o(\pi) q^{\sigma(\pi)} = \sum_{k=0}^{\infty} (-a)^k q^{k^2}, \tag{3.3}$$

which is the statement of Theorem 2.

Even though the weights in Theorems 1 and 2 are different in general, when $a = 1$ they are identical and so, as was pointed out in Sect. 1, both Theorems 1 and 2 yield Theorem 3 as special case.

In Andrews [6; pp. 156–157] the case $a = -1$ of (1.6) is proved, namely

$$\sum_{n=0}^{\infty} q^{2n}(q^{2n+2}; q^2)_{\infty}(-q^{2n+1}; q^2)_{\infty} = \sum_{k=0}^{\infty} q^{k^2}, \tag{3.4}$$

and shown that (3.4) is combinatorially equivalent to

Andrews' Theorem *Let $\varepsilon_e(n)$ (resp. $\varepsilon_o(n)$) denote the number of partitions of n into distinct non-negative parts with smallest part even such that the number of parts is even (resp. odd). Then*

$$\varepsilon_e(n) - \varepsilon_o(n) = 1 \quad \text{if } n = k^2, \quad \text{and} \quad 0 \quad \text{otherwise}. \tag{3.5}$$

This closely resembles our Theorem 3 but is not the same result. In [3] we show how Theorem 3 is equivalent to Andrews' theorem, in the sense that each can be derived from the other. The idea is similar to how we showed above that in (1.6) the real contribution comes from partitions in $P_{d,o}$ and not from partitions in $P_{d,e}$.

4 Gauss' identity and a new weighted partition theorem

A celebrated theta function identity of Gauss (see [6]) is

$$\prod_{m=0}^{\infty} \frac{(1 - q^{2m})}{(1 - q^{2m-1})} = \sum_{k=0}^{\infty} q^{k(k+1)/2}. \tag{4.1}$$

We will first show that (4.1) follows from (1.6) and use this approach to deduce a new weighted partition identity (Theorem 4 below).

In (1.6) replace a by aq to get

$$\sum_{n=0}^{\infty} q^{2n}(q^{2n+2}; q^2)_{\infty}(aq^{2n+2}; q^2)_{\infty} = \sum_{k=0}^{\infty}(-1)^k a^k q^{k^2+k}.$$

If we replace q^2 by q in the above equation, we get

$$\sum_{n=0}^{\infty} q^n(q^{n+1})_{\infty}(aq^{n+1})_{\infty} = \sum_{k=0}^{\infty}(-1)^k a^k q^{k(k+1)/2}. \tag{4.2}$$

Now choose $a = -1$ in (4.2) and note that it becomes

$$\sum_{n=0}^{\infty} q^n(q^{2n+2}; q^2)_{\infty} = \sum_{k=0}^{\infty} q^{k(k+1)/2}. \tag{4.3}$$

Note that the left hand side of (4.3) is

$$(q^2; q^2)_{\infty} \sum_{n=0}^{\infty} \frac{q^n}{(q^2; q^2)_n} = \frac{(q^2; q^2)_{\infty}}{(q; q^2)_{\infty}}. \tag{4.4}$$

Gauss identity (4.1) follows from (4.3) and (4.4).

Next we will use this approach to Gauss' identity to obtain a new weighted partition theorem.

In (4.2) replace a by $-a$ and divide both sides by $(q)_{\infty}$ to get

$$\sum_{n=0}^{\infty} \frac{q^n}{(q)_n} \cdot (-aq^{n+1})_{\infty} = \frac{1}{(q)_{\infty}} \sum_{k=0}^{\infty} a^k q^{k(k+1)/2}. \tag{4.5}$$

We will now give a combinatorial proof of (4.5).

First we rewrite the right hand side of (4.5) as

$$\left\{ \sum_{n=0}^{\infty} p(n)q^n \right\}\left\{ \sum_{k=0}^{\infty} a^k q^{k(k+1)/2} \right\} = \sum_{m=0}^{\infty} q^m \sum_k a^k p(m - T_k), \tag{4.6}$$

where $p(n)$ is the number of (unrestricted) partitions of n and $T_k = k(k+1)/2$ is the k-th triangular number. Thus the right hand side of (4.5) is the generating function of the weighted sum (the weights being a^k) of the partition function over shifted values $p(m - T_k)$.

Now given any partition π of an integer n, consider the *maximal chain* of distinct parts as consecutive parts starting from the largest part and descending downwards. Let us call this maximal chain as a *string*. The string is empty if the largest part repeats. Note that left hand side of (4.5) is the generating function of bi-partitions (π_1, π_2), where π_1 is an ordinary partition, and π_2 is a partition into distinct parts such that each part of π_2 is greater that each part of π_1, and the power of a is the number of parts of π_2. We will now interpret this decomposition using strings.

Now given an unrestricted partition π, suppose it has a string of length ℓ. Then for any $k \leq \ell$, we can separate out the k largest parts of π to form a partition π_4, and we denote the partition from the remaining parts as π_3. Note that π_4 is a partition into distinct parts, and that every part of π_4 is greater than every part of π_3. There are several ways of decomposing π in the form (π_3, π_4) (the number of such decompositions of π is $\ell + 1$), and the left hand side of (4.5) precisely represents this decomposition process.

Finally, given an unrestricted partition π of an integer m which contains a string of length ℓ, choose any $k \leq \ell$, and subtract $k - 1$ from the largest part, $k - 2$ from the second largest part, $k - 3$ from the next largest part, etc., and finally 0 from the k-th largest part. We call this *Eulerian subtraction*. This yields a partition of $m - T_k$ into at least k parts. Conversely, given a partition of any integer into at least k parts, by adding $k - 1$ to the largest part, $k - 2$ to the second largest, ..., and 0 to the k-th largest part, we get a partition in which the string length is $\geq k$. So while the left hand side of (4.6) keeps track of the splitting of a partition using strings, the right hand side deals with the sum of the partition function over shifts by triangular numbers, and the two are equal by Eulerian subtraction. This proves (4.6) and (4.5).

Now we consider the special case $a = 1$ in (4.5), but we will not rewrite the left hand side of (4.5) in the form (4.3), but keep it as it is and interpret it in terms of strings as above. When $a = 1$ the triangular series on the right in (4.5) becomes the product in (4.1). So when $a = 1$, we get from (4.5)

$$\sum_{n=0}^{\infty} \frac{q^n}{(q)_n}(-q^{n+1})_\infty = \frac{1}{(q)_\infty}\frac{(q^2;q^2)_\infty}{(q;q^2)_\infty} = \frac{(-q)_\infty}{(q;q^2)_\infty} = (-q^2;q^2)_\infty\frac{(-q;q^2)_\infty}{(q;q^2)_\infty}.$$
(4.7)

Observe that

$$\frac{1+q^{2m-1}}{1-q^{2m-1}} = 1 + \frac{2q^{2m-1}}{1-q^{2m-1}} = 1 + 2(q^{2m-1} + q^{2(2m-1)} + q^{3(2m-1)} + \cdots).$$ (4.8)

From (4.8) we see that the product on the right in (4.7) is the generating function for partitions in which the even parts do not repeat, but these partitions π^* are counted with weight $2^{\nu_{d,o}(\pi^*)}$, where $\nu_{d,o}(\pi^*)$ is the number of different odd parts of π^*. So (4.5), (4.7), and the above arguments yield the following weighted partition result:

Theorem 4 *Suppose π is an unrestricted partition which has a string of length ℓ. Define the weight $\omega_s(\pi)$ to be $\ell + 1$.*

Let G denote the set of all partitions in which the even parts do not repeat. For $\pi^* \in G$, define its weight to be $2^{\nu_{d,o}(\pi^*)}$, where $\nu_{d,o}(\pi^*)$ is the number of different odd parts of π^*. Then

$$\sum_{\sigma(\pi)=n} \omega_s(\pi) = \sum_k p(n - T_k) = \sum_{\pi^* \in G, \sigma(\pi^*)=n} 2^{\nu_{d,o}(\pi)}.$$ (4.9)

Remarks (1) In [1] we established that for $g(n)$, the number of partitions of n in which the even parts do not repeat, the following weighted partition identity holds:

Let D denote the set of partitions into distinct parts. For $\pi_d \in D$, $\pi_d : b_1 + b_2 + \cdots + b_v$, let there be k gaps $b_i - b_{i+1} \geq 2$, with b_{v+1} defined to be 0. Define the weight $\omega_d(\pi_d) = 2^k$. Then

$$g(n) = \sum_{\pi_d \in D; \sigma(\pi_d)=n} \omega_d(\pi). \tag{4.10}$$

(2) Instead of partitions with non-repeating even parts, consider the set A of partitions with non-repeating odd parts. The generating function of partitions $\tilde{\pi} \in A$ is

$$\prod_{m=1}^{\infty} \frac{1+q^{2m-1}}{1-q^{2m}}.$$

Observe that one way to decompose the generating function of the partition function $p(n)$ is

$$\sum_{n=0}^{\infty} p(n)q^n = \frac{1}{(q)_\infty} = \frac{(-q)_\infty}{(q^2;q^2)_\infty} = (-q;q^2)_\infty \frac{(-q^2;q^2)_\infty}{(q^2;q^2)_\infty}. \tag{4.11}$$

Now interpreting

$$\frac{1+q^{2m}}{1-q^{2m}}$$

as in (4.8), we get the following companion to Theorem 4 which was stated in [2]:

Theorem A *Let A denote the set of partitions with non-repeating odd parts. For $\tilde{\pi} \in A$, let its weight be defined as $2^{v_{d,e}(\tilde{\pi})}$, where $v_{d,e}(\tilde{\pi})$ is the number of different even parts of $\tilde{\pi}$. Then*

$$p(n) = \sum_{\tilde{\pi} \in A, \sigma(\tilde{\pi})=n} 2^{v_{d,e}(\tilde{\pi})}. \tag{4.12}$$

In a subsequent paper [4], we will utilize the link between $p(n)$ and the partitions $\pi^* \in G$ in (4.9) to construct a nested sequence of decreasing sets of partitions converging to the set of partitions into odd parts and study the weighted partition theorems connected with this nested sequence. In view of Euler's fundamental observation equating partitions into odd parts with partitions into distinct parts, it is natural to ask whether there is a similar nested sequence of decreasing sets starting with unrestricted partitions and ending with partitions into distinct parts. The answer is yes, and we produce this second nested sequence (see [4]) by utilizing the link established in (4.12) between $p(n)$ and the partitions $\tilde{\pi} \in A$. This produces another sequence of weighted partition theorems.

5 Analytic representation of Theorem 2 and consequences

The most transparent analytic representation of Theorem 2 is

$$\sum_{n=1}^{\infty} -aq^{2n-1}(q^{2n};q^2)_\infty(aq^{2n+1};q^2)_\infty = \sum_{k=1}^{\infty}(-a)^k q^{k^2}, \tag{5.1}$$

and not (1.6). The left hand side can be rewritten as

$$-(aq;q^2)_\infty \sum_{n=1}^{\infty} \frac{aq^{2n-1}}{(aq;q^2)_n}\cdot(q^{2n};q^2)_\infty. \tag{5.2}$$

Recall the well known theta function identity

$$\sum_{k=-\infty}^{\infty}(-1)^k q^{k^2} = \frac{(q)_\infty}{(-q)_\infty}. \tag{5.3}$$

So when $a = 1$, we can use (5.3) to combine (5.1) and (5.3) into

$$\frac{1}{2}\left\{\frac{(q)_\infty}{(-q)_\infty} - 1\right\} = -(q;q^2)_\infty \sum_{n=1}^{\infty} \frac{q^{2n-1}}{(q;q^2)_n}(q^{2n};q^2)_\infty. \tag{5.4}$$

Now Euler's celebrated result equating partitions into odd parts with partitions into distinct parts is the identity

$$(-q)_\infty = \frac{1}{(q;q^2)_\infty}. \tag{5.5}$$

In view of (5.5), we may rewrite (5.4) as

$$\frac{(-q)_\infty - (q)_\infty}{2} = \sum_{n=1}^{\infty} \frac{q^{2n-1}}{(q;q^2)_n}(q^{2n};q^2)_\infty. \tag{5.6}$$

There is the well known identity

$$(-zq)_\infty = \sum_{n=0}^{\infty} \frac{z^n q^{n(n+1)/2}}{(q)_n}, \tag{5.7}$$

and the left hand side of (5.6) is simply the odd part of the expression in (5.7) as a function of z and evaluated at $z = 1$. Thus (5.7) implies that (5.6) is equivalent to

$$\sum_{n=1}^{\infty} \frac{q^{2n^2-n}}{(q)_{2n-1}} = \sum_{n=1}^{\infty} \frac{q^{2n-1}}{(q;q^2)_n}(q^{2n};q^2)_\infty. \tag{5.8}$$

We now divide both sides of (5.8) by $(q^2;q^2)_\infty$ to write it as

$$\frac{1}{(q^2;q^2)_\infty} \sum_{n=1}^{\infty} \frac{q^{2n^2-n}}{(q)_{2n-1}} = \sum_{n=1}^{\infty} \frac{q^{2n-1}}{(q)_{2n-1}}. \tag{5.9}$$

We can provide a simple proof of (5.9) by interpreting both sides combinatorially: The right hand side is the generating function of partitions π into an odd number of parts, by interpreting $2n - 1$ as representing the number of parts. Now write

$$\pi = \sum_{i \geq 1} f_i . i, \tag{5.10}$$

where i is an integer (part) and $f_i \geq 0$ is the frequency of its occurrence. We know that the total number of parts

$$\sum_i f_i = odd. \tag{5.11}$$

Next, write

$$f_i = 2[f_i/2] + \delta_i, \tag{5.12}$$

where $[x]$ is the greatest integer $\leq x$, and δ_i is either 0 or 1. Observe from (5.11) and (5.12) that

$$\sum_{i \geq 1} \delta_i = odd, \qquad \delta_i = 0 \quad \text{or} \quad 1. \tag{5.13}$$

We may use (5.12) to write

$$\pi = \sum_{i \geq 1} [f_i/2] 2i + \sum_{i \geq 1} \delta_i . i \tag{5.14}$$

and interpret this as the decomposition of π in the form $(\pi_2, \pi_{1,d})$, where π_2 is a partition into even parts, and using (5.14) interpret $\pi_{1,d}$ as a partition into an odd number of distinct parts. The left hand side of (5.9) is precisely the generating function of such bipartitions $(\pi_2, \pi_{1,d})$, and so this proves (5.9) and therefore (5.1) as well.

The purpose of the above arguments is not just to prove (5.1) but also to deduce some new partition theorems as we show presently.

We go back to (5.6) and rewrite it as

$$(-q)_\infty = (q)_\infty + 2 \sum_{n=1}^{\infty} \frac{q^{2n-1}}{(q;q^2)_n} (q^{2n};q^2)_\infty. \tag{5.15}$$

This yields very interesting partition results. To begin with, define S to be the set of partitions in which the odd parts may repeat, an odd part must occur, the evens should not repeat, and every even part is greater than every odd part. Next let $S(n)$ denote the number of partitions π of n with $\pi \in S$. Also let $S_e(n)$ (resp. $S_o(n)$) denote the number of partitions of the type enumerated by $S(n)$ but with an even (resp. odd) number of even parts. With these definitions we observe first that the series on the right hand side of (5.16) is the generating function of $S_e(n) - S_o(n)$.

Let us write the celebrated Euler's pentagonal numbers theorem as

$$(q)_\infty = \sum_{n=0}^{\infty} \varepsilon_n, \tag{5.16}$$

where

$$\varepsilon_n = (-1)^k, \quad \text{if } n = \frac{3k^2 \pm k}{2}, \quad \text{and} \quad 0 \quad \text{otherwise.} \tag{5.17}$$

With $(-q)_\infty$ interpreted as the generating function of $P_d(n)$, the number of partitions of n into distinct parts, we have

Theorem 5 *For $n \geq 1$*

$$P_d(n) = \varepsilon_n + 2(S_e(n) - S_o(n)).$$

It is of course very well known that $P_d(n)$ is odd precisely at the pentagonal numbers and even elsewhere; Franklin's famous combinatorial proof of the pentagonal numbers theorem tells us how to split the partitions enumerated by $P_d(n)$ into two equal subsets (except at the pentagonal numbers), and so $P_d(n)$ is seen as twice the number of partitions in each of these subsets. Theorem 5 is interesting because it relates the parity of $P_d(n)$ with *another* set of partitions. Theorem 5 actually has a very interesting consequence.

Since Euler's identity (5.5) equates partitions into distinct parts with partitions into odd parts, it is obvious that $S(n) \geq P_d(n)$. But note the stronger inequality

$$S_e(n) \geq P_d(n), \tag{5.18}$$

because from among the partitions enumerated by $S_e(n)$, those which do not have even parts at all are precisely the partitions of n into odd parts and these are equal to $P_d(n)$ in number by Euler's identity. Next rewrite Theorem 5 as

$$S_e(n) - S_o(n) = \frac{P_d(n)}{2} - \frac{\varepsilon_n}{2}. \tag{5.19}$$

From (5.18) and (5.19) we see that

$$S_o(n) \geq \frac{P_d(n)}{2} + \frac{\varepsilon_n}{2}. \tag{5.20}$$

Finally, since

$$S(n) = S_e(n) + S_o(n)$$

by definition, we combine (5.18) and (5.20) to deduce

Theorem 6 *For $n \geq 1$*

$$S(n) \geq \frac{3}{2} P_d(n) + \frac{\varepsilon_n}{2}.$$

6 A companion result

In [3] the method employed to prove (1.1) was used to obtain a companion identity, namely,

$$\sum_{n=1}^{\infty} \frac{(-a)^n q^{n(n+1)/2}(-q)_n}{(aq^2; q^2)_{n+1}} = \sum_{k=0}^{\infty} (-a)^k q^{k^2}. \tag{6.1}$$

It was shown in [3] that even though the series on the left in (6.1) is different from the one on the left in (1.1), the combinatorial interpretation of both identities are the same, namely Theorem 1. In a similar sense, identities (1.6) and (5.1) are companions because they both are analytic representations of Theorem 2. We now ask what happens if instead of applying the ideas of Sect. 5 to analyze the odd part of $(-q)_\infty$ as on the left of (5.6), we now consider the even part of $(-q)_\infty$, namely

$$\frac{(-q)_\infty + (q)_\infty}{2}.$$

Analogous to (5.6) and (5.8), we have the identity

$$\sum_{n=0}^{\infty} \frac{q^{2n^2+n}}{(q)_{2n}} = \frac{(-q)_\infty + (q)_\infty}{2} = \sum_{n=0}^{\infty} \frac{q^{2n}}{(q; q^2)_n} (q^{2n+2}; q^2)_\infty. \tag{6.2}$$

Let us rewrite (6.2) as

$$(-q)_\infty = -(q)_\infty + 2\left\{ (q^2; q^2)_\infty + \sum_{n=1}^{\infty} \frac{q^{2n}}{(q; q^2)_n} (q^{2n+2}; q^2)_\infty \right\}. \tag{6.3}$$

Now let T denote the set of partitions in which the odd parts may repeat, the evens should not repeat, even parts exist, and every even part is greater that every odd part. Let $T(n)$ denote the number of partitions π of n with $\pi \in T$. Next let $T_e(n)$ (resp. $T_o(n)$) denote the number of partitions of n of the type enumerated by $T(n)$ but with an even (resp. odd) number of even parts. Then the infinite sum inside the brackets on the right in (6.3) is the generating function of $T_o(n) - T_e(n)$, namely

$$\sum_{n=1}^{\infty} \{T_o(n) - T_e(n)\} q^n \tag{6.4}$$

and *not* of $T_e(n) - T_o(n)$, because q^{2n} has a $+$ sign attached to it! Now in partitions enumerated by $T(n)$, odd parts may or may not occur, but even parts have to occur. Suppose we consider partitions $\pi_0 \in T$ without any odd parts. Call the set of such partitions T^*. Let $T^*(n)$, $T_e^*(n)$, $T_o^*(n)$ be partition functions having analogous interpretations. Observe that T^* in simply the set of non-empty partitions into distinct even parts, and so

$$(q^2; q^2)_\infty = 1 + \sum \{T_e^*(n) - T_o^*(n)\} q^n. \tag{6.5}$$

Note that T^* is a subset of T, and $\tilde{T} = T - T^*$ is the set of partitions in T having an odd part as well. From (6.4) and (6.5) we see that the partitions in T which do not have any odd parts are actually counted with opposite signs and so the net contribution due these partitions in (6.3) is 0. Thus (6.3) can be rewritten as

$$(-q)_\infty = -(q)_\infty + 2\left\{1 + \sum_{n=1}^\infty \{\tilde{T}_o(n) - \tilde{T}_e(n)\}q^n\right\}. \tag{6.6}$$

Thus analogous to Theorem 5, we have

Theorem 7 *For $n \geq 1$*

$$P_d(n) = -\varepsilon_n + 2(\tilde{T}_o(n) - \tilde{T}_e(n)).$$

The main difference here is that we do not have an inequality like (5.19) and so we do not (yet) have a result analogous to Theorem 6.

7 Odd-even splits and Euler's trick

Euler's proof of the fundamental identity (5.5) utilizes the simple observation

$$1 + x = \frac{1 - x^2}{1 - x}. \tag{7.1}$$

With this trick applied to every factor of the form $(1 + q^m)$, we have

$$(-q)_\infty = \frac{(q^2; q^2)_\infty}{(q)_\infty} = \frac{1}{(q; q^2)_\infty}. \tag{7.2}$$

Since the refined generating function $(-zq)_\infty$ for partitions into distinct parts admits the series representation as in (5.7), we will now see what happens if Euler's trick is applied to $(-zq)_\infty$. We get

$$(-zq)_\infty = \frac{(z^2q^2; q^2)_\infty}{(zq)_\infty}. \tag{7.3}$$

There is no cancellation in (7.3) as in (7.2) which is perhaps why the application of Euler's trick to $(-zq)_\infty$ has not attracted attention. However, if we cast (7.3) in the form

$$\frac{1}{(z^2q^2; q^2)_\infty} \cdot (-zq)_\infty = \frac{1}{(zq)_\infty} \tag{7.4}$$

and use (5.7) to rewrite this as

$$\frac{1}{(z^2q^2; q^2)_\infty} \sum_{n=0}^\infty \frac{z^n q^{n(n+1)/2}}{(q)_n} = \sum_{n=0}^\infty \frac{z^n q^n}{(q)_n}, \tag{7.5}$$

we can say something interesting about the odd and even parts of the two sides. More precisely, since $(z^2q^2; q^2)_\infty$ is an even function of z, the odd and even parts of the two sides of (7.5) could be compared by concentrating on the odd and even parts of the two series there. That is, we have

$$\frac{1}{(z^2q^2; q^2)_\infty} \sum_{n=1}^{\infty} \frac{z^{2n-1}q^{2n^2-n}}{(q)_{2n-1}} = \sum_{n=1}^{\infty} \frac{z^{2n-1}q^{2n-1}}{(q)_{2n-1}}, \tag{7.6}$$

and

$$\frac{1}{(z^2q^2; q^2)_\infty} \sum_{n=0}^{\infty} \frac{z^{2n}q^{2n^2+n}}{(q)_{2n}} = \sum_{n=1}^{\infty} \frac{z^{2n}q^{2n}}{(q)_{2n}}. \tag{7.7}$$

Note that $z = 1$ in (7.6) is precisely (5.9) which is equivalent to (5.16). Similarly, $z = 1$ in (7.7) is (6.2), which is equivalent to (6.3). Thus the identities (5.15) and (6.3) from which we deduced Theorems 5 and 7 are emerging from the odd-even split underlying Euler's method to prove (5.5).

In a subsequent paper [4] we will consider variations in applying Euler's trick and generate an infinite sequence of weighted partition identities connecting unrestricted partitions, partitions into distinct parts, and partitions into odd parts.

Note added in proof I recently conjectured that the weighted partition function $p_\omega(n)$, which is the sum of the weights $\omega_s(\pi)$ taken over all partitions π of n as in Theorem 4, has the very interesting property that for any integer k, $p_\omega(n)$ is almost always a multiple of 2^k. Note that the generating function of $p_\omega(n)$ is

$$(-q)^2_\infty.$$

The conjecture can be proved for $k \leq 3$ by observing that $(-q)_\infty$ is the generating function of partitions into distinct parts, and by appealing to the property that the number of partitions of an integer n into distinct parts is odd precisely when n is a pentagonal number. The theory of modular forms would be needed to prove the conjecture for all $k \geq 1$; Ken Ono (private communication) has indicated that using modular forms, a stronger form of this conjecture can be proved.

Alladi and Gordon (Manuscr. Math. 79:113–126, 1993) introduced a generalization of the celebrated partition theorem of Schur by studying the expansion of the product $(-aq)_\infty(-bq)_\infty$. The generating function of $p_\omega(n)$ is the special case $a = b = 1$ of this product. In a subsequent paper we will investigate connections between the generalized Schur theorem, $p_\omega(n)$, and partitions into distinct parts, as well as the divisibility of $p_\omega(n)$ by powers of 2.

References

1. Alladi, K.: Partition identities involving gaps and weights. Trans. Am. Math. Soc. **349**, 5001–5019 (1997)
2. Alladi, K.: Partition identities involving gaps and weights—II. Ramanujan J. **2**, 21–37 (1998)
3. Alladi, K.: A partial theta identity of Ramanujan and its number theoretic interpretation. Ramanujan J. (to appear)
4. Alladi, K.: Variations on an idea of Euler—infinite sequences of weighted partition identities (in preparation)
5. Andrews, G.E.: Problem 5865. Am. Math. Mon. **79**, 668 (1972)

6. Andrews, G.E.: The Theory of Partitions. Encyclopedia of Math. and Its Applications, vol. 2. Addison-Wesley, Reading (1976)
7. Andrews, G.E.: Private communication (2008)
8. Andrews, G.E., Berndt, B.C.: Ramanujan's Lost Notebook, Part II. Springer, New York (2009)
9. Berndt, B.C., Kim, B., Yee, A.J.: Ramanujan's Lost Notebook: Combinatorial proofs of identities associated with Heine's transformation or partial theta functions. J. Comb. Theory Ser. A (to appear)
10. Ramanujan, S.: The Lost Notebook and Other Unpublished Papers. Narosa, New Delhi (1987)

Ramanujan J (2010) 23: 243–251
DOI 10.1007/s11139-009-9178-9

New identities involving sums of the tails related to real quadratic fields

Kathrin Bringmann · Ben Kane

To George Andrews, who has been a great inspiration, on the occasion of his 70th birthday

Received: 3 May 2009 / Accepted: 20 May 2009 / Published online: 31 August 2010
© Springer Science+Business Media, LLC 2010

Abstract In previous work, the authors discovered new examples of q-hypergeometric series related to the arithmetic of $\mathbb{Q}(\sqrt{2})$ and $\mathbb{Q}(\sqrt{3})$. Building on this work, we construct in this paper sum of the tails identities for which some which some of these functions occur as error terms. As an application, we obtain formulas for the generating function of a certain zeta functions for real quadratic fields at negative integers.

Keywords q-hypergeometric series · Sum of tails · Modular form · Real quadratic fields

Mathematics Subject Classification (2000) 11P81 · 11E16 · 05A17

1 Introduction

Andrews, Dyson and Hickerson [2] investigated the function

$$\sigma(q) := 1 + \sum_{n \geq 1} \frac{q^{\frac{n(n+1)}{2}}}{(-q)_n} = 1 + q - q^2 + 2q^3 + \cdots + 4q^{45} + \cdots$$

which first appeared in Ramanujan's lost notebook [7]. We denote as usual $(a)_n := (a;q)_n := \prod_{m=0}^{n-1}(1 - aq^m)$. By showing that the Fourier coefficients of $\sigma(q)$ are

The first author was partially supported by NSF grant DMS-0757907 and the Alfried Krupp prize.

K. Bringmann (✉) · B. Kane
Mathematical Institute, University of Cologne, Weyertal 86-90, 50931 Cologne, Germany
e-mail: kbringma@math.uni-koeln.de

B. Kane
e-mail: bkane@math.uni-koeln.de

connected to the arithmetic of $\mathbb{Q}(\sqrt{6})$, the authors of [2] were able to prove that σ is lacunary, i.e. its coefficients are almost always zero, and yet attains every integer infinitely many times.

Since Andrews, Dyson and Hickerson's investigation, the function $\sigma(q)$ has shown up in a variety of settings and is related to the theory of automorphic forms in many interesting ways, as described in [4]. Here we consider one such type of result involving sum of tails identities. Such identities were first studied by Zagier [8] in his work on Vassiliev invariants, who showed that

$$\sum_{n\geq 0}((q)_\infty - (q)_n) = (q)_\infty D_1(q) + E_1(q),$$

where

$$D_1(q) := -\frac{1}{2} + \sum_{n\geq 1} d(n)q^n, \qquad E_1(q) := \frac{1}{2}\sum_{n\geq 1}\left(\frac{12}{n}\right)nq^{\frac{n^2-1}{24}}.$$

Here $d(n)$ denotes the number of divisors of n and $\left(\frac{\bullet}{n}\right)$ is the Kronecker symbol. In [3], Andrews, Jimenez-Urroz, and Ono obtained a number of related identities, which have the form

$$\sum_{n\geq 0}(F(q) - F_n(q)) = F(q)D(q) + E(q), \tag{1.1}$$

where F is a (modular) infinite product, $F_n \to F$, and D is a divisor function. Since the coefficients of E grow much slower than those of FD, the function E may be considered as an "error series". Of particular interest are special cases of (1.1) in which the "error series" is $\sigma(q)$, for example the following formula which can be found in Ramanujan's lost notebook

$$\sum_{n\geq 0}((-q)_\infty - (-q)_n) = (-q)_\infty D_1(q) + \frac{1}{2}\sigma(q). \tag{1.2}$$

In [4], the authors discovered 8 additional examples of q-hypergeometric series related to the arithmetic of $\mathbb{Q}(\sqrt{2})$ and $\mathbb{Q}(\sqrt{3})$ including the functions

$$f(q) := \sum_{n\geq 0}\frac{(q)_{2n}}{(-q)_{2n+1}}q^n, \qquad h(q) := \sum_{n\geq 0}\frac{(q)_{2n+1}}{(-q)_{2n+2}}q^{n+1}.$$

Using the theory of Bailey pairs, it was shown in [4] that

$$qf(q^2) = \sum_{\mathfrak{a}\subset O_K}\left(\frac{-4}{\mathcal{N}(\mathfrak{a})}\right)q^{\mathcal{N}(\mathfrak{a})}, \tag{1.3}$$

$$h(q) = -\sum_{\mathfrak{a}\subset O_K}(-1)^{\mathcal{N}(\mathfrak{a})}q^{\mathcal{N}(\mathfrak{a})}, \tag{1.4}$$

where $K := \mathbb{Q}(\sqrt{2})$, O_K is the ring of integers of K, the sum runs over all ideals of O_K, and $\mathcal{N}$ denotes the norm of an ideal. As in the case of σ, the identities in (1.3) and (1.4) imply arithmetic information about the coefficients of f and h including lacunary behavior.

In this note we consider sums of tails identities resembling (1.2) which involve the functions f and h as "error series". To state our results, for $i \in \{0, 1\}$ we define

$$\theta_i(q) := \sum_{n \in \mathbb{Z}} q^{\frac{1}{2}(2n+i)^2}$$

and, denoting as usual the q-binomial coefficients by $\begin{bmatrix} A \\ B \end{bmatrix} := \frac{(q)_A}{(q)_B (q)_{A-B}}$, we let its "finite companion" be given by

$$\theta_{i,n}(q) := (q)_n \sum_{j \in \mathbb{Z}} \begin{bmatrix} 2n + 1 \\ n - 2j \end{bmatrix} q^{\frac{1}{2}(2j+i)^2}.$$

We note that $\lim_{n \to \infty} \theta_{i,n}(q) = \theta_i(q)$ and that the theta series $\theta_i(q^2)$ are modular forms which can be expressed as an infinite product, namely they can be written $\theta_0(q^{\frac{1}{2}}) = \frac{(q^2;q^2)_\infty^5}{(q)_\infty^2 (q^4;q^4)_\infty^2}$ and $\theta_1(q^2) = 2q \frac{(q^{16};q^{16})_\infty^2}{(q^8;q^8)_\infty}$. Moreover, we define

$$D_2(q) := \sum_{n=1}^{\infty} d_o(n)q^n,$$

where $d_o(n)$ counts the number of odd divisors of n.

Theorem 1.1 *We have the sum of tails identities*

$$\sum_{n \geq 0} \left(\frac{\theta_0(q)}{(q)_\infty(q;q^2)_\infty} - \frac{\theta_{0,n}(q)}{(q)_n(q;q^2)_{n+1}} \right) = 2 \frac{\theta_0(q)}{(q)_\infty(q;q^2)_\infty} D_2(q) - h(q), \qquad (1.5)$$

$$\sum_{n \geq 0} \left(\frac{\theta_1(q)}{(q)_\infty(q;q^2)_\infty} - \frac{\theta_{1,n}(q)}{(q)_n(q;q^2)_{n+1}} \right) = 2 \frac{\theta_1(q)}{(q)_\infty(q;q^2)_\infty} D_2(q) + q^{\frac{1}{2}} f(q). \quad (1.6)$$

As an application, we next consider formulas for the generating function of the following zeta functions for real quadratic fields at negative integers, paralleling the results obtained in [8] and [3]. For this define the usual Hecke L-function $\zeta_K(s)$ for K and a twisted zeta-function $\widetilde{\zeta_K}(s)$ by

$$\zeta_K(s) := \sum_{\mathfrak{a} \subset O_K} \mathcal{N}(\mathfrak{a})^{-s}, \qquad \widetilde{\zeta_K}(s) := \sum_{\mathfrak{a} \subset O_K} \left(\frac{-4}{\mathcal{N}(\mathfrak{a})} \right) \mathcal{N}(\mathfrak{a})^{-s}.$$

Theorem 1.1 will give us the following formulas from which $\zeta_K(s)$ and $\widetilde{\zeta_K}(s)$ may be calculated at negative integers.

Theorem 1.2 *As a power series in t, we have the equations*

$$-\sum_{n \geq 0} \frac{\theta_{0,n}(-e^{-t})}{(-e^{-t}; -e^{-t})_n(-e^{-t}; e^{-2t})_{n+1}} = \sum_{n \geq 1} (-1)^n \zeta_K(-n) \frac{t^n}{n!}, \qquad (1.7)$$

$$\sum_{n\geq0}\frac{\theta_{1,n}(-e^{-t})}{(-e^{-t};-e^{-t})_n(-e^{-t};e^{-2t})_{n+1}} = \sum_{n\geq1}\left(-\frac{1}{2}\right)^n \widetilde{\zeta_K}(-n)\frac{t^n}{n!}. \qquad (1.8)$$

We note that similarly we could also consider the limiting behavior at other odds roots of unity.

This paper is organized as follows. In Sect. 2, we first introduce an auxiliary function with an extra parameter, which is in special cases related to f and h, and thus also to the arithmetic of K. This auxiliary function also has some nice combinatorial meaning. We then show that this function is the "error series" in a sum of tails identity. In Sect. 3, we use the results from Sect. 2 to prove Theorem 1.1. We conclude Sect. 3 with another sums of tails result involving h, but where $\widetilde{h}(q)$ is not the "error series." Finally, we will establish the formulas for $\zeta_K(s)$ and $\widetilde{\zeta_K}(s)$ at negative integers in Sect. 4.

2 An auxiliary function

We will first investigate the properties of an auxiliary function which we will then bootstrap to obtain the main results for f and h. To this end, we define

$$g_z(q) := \sum_{n\geq0}\frac{(q)_n}{(-q)_{n+1}}z^{-n}q^{\frac{n}{2}}.$$

We will see in (2.6) that $z^{-1}qg_z(q^2)$ is the generating function for a natural combinatorial object. For this recall that an *overpartition* of the integer n is a partition, where the last occurrence of each part may be overlined. Denote by P the set of overpartitions into odd parts where the largest part must be overlined. If we denote the largest part by $L(\Lambda)$ and the number of parts by $M(\Lambda)$, then the coefficient of $z^{-m}q^n$ in $z^{-1}qg_z(q^2)$ is the number of overpartitions $\Lambda \in P$ of n with $M(\Lambda) = m$, weighted by $(\frac{-4}{L(\Lambda)})$.

We will particularly be interested in the two specializations $z = \pm1$. For the specialization $z = 1$, the n-th Fourier coefficient of $qg_{+1}(q^2)$ equals

$$\#\{\Lambda \in P : L(\Lambda) \equiv 1 \ (\mathrm{mod}\ 4)\} - \#\{\Lambda \in P : L(\Lambda) \equiv 3 \ (\mathrm{mod}\ 4)\},$$

while the n-th coefficient of $qg_{-1}(q^2)$ equals

$$\#\{\Lambda \in P : L(\Lambda) \equiv 2M(\Lambda)+1 \ (\mathrm{mod}\ 4)\} - \#\{\Lambda \in P : L(\Lambda) \not\equiv 2M(\Lambda)+1 \ (\mathrm{mod}\ 4)\}.$$

However, we will see that the coefficients of $qg_{\pm1}(q^2)$ will always be equal up to absolute value, due to the following relations with $f(q)$ and $h(q)$:

$$\frac{1}{2}(g_{+1}(q) + g_{-1}(q)) = f(q),$$

$$\frac{1}{2}(g_{+1}(q) - g_{-1}(q)) = q^{-\frac{1}{2}}h(q).$$

Taking $q \rightarrow q^2$ and adding the two above equations relates the coefficients of $g_1(q)$ to the arithmetic of O_K by

$$qg_{+1}(q^2) = qf(q^2) + h(q^2) = \sum_{\mathfrak{a} \subset O_K} \left[\left(\frac{-4}{\mathcal{N}(\mathfrak{a})} \right) - \frac{1 + (-1)^{\mathcal{N}(\mathfrak{a})}}{2} i^{\mathcal{N}(\mathfrak{a})} \right] q^{\mathcal{N}(\mathfrak{a})}, \tag{2.1}$$

which follows from (1.3) and (1.4). Here we have used the fact that there is a unique ideal of O_K of norm 2, so $h(q^2)$ is precisely the generating function for the elements of norm $2n$, weighted by $-i^{2n}$. Subtracting the two equations yields

$$qg_{-1}(q^2) = qf(q^2) - h(q^2) = \sum_{\mathfrak{a} \subset O_K} \left[\left(\frac{-4}{\mathcal{N}(\mathfrak{a})} \right) + \frac{1 + (-1)^{\mathcal{N}(\mathfrak{a})}}{2} i^{\mathcal{N}(\mathfrak{a})} \right] q^{\mathcal{N}(\mathfrak{a})}. \tag{2.2}$$

Remark Since $\left(\frac{-4}{2n} \right) = 0$ and $1 + (-1)^{2n+1} = 0$, it follows that the absolute value of the n-th Fourier coefficient of both $qg_{+1}(q^2)$ and $qg_{-1}(q^2)$ equals the number of ideals of O_K of norm n. Hence, although the coefficients of these series are not multiplicative, they are multiplicative up to ± 1.

We next show a sum of the tail-identity involving the functions g_z. For this, we let

$$\mathcal{D}_z(q) := \sum_{n \geq 1} \frac{z^{-1} q^{n+\frac{1}{2}}}{1 - z^{-1} q^{n+\frac{1}{2}}} - \sum_{n \geq 0} \frac{zq^{n-\frac{1}{2}}}{1 - zq^{n-\frac{1}{2}}}.$$

Theorem 2.1 *We have*

$$\sum_{n \geq 0} \left(\frac{(zq^{\frac{1}{2}})_\infty}{(-zq^{\frac{1}{2}})_\infty} - \frac{(zq^{\frac{1}{2}})_n}{(-zq^{\frac{1}{2}})_{n+1}} \right) = \frac{(zq^{\frac{1}{2}})_\infty}{(-zq^{\frac{1}{2}})_\infty} (\mathcal{D}_z(q) - 1 + 2D_2(q)) - z^{-1} q^{\frac{1}{2}} g_z(q). \tag{2.3}$$

This immediately implies the following corollary.

Corollary 2.2 *The following equations hold*

$$\sum_{n \geq 0} \left(\frac{(q^{\frac{1}{2}})_\infty}{(-q^{\frac{1}{2}})_\infty} - \frac{(q^{\frac{1}{2}})_n}{(-q^{\frac{1}{2}})_{n+1}} \right) = 2 \frac{(q^{\frac{1}{2}})_\infty}{(-q^{\frac{1}{2}})_\infty} D_2(q) - q^{\frac{1}{2}} g_1(q), \tag{2.4}$$

$$\sum_{n \geq 0} \left(\frac{(-q^{\frac{1}{2}})_\infty}{(q^{\frac{1}{2}})_\infty} - \frac{(-q^{\frac{1}{2}})_n}{(q^{\frac{1}{2}})_{n+1}} \right) = 2 \frac{(-q^{\frac{1}{2}})_\infty}{(q^{\frac{1}{2}})_\infty} D_2(q) + q^{\frac{1}{2}} g_{-1}(q). \tag{2.5}$$

Proof From Theorem 1 of [3] with $a = -zq^{\frac{1}{2}}$ and $t = zq^{-\frac{1}{2}}$, one may easily deduce that

$$\sum_{n\geq 1}\left(\frac{(zq^{-\frac{1}{2}})_\infty}{(-zq^{\frac{1}{2}})_\infty} - \frac{(zq^{-\frac{1}{2}})_n}{(-zq^{\frac{1}{2}})_n}\right) = \frac{(zq^{-\frac{1}{2}})_\infty}{(-zq^{\frac{1}{2}})_\infty}(\mathcal{D}_z(q) - 1 + 2D_2(q))$$

$$+ 1 + \sum_{n\geq 1}\frac{(-z^{-1}q^{\frac{1}{2}})_n}{(z^{-1}q^{\frac{3}{2}})_n}(-1)^n q^n.$$

Dividing both sides by $1 - zq^{-\frac{1}{2}} = -zq^{-\frac{1}{2}}(1 - z^{-1}q^{\frac{1}{2}})$ and shifting the sum of tails gives

$$\sum_{n\geq 0}\left(\frac{(zq^{\frac{1}{2}})_\infty}{(-zq^{\frac{1}{2}})_\infty} - \frac{(zq^{\frac{1}{2}})_n}{(-zq^{\frac{1}{2}})_{n+1}}\right) = \frac{(zq^{\frac{1}{2}})_\infty}{(-zq^{\frac{1}{2}})_\infty}(\mathcal{D}_z(q) - 1$$

$$+ 2D_2(q)) - z^{-1}q^{\frac{1}{2}}\sum_{n\geq 0}\frac{(-z^{-1}q^{\frac{1}{2}})_n}{(z^{-1}q^{\frac{1}{2}})_{n+1}}(-1)^n q^n.$$

Using Heine's second transformation (cf. [1, p. 11, (2.15)]) with $a = b = q$, $c = -q^2$, and $z = z^{-1}q^{\frac{1}{2}}$ gives that

$$g_z(q) = \sum_{n\geq 0}\frac{(-z^{-1}q^{\frac{1}{2}})_n}{(z^{-1}q^{\frac{1}{2}})_{n+1}}(-1)^n q^n. \tag{2.6}$$

This gives the claim. $\qquad\square$

3 Proof of the main result

In this section, we will use Theorem 2.2 to obtain the desired sum of tails results for $f(q)$ and $h(q)$. Since the proof of both identities are quite similar, we only prove the first here. We first define

$$F(q) := \frac{(q^{\frac{1}{2}})_\infty}{(-q^{\frac{1}{2}})_\infty} + \frac{(-q^{\frac{1}{2}})_\infty}{(q^{\frac{1}{2}})_\infty}$$

and its "finite companion"

$$F_n(q) := \frac{(q^{\frac{1}{2}})_n}{(-q^{\frac{1}{2}})_{n+1}} + \frac{(-q^{\frac{1}{2}})_n}{(q^{\frac{1}{2}})_{n+1}}.$$

Specializing equation (2.3) once with $z = 1$ and once $z = -1$ and then summing gives

$$\sum_{n\geq 0}(F(q) - F_n(q)) = 2F(q)D_2(q) - 2h(q).$$

Thus it remains to compute $F(q)$ and $F_n(q)$. Using the Jacobi triple product identity (cf. [6, p. 17]) one obtains

$$F(q^2) = \frac{1}{(q^2; q^4)_\infty (q^2; q^2)_\infty} \left(\frac{(q)_\infty^2}{(q^2; q^2)_\infty} + \frac{(q^2; q^2)_\infty^5}{(q^4; q^4)_\infty^2 (q)_\infty^2} \right)$$

$$= \frac{1}{(q^2; q^4)_\infty (q^2; q^2)_\infty} \left(\sum_{n \in \mathbb{Z}} (-1)^n q^{n^2} + \sum_{n \in \mathbb{Z}} q^{n^2} \right).$$

Thus

$$F(q) = \frac{2}{(q; q^2)_\infty (q)_\infty} \theta_0(q).$$

To compute $F_n(q)$ we first rewrite

$$F_n(q) = \frac{1}{(q; q^2)_{n+1}} ((q^{\frac{1}{2}})_n^2 (1 - q^{n+\frac{1}{2}}) + (-q^{\frac{1}{2}})_n^2 (1 + q^{n+\frac{1}{2}}))$$

and then use McMahon's finite version of the Jacobi triple product identity [5, vol. 2, Sect. 323] with $x = q^{-\frac{1}{2}}$ for the first summand and $x = -q^{-\frac{1}{2}}$ for the second summand. This gives

$$F_n(q) = \frac{1}{(q; q^2)_{n+1}} \left((1 - q^{n+\frac{1}{2}}) \sum_{j \in \mathbb{Z}} \begin{bmatrix} 2n \\ n+j \end{bmatrix} (-1)^j q^{\frac{j^2}{2}} \right.$$

$$\left. + (1 + q^{n+\frac{1}{2}}) \sum_{j \in \mathbb{Z}} \begin{bmatrix} 2n \\ n+j \end{bmatrix} q^{\frac{j^2}{2}} \right)$$

$$= \frac{2}{(q; q^2)_{n+1}} \left(\sum_{j \in \mathbb{Z}} \begin{bmatrix} 2n \\ n+2j \end{bmatrix} q^{2j^2} + q^{n+\frac{1}{2}} \sum_{j \in \mathbb{Z}} \begin{bmatrix} 2n \\ n+2j+1 \end{bmatrix} q^{2j^2+2j+\frac{1}{2}} \right)$$

$$= \frac{2}{(q; q^2)_{n+1}} \sum_{j \in \mathbb{Z}} \frac{(q)_{2n}}{(q)_{n+2j+1}(q)_{n-2j}} q^{2j^2}$$

$$\times (1 - q^{n+2j+1} + q^{n+2j+1}(1 - q^{n-2j}))$$

$$= \frac{2}{(q; q^2)_{n+1}} \sum_{j \in \mathbb{Z}} \begin{bmatrix} 2n+1 \\ n-2j \end{bmatrix} q^{2j^2} = \frac{2}{(q; q^2)_{n+1}(q)_n} \theta_{0,n}(\tau).$$

There is additionally another sum of tails identity related to $g_z(q)$. In this case $g_z(q)$ does not play the role of the "error series" but rather as part of the "divisor function".

Theorem 3.1 *We have the sum of tails identity*

$$\sum_{n \geq 0} \left(\frac{(-q)_\infty}{(q)_\infty} - \frac{(-q)_n}{(q)_n(1 - q^{2n+1})} \right) = \frac{(-q)_\infty}{(q)_\infty} (2D_2(q) - h(q)).$$

Proof From Theorem 2 of [3] with $a = -q$, $b = z^{-1}q^{\frac{1}{2}}$, and $c = z^{-1}q^{\frac{3}{2}}$, one obtains

$$\sum_{n\geq 0}\left(\frac{(-q)_\infty}{(q)_\infty} - \frac{(-q)_n}{(q)_n(1 - z^{-1}q^{n+\frac{1}{2}})}\right)$$

$$= \frac{(-q)_\infty}{(q)_\infty}\left(2\sum_{n\geq 1}\frac{q^n}{1 - q^{2n}} - \sum_{n\geq 1}\frac{(q)_n}{(-q)_n(1 - q^n)}z^{-n}q^{\frac{n}{2}}\right)$$

$$= \frac{(-q)_\infty}{(q)_\infty}(2D_2(q) - z^{-1}q^{\frac{1}{2}}g_z(q)).$$

The result then follows after summing the terms with $z = 1$ and $z = -1$ and dividing by 2.   □

In addition to these sum of tails identities, the auxiliary function $g_z(q)$ also allows us to deduce new combinatorial interpretations for $f(q)$ and $h(q)$ in terms of overpartitions. Using Heine's first transformation (cf. [1, p. 10, (2.11)]) with $a = b = q$, $c = -q^2$, and $z = z^{-1}q^{\frac{1}{2}}$ gives

$$g_z(q) = \frac{(q)_\infty}{(-q)_\infty}\sum_{n\geq 0}\frac{(-q)_n}{(q)_n(1 - z^{-1}q^{n+\frac{1}{2}})}q^n.$$

After pulling the infinite product inside the sum, this immediately gives

$$f(q) = \sum_{n\geq 0}\frac{(q^{n+1})_n(q^{2n+2})_\infty}{(-q^{n+1})_\infty}q^n, \qquad h(q) = \sum_{n\geq 0}\frac{(q^{n+1})_n(q^{2n+2})_\infty}{(-q^{n+1})_\infty}q^{2n+1}.$$

To describe the new combinatorial interpretation for the coefficients of f and h, let $n = s(\Lambda)$ denote the smallest part of the overpartition Λ. Then $-f(q)$ is the generating function for overpartitions where $s(\Lambda)$ is overlined and occurs exactly once and $2s(\Lambda) + 1$ cannot be overlined, weighted by $(-1)^{M(\Lambda)}$.

Similarly, writing $2n + 1 = n + (n + 1)$, we see that $h(q)$ is the generating function for overpartitions for which $s(\Lambda)$ is overlined and occurs precisely once, a non-overlined part of size $s(\Lambda) + 1$ must occur, and $2s(\Lambda) + 1$ cannot be overlined, again weighted by $(-1)^{M(\Lambda)}$.

4 Values of zeta functions at negative integers

We will use Theorem 1.1 to establish Theorem 1.2 here.

Proof of Theorem 1.2 This proof will follow analogously to that of Zagier [8, Theorem 3] so we will give a brief argument for (1.7) and leave the details of (1.8) to the reader. We first note that

$$\frac{\theta_0(q)}{(q)_\infty(q;q^2)_\infty} = \frac{(-q)_\infty^2(-q^2;q^2)_\infty^3}{(-q^4;q^4)_\infty^2}$$

and hence vanishes to infinite order as $q \to -1$. We next replace q by $-e^{-t}$ and define $a(n)$ and $b(n)$ by the asymptotic expansions as $t \to 0$ given by

$$-\sum_{n \geq 0} \frac{\theta_{0,n}(-e^{-t})}{(-e^{-t}; -e^{-t})_n(-e^{-t}; e^{-2t})_{n+1}} = \sum_{n \geq 0} a(n)t^n,$$

$$-h(-e^{-t}) := \sum_{\mathbf{a} \subset O_K} e^{-t \mathcal{N}(\mathbf{a})} \sim \sum_{n \geq 0} b(n)t^n.$$

By Theorem 1.1 we have $a(n) = b(n)$ for every $n \geq 0$. We then integrate to obtain

$$-\int_0^\infty h(-e^{-t})t^{s-1}dt = \sum_{\mathbf{a} \subset O_K} \int_0^\infty e^{-t \mathcal{N}(\mathbf{a})}t^{s-1}dt$$

$$= \Gamma(s) \sum_{\mathbf{a} \subset O_K} \mathcal{N}(\mathbf{a})^{-s} = \Gamma(s)\zeta_K(s).$$

The residue at $s = -n$ is hence given by $\frac{(-1)^n}{n!}\zeta_K(-n)$. However, for any N fixed we have

$$-\int_0^\infty h(-e^{-t})t^{s-1}dt = \int_0^\infty \left(\sum_{0 \leq n < N} b(n)t^n + O(t^N) \right)t^{s-1}dt$$

$$= \sum_{0 \leq n < N} \frac{b(n)}{s+n} + H(s),$$

where $H(s)$ is holomorphic for $\mathrm{Re}(s) > -N$, so that for $n < N$ the residue at $s = -n$ is $b(n)$. $\qquad\square$

Acknowledgements The authors thank George Andrews and Jeremy Lovejoy for fruitful conversations. Moreover, they thank Ken Ono for helpful comments on an earlier version of the paper.

References

1. Andrews, G.: q-Series: Their Development and Application in Analysis Number Theory, Combinatorics, Physics, and Computer Algebra. CBMS Regional Conference Series in Mathematics, vol. 66, p. 130. Am. Math. Soc., Providence (1986)
2. Andrews, G., Dyson, F., Hickerson, D.: Partitions and indefinite quadratic forms. Invent. Math. **91**, 391–407 (1988)
3. Andrews, G., Jimenez-Urroz, J., Ono, K.: q-series identities and values of certain L-functions. Duke Math. J. **108**, 395–419 (2001)
4. Bringmann, K., Kane, B.: Multiplicative q-hypergeometric series arising from real quadratic fields. Trans. Am. Math. Soc. (2010, accepted). arxiv:0812.4397
5. MacMahon, P.A.: Combinatory Analysis, vols. I, II. Cambridge University Press, Cambridge (1915). Reprinted by Chelsea, New York (1960)
6. Ono, K.: Web of Modularity: Arithmetic of the Coefficients of Modular Forms and q-Series. CBMS Regional Conference Series in Mathematics, vol. 102. Am. Math. Soc., Providence (2003)
7. Ramanujan, S.: The Lost Notebook and Other Unpublished Papers. Narosa, New Delhi (1988)
8. Zagier, D.: Vassiliev invariants and a strange identity related to the Dedekind eta-function. Topology **40**, 945–960 (2001)

Ramanujan J (2010) 23: 253–264
DOI 10.1007/s11139-009-9184-y

Rademacher-type formulas for restricted partition and overpartition functions

Andrew V. Sills

Dedicated to George Andrews on the occasion of his seventieth birthday

Received: 18 May 2009 / Accepted: 9 June 2009 / Published online: 11 May 2010
© Springer Science+Business Media, LLC 2010

Abstract A collection of Hardy-Ramanujan-Rademacher type formulas for restricted partition and overpartition functions is presented, framed by several biographical anecdotes.

Keywords Partitions · Circle method · Rogers-Ramanujan identities

Mathematics Subject Classification (2000) 11P82 · 11P85 · 05A19

1 Introduction

When George Andrews matriculated in the Ph.D. program at the University of Pennsylvania in the fall of 1961, his intention was to specialize in geometric number theory. He had been attracted to Penn's graduate program in part because the 1961–1962 academic year had been designated a special year in number theory there. The academic year culminated in a celebration of the seventieth birthday of Professor Hans Rademacher.

Rademacher taught Andrews in his analytic number theory class that year, and there Andrews was introduced to the theory of partitions. A partition λ of an integer n is a weakly decreasing finite sequence of positive integers $(\lambda_1, \lambda_2, \dots, \lambda_s)$ whose sum is n. Each λ_i is called a 'part' of the partition λ. The theory of integer partitions began with Euler [22], who introduced generating functions to study $p(n)$, the number of partitions of n, and found that the generating function for $p(n)$ was representable as

A.V. Sills (✉)
Department of Mathematical Sciences, Georgia Southern University, Statesboro,
GA 31407-8093, USA
e-mail: ASills@GeorgiaSouthern.edu

an elegant infinite product:

$$\sum_{n=0}^{\infty} p(n)x^n = \prod_{m \geq 1} \frac{1}{1 - x^m}. \tag{1.1}$$

The "circle method" was created by Hardy and Ramanujan and later improved by Rademacher, in connection with the study of the function $p(n)$, the number of partitions of the integer n. The circle method has proved to be one of the most useful tools in the history of analytic number theory. Expositions of the circle method may be found in [3, 10, 52–54].

Rademacher's formula for $p(n)$ is given by

$$p(n) = \frac{1}{\pi\sqrt{2}} \sum_{k=1}^{\infty} \sqrt{k} \sum_{\substack{0 \leq h < k \\ (h,k)=1}} \omega(h,k) e^{-2\pi i n h/k} \frac{d}{dn} \left(\frac{\sinh(\frac{\pi}{k}\sqrt{\frac{2}{3}(n - \frac{1}{24})})}{\sqrt{n - \frac{1}{24}}} \right), \tag{1.2}$$

where $\omega(h,k)$ is a $24k$th root of unity that frequently occurs in the study of modular forms and is given by

$$\omega(h,k) = \begin{cases} (\frac{-k}{h}) \exp(-\pi i \{ \frac{1}{4}(2 - hk - h) + \frac{1}{12}(k - \frac{1}{k})(2h - H + h^2 H) \}), & \text{if } 2 \nmid h, \\ (\frac{-h}{k}) \exp(-\pi i \{ \frac{1}{4}(k - 1) + \frac{1}{12}(k - \frac{1}{k})(2h - H + h^2 H) \}), & \text{if } 2 \nmid k, \end{cases}$$

$(\frac{a}{b})$ is the Legendre-Jacobi symbol, and H is any solution of the congruence

$$hH \equiv -1 \pmod{k}.$$

Andrews reports [64] that the formula for $p(n)$

... was a revolutionary and surprising achievement. The form of this formula is even more stunning. It involves transcendental numbers and expressions that seem to be totally unrelated that might be appropriate, say, in a course on engineering or theoretical physics, but for actually counting how many ways you can add up sums to get a particular number, they seem absolutely incredible. In fact, I was *stunned* the first time I saw this formula. I could not *believe* it, and the experience of seeing it explained and understanding how it took shape really, I think, convinced me that this was the area of mathematics that I wanted to pursue.

Many practitioners, including a number of Ph.D. students and postdocs who worked under Rademacher, have used the circle method to study various restricted partition functions, often associated with sets of partitions enumerated in famous theorems. These practitioners included Grosswald [24, 25], Haberzetle [26], Hagis [27–35], Hua [38], Iseki [39–41], Lehner [42], Livingood [43], Niven [51], and Subramanyasastri [63].

Let us consider several examples.

Theorem 1 (Euler, 1748) *Let $q(n)$ denote the number of partitions of n into odd parts. Let $r(n)$ denote the number of partitions of n into distinct parts. Then $q(n) = r(n)$ for all integers n.*

Theorem 2 (Hagis, 1963)

$$q(n) = \frac{\pi}{\sqrt{24n+1}} \sum_{\substack{k \geq 1 \\ 2 \nmid k}} \frac{1}{k} \sum_{\substack{0 \leq h < k \\ (h,k)=1}} e^{-2\pi nh/k} \frac{\omega(h,k)}{\omega(2h,k)} I_1\left(\frac{\pi\sqrt{24n+1}}{6\sqrt{2}k}\right), \qquad (1.3)$$

where

$$I_\nu(z) := \sum_{r=0}^{\infty} \frac{(\frac{1}{2}z)^{\nu+2r}}{r!\,\Gamma(\nu+r+1)} \qquad (1.4)$$

is the Bessel function of purely imaginary argument.

Theorem 3 (Schur, 1926) *Let $s(n)$ denote the number of partitions of n into parts congruent to $\pm 1 \pmod 6$. Let $t(n)$ denote the number of partitions λ of n where $\lambda_i - \lambda_{i+1} \geq 3$ and $\lambda_i - \lambda_{i+1} > 3$ if $3 \mid \lambda_i$. Then $s(n) = t(n)$ for all n.*

Theorem 4 (Niven, 1940)

$$s(n) = \frac{\pi}{\sqrt{36n-3}} \sum_{d \mid 6} \sqrt{(d-2)(d-3)} \sum_{\substack{k \geq 1 \\ (k,6)=d}} \frac{1}{k}$$

$$\times \sum_{\substack{0 \leq h < k \\ (h,k)=1}} e^{-2\pi nh/k} \frac{\omega(h,k)\omega(6h/d,k/d)}{\omega(\frac{2h}{(d,2)},\frac{k}{(d,2)})\omega(\frac{3h}{(d,3)},\frac{k}{(d,3)})}$$

$$\times I_1\left(\frac{\pi\sqrt{d(12n-1)}}{3\sqrt{6}k}\right). \qquad (1.5)$$

Recently, the author found [59]

$$\bar{p}(n) = \frac{1}{2\pi} \sum_{\substack{k \geq 1 \\ 2 \nmid k}} \sqrt{k} \sum_{\substack{0 \leq h < k \\ (h,k)=1}} \frac{\omega(h,k)^2}{\omega(2h,k)} e^{-2\pi inh/k} \frac{d}{dn}\left(\frac{\sinh(\frac{\pi\sqrt{n}}{k})}{\sqrt{n}}\right) \qquad (1.6)$$

and

$$pod(n) = \frac{2}{\pi\sqrt{6}} \sum_{d \mid 4} \sqrt{(d-2)(5d-17)} \sum_{\substack{k \geq 1 \\ (k,4)=d}} \sqrt{k}$$

$$\times \sum_{\substack{0 \leq h < k \\ (h,k)=1}} \frac{\omega(h,k)\,\omega(4h/d,k/d)}{\omega(\frac{2h}{(d,2)},\frac{k}{(d,2)})} e^{-2\pi i n h/k}$$

$$\times \frac{d}{dn}\left(\frac{\sinh(\frac{\pi\sqrt{d(8n-1)}}{4k})}{\sqrt{8n-1}}\right), \tag{1.7}$$

where $pod(n)$ denotes the number of partitions of n where no odd part is repeated, and $\bar{p}(n)$ denotes the number of overpartitions of n. An *overpartition* of n is a finite weakly decreasing sequence of positive integers where the last occurrence of a given part may or may not be overlined. Thus the eight overpartitions of 3 are (3), $(\bar{3})$, $(2,1)$, $(\bar{2},1)$, $(2,\bar{1})$, $(\bar{2},\bar{1})$, $(1,1,1)$, $(1,1,\bar{1})$. Overpartitions were introduced by S. Corteel and J. Lovejoy in [19] and have been studied extensively by them and others including Bringmann, Chen, Fu, Goh, Hirschhorn, Hitczenko, Lascoux, Mahlburg, Robbins, Rødseth, Sellers, Yee, and Zho [12, 16–21, 23, 36, 37, 44–50, 55, 56].

Recently, Bringmann and Ono [13] have given exact formulas for the coefficients of all harmonic Maass forms of weight $\leq \frac{1}{2}$. All of the generating functions considered herein are weakly holomorphic forms of weight either 0 or $-\frac{1}{2}$, and thus they are harmonic Maass forms of weight $\leq \frac{1}{2}$. Accordingly, all of the exact formulas for restricted partition and overpartition functions presented here could be derived from the general theorem in [13].

In this article, we will present several anecdotes from the professional life of George Andrews, and present some new Rademacher type formulas related to the events described.

2 Identities in the lost notebook

Certainly one of the most exciting incidents of George Andrews' professional life was his unearthing of Ramanujan's lost notebook at the Wren Library at Trinity College, Cambridge University in 1976 (see [5, pp. 5–6, §1.5] and [7, p. 1 ff] for a full account). As is now well known, the lost notebook contains many identities of the Rogers-Ramanujan type. Many of the infinite products appearing in these identities are easily identified as generating functions for certain restricted classes of partitions or overpartitions. The methods of Rademacher may be applied to find explicit formulas for the coefficients appearing in the series expansions of these generating functions.

Below is a list of some of the Rogers-Ramanujan type identities which appear in the lost notebook. Some of these identities also appear in Slater [57]. Specifically, (2.1) is Slater's (6); (2.3) is Slater's (12); (2.5) is Slater's (22); (2.6) is Slater's (25); (2.7) is Slater's (28); (2.8) is Slater's (29); and (2.11) is Slater's (50).

The standard abbreviations

$$(a;b)_j = \prod_{i=0}^{j-1}(1 - ab^i), \qquad (a)_j := (a;x)_j$$

will be used. Note that $(a; b)_0 = 1$. Here and throughout, we assume $|x| < 1$ to guarantee convergence

$$\sum_{n=0}^{\infty} \frac{x^{n^2}(-1)_n}{(x)_n(x; x^2)_n} = \prod_{m=1}^{\infty} \frac{(1 + x^{3m-2})(1 + x^{3m-1})}{(1 - x^{3m-2})(1 - x^{3m-1})} \qquad [8, \text{Ent } 4.2.8], \quad (2.1)$$

$$\sum_{n=0}^{\infty} \frac{x^{n^2}(-x)_n}{(x)_n(x; x^2)_{n+1}} = \prod_{m=1}^{\infty} \frac{(1 + x^{3m-2})(1 + x^{3m-1})}{(1 - x^{3m-2})(1 - x^{3m-1})} \qquad [8, \text{Ent } 4.2.9], \quad (2.2)$$

$$\sum_{n=0}^{\infty} \frac{x^{n(n+1)/2}(-1)_n}{(x)_n} = \prod_{m=1}^{\infty} \frac{1 + x^{2m-1}}{1 - x^{2m-1}} \qquad [8, \text{Ent } 1.7.14], \qquad\qquad (2.3)$$

$$\sum_{n=0}^{\infty} \frac{x^{n^2}(-x^2; x^2)_n}{(x)_{2n+1}} = \prod_{m=1}^{\infty} \frac{1 + x^{2m-1}}{1 - x^{2m-1}} \qquad [8, \text{Ent } 1.7.13]. \qquad\qquad (2.4)$$

$$\sum_{n=0}^{\infty} \frac{x^{n(n+1)}(-x)_n}{(x; x^2)_{n+1}(x)_n} = \prod_{m=1}^{\infty} \frac{(1 - x^{6m})(1 - x^{6m-1})(1 - x^{6m-5})}{(1 - x^m)(1 - x^{2m-1})}$$

$$[8, \text{Ent } 4.2.12], \qquad\qquad (2.5)$$

$$\sum_{n=0}^{\infty} \frac{x^{n^2}(-x; x^2)_n}{(x^4; x^4)_n} = \prod_{m=1}^{\infty} \frac{(1 - x^{3m})(1 - x^{12m})}{(1 - x^{6m-5})(1 - x^{6m-1})(1 - x^{4m})}$$

$$[8, \text{Ent } 4.2.7], \qquad\qquad (2.6)$$

$$\sum_{n=0}^{\infty} \frac{x^{n(n+1)}(-x^2; x^2)_n}{(x)_{2n+1}} = \prod_{m=1}^{\infty} \frac{(1 - x^{12m})(1 - x^{12m-9})(1 - x^{12m-3})}{1 - x^m}$$

$$[8, \text{Ent } 4.3.12], \qquad\qquad (2.7)$$

$$\sum_{n=0}^{\infty} \frac{x^{n^2}(-x; q^2)_n}{(x)_{2n}} = \prod_{m=1}^{\infty} \frac{(1 - x^{6m})(1 - x^{12m-6})}{1 - x^m} \qquad [8, \text{Ent. } 5.2.3], \quad (2.8)$$

$$\sum_{n=0}^{\infty} \frac{x^{n(n+1)/2}(-x^2; x^2)_n}{(x)_n(x; x^2)_{n+1}} = \prod_{m=1}^{\infty} \frac{1 + x^m}{(1 - x^{2m-1})(1 - x^{8m-4})} \qquad [8, \text{Ent. } 1.7.5], \quad (2.9)$$

$$\sum_{n=0}^{\infty} \frac{x^{n(n+1)/2}(-1; x^2)_n}{(x)_n(x; x^2)_n} = \prod_{m=1}^{\infty} \frac{(1 - x^{4m})(1 - x^{8m-4})(1 + x^m)}{1 - x^m}$$

$$[8, \text{Ent. } 1.7.4], \qquad\qquad (2.10)$$

$$\sum_{n=0}^{\infty} \frac{x^{n(n+2)}(-x; x^2)_n}{(x)_{2n+1}} = \prod_{m=1}^{\infty} \frac{(1 - x^{12m})(1 - x^{12m-10})(1 - x^{12m-9})}{1 - x^m}$$

$$[8, \text{Ent. } 3.4.4]. \qquad\qquad (2.11)$$

Let us denote the cöefficient of x^n in the power series expansion of equation (j) above by $R_j(n)$. The following combinatorial interpretations are then immediate:

- $R_{2.1}(n) = R_{2.2}(n) =$ the number of overpartitions of n into nonmultiples of 3.
- $R_{2.3}(n) = R_{2.4}(n) =$ the number of overpartitions of n with only odd parts.
- $R_{2.5}(n) =$ the number of overpartitions of n where nonoverlined parts are congruent to $\pm 2, 3 \pmod 6$.
- $R_{2.7}(n) =$ the number of partitions of n into parts not congruent to $0, \pm 3 \pmod{12}$.
- $R_{2.9}(n) =$ the number of overpartitions of n where the nonoverlined parts are odd or congruent to $4 \pmod 8$.
- $R_{2.11}(n) =$ the number of partitions of n into parts not congruent to $0, \pm 2 \pmod{12}$.

The circle method yields the following formulas, which are believed to be new to the literature. It could be argued that a number of them capture much of the elegance of the formula for $p(n)$. They were found with the aid of *Mathematica* program written by the author. For a discussion of the automation of certain key steps of the circle method, along with additional examples of Rademacher type formulas for restricted partition and overpartition functions, please see [60]. As noted earlier, they could also be derived using the results of Bringmann and Ono [13]

$$R_{2.1}(n) = \frac{\pi}{3\sqrt{2n}} \sum_{\substack{k \geq 1 \\ 2 \nmid k, 3 \nmid k}} \frac{1}{k} \sum_{\substack{0 \leq h < k \\ (h,k)=1}} e^{-2\pi i n h/k} \frac{\omega(h,k)^2 \omega(6h,k)}{\omega(2h,k)\omega(3h,k)^2} I_1\left(\frac{\pi\sqrt{2n}}{k\sqrt{3}}\right),$$

(2.12)

$$R_{2.3}(n) = \frac{\pi}{4\sqrt{n}} \sum_{\substack{k \geq 1 \\ 2 \nmid k}} \frac{1}{k} \sum_{\substack{0 \leq h < k \\ (h,k)=1}} e^{-2\pi i n h/k} \frac{\omega(h,k)^2 \omega(4h,k)}{\omega(2h,k)^3} I_1\left(\frac{\pi\sqrt{n}}{k\sqrt{2}}\right),$$

(2.13)

$$R_{2.5}(n) = \frac{\pi}{2\sqrt{18n+6}} \sum_{\substack{k \geq 1 \\ 2 \nmid k}} \frac{\sqrt{(k,6)}}{k} \sum_{\substack{0 \leq h < k \\ (h,k)=1}} e^{-2\pi i n h/k} \frac{\omega(h,k)\omega\left(\frac{3h}{(k,3)}, \frac{k}{(k,3)}\right)}{\omega\left(\frac{6h}{(k,6)}, \frac{k}{(k,6)}\right)}$$

$$\times I_1\left(\frac{\pi\sqrt{6n+2}}{3k}\right),$$

(2.14)

$$R_{2.6}(n) = \frac{\pi}{3\sqrt{264n-33}} \sum_{d \in \{1,4,12\}} \sqrt{d^2+83d+48} \sum_{\substack{k \geq 1 \\ (k,12)=d}} \frac{1}{k}$$

$$\times \sum_{\substack{0 \leq h < k \\ (h,k)=1}} e^{-2\pi i n h/k} \frac{\omega(h,k)\omega\left(\frac{4h}{(d,4)}, \frac{k}{(d,4)}\right)\omega\left(\frac{6h}{(d,6)}, \frac{k}{(d,6)}\right)}{\omega\left(\frac{3h}{(d,3)}, \frac{k}{(d,3)}\right)^2 \omega\left(\frac{2h}{(d,2)}, \frac{k}{(d,2)}\right)}$$

$$\times I_1\left(\pi \frac{\sqrt{(16d-d^2-12)(8n-1)}}{12k}\right),$$

(2.15)

$$R_{2.7}(n) = \frac{\pi}{4\sqrt{90n+30}} \sum_{d|6} \sqrt{(d-3)(9d^2-52d+28)} \sum_{\substack{k\geq 1 \\ (k,12)=d}} \frac{1}{k}$$

$$\times \sum_{\substack{0\leq h<k \\ (h,k)=1}} e^{-2\pi ihn/k} \frac{\omega(h,k)\omega(\frac{6h}{(d,6)}, \frac{k}{(d,6)})}{\omega(\frac{3h}{(d,3)}, \frac{k}{(d,3)})\omega(\frac{12h}{(d,12)}, \frac{k}{(d,12)})}$$

$$\times I_1\left(\frac{\pi\sqrt{(8+8d-d^2)(3n+1)}}{3k\sqrt{10}}\right), \tag{2.16}$$

$$R_{2.8}(n) = \frac{\pi}{3\sqrt{264n-11}} \sum_{d\in\{1,4,12\}} \sqrt{2d^2+d+96} \sum_{\substack{k\geq 1 \\ (k,12)=d}} \frac{1}{k}$$

$$\times \sum_{\substack{0\leq h<k \\ (h,k)=1}} e^{-2\pi ihn/k} \frac{\omega(h,k)\omega(\frac{12h}{d}, \frac{k}{d})}{\omega(\frac{6h}{(d,6)}, \frac{k}{(d,6)})^2}$$

$$\times I_1\left(\frac{\pi\sqrt{(84+16d-d^2)(24n-1)}}{12k\sqrt{33}}\right), \tag{2.17}$$

$$R_{2.9}(n) = \frac{\pi\sqrt{3}}{4\sqrt{8n+2}} \sum_{\substack{k\geq 1 \\ 2\nmid k}} \frac{1}{k} \sum_{\substack{0\leq h<k \\ (h,k)=1}} e^{-2\pi ihn/k} \frac{\omega(h,k)^2\omega(4h,k)}{\omega(2h,k)^2\omega(8h,k)}$$

$$\times I_1\left(\frac{\pi\sqrt{12n+3}}{4k}\right), \tag{2.18}$$

$$R_{2.10}(n) = \frac{\pi\sqrt{3}}{8\sqrt{n}} \sum_{\substack{k\geq 1 \\ 2\nmid k}} \frac{1}{k} \sum_{\substack{0\leq h<k \\ (h,k)=1}} e^{-2\pi ihn/k} \frac{\omega(h,k)^2\omega(8h,k)}{\omega(4h,k)^2\omega(2h,k)} I_1\left(\frac{\pi\sqrt{3n}}{2k}\right), \tag{2.19}$$

$$R_{2.11}(n) = \frac{\pi}{6\sqrt{24n+15}} \sum_{d=1}^{4} \sqrt{(2-d)(7d^2-46d+48)} \sum_{\substack{k\geq 1 \\ (k,12)=d}} \frac{1}{k}$$

$$\times \sum_{\substack{0\leq h<k \\ (h,k)=1}} e^{-2\pi ihn/k} \frac{\omega(h,k)\omega(\frac{4h}{(d,4)}, \frac{k}{(d,4)})\omega(\frac{6h}{(d,6)}, \frac{k}{(d,6)})}{\omega(\frac{2h}{(d,2)}, \frac{k}{(d,2)})\omega(\frac{12h}{d}, \frac{k}{d})^2}$$

$$\times I_1\left(\frac{\pi\sqrt{(8n+5)(d^2-4d+12)}}{12k}\right). \tag{2.20}$$

3 Capparelli's conjecture

The year 1992 marked the one hundredth anniversary of the birth of Rademacher, and on July 21–25 of that year a conference honoring the memory of Rademacher was held at Penn State, and George Andrews was of course one of the conference organizers. On the first day of the conference, James Lepowsky of Rutgers gave a talk in which he mentioned that his student Stefano Capparelli had conjectured the following partition identity [14] as a result of his studies of the standard level 3 modules associated with the Lie algebra $A_2^{(2)}$:

Theorem 5 (Capparelli)

- *Let $C(n)$ denote the number of partitions of n into parts $\equiv \pm 2, \pm 3$ (mod 12).*
- *Let $D(n)$ denote the number of partitions $\lambda = (\lambda_1, \lambda_2, \dots)$ of n such that*
  - $\lambda_j - \lambda_{j+1} \geq 2$,
  - $\lambda_j - \lambda_{j+1} = 2$ *only if $\lambda_j \equiv 1$ (mod 3), and*
  - $\lambda_j - \lambda_{j+1} = 3$ *only if λ_j is a multiple of 3.*
- *Then $C(n) = D(n)$ for all n.*

This identity is clearly similar in the spirit of those in the classical literature such as Schur's identity (our Theorem 3), yet was new. Needless to say, Andrews and others at the conference were quite intrigued by the conjecture. Andrews worked intently for the next several evenings, and was able to find a proof [6] of the identity in time to present it as his talk on the last day of the conference. Of this proof, Andrews wrote [1, p. 505], "In my proof of Capparelli's conjecture, I was completely guided by the Wilf-Zeilberger method, even if I didn't use Doron's program explicitly. I couldn't have produced my proof without knowing the principle behind 'WZ.'" Although Andrews' WZ-inspired proof (see [62, 65–68]) was the first proof of the Capparelli conjecture, Lie theoretic proofs were later found by Tamba and Xie [61] and Capparelli himself [15].

The generating function for the partitions enumerated by the $C(n)$ in Capparelli's identity is

$$\sum_{n=0}^{\infty} C(n)x^n = \prod_{m \geq 1} \frac{1}{(1 - x^{12m-10})(1 - x^{12m-9})(1 - x^{12m-3})(1 - x^{12m-2})},$$

and indeed the Rademacher method may be applied to find an explicit formula for $C(n)$

$$C(n) = \frac{\pi}{\sqrt{24n-1}} \sum_{\substack{d \in \{1,2,3,12\}}} \sqrt{12 + 308d + 12d^2 - 2d^3} \sum_{\substack{k \geq 1 \\ (k,12)=d}} \frac{1}{k}$$

$$\times \sum_{\substack{0 \leq h < k \\ (h,k)=1}} e^{-2\pi nh/k} \frac{\omega(\frac{12h}{d}, \frac{k}{d})\omega(\frac{3h}{(d,3)}, \frac{k}{(d,3)})\omega(\frac{2h}{(d,2)}, \frac{k}{(d,2)})}{\omega(\frac{6h}{(d,6)}, \frac{k}{(d,6)})^2 \omega(\frac{4h}{(d,4)}, \frac{k}{(d,4)})}$$

$$\times I_1\left(\frac{\pi\sqrt{(24n-1)(201 - 231d + 91d^2 - 6d^3)}}{6\sqrt{165}k}\right). \tag{3.1}$$

4 The Bailey chain

Of course, Andrews has contributed a large number of important and useful discoveries to the body of mathematical knowledge. One of this author's favorites is the *Bailey chain*, i.e. the realization that the Bailey lemma is self-replicating and therefore any Bailey pair implies infinitely many others. In particular, every Rogers-Ramanujan type identity is automatically part of an infinite family (see [4, 5]).

The Bailey chain provides an explanation and a context for many infinite family q-series identities and their combinatorial counterparts. For example, David Bressoud's identity [11, p. 15, Eq. (3.4) with $k = r$]

$$\sum_{n_1, n_2, \ldots, n_{r-1} \geq 0} \frac{x^{N_1^2 + N_2^2 + \cdots + N_{r-1}^2}}{(x)_{n_1} (x)_{n_2} \cdots (x)_{n_{r-2}} (x^2; x^2)_{n_{r-1}}} = \prod_{m=1}^{\infty} \frac{(1 - x^{2rm-r})(1 - x^{rm})}{1 - x^m},$$

(4.1)

where $N_j := n_j + n_{j+1} + \cdots + n_{r-1}$ and $r \geqq 2$, follows from inserting the Bailey pair

$$\alpha_n(a, x) = \frac{(-1)^n x^{n^2} (1 - ax^{2n})(a^2; x^2)_n}{(1 - a)(x^2; x^2)_n}, \qquad \beta_n(a, x) = \frac{1}{(x^2; x^2)_n}$$

into a certain limiting case of the Bailey chain [5, p. 30, Theorem 3.5], setting $a = 1$, and then applying Jacobi's triple product identity [5, p. 63, Eq. (7.1)]. Although Bressoud's combinatorial counterpart to [11, p. 15, Eq. (3.4)] excludes the special case with $k = r$ (our (4.1) above), the author provided a combinatorial interpretation [58, p. 315, Theorem 6.9], which we recall here:

Theorem 6 *For $r \geq 2$, let $B_r(n)$ denote the number of partitions $\lambda = (\lambda_1, \lambda_2, \ldots)$ of n such that*

- *1 appears as a part less than r times,*
- *$\lambda_j - \lambda_{j+r-1} \geqq 2$, and*
- *if $\lambda_j - \lambda_{j+r-2} \leqq 1$, then $\sum_{h=0}^{r-2} \lambda_{j+h} \equiv (r-1) \pmod 2$.*

For $r \geq 3$, let $A_r(n)$ denote the number of partitions of n such that

- *no part is a multiple of r,*
- *for any nonnegative integer j, either $rj + 1$ or $r(j+1) - 1$, but not both, may appear as parts,*

and let $A_2(n)$ denote the number of partitions of n into distinct odd parts. Then $A_r(n) = B_r(n)$ for all integers n.

Remark 1 The combinatorial interpretation of the $A_r(n)$ was facilitated by ideas advanced by Andrews and Lewis [9].

We conclude with a Rademacher-type formula for the $A_r(n)$:

$$A_r(n) = \frac{2\pi\sqrt{2}}{\sqrt{24n-1}} \sum_{d|r} \frac{(d,r)}{\sqrt{dr}} \chi(2r + d^2 > 4(d,r)^2) \sum_{\substack{k \geq 1 \\ (h,k)=d}} k^{-1}$$

$$\times \sum_{\substack{0 \leq h < k \\ (h,k)=1}} e^{-2\pi i n h/k} \frac{\omega(h,k)\omega(2rh/d, k/d)}{\omega\left(\frac{rh}{(d,r)}, \frac{k}{(d,r)}\right)}$$

$$\times I_1\left(\frac{\pi}{6k} \sqrt{\frac{(24n-1)\left(2r + d^2 - 4(d,r)^2\right)}{2r}}\right), \tag{4.2}$$

where

$$\chi(P) = \begin{cases} 1 & \text{if } P \text{ is true, and} \\ 0 & \text{if } P \text{ is false.} \end{cases}$$

Acknowledgements The author thanks his thesis advisor, George Andrews, for all his help, encouragement, and kindness over many years. The author also thanks the anonymous referee for helping him to correct a potentially misleading statement in the original version.

Note added in proof Shortly after the Rademacher Centenary Conference, Alladi, Andrews, and Gordon obtained a generalization and refinement of Capparelli's partition theorem for which they gave a combinatorial bijective proof. Capparelli had actually stated a companion result to Theorem 5 above. Both results of Capparelli in a refined form follow from the generalization of Alladi-Andrews-Gordon. See [2].

References

1. 1998 Steele Prizes: Not. Am. Math. Soc. **45**, 504–508 (1998)
2. Alladi, K., Andrews, G.E., Gordon, B.: Refinements and generalizations of Capparelli's conjecture on partitions. J. Algebra **174**, 636–658 (1995)
3. Andrews, G.E.: The Theory of Partitions. Encyclopedia of Mathematics and its Applications, vol. 2. Addison-Wesley, Reading (1976). Reissued, Cambridge (1998)
4. Andrews, G.E.: Multiple series Rogers-Ramanujan type identities. Pac. J. Math. **114**, 267–283 (1984)
5. Andrews, G.E.: q-Series: Their Development and Application in Analysis, Number Theory, Combinatorics, Physics, and Computer Algebra. Regional Conference Series in Mathematics, vol. 66. Am. Math. Soc., Providence (1986)
6. Andrews, G.E.: Schur's Theorem, Capparelli's conjecture and q-trinomial coefficients. In: Proc. Rademacher Centenary Conf., 1992. Contemp. Math., vol. 166, pp. 141–154. Am. Math. Soc., Providence (1994)
7. Andrews, G.E., Berndt, B.C.: Ramanujan's Lost Notebook Part I. Springer, New York (2005)
8. Andrews, G.E., Berndt, B.C.: Ramanujan's Lost Notebook Part II. Springer, New York (2009)
9. Andrews, G.E., Lewis, R.P.: An algebraic identity of F.H. Jackson and its implication for partitions. Discrete Math. **232**, 77–83 (2001)
10. Apostol, T.M.: Modular Functions and Dirichlet Series in Number Theory, 2nd edn. Graduate Texts in Mathematics, vol. 41. Springer, New York (1990)
11. Bressoud, D.M.: Analytic and combinatorial generalizations of the Rogers-Ramanujan identities. Mem. Am. Math. Soc. **24**(227), 1–54 (1980)
12. Bringmann, K., Lovejoy, J.: Dyson's rank, overpartitions, and weak Maass forms. Int. Math. Res. Not. IMRN **19**, 1–34 (2007)

13. Bringmann, K., Ono, K.: Coefficients of harmonic Maass forms. In: Proceedings of the 2008 University of Florida Conference on Partitions, q-series, and Modular Forms. Developments in Mathematics Series. Springer, New York (to appear)
14. Capparelli, S.: Vertex operator relations for ane Lie algebras and combinatorial identities. Ph.D. thesis, Rutgers (1988)
15. Capparelli, S.: A construction of the level 3 modules for the ane algebra $A_2^{(2)}$ and a new combinatorial identity of the Rogers-Ramanujan type. Trans. Am. Math. Soc. **348**(2), 481–501 (1996)
16. Chen, W.Y.C., Zhao, J.J.Y.: The Gaussian coefficients and overpartitions. Discrete Math. **305**, 350–353 (2005)
17. Corteel, S., Goh, W.M.Y., Hitczenko, P.: A local limit theorem in the theory of overpartitions. Algorithmica **46**, 329–343 (2006)
18. Corteel, S., Hitczenko, P.: Multiplicity and number of parts in overpartitions. Ann. Comb. **8**, 287–301 (2004)
19. Corteel, S., Lovejoy, J.: Overpartitions. Trans. Am. Math. Soc. **356**, 1623–1635 (2004)
20. Corteel, S., Lovejoy, J., Yee, A.J.: Overpartitions and generating functions for generalized Frobenius partitions. In: Mathematics and Computer Science. III. Trends Math., pp. 15–24. Birkhäuser, Basel (2004)
21. Corteel, S., Mallet, O.: Overpartitions, lattice paths, and Rogers-Ramanujan identities. J. Comb. Theory Ser. A **114**, 1407–1437 (2007)
22. Euler, L.: Introductio in Analysin Infinatorum. Marcum-Michaelem Bousquet, Lausanne (1748)
23. Fu, A.M., Lascoux, A.: q-identities related to overpartitions and divisor functions. Electron. J. Comb. **12**, #R38 (2005). 7 pp.
24. Grosswald, E.: Some theorems concerning partitions. Trans. Am. Math. Soc. **89**, 113–128 (1958)
25. Grosswald, E.: Partitions into prime powers. Mich. Math. J. **7**, 97–122 (1960)
26. Haberzetle, M.: On some partition functions. Am. J. Math. **63**, 589–599 (1941)
27. Hagis, P.: A problem on partitions with a prime modulus $p \geq 3$. Trans. Am. Math. Soc. **102**, 30–62 (1962)
28. Hagis, P.: Partitions into odd summands. Am. J. Math. **85**, 213–222 (1963)
29. Hagis, P.: On a class of partitions with distinct summands. Trans. Am. Math. Soc. **112**, 401–415 (1964)
30. Hagis, P.: Partitions into odd and unequal parts. Am. J. Math. **86**, 317–324 (1964)
31. Hagis, P.: Partitions with odd summands-some comments and corrections. Am. J. Math. **87**, 218–220 (1965)
32. Hagis, P.: A correction of some theorems on partitions. Trans. Am. Math. Soc. **118**, 550 (1965)
33. Hagis, P.: On partitions of an integer into distinct odd summands. Am. J. Math. **87**, 867–873 (1965)
34. Hagis, P.: Some theorems concerning partitions into odd summands. Am. J. Math. **88**, 664–681 (1966)
35. Hagis, P.: Partitions with a restriction on the multiplicity of summands. Trans. Am. Math. Soc. **155**, 375–384 (1971)
36. Hirschhorn, M.D., Sellers, J.A.: Arithmetic relations for overpartitions. J. Comb. Math. Comb. Comput. **53**, 65–73 (2005)
37. Hirschhorn, M.D., Sellers, J.A.: Arithmetic properties of overpartitions into odd parts. Ann. Comb. **10**, 353–367 (2006)
38. Hua, L.K.: On the number of partitions into unequal parts. Trans. Am. Math. Soc. **51**, 194–201 (1942)
39. Iseki, S.: A partition function with some congruence condition. Am. J. Math. **81**, 939–961 (1959)
40. Iseki, S.: On some partition functions. J. Math. Soc. Jpn. **12**, 81–88 (1960)
41. Iseki, S.: Partitions in certain arithmetic progressions. Am. J. Math. **83**, 243–264 (1961)
42. Lehner, J.: A partition function connected with the modulus five. Duke Math. J. **8**, 631–655 (1941)
43. Livingood, J.: A partition function with prime modulus $p > 3$. Am. J. Math. **67**, 194–208 (1945)
44. Lovejoy, J.: Gordon's theorem for overpartitions. J. Comb. Theory Ser. A **103**, 393–401 (2003)
45. Lovejoy, J.: Overpartitions and real quadratic fields. J. Number Theory **106**, 178–186 (2004)
46. Lovejoy, J.: Overpartition theorems of the Rogers-Ramanujan type. J. Lond. Math. Soc. **69**(2), 562–574 (2004)
47. Lovejoy, J.: A theorem on seven-colored overpartitions and its applications. Int. J. Number Theory **1**, 215–224 (2005)
48. Lovejoy, J.: Rank and conjugation for the Frobenius representation of an overpartition. Ann. Comb. **9**, 321–334 (2005)
49. Lovejoy, J.: Partitions and overpartitions with attached parts. Arch. Math. (Basel) **88**, 316–322 (2007)
50. Mahlburg, K.: The overpartition function modulo small powers of 2. Discrete Math. **286**(3), 263–267 (2004)

51. Niven, I.: On a certain partition function. Am. J. Math. **62**, 353–364 (1940)
52. Rademacher, H.: On the partition function $p(n)$. Proc. Lond. Math. Soc. **43**(2), 241–254 (1937)
53. Rademacher, H.: On the expansion of the partition function in a series. Ann. Math. **44**(2), 416–422 (1943)
54. Rademacher, H.: Topics in Analytic Number Theory. Die Grundelhren der mathematischen Wissenschaften, Bd. 169. Springer, Berlin (1973)
55. Robbins, N.: Some properties of overpartitions. JP J. Algebra Number Theory Appl. **3**, 395–404 (2003)
56. Rødseth, Ø., Sellers, N.: On m-ary overpartitions. Ann. Comb. **9**, 345–353 (2005)
57. Slater, L.J.: Further identities of the Rogers-Ramanujan type. Proc. Lond. Math. Soc. **54**, 147–167 (1952)
58. Sills, A.V.: Identities of the Rogers-Ramanujan-Slater type. Int. J. Number Theory **3**, 293–323 (2007)
59. Sills, A.V.: A Rademacher type formula for overpartitions. Preprint (2009)
60. Sills, A.V.: Towards an automation of the circle method. Preprint (2009)
61. Tamba, M., Xie, C.: Level three standard modules for $A_2^{(2)}$ and combinatorial identities. J. Pure Appl. Algebra **105**(1), 53–92 (1995)
62. Petkovšek, M., Wilf, H.S., Zeilberger, D.: $A = B$. A.K. Peters, Wellesley (1996)
63. Subramanyasastri, V.V.: Partitions with congruence conditions. J. Indian Math. Soc. **11**, 55–80 (1972)
64. Sykes, C. (writer, producer, director): The Man Who Loved Numbers, NOVA documentary, PBS (WGBH Boston) original airdate: March 22, 1988
65. Wilf, H.S., Zeilberger, D.: Rational functions certify combinatorial identities. J. Am. Math. Soc. **3**, 147–158 (1990)
66. Wilf, H.S., Zeilberger, D.: An algorithmic proof theory for hypergeometric (ordinary and 'q') multi-sum/integral identities. Invent. Math. **108**, 575–633 (1992)
67. Zeilberger, D.: A fast algorithm for proving termination hypergeometric identities. Discrete Math. **80**, 207–211 (1990)
68. Zeilberger, D.: The method of creative telescoping. J. Symb. Comput. **11**, 195–204 (1991)

Ramanujan J (2010) 23: 265–295
DOI 10.1007/s11139-009-9207-8

Bijective proofs using two-line matrix representations for partitions

Eduardo H.M. Brietzke · José Plínio O. Santos · Robson da Silva

Dedicated to George Andrews for his 70th birthday

Received: 1 June 2009 / Accepted: 5 October 2009 / Published online: 21 April 2010
© Springer Science+Business Media, LLC 2010

Abstract In this paper, we present bijective proofs of several identities involving partitions by making use of a new way for representing partitions as two-line matrices. We also apply these ideas to give a combinatorial proof for an identity related to three-quadrant Ferrers graphs.

Keywords Partitions · Combinatorial identities · Lebesgue identity · Ferrers graph

Mathematics Subject Classification (2000) Primary 11P81 · Secondary 05A19

1 Introduction

In this paper, we present a number of results by making use of a new way of representing partitions as two-line matrices introduced in [9]. As one will see, one of the new notations for unrestricted partitions has more explicit information on the conjugate partition than what is given by the well-known Frobenius' symbol.

In Sect. 2, we present basically three distinct notations for unrestricted partitions that are given in [9]. In Sect. 3, we construct a bijection between the set of partitions into distinct odd parts with no parts equal to 1 and the set of self-conjugate

E.H.M. Brietzke
Instituto de Matemática, UFRGS C.P. 15080, 90509-900 Porto Alegre, RS, Brazil
e-mail: brietzke@mat.ufrgs.br

J.P.O. Santos (✉)
IMECC-UNICAMP, C.P. 6065, 13084-970 Campinas, SP, Brazil
e-mail: josepli@ime.unicamp.br

R. da Silva
ICE-UNIFEI, C.P. 50, 37500-903 Itajubá, MG, Brazil
e-mail: rsilva@unifei.edu.br

three-quadrant Ferrers graphs. In Sect. 4, we present combinatorial proofs for three identities and we establish bijections between several classes of partitions. One of the results resembles the Euler Pentagonal Number Theorem, and another one is related to Ramanujan's partial theta function. In the final section, we provide a new bijective proof for the Lebesgue Identity based on the two-line matrix representation for partitions.

2 Representation of partitions by two-line matrices

A very well-known representation of a partition by a two-line matrix is by means of the Frobenius' symbol. The purpose of this section is to recall three different ways of representing partitions given in [9]. The main motivation for these representations comes from results obtained in [9].

Theorem A (Theorem 8, [9]) *The number of unrestricted partitions of n is equal to the number of two-line matrices of the form*

$$\begin{pmatrix} c_1 & c_2 & c_3 & \cdots & c_k \\ d_1 & d_2 & d_3 & \cdots & d_k \end{pmatrix}, \tag{2.1}$$

where

$$c_k = 0, \qquad d_k \neq 0,$$
$$c_t = c_{t+1} + d_{t+1}, \quad \text{for any } t < k, \tag{2.2}$$
$$n = \sum c_t + \sum d_t.$$

A natural bijection between the two sets is given in [9]. Indeed, there are two different natural bijections between unrestricted partitions and two-line matrices satisfying (2.2). Perhaps the best way to describe them is by an example.

First bijection The number k of columns of the matrix corresponds to the number of parts of the partition.

Consider, for example, the partition $\lambda = (6, 5, 2, 2)$ of 15. We associate to λ a 2×4 matrix A of the form given in Theorem A in such a way that the sums of the entries in each column of A are the parts of λ. We have no choice for the fourth column, but to pick $c_4 = 0$ and $d_4 = 2$. Since c_3 must be 2 and the entries of the third column must add up to 2, then $d_3 = 0$. By the same argument, we must have $c_2 = 2$ and $d_2 = 3$. Also, $c_1 = 5$ and $d_1 = 1$. The representation is

$$\lambda = 6 + 5 + 2 + 2 = \begin{pmatrix} 5 & 2 & 2 & 0 \\ 1 & 3 & 0 & 2 \end{pmatrix}.$$

To go from the matrix to the partition, we only have to add the entries in each column. As explained in [9], the second line of the matrix provides a complete description of the partition λ' conjugate to λ. In the above example, the second row $(1, 3, 0, 2)$ indicates that λ' contains one 1, three 2s, no 3s, and two 4s.

Second bijection The number k of columns of the matrix corresponds to the largest part of the partition.

To any positive integer j, we associate a $2 \times j$ matrix with the first $j - 1$ entries equal to 1 in the first row, the last entry in the second row equal to 1, and the remaining entries equal to 0. For example, we have

$$1 = \begin{pmatrix} 0 \\ 1 \end{pmatrix}, \quad 2 = \begin{pmatrix} 1 & 0 \\ 0 & 1 \end{pmatrix}, \quad 3 = \begin{pmatrix} 1 & 1 & 0 \\ 0 & 0 & 1 \end{pmatrix}, \quad \ldots$$

If $j \le k$, given a $2 \times j$ matrix A and a $2 \times k$ matrix B, we perform the sum $A + B$ as if $k - j$ zero columns were added to the right of A. To the partition $\lambda = (6, 5, 2, 2)$ considered before, for instance, we associate the matrix

$$6 + 5 + 2 + 2 = \begin{pmatrix} 1 & 1 & 1 & 1 & 1 & 0 \\ 0 & 0 & 0 & 0 & 0 & 1 \end{pmatrix} + \begin{pmatrix} 1 & 1 & 1 & 1 & 0 & 0 \\ 0 & 0 & 0 & 0 & 1 & 0 \end{pmatrix}$$

$$+ \begin{pmatrix} 1 & 0 & 0 & 0 & 0 & 0 \\ 0 & 1 & 0 & 0 & 0 & 0 \end{pmatrix} + \begin{pmatrix} 1 & 0 & 0 & 0 & 0 & 0 \\ 0 & 1 & 0 & 0 & 0 & 0 \end{pmatrix}$$

$$= \begin{pmatrix} 4 & 2 & 2 & 2 & 1 & 0 \\ 0 & 2 & 0 & 0 & 1 & 1 \end{pmatrix}.$$

Note that if we now add up the columns of the obtained matrix, we get the conjugate partition and, as before, the second line contains a description of the given partition, saying that there are no 1s, two 2s, no 3s, no 4s, one 5, and one 6.

We now present two new bijections between unrestricted partitions of n and certain classes of two-line matrices. It is shown in [9] that these sets have the same number of elements, but a natural bijection between them is not presented there.

In [9], for positive integers k and j, a set of nonnegative integers is defined by

$$A_{k,j} = \{ck + dj \,|\, c, d \ge 0\}.$$

Theorem B (Theorem 11, [9]) *Let $f(n)$ be the number of partitions of n of the form $\lambda_1 + \lambda_2 + \cdots + \lambda_s$, with $\lambda_r \in A_{k,j}$, $c_s \ne 0$ and such that for $c_t k + d_t j$ and $c_{t+1} k + d_{t+1} j$ consecutive parts, $c_t \ge 2 + c_{t+1} + d_{t+1}$. Then,*

$$\sum_{n=0}^{\infty} f(n) q^n = \sum_{n=0}^{\infty} \frac{q^{kn^2}}{(q^j; q)_n (q^k; q)_n}. \tag{2.3}$$

In the particular case $k = j = 1$, denoting by $p(n)$ the number of partitions of n, since

$$\sum_{n=0}^{\infty} \frac{q^{n^2}}{(q; q)_n^2} = \prod_{n=1}^{\infty} \frac{1}{(1 - q^n)} = \sum_{n=0}^{\infty} p(n) q^n, \tag{2.4}$$

the following corollary is obtained.

Corollary C (Corollary 12, [9]) *The number of unrestricted partitions of n is equal to the number of two-line matrices of the form (2.1) where*

$$c_k \neq 0,$$

$$c_t \geq 2 + c_{t+1} + d_{t+1}, \quad \text{for any } t < k, \tag{2.5}$$

$$n = \sum c_t + \sum d_t.$$

We now construct the first new bijection for the set of unrestricted partitions, namely a bijection between the set of matrices of the form (2.1) satisfying (2.5) and the set of unrestricted partitions of n. This provides a direct proof of Corollary C.

Construction of the first bijection We now construct a natural bijection between the set of unrestricted partitions of n and the set of matrices of the form (2.1) with non-negative integer entries satisfying (2.5).

It suffices, for each $k \geq 1$ fixed, to establish a bijection between the set of unrestricted partitions of n with Durfee square with side k and the set of matrices described above with the additional requirement that the matrix should have k columns.

For instance, a partition $\lambda = (\lambda_1, \lambda_2, \ldots, \lambda_r)$ (as usual we suppose $\lambda_t \geq \lambda_{t+1}$) with Durfee square of side 3 is completely characterized once we declare how many times each one of the numbers 1, 2, and 3 appears as a part of λ, and if we also know the three largest parts λ_1, λ_2, and λ_3 of λ. We have to associate to λ a 2×3 matrix that gives a clue of these six numbers. Let

$$\begin{pmatrix} c_1 & c_2 & c_3 \\ d_1 & d_2 & d_3 \end{pmatrix} \tag{2.6}$$

be such a matrix. By (2.5), we can write

$$c_3 = 1 + j_3, \tag{2.7}$$

$$c_2 = 3 + j_2 + j_3 + d_3, \tag{2.8}$$

$$c_1 = 5 + j_1 + j_2 + j_3 + d_2 + d_3 \tag{2.9}$$

with $j_1, j_2, j_3 \geq 0$. Hence, the matrix (2.6) may be rewritten as

$$\begin{pmatrix} 5 + j_1 + j_2 + j_3 + d_2 + d_3 & 3 + j_2 + j_3 + d_3 & 1 + j_3 \\ d_1 & d_2 & d_3 \end{pmatrix} \tag{2.10}$$

or, still, as the sum

$$\begin{pmatrix} 5 & 3 & 1 \\ 0 & 0 & 0 \end{pmatrix} + \begin{pmatrix} 0 & 0 & 0 \\ d_1 & 0 & 0 \end{pmatrix} + \begin{pmatrix} d_2 & 0 & 0 \\ 0 & d_2 & 0 \end{pmatrix} + \begin{pmatrix} d_3 & d_3 & 0 \\ 0 & 0 & d_3 \end{pmatrix}$$

$$+ \begin{pmatrix} j_1 & 0 & 0 \\ 0 & 0 & 0 \end{pmatrix} + \begin{pmatrix} j_2 & j_2 & 0 \\ 0 & 0 & 0 \end{pmatrix} + \begin{pmatrix} j_3 & j_3 & j_3 \\ 0 & 0 & 0 \end{pmatrix}. \tag{2.11}$$

From the above discussion, it follows that the partition λ can be characterized by the numbers

$d_1 \longrightarrow$ the number of times that 1 is a part of λ

$d_2 \longrightarrow$ the number of times that 2 is a part of λ

$d_3 \longrightarrow$ the number of times that 3 is a part of λ, not counting here eventual parts equal to 3 contained in the Durfee square

$j_1 \longrightarrow \lambda_1 - \lambda_2$, i.e., the number of times 1 is a part of the conjugate partition λ'

$j_2 \longrightarrow \lambda_2 - \lambda_3$, i.e., the number of times 2 is a part of λ'

$j_3 \longrightarrow \lambda_3 - 3$, i.e., the number of units that λ_3 exceeds the side of the Durfee square.

Note that the first three parts are completely determined: $\lambda_3 = 3 + j_3$, $\lambda_2 = 3 + j_2 + j_3$, and $\lambda_1 = 3 + j_1 + j_2 + j_3$.

Example 1 Consider the partition $\lambda = (5, 4, 4, 2, 2, 1)$ of $n = 18$.

We decompose the Ferrers' graph of λ as shown in the picture, taking into account the size of the Durfee square and the number of times 1, 2, and 3 are parts of λ and of the conjugate partition λ'. In the present example, $d_1 = 1$, $d_2 = 2$, $d_3 = 0$, $j_1 = 1$, $j_2 = 0$, and $j_3 = 1$. This decomposition suggests the sum of matrices below:

$$\begin{pmatrix} 5 & 3 & 1 \\ 0 & 0 & 0 \end{pmatrix} \quad \text{from the Durfee square,}$$

$$\begin{pmatrix} 0 & 0 & 0 \\ 1 & 0 & 0 \end{pmatrix} + \begin{pmatrix} 2 & 0 & 0 \\ 0 & 2 & 0 \end{pmatrix} = \begin{pmatrix} 2 & 0 & 0 \\ 1 & 2 & 0 \end{pmatrix} \quad \text{because 1 appears once and 2}$$

appears twice as parts of λ and

$$\begin{pmatrix} 1 & 0 & 0 \\ 0 & 0 & 0 \end{pmatrix} + \begin{pmatrix} 1 & 1 & 1 \\ 0 & 0 & 0 \end{pmatrix} = \begin{pmatrix} 2 & 1 & 1 \\ 0 & 0 & 0 \end{pmatrix} \quad \text{since 3 is twice part of } \lambda', \text{ but}$$

appears only once outside of the Durfee square and 1 is part of λ' only once.

Hence, the matrix corresponding to the partition λ is

$$\begin{pmatrix} 5 & 3 & 1 \\ 0 & 0 & 0 \end{pmatrix} + \begin{pmatrix} 2 & 0 & 0 \\ 1 & 2 & 0 \end{pmatrix} + \begin{pmatrix} 2 & 1 & 1 \\ 0 & 0 & 0 \end{pmatrix} = \begin{pmatrix} 9 & 4 & 2 \\ 1 & 2 & 0 \end{pmatrix}.$$

There is an easier way of obtaining the matrix-representation of the partition. We do not need to write out first the sum of matrices as done above. Indeed, in the example above, note that the sums of columns, namely 10, 6, and 2, are the lengths of the hooks shown in the picture to the left. In the second row, the entries 1 and 2 indicate the vertical displacement of the corresponding hook with respect to the previous one, and 0 indicates the vertical displacement of the last hook with respect to the bottom of the Durfee square.

The matrix $\left(\begin{smallmatrix} 5 & 3 & 1 \\ 0 & 0 & 0 \end{smallmatrix}\right)$ corresponds to the Durfee square. Note that the entries 5, 3, and 1 of the first row are precisely the numbers of elements each hook has in common with the Durfee square.

Example 2 Determine the partition represented by the matrix $\left(\begin{smallmatrix} 15 & 9 & 5 & 1 \\ 2 & 3 & 0 & 1 \end{smallmatrix}\right)$.

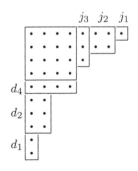

The Durfee square has side four and the length of the hooks are $15 + 2 = 17$, $9 + 3 = 12$, $5 + 0 = 5$, and $1 + 1 = 2$. In addition, the second row of the matrix indicates that the first hook starts two units below the second hook which, in turn, initiates three units below the third hook, the third hook begins at the same level as the fourth one, and this last hook begins one unit below the bottom of the Durfee square. This is all we need to know to determine the Ferrers' graph of the partition, which is shown in the picture to the right. The corresponding partition is $\lambda = (8, 7, 5, 4, 4, 2, 2, 2, 1, 1)$.

We now undertake the construction of the second bijection between unrestricted partitions and two-line matrices. Again, our motivation is a result presented in [9].

Theorem D (Theorem 9, [9]) *The generating function for matrices of the form*

$$\begin{pmatrix} c_1 \times k & c_2 \times k & \cdots & c_s \times k \\ d_1 \times j & d_2 \times j & \cdots & d_s \times j \end{pmatrix},$$

where $d_t \neq 0$, $c_t \geq 1 + c_{t+1} + d_{t+1}$ and the sum of all entries is equal to n, is given by the LHS of the identity

$$\sum_{n=0}^{\infty} \frac{q^{kn^2+(j-k)n}}{(q^j;q^k)_n(q^k;q^k)_n} = \frac{1}{(q^j;q^k)_\infty}. \tag{2.12}$$

In the particular case $k = j = 1$, using (2.4), the following result follows.

Theorem E (Theorem 10, [9]) *The number of unrestricted partitions of n is equal to the number of two-line matrices of the form*

$$\begin{pmatrix} c_1 & c_2 & \cdots & c_s \\ d_1 & d_2 & \cdots & d_s \end{pmatrix} \tag{2.13}$$

where

$$d_t \neq 0,$$

$$c_t \geq 1 + c_{t+1} + d_{t+1}, \tag{2.14}$$

$$n = \sum c_t + \sum d_t.$$

In [9], the equinumerability expressed in Theorem E is established by the generating function technique, but a natural bijection between the two sets is not presented.

Construction of the second bijection We now construct a natural bijection between the set of unrestricted partitions of n and the set of matrices of the form (2.13) with non-negative integer entries satisfying (2.14). We rewrite the condition $c_t \geq 1 + c_{t+1} + d_{t+1}$ as

$$c_t = 1 + j_t + c_{t+1} + d_{t+1} \quad \forall t < k, \tag{2.15}$$

$$c_k = j_k, \tag{2.16}$$

$$j_t \geq 0. \tag{2.17}$$

In this case, we also take $s = k$ and define a bijection between the set of unrestricted partitions of n with Durfee square of side k and the set of matrices of the form (2.13) satisfying (2.15), (2.16) and (2.17), in addition to (2.14), with k columns.

For example, if $k = 3$ then

$$\begin{pmatrix} c_1 & c_2 & c_3 \\ d_1 & d_2 & d_3 \end{pmatrix} = \begin{pmatrix} 2 + j_1 + j_2 + j_3 + d_2 + d_3 & 1 + j_2 + j_3 + d_3 & j_3 \\ d_1 & d_2 & d_3 \end{pmatrix}$$

$$= \begin{pmatrix} 2 & 1 & 0 \\ 0 & 0 & 0 \end{pmatrix} + \begin{pmatrix} 0 & 0 & 0 \\ d_1 & 0 & 0 \end{pmatrix} + \begin{pmatrix} d_2 & 0 & 0 \\ 0 & d_2 & 0 \end{pmatrix} + \begin{pmatrix} d_3 & d_3 & 0 \\ 0 & 0 & d_3 \end{pmatrix}$$

$$+ \begin{pmatrix} j_1 & 0 & 0 \\ 0 & 0 & 0 \end{pmatrix} + \begin{pmatrix} j_2 & j_2 & 0 \\ 0 & 0 & 0 \end{pmatrix} + \begin{pmatrix} j_3 & j_3 & j_3 \\ 0 & 0 & 0 \end{pmatrix}. \tag{2.18}$$

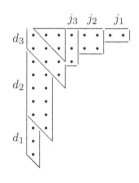

Compare (2.18) with the picture to the left. The entries d_1, d_2, and d_3 in the second row of the matrix are all non-vanishing and indicate the sizes of the subsets of the Ferrers' graph in the picture. The elements in the first row of the matrix, in turn, indicate the number of units still necessary to complete the hooks. The j_t's can vanish, but $d_t \neq 0$ for any t.

In the picture to the left, we show the example of the partition $\lambda = (8, 6, 4, 3, 2, 2, 2, 2, 1, 1)$ for which $d_1 = 3$, $d_2 = 5$, $d_3 = 2$, $j_1 = 2$, $j_2 = 2$, and $j_3 = 1$. The matrix associated to this partition is

$$\begin{pmatrix} 14 & 6 & 1 \\ 3 & 5 & 2 \end{pmatrix}.$$

The correspondence described by (2.18) or, alternatively, in a pictorial manner in the above picture, provides a new representation of unrestricted partitions by matrices. As with the representation obtained previously from (2.10), the number of columns in the matrix is equal to the side of the Durfee square of the partition.

Note that by composition, we obtain a bijection between the set of matrices of the form (2.1) satisfying (2.5) and those satisfying (2.14). The bijection simply takes

$$\begin{pmatrix} c_1 & c_2 & \cdots & c_k \\ d_1 & d_2 & \cdots & d_k \end{pmatrix} \longmapsto \begin{pmatrix} c_1 - 1 & c_2 - 1 & \cdots & c_k - 1 \\ d_1 + 1 & d_2 + 1 & \cdots & d_k + 1 \end{pmatrix}.$$

3 A combinatorial proof of an identity involving three-quadrant Ferrers graphs

In [3], G.E. Andrews presented the three-quadrant Ferrers graphs as an extension of the two-quadrant Ferrers graphs. The three-quadrant Ferrers graphs of a positive integer n are constructed by placing n points in the first, second and fourth quadrants of the plane observing that each point must have at least one positive coordinate. We also require that the points on the x-axis and on the y-axis form the longest row and the tallest column among all rows and all columns in the set with positive x-coordinates and y-coordinates, respectively.

For example,

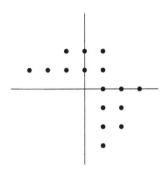

We recall from [3] that the self-conjugate three-quadrant Ferrers graphs are the ones that are unchanged by the mapping $(x, y) \mapsto (y, x)$. Let $s_3(n)$ denote the number of self-conjugate three-quadrant Ferrers graphs of n.

For example, the next two three-quadrant Ferrers graphs are self-conjugate:

In [3], the following identity is proved

$$\sum_{n=0}^{\infty} s_3(n) q^n = \prod_{n=0}^{\infty} \frac{(1 + q^{2n+3})}{(1 - q^{2n+2})}. \tag{3.1}$$

The RHS of (3.1) is the generating function for partitions in which the odd parts are distinct and greater than 1. We present in this section a bijective proof of (3.1) by using a two-line matrix representation for these partitions.

We now recall from [9] and [8] that we can represent a partition of n in which the odd parts are distinct and greater than 1 as a two-line matrix of the form

$$\begin{pmatrix} c_1 & c_2 & \cdots & c_s \\ d_1 & d_2 & \cdots & d_s \end{pmatrix}, \qquad (3.2)$$

where c_t and d_t satisfy $c_s = 2$ or 3, $d_t \equiv 0 \pmod{2}$ and

- if $c_t \equiv 0 \pmod 2$ and $c_{t+1} \equiv 0 \pmod 2$, then $c_t = c_{t+1} + d_{t+1}$;
- if $c_t \equiv 0 \pmod 2$ and $c_{t+1} \equiv 1 \pmod 2$, then $c_t = 1 + c_{t+1} + d_{t+1}$;
- if $c_t \equiv 1 \pmod 2$ and $c_{t+1} \equiv 0 \pmod 2$, then $c_t = 1 + c_{t+1} + d_{t+1}$;
- if $c_t \equiv 1 \pmod 2$ and $c_{t+1} \equiv 1 \pmod 2$, then $c_t = 2 + c_{t+1} + d_{t+1}$,

with the sum of all entries equal to n. This same class of matrices is used in Sect. 4 with a different purpose.

We associate to each two-line matrix

$$\begin{pmatrix} c_1 & c_2 & \cdots & c_s \\ d_1 & d_2 & \cdots & d_s \end{pmatrix},$$

a partition $\lambda_1 + \cdots + \lambda_s$ by just adding up the elements of the columns: $\lambda_t = c_t + d_t$. For example, the matrix $\begin{pmatrix} 3 & 2 & 2 \\ 2 & 0 & 0 \end{pmatrix}$ is associated to $5 + 2 + 2$.

As the entries in the second line are even numbers, the parity of each part is determined by the corresponding entry of the first line. Then, it is impossible to have two equal consecutive odd parts $\lambda_t = c_t + d_t$ and $\lambda_{t+1} = c_{t+1} + d_{t+1}$, since we have $c_t = 2 + c_{t+1} + d_{t+1}$ in this case. As we begin with $c_s = 2$ or 3, each part, obtained by adding up the entries in corresponding column, is grater than 1.

We describe next how to go from a partition where the odd parts are distinct and greater than 1 to a two-line matrix satisfying the conditions above:

1. We start at the end of the first line, placing 2 or 3 if λ_s is even or odd, respectively. Then, we complete the column with an even number d_s such that the sum of the entries of this column is equal to λ_s. So the last column is uniquely determined.
2. In order to create the column just before the last one, we must observe the parity of λ_{s-1} and λ_s because c_{s-1} and c_s have to satisfy the conditions above. For example, if c_{s-1} and c_s are both odd, then $c_{s-1} = 2 + c_s + d_s$, and we complete this column by choosing an even number d_{s-1} such that $\lambda_{s-1} = c_{s-1} + d_{s-1}$. It is not difficult to see that there is only one way to fill the column up.
3. To build the previous column, we observe the parity of λ_{t-1} and λ_t and follow the procedure described in item 2.

For example, Table 1 presents the matrices and the partitions having odd parts distinct and greater than 1 for $n = 8$.

3.1 The bijective proof

It is convenient to rewrite the matrices of the form (3.2) as

$$\begin{pmatrix} c_1 & c_2 & \cdots & c_s \\ 2d_1 & 2d_2 & \cdots & 2d_s \end{pmatrix}, \qquad (3.3)$$

Table 1 The matrices and partitions for $n = 8$

$$\begin{pmatrix} 2 \\ 6 \end{pmatrix} \qquad 8$$

$$\begin{pmatrix} 2 & 2 \\ 4 & 0 \end{pmatrix} \qquad 6 + 2$$

$$\begin{pmatrix} 5 & 3 \\ 0 & 0 \end{pmatrix} \qquad 5 + 3$$

$$\begin{pmatrix} 4 & 2 \\ 0 & 2 \end{pmatrix} \qquad 4 + 4$$

$$\begin{pmatrix} 2 & 2 & 2 \\ 2 & 0 & 0 \end{pmatrix} \qquad 4 + 2 + 2$$

$$\begin{pmatrix} 2 & 2 & 2 & 2 \\ 0 & 0 & 0 & 0 \end{pmatrix} \qquad 2 + 2 + 2 + 2$$

where c_t and d_t satisfy $c_s = 2$ or 3 and

- if $c_t \equiv 0 \pmod 2$ and $c_{t+1} \equiv 0 \pmod 2$, then $c_t = c_{t+1} + 2d_{t+1}$;
- if $c_t \equiv 0 \pmod 2$ and $c_{t+1} \equiv 1 \pmod 2$, then $c_t = 1 + c_{t+1} + 2d_{t+1}$;
- if $c_t \equiv 1 \pmod 2$ and $c_{t+1} \equiv 0 \pmod 2$, then $c_t = 1 + c_{t+1} + 2d_{t+1}$;
- if $c_t \equiv 1 \pmod 2$ and $c_{t+1} \equiv 1 \pmod 2$, then $c_t = 2 + c_{t+1} + 2d_{t+1}$.

Due to the restrictions on c_t, we have $c_t = i_t + c_{t+1} + 2d_{t+1}$, where $i_t \in \{0, 1, 2\}$, for $t = 1, 2, \ldots, s - 1$ and $c_s = 2 + i_s$, where $i_s = 0, 1$. We can split (3.3) as the sum

$$\begin{pmatrix} 2 + i_1 + \cdots + i_s + 2d_2 + \cdots + 2d_s & \cdots & 2 + i_{s-1} + i_s + 2d_s & 2 + i_s \\ 2d_1 & \cdots & 2d_{s-1} & 2d_s \end{pmatrix}$$

$$= \begin{pmatrix} 2 + i_1 + \cdots + i_s & \cdots & 2 + i_{s-1} + i_s & 2 + i_s \\ 0 & \cdots & 0 & 0 \end{pmatrix}$$

$$+ 2 \begin{pmatrix} d_2 + \cdots + d_s & \cdots & d_s & 0 \\ d_1 & \cdots & d_{s-1} & d_s \end{pmatrix}.$$

Note that the numbers $2 + i_1 + \cdots + i_s, \ldots, 2 + i_{s-1} + i_s, 2 + i_s$ form a nonincreasing sequence ending in either 2 or 3, and that there is no gap in the odd numbers in this sequence (if $2k + 1$ is in the sequence, then so is every odd integer between 3 and $2k + 1$).

For example, the matrix

$$\begin{pmatrix} 26 & 22 & 19 & 11 & 8 & 3 \\ 2 & 4 & 2 & 6 & 2 & 4 \end{pmatrix}$$

can be split as

$$\begin{pmatrix} 8 & 8 & 7 & 5 & 4 & 3 \\ 0 & 0 & 0 & 0 & 0 & 0 \end{pmatrix} + 2 \begin{pmatrix} 9 & 7 & 6 & 3 & 2 & 0 \\ 1 & 2 & 1 & 3 & 1 & 2 \end{pmatrix}. \qquad (3.4)$$

We will use this example to show how to associate a matrix like (3.3) and a three-quadrant Ferrers graph. The arguments in this example can be directly extended to the general case. We associate the matrix

$$\begin{pmatrix} 9 & 7 & 6 & 3 & 2 & 0 \\ 1 & 2 & 1 & 3 & 1 & 2 \end{pmatrix}$$

to the partition having one part equal to 1, two parts equal to 2, one part equal to 3, three parts equal to 4, one part equal to 5, and two parts equal to 6 $(6 + 6 + 5 + 4 + 4 + 4 + 3 + 2 + 2 + 1)$ whose Ferrers graph is

The pair of Ferrers graphs obtained from the second matrix in (3.4) can be placed in the plane in the following way

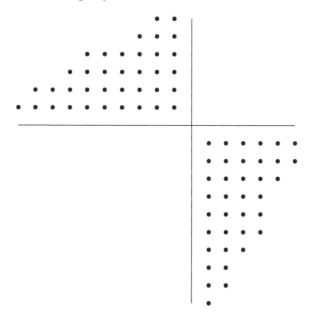

Now we describe how to place the contribution from the first matrix in (3.4) in the plane in order to obtain a self-conjugate three-quadrant Ferrers graph. To do this, we consider first the odd numbers in the first line: 7, 5, and 3. These numbers are represented in the first quadrant of the plane in the following way

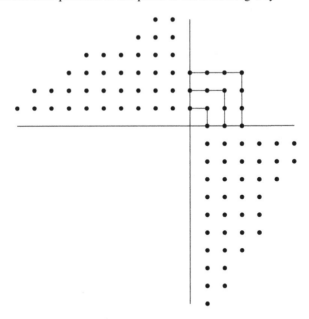

The even numbers, namely 8, 8 and 4, are represented in the first quadrant after the odd numbers in a non increasing sequence as follows

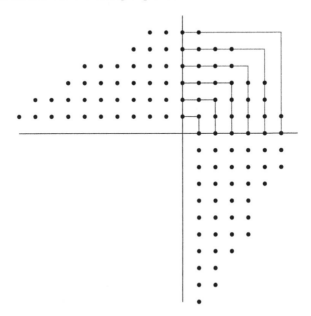

Note that the first even number after the greatest odd number, if there is one, is 1 plus the greatest odd number. Therefore, the construction of the three-quadrant

Ferrers graph described above can always be done. It is clear that this procedure can be reversed in order to obtain the two-line matrix representing a partition in which the odd parts are distinct and greater than 1.

The same argument applies in general and we have a bijection. Hence, identity (3.1) is proved.

4 Three identities and related bijections between certain classes of partitions

In this section, we prove three partition identities ((4.1), (4.14), and (4.26) below) using combinatorial arguments and we establish bijections between several classes of partitions related to these identities. Our main tool is constructing bijections between certain classes of partitions and sets of two-line matrices. As an application of our results proved by combinatorial arguments, we obtain a straightforward proof of an identity about Ramanujan's ψ function (see (4.29) below), which is similar to dozens of identities for partial theta functions obtained by Ramanujan in his notebooks (see, for example, [2] and Chap. 9 of [4]).

Our first objective is to study the identity

$$1 + \sum_{k=1}^{\infty} \frac{(1+q)(1+q^3)\cdots(1+q^{2k-1})}{(1-q^2)(1-q^4)\cdots(1-q^{2k})} q^{2k} = \prod_{n=1}^{\infty} \frac{1+q^{2n+1}}{1-q^{2n}}. \tag{4.1}$$

We provide two combinatorial proofs for identity (4.1). As a matter of fact, we consider a slight improvement of (4.1).

Theorem 4.1 *The following identity holds*

$$1 + \sum_{k=1}^{\infty} \frac{(1+zq)(1+zq^3)\cdots(1+zq^{2k-1})}{(1-q^2)(1-q^4)\cdots(1-q^{2k})} q^{2k} = \prod_{n=1}^{\infty} \frac{1+zq^{2n+1}}{1-q^{2n}}. \tag{4.2}$$

Proof We introduce two classes of partitions

$\mathcal{P}_n$, the set of all partitions λ of n such that

- the odd parts of λ are distinct;
- the largest part of λ is even;

and

$\mathcal{Q}_n$, the set of all partitions μ of n such that

- the odd parts of μ are distinct;
- 1 is not a part of μ.

Identity (4.1) is equivalent to the statement that, for any n, $\mathcal{P}_n$ and $\mathcal{Q}_n$ have the same number of elements, or in other words, there is a bijection between these two classes of partitions. Identity (4.2) suggests that it is possible to find such a bijection that preserves the number of odd parts of the partitions.

Denote by $\mathcal{P}_{n,k,m}$ the subset of all $\lambda \in \mathcal{P}_n$ such that λ contains exactly m odd parts and its largest part is $a(\lambda) = 2k$. Consider also the subset $\mathcal{Q}_{n,k,m}$ of all $\mu \in \mathcal{Q}_n$ such that μ has exactly m odd parts and its largest part is $a(\mu) = 2k$ or $2k + 1$.

The coefficients of $z^m q^n$ on the left-hand-side and on the right-hand-side of (4.2) are, respectively, the number of elements of the sets

$$\mathcal{P}_{n,m} := \bigcup_k \mathcal{P}_{n,k,m} \quad \text{and} \quad \mathcal{Q}_{n,m} := \bigcup_k \mathcal{Q}_{n,k,m}. \tag{4.3}$$

Hence, to prove (4.2), it suffices to construct a bijection between the two classes of partitions $\mathcal{P}_{n,k,m}$ and $\mathcal{Q}_{n,k,m}$.

We now define a bijection $\varphi : \mathcal{P}_{n,k,m} \to \mathcal{Q}_{n,k,m}$. Given any partition $\lambda \in \mathcal{P}_{n,k,m}$, we define $\varphi(\lambda) \in \mathcal{Q}_{n,k,m}$ by:

– if 1 is not a part of λ, $\varphi(\lambda) = \lambda$;
– if 1 is a part of λ, $\varphi(\lambda)$ is the partition obtained by removing 1 from λ and adding one unit to its largest part $a(\lambda)$. If λ contains more than one copy of $a(\lambda)$, we add 1 to one of them only. Since $a(\lambda)$ is even, the number of odd parts is not altered.

Then, φ is a bijection with an inverse φ^{-1} defined by:

– if $\mu \in \mathcal{Q}_{n,k,m}$ is such that its largest part is even, then $\varphi^{-1}(\mu) = \mu$;
– if $\mu \in \mathcal{Q}_{n,k,m}$ and its largest part $a(\mu)$ is odd, $\varphi^{-1}(\mu)$ is obtained by removing from μ one part equal to $a(\mu)$ and adding one part equal to $a(\mu) - 1$ and one part equal to 1. $\qquad\Box$

Of course, φ defined above can be extended to bijections from $\varphi : \mathcal{P}_{n,m} \to \mathcal{Q}_{n,m}$ or from $\varphi : \mathcal{P}_n \to \mathcal{Q}_n$.

Example For $n = 11$, the two classes $\mathcal{P}_n = \mathcal{P}_{n,1}$ and $\mathcal{Q}_n = \mathcal{Q}_{n,1}$ defined in (4.3) and the correspondence φ constructed in Theorem 4.1 are given by

$$
\begin{array}{llll}
(10,1) & \longmapsto & (11), & (6,2,2,1) \longmapsto (7,2,2), \\
(8,3) & \longmapsto & (8,3), & (4,4,3) \longmapsto (4,4,3), \\
(8,2,1) & \longmapsto & (9,2), & (4,4,2,1) \longmapsto (5,4,2), \\
(6,5) & \longmapsto & (6,5), & (4,3,2,2) \longmapsto (4,3,2,2), \\
(6,4,1) & \longmapsto & (7,4), & (4,2,2,2,1) \longmapsto (5,2,2,2), \\
(6,3,2) & \longmapsto & (6,3,2), & (2,2,2,2,2,1) \longmapsto (3,2,2,2,2).
\end{array}
$$

We now present another proof for Identity (4.2) based on a two-line matrix representation for partitions. The idea of the proof is to introduce a certain class of two-line matrices and to interpret it in terms of partitions in two different ways. Consider the set $\mathcal{M}^{(2,3)}_{n,k,m}$ of $2 \times k$ matrices of the form

$$A = \begin{pmatrix} c_1 & c_2 & \cdots & c_k \\ d_1 & d_2 & \cdots & d_k \end{pmatrix} \tag{4.4}$$

with non-negative integer entries, having exactly m odd entries in the first row and satisfying

$$c_k \in \{2,3\}, \qquad d_t \equiv 0 \pmod 2, \quad \forall t; \tag{4.5}$$

$$c_t \equiv 0 \pmod 2, c_{t+1} \equiv 0 \pmod 2 \Longrightarrow c_t = c_{t+1} + d_{t+1}; \tag{4.6}$$

$$c_t \equiv 0 \pmod 2, c_{t+1} \equiv 1 \pmod 2 \Longrightarrow c_t = 1 + c_{t+1} + d_{t+1}; \tag{4.7}$$

$$c_t \equiv 1 \pmod 2, c_{t+1} \equiv 0 \pmod 2 \Longrightarrow c_t = 1 + c_{t+1} + d_{t+1}; \tag{4.8}$$

$$c_t \equiv 1 \pmod 2, c_{t+1} \equiv 1 \pmod 2 \Longrightarrow c_t = 2 + c_{t+1} + d_{t+1}; \tag{4.9}$$

$$\sum c_t + \sum d_t = n. \tag{4.10}$$

This class of matrices was introduced in [8].

We now introduce a notation needed in the next theorem. Let $A \in \mathcal{M}_{n,k,m}^{(2,3)}$. If, for example, $k = 4$ and

$$A = \begin{pmatrix} c_1 & c_2 & c_3 & c_4 \\ d_1 & d_2 & d_3 & d_4 \end{pmatrix} \in \mathcal{M}_{n,4,m}^{(2,3)}$$

then considering the restrictions (4.5)–(4.10) it is not difficult to verify that we can write in a unique manner $c_4 = 2 + j_4$, $c_3 = 2 + d_4 + j_3 + 2j_4$, $c_2 = 2 + d_3 + d_4 + j_2 + 2j_3 + 2j_4$, $c_1 = 2 + d_2 + d_3 + d_4 + j_1 + 2j_2 + 2j_3 + 2j_4$, with $j_t \in \{0,1\}$, i.e., A is of the form

$$A = \begin{pmatrix} 2 + d_2 + d_3 + d_4 + j_1 + 2j_2 + 2j_3 + 2j_4 & 2 + d_3 + d_4 + j_2 + 2j_3 + 2j_4 \\ d_1 & d_2 \end{pmatrix}$$

$$\begin{pmatrix} 2 + d_4 + j_3 + 2j_4 & 2 + j_4 \\ d_3 & d_4 \end{pmatrix}$$

with $j_t \in \{0,1\}$. In the general case of a $2 \times k$ matrix $A \in \mathcal{M}_{n,k,m}^{(2,3)}$, we can express in a unique way

$$c_1 = 2 + d_2 + d_3 + \cdots + d_k + j_1 + 2j_2 + \cdots + 2j_k,$$

$$c_2 = 2 + d_3 + d_4 + \cdots + d_k + j_2 + 2j_3 + \cdots + 2j_k,$$

$$\vdots \tag{4.11}$$

$$c_{k-1} = 2 + d_k + j_{k-1} + 2j_k,$$

$$c_k = 2 + j_k,$$

where $j_t \in \{0,1\}$.

Theorem 4.2 *For any (n,k,m), with $0 \le m \le n$ and $k \ge 1$, there is a bijection*

$$\psi : \mathcal{M}_{n,k,m}^{(2,3)} \longrightarrow \mathcal{P}_{n,k,m},$$

where $\mathcal{P}_{n,k,m}$ is the set of all partitions λ of n such that

– λ has exactly m odd parts and the odd parts are distinct;
– the largest part of λ is $a(\lambda) = 2k$.

Precisely, given $A \in \mathcal{M}_{n,k,m}^{(2,3)}$, $\lambda = \psi(A)$ is the partition containing:

– $\frac{1}{2}d_t$ parts equal to $2t$, for any $t \in \{1, 2, \ldots, k\}$;
– j_t parts equal to $2t - 1$, for any $t \in \{1, 2, \ldots, k\}$;
– one part equal to $2k$,

where the $j_t \in \{0, 1\}$ are defined by (4.11).

Proof Let $j_t \in \{0, 1\}$ as above. From (4.10) and (4.11), it follows that

$$n = 2k + (d_1 + 2d_2 + 3d_3 + \cdots + kd_k) + (j_1 + 3j_2 + 5j_3 + \cdots + (2k-1)j_k). \quad (4.12)$$

We define a partition $\lambda = \psi(A)$ containing:

– $\frac{1}{2}d_t$ parts equal to $2t$, for any $t \in \{1, 2, \ldots, k\}$;
– j_t parts equal to $2t - 1$, for any $t \in \{1, 2, \ldots, k\}$;
– one part equal to $2k$.

Note that, for any $t \le k$,

$$j_t = \begin{cases} 0, & \text{if } c_t \text{ is even,} \\ 1, & \text{if } c_t \text{ is odd.} \end{cases}$$

Therefore, the number of odd parts in λ is equal to the number of odd elements in the first row of A.

It is easy to see that ψ defined above is a bijection between the set of matrices $\mathcal{M}_{n,k,m}^{(2,3)}$ and the set of partitions $\mathcal{P}_{n,k,m}$. $\qquad \square$

Denote by $\tilde{\mathcal{Q}}_{n,k,m}$ ($0 \le m \le n$, $k \ge 1$) the set of partitions μ of n such that

– μ contains exactly k parts, m of these being odd;
– the odd parts of μ are distinct;
– 1 is not a part of μ.

Note that, for $\mathcal{Q}_{n,m}$ defined by (4.3), we have

$$\mathcal{Q}_{n,m} = \bigcup_k \mathcal{Q}_{n,k,m} = \bigcup_k \tilde{\mathcal{Q}}_{n,k,m}. \quad (4.13)$$

Our next result establishes a bijection between $\tilde{\mathcal{Q}}_{n,k,m}$ and the same class of matrices as before.

Theorem 4.3 *There is a natural bijection* $\theta : \mathcal{M}_{n,k,m}^{(2,3)} \to \tilde{\mathcal{Q}}_{n,k,m}$. *Given any matrix* $A \in \mathcal{M}_{n,k,m}^{(2,3)}$ *of the form* (4.4), *the corresponding partition* $\mu = \theta(A)$ *is obtained adding up the entries in each column of* A,

$$\mu = (c_1 + d_1, c_2 + d_2, \ldots, c_k + d_k).$$

Corollary 4.4 *Identity* (4.2) *holds true.*

Proof As mentioned above, the cardinalities of the sets $\mathcal{P}_{n,m}$ and $\mathcal{Q}_{n,m}$ defined by (4.3) are, respectively, equal to the coefficients of $z^m q^n$ on left-hand-side and on the right-hand-side of (4.2). Since, by Theorem 4.2 and Theorem 4.3, $\mathcal{P}_{n,k,m}$ and $\tilde{\mathcal{Q}}_{n,k,m}$ are both in one-to-one correspondence with the same set of matrices $\mathcal{M}_{n,k,m}^{(2,3)}$, using (4.13) we obtain identity (4.2). $\qquad\square$

Application We can construct a bijection $\theta \circ \psi^{-1} : \mathcal{P}_{n,k,m} \to \mathcal{Q}_{n,k,m}$, or $\theta \circ \psi^{-1} : \mathcal{P}_{n,m} \to \mathcal{Q}_{n,m}$, if we want, which is different from the bijection constructed in the proof of Theorem 4.1. For example, if $n = 11$ the bijection $\theta \circ \psi^{-1}$ is described below:

$$(2,2,2,2,2,1) \longmapsto \begin{pmatrix} 3 \\ 8 \end{pmatrix} \longmapsto (11),$$

$$(4,3,2,2) \longmapsto \begin{pmatrix} 4 & 3 \\ 4 & 0 \end{pmatrix} \longmapsto (8,3),$$

$$(4,4,3) \longmapsto \begin{pmatrix} 6 & 3 \\ 0 & 2 \end{pmatrix} \longmapsto (6,5),$$

$$(4,2,2,2,1) \longmapsto \begin{pmatrix} 3 & 2 \\ 6 & 0 \end{pmatrix} \longmapsto (9,2),$$

$$(4,4,2,1) \longmapsto \begin{pmatrix} 5 & 2 \\ 2 & 2 \end{pmatrix} \longmapsto (7,4),$$

$$(6,2,2,1) \longmapsto \begin{pmatrix} 3 & 2 & 2 \\ 4 & 0 & 0 \end{pmatrix} \longmapsto (7,2,2),$$

$$(6,4,1) \longmapsto \begin{pmatrix} 5 & 2 & 2 \\ 0 & 2 & 0 \end{pmatrix} \longmapsto (5,4,2),$$

$$(6,3,2) \longmapsto \begin{pmatrix} 4 & 3 & 2 \\ 2 & 0 & 0 \end{pmatrix} \longmapsto (6,3,2),$$

$$(6,5) \longmapsto \begin{pmatrix} 4 & 4 & 3 \\ 0 & 0 & 0 \end{pmatrix} \longmapsto (4,4,3),$$

$$(8,2,1) \longmapsto \begin{pmatrix} 3 & 2 & 2 & 2 \\ 2 & 0 & 0 & 0 \end{pmatrix} \longmapsto (5,2,2,2),$$

$$(8,3) \longmapsto \begin{pmatrix} 4 & 3 & 2 & 2 \\ 0 & 0 & 0 & 0 \end{pmatrix} \longmapsto (4,3,2,2),$$

$$(10,1) \longmapsto \begin{pmatrix} 3 & 2 & 2 & 2 & 2 \\ 0 & 0 & 0 & 0 & 0 \end{pmatrix} \longmapsto (3,2,2,2,2).$$

We now study another identity (see (4.14) below) that, surprisingly, has the same right-hand-side as (4.1).

Theorem 4.5 *The following identity holds*

$$\frac{1}{1-q^2} + \sum_{k=2}^{\infty} \frac{(1+q)(1+q^3)\cdots(1+q^{2k-3})}{(1-q^2)(1-q^4)\cdots(1-q^{2k})} q^{2k-1} = \prod_{n=1}^{\infty} \frac{1+q^{2n+1}}{1-q^{2n}}, \qquad (4.14)$$

or, multiplying both sides by $1+q$,

$$\frac{1}{1-q} + \sum_{k=2}^{\infty} \frac{(1+q)^2(1+q^3)\cdots(1+q^{2k-3})}{(1-q^2)(1-q^4)\cdots(1-q^{2k})} q^{2k-1} = \prod_{n=1}^{\infty} \frac{1+q^{2n-1}}{1-q^{2n}}. \qquad (4.15)$$

The proof of (4.14) depends on a relationship between two classes of partitions established in Theorem 4.6 below.

In addition to $\mathcal{Q}_n$, the set of partitions of n such that the odd parts are distinct and 1 is not a part, introduced in the proof of Theorem 4.1, we consider the following class of partitions:

$\mathcal{R}_n$, the set of all partitions λ of n such that

- the odd parts of λ are distinct;
- λ contains at least one odd part;
- if $a_o(\lambda)$ denotes the largest odd part, then $a_o(\lambda) + 1 \geq$ any even part.

Our next result resembles the Euler Pentagonal Number Theorem.

Theorem 4.6 *Let $\mathcal{R}_n$ and $\mathcal{Q}_n$ be the two classes of partitions defined above. Then,*

$$|\mathcal{Q}_n| - |\mathcal{R}_n| = (-1)^n. \qquad (4.16)$$

Proof of Theorem 4.6 We construct a correspondence ϕ which associates $\phi(\lambda) \in \mathcal{Q}_n$ to any given $\lambda \in \mathcal{R}_n$:

- if 1 is a part of λ, we define $\phi(\lambda)$ as the partition obtained by removing 1 from λ and adding 1 to its largest odd part;
- if 1 is not a part of λ, we take $\phi(\lambda) = \lambda$.

Then, ϕ is a one-to-one, with an inverse ϕ^{-1} defined by:

- if $\mu \in \mathcal{Q}_n$ is such that its largest odd part $a_o(\mu)$ and largest even part $a_e(\mu)$ satisfy $a_o(\mu) + 1 \geq a_e(\mu)$, then $\phi^{-1}(\mu) = \mu$;
- if $\mu \in \mathcal{Q}_n$ contains no odd parts or if $a_o(\mu) + 1 < a_e(\mu)$, we construct $\phi^{-1}(\mu)$ by removing from μ one part equal to $a_e(\mu)$ and add one part equal to $a_e(\mu) - 1$ and one part equal to 1.

The correspondence ϕ defines a bijection except for the following two special features:

- when n is even, there is one element $\mu = (2, 2, \ldots, 2) \in \mathcal{Q}_n$ without a corresponding $\phi^{-1}(\lambda) \in \mathcal{R}_n$;
- when n is odd, there is one element $\lambda = (2, 2, \ldots, 2, 1) \in \mathcal{R}_n$ without a corresponding $\phi(\lambda) \in \mathcal{Q}_n$.

Therefore, (4.16) holds.   □

Proof of Theorem 4.5 It is easy to see that the generating function of $\mathcal{R}_n$ is

$$\frac{1}{1-q^2}q + \sum_{k=2}^{\infty} \frac{(1+q)(1+q^3)\cdots(1+q^{2k-3})}{(1-q^2)(1-q^4)\cdots(1-q^{2k})}q^{2k-1}. \tag{4.17}$$

From (4.16) and (4.17), we obtain

$$\frac{1}{1+q} + \frac{1}{1-q^2}q + \sum_{k=2}^{\infty} \frac{(1+q)(1+q^3)\cdots(1+q^{2k-3})}{(1-q^2)(1-q^4)\cdots(1-q^{2k})}q^{2k-1} = \prod_{n=1}^{\infty} \frac{1+q^{2n+1}}{1-q^{2n}},$$

from which (4.14) follows immediately. □

In order to construct a bijection between certain classes of partitions, we consider the set $\mathcal{M}_{n,k,m}^{(1)}$ of all $2 \times k$ matrices of the form (4.4) with non-negative integer entries, having exactly m odd entries in the first row and satisfying

$$c_k = 1, \qquad d_t \equiv 0 \ (\mathrm{mod}\ 2), \quad \forall t; \tag{4.18}$$

$$c_t \equiv 0 \ (\mathrm{mod}\ 2), c_{t+1} \equiv 0 \ (\mathrm{mod}\ 2) \Longrightarrow c_t = c_{t+1} + d_{t+1}; \tag{4.19}$$

$$c_t \equiv 0 \ (\mathrm{mod}\ 2), c_{t+1} \equiv 1 \ (\mathrm{mod}\ 2) \Longrightarrow c_t = 1 + c_{t+1} + d_{t+1}; \tag{4.20}$$

$$c_t \equiv 1 \ (\mathrm{mod}\ 2), c_{t+1} \equiv 0 \ (\mathrm{mod}\ 2) \Longrightarrow c_t = 1 + c_{t+1} + d_{t+1}; \tag{4.21}$$

$$c_t \equiv 1 \ (\mathrm{mod}\ 2), c_{t+1} \equiv 1 \ (\mathrm{mod}\ 2) \Longrightarrow c_t = 2 + c_{t+1} + d_{t+1}; \tag{4.22}$$

$$\sum c_t + \sum d_t = n. \tag{4.23}$$

Let $A \in \mathcal{M}_{n,k,m}^{(1)}$. As before, if, for example, $k = 4$, then it is easy to verify that A is expressed in a unique way as

$$A = \begin{pmatrix} 2 + d_2 + d_3 + d_4 + j_1 + 2j_2 + 2j_3 & 2 + d_3 + d_4 + j_2 + 2j_3 & 2 + d_4 + j_3 & 1 \\ d_1 & d_2 & d_3 & d_4 \end{pmatrix},$$

with $j_t \in \{0, 1\}$.

In general, for any $A \in \mathcal{M}_{n,k,m}^{(1)}$, we can write in a unique way

$$c_1 = 2 + d_2 + d_3 + \cdots + d_k + j_1 + 2j_2 + \cdots + 2j_{k-1},$$
$$c_2 = 2 + d_3 + d_4 + \cdots + d_k + j_2 + 2j_3 + \cdots + 2j_{k-1},$$
$$\vdots \tag{4.24}$$
$$c_{k-1} = 2 + d_k + j_{k-1},$$
$$c_k = 1,$$

where $j_t \in \{0, 1\}$.

Consider also the set $\mathcal{R}_{n,k,m}$ of partitions λ of n, having exactly m odd parts, such that the odd parts are distinct, the largest odd part is $a_o(\lambda) = 2k - 1$, and any even part is less than or equal to $2k$.

Theorem 4.7 *For any n, k, and m, with $1 \le m \le n$ and $k \ge 1$, there is a bijection $\zeta : \mathcal{M}_{n,k,m}^{(1)} \to \mathcal{R}_{n,k,m}$. Precisely, given $A \in \mathcal{M}_{n,k,m}^{(1)}$, $\lambda = \zeta(A)$ is the partition containing*

- *$\frac{1}{2}d_t$ parts equal to $2t$, for any $t \in \{1, 2, \ldots, k\}$;*
- *j_t parts equal to $2t - 1$, for any $t \in \{1, 2, \ldots, k - 1\}$;*
- *one part equal to $2k - 1$,*

where the $j_t \in \{0, 1\}$ are given by (4.24).

Remark In the case $k = 1$ (hence, n is necessarily odd) λ contains one part equal to 1 and $\frac{1}{2}d_1$ parts equal to 2.

Proof From (4.23) and (4.24), it follows that

$$n = (2k - 1) + (d_1 + 2d_2 + 3d_3 + \cdots + kd_k) + \left(j_1 + 3j_2 + 5j_3 + \cdots + (2k - 3)j_{k-1}\right).$$
$$(4.25)$$

To the matrix A we associate a partition λ containing:

- *$\frac{1}{2}d_t$ parts equal to $2t$, for any $t \in \{1, 2, \ldots, k\}$;*
- *j_t parts equal to $2t - 1$, for any $t \in \{1, 2, \ldots, k - 1\}$;*
- *one part equal to $2k - 1$.*

Note that, since $j_t = 1$ if c_t is odd and $j_t = 0$ if c_t is even, for any $t \le k - 1$, it follows that the number of odd parts in λ is equal to the number of odd elements in the first row of A. This completes the argument. □

Remark In practice, to deal with the inverse correspondence ζ^{-1}, it is useful to identify integers with matrices as described below. For simplicity, we consider the case $k = 4$:

$$1 = \begin{pmatrix} 1 & 0 & 0 & 0 \\ 0 & 0 & 0 & 0 \end{pmatrix}, \qquad 4 = \begin{pmatrix} 2 & 0 & 0 & 0 \\ 0 & 2 & 0 & 0 \end{pmatrix}, \qquad 7 = \begin{pmatrix} 2 & 2 & 2 & 1 \\ 0 & 0 & 0 & 0 \end{pmatrix},$$

$$2 = \begin{pmatrix} 0 & 0 & 0 & 0 \\ 2 & 0 & 0 & 0 \end{pmatrix}, \qquad 5 = \begin{pmatrix} 2 & 2 & 1 & 0 \\ 0 & 0 & 0 & 0 \end{pmatrix}, \qquad 8 = \begin{pmatrix} 2 & 2 & 2 & 0 \\ 0 & 0 & 0 & 2 \end{pmatrix}.$$

$$3 = \begin{pmatrix} 2 & 1 & 0 & 0 \\ 0 & 0 & 0 & 0 \end{pmatrix}, \qquad 6 = \begin{pmatrix} 2 & 2 & 0 & 0 \\ 0 & 0 & 2 & 0 \end{pmatrix},$$

For example, the partition $\lambda = (8, 7, 4, 4, 4, 3, 1)$ corresponds to the matrix

$$\begin{pmatrix} 2 & 2 & 2 & 0 \\ 0 & 0 & 0 & 2 \end{pmatrix} + \begin{pmatrix} 2 & 2 & 2 & 1 \\ 0 & 0 & 0 & 0 \end{pmatrix} + \begin{pmatrix} 6 & 0 & 0 & 0 \\ 0 & 6 & 0 & 0 \end{pmatrix} + \begin{pmatrix} 2 & 1 & 0 & 0 \\ 0 & 0 & 0 & 0 \end{pmatrix}$$

$$+ \begin{pmatrix} 1 & 0 & 0 & 0 \\ 0 & 0 & 0 & 0 \end{pmatrix} = \begin{pmatrix} 13 & 5 & 4 & 1 \\ 0 & 6 & 0 & 2 \end{pmatrix}.$$

In the case of the correspondence ψ considered in Theorem 4.2, the identification is the same, except for the integer $2k$. For example, for $k = 4$, remember that the largest part of any $\mu \in \mathcal{P}_{n,4,m}$ is $2k = 8$. The only difference is that the first 8 is identified with the matrix

$$\begin{pmatrix} 2 & 2 & 2 & 2 \\ 0 & 0 & 0 & 0 \end{pmatrix}$$

and each additional 8 is identified with

$$\begin{pmatrix} 2 & 2 & 2 & 0 \\ 0 & 0 & 0 & 2 \end{pmatrix}.$$

We now introduce another class of partitions. Let $\mathcal{S}_{n,k,m}$ denote the set of all partitions π of n such that

- π has exactly k parts and m of these parts are odd;
- the odd parts of π are distinct;
- the smallest part of π is odd.

Theorem 4.8 *There is a natural bijection* $\xi : \mathcal{M}^{(1)}_{n,k,m} \to \mathcal{S}_{n,k,m}$. *Given any matrix* $A \in \mathcal{M}^{(1)}_{n,k,m}$, *the corresponding partition* $\pi = \xi(A)$ *is defined as*

$$\pi = (c_1 + d_1, c_2 + d_2, \ldots, c_k + d_k),$$

i.e., by adding up the entries in each column of A.

Proof The proof is straightforward. □

Corollary 4.9 *There is a bijection* $\xi \circ \zeta^{-1} : \mathcal{R}_{n,k,m} \to \mathcal{S}_{n,k,m}$.

Proof Combining the bijections given in Theorem 4.7 and in Theorem 4.8, one obtains a bijection from $\xi \circ \zeta^{-1} : \mathcal{R}_{n,k,m} \to \mathcal{S}_{n,k,m}$. □

Example Let $n = 11$. For $m = k = 1$, we have

$$(2, 2, 2, 2, 2, 1) \longmapsto \begin{pmatrix} 1 \\ 10 \end{pmatrix} \longmapsto (11).$$

For $m = 1$ and $k = 2$, we have

$$(3, 2, 2, 2, 2) \longmapsto \begin{pmatrix} 2 & 1 \\ 8 & 0 \end{pmatrix} \longmapsto (10, 1),$$

$$(4, 3, 2, 2) \longmapsto \begin{pmatrix} 4 & 1 \\ 4 & 2 \end{pmatrix} \longmapsto (8, 3),$$

$$(4, 4, 3) \longmapsto \begin{pmatrix} 6 & 1 \\ 0 & 4 \end{pmatrix} \longmapsto (6, 5).$$

For $m = 1$ and $k = 3$, we have

$$(5, 2, 2, 2) \longmapsto \begin{pmatrix} 2 & 2 & 1 \\ 6 & 0 & 0 \end{pmatrix} \longmapsto (8, 2, 1),$$

$$(6,5) \quad \longmapsto \quad \begin{pmatrix} 4 & 4 & 1 \\ 0 & 0 & 2 \end{pmatrix} \quad \longmapsto \quad (4,4,3),$$

$$(5,4,2) \quad \longmapsto \quad \begin{pmatrix} 4 & 2 & 1 \\ 2 & 2 & 0 \end{pmatrix} \quad \longmapsto \quad (6,4,1),$$

$$(5,3,2,1) \quad \longmapsto \quad \begin{pmatrix} 5 & 3 & 1 \\ 2 & 0 & 0 \end{pmatrix} \quad \longmapsto \quad (7,3,1).$$

As an application of Theorem 4.8, we obtain the following identity.

Theorem 4.10 *The following identity holds*

$$\frac{q}{1-q^2} + \sum_{n=2}^{\infty} \frac{(1+q)(1+q^3)\cdots(1+q^{2n-3})}{(1-q^2)(1-q^4)\cdots(1-q^{2n})} q^{2n-1} = \sum_{n=1}^{\infty} q^{2n-1} \prod_{i=0}^{\infty} \frac{1+q^{2n+2i+1}}{1-q^{2n+2i}}.$$

$$(4.26)$$

Remark Indeed a slightly stronger version of (4.26) holds true, namely

$$\frac{zq}{1-q^2} + \sum_{k=2}^{\infty} \frac{(1+zq)(1+zq^3)\cdots(1+zq^{2k-3})}{(1-q^2)(1-q^4)\cdots(1-q^{2k})} zq^{2k-1}$$

$$= \sum_{k=1}^{\infty} zq^{2k-1} \prod_{i=0}^{\infty} \frac{1+zq^{2k+2i+1}}{1-q^{2k+2i}}.$$

$$(4.27)$$

Proof The proof follows from Theorem 4.8 and Corollary 4.9. On the one hand, the coefficient of $z^m q^n$ in the expansion of $\frac{(1+zq)(1+zq^3)\cdots(1+zq^{2k-3})}{(1-q^2)(1-q^4)\cdots(1-q^{2k})} zq^{2k-1}$ is the number of elements of $\mathcal{R}_{n,k,m}$. Therefore, the coefficient of $z^m q^n$ on the left-hand-side of (4.27) is the number of elements in the union $\bigcup_k \mathcal{R}_{n,k,m}$.

On the other hand, if we denote by $\mathcal{S}_{n,m}$ the set of all partitions π of n such that

- the odd parts of π are distinct;
- π has exactly m odd parts;
- the smallest part of π is odd,

then the class $\mathcal{S}_{n,k,m}$, defined immediately before the statement of Theorem 4.8, consists of the partitions $\pi \in \mathcal{S}_{n,m}$ with exactly k parts. We consider also the class

$$\tilde{\mathcal{S}}_{n,k,m} = \{\pi \in \mathcal{S}_{n,m} \mid \text{the smallest part of } \pi \text{ is } 2k-1\}.$$

Clearly,

$$\mathcal{S}_{n,m} = \bigcup_k \mathcal{S}_{n,k,m} = \bigcup_k \tilde{\mathcal{S}}_{n,k,m},$$

with the above unions being of pairwise disjoint sets.

The coefficient of $z^m q^n$ in the expansion of $zq^{2k-1} \prod_{i=0}^{\infty} \frac{1+zq^{2k+2i+1}}{1-q^{2k+2i}}$ is the number of elements of $\tilde{\mathcal{S}}_{n,k,m}$. Hence, the coefficient of $z^m q^n$ on the right-hand-side of (4.27) is the number of elements in the union $\mathcal{S}_{n,m} = \bigcup_k \mathcal{S}_{n,k,m} = \bigcup_k \tilde{\mathcal{S}}_{n,k,m}$. Since, by Corollary 4.9 of Theorem 4.8, $\mathcal{R}_{n,k,m}$ and $\mathcal{S}_{n,k,m}$ have the same number of elements, (4.27) follows. $\square$

It is important to point out that as a consequence of Theorem 4.10, that has been proved by a combinatorial argument, we can easily prove a well-known identity about Ramanujan's partial theta function

$$\psi(q) = \sum_{n=0}^{\infty} q^{\frac{n(n+1)}{2}}. \tag{4.28}$$

Theorem 4.11 *The function ψ, defined by (4.28), satisfies the identity*

$$1 + q + \sum_{n=1}^{\infty} \frac{(1-q^2)(1-q^4)\cdots(1-q^{2n})}{(1-q^3)(1-q^5)\cdots(1-q^{2n+1})} q^{2n+1} = \sum_{n=0}^{\infty} q^{\frac{n(n+1)}{2}}. \tag{4.29}$$

Proof From (4.26), it follows that

$$\frac{q}{1-q^2} + \sum_{n=2}^{\infty} \frac{(1+q)(1+q^3)\cdots(1+q^{2n-3})}{(1-q^2)(1-q^4)\cdots(1-q^{2n})} q^{2n-1} = \left(\prod_{i=1}^{\infty} \frac{1+q^{2i+1}}{1-q^{2i}}\right) \cdot U(q),$$

where $U(q)$ is defined by

$$U(q) := q + \sum_{n=2}^{\infty} \frac{(1-q^2)(1-q^4)\cdots(1-q^{2n-2})}{(1+q^3)(1+q^5)\cdots(1+q^{2n-1})} q^{2n-1}.$$

Using (4.14), we obtain

$$\left(\prod_{i=1}^{\infty} \frac{1+q^{2i+1}}{1-q^{2i}} - \frac{1}{1+q}\right) = \left(\prod_{i=1}^{\infty} \frac{1+q^{2i+1}}{1-q^{2i}}\right) \cdot U(q). \tag{4.30}$$

To prove (4.29), it suffices to verify that

$$1 - U(-q) = \psi(q) = \sum_{n=0}^{\infty} q^{\frac{n(n+1)}{2}}. \tag{4.31}$$

Using the notation

$$(a; q)_{\infty} := \prod_{i=0}^{\infty} (1 - aq^i), \quad |q| < 1,$$

we can rephrase (4.30) as

$$\frac{(-q^3; q^2)_{\infty}}{(q^2; q^2)_{\infty}} - \frac{1}{1+q} = \frac{(-q^3; q^2)_{\infty}}{(q^2; q^2)_{\infty}} \cdot (1 - \psi(-q)),$$

which is easily seen to be true, using (see (1.3.4), page 11, in [5])

$$\psi(q) = \frac{(q^2; q^2)_\infty}{(q; q^2)_\infty}.$$

□

Remark Identity (4.2), and hence Identity (4.1) as well, is a particular case of the following identity (see (1.5) in [1])

$$1 + \sum_{k=0}^{\infty} \frac{(1+a)(1+aq)\cdots(1+aq^{k-1})}{(1-q)(1-q^2)\cdots(1-q^k)} z^k q^k = \prod_{j=1}^{\infty} \frac{1+azq^j}{1-zq^j}. \tag{4.32}$$

Indeed, taking $z = 1$ in (4.32) and then replacing q by q^2 and a by z, one obtains (4.2) as a particular case.

In [1], (4.32) is proved by a combinatorial argument, but the proof is more involved than our proof of (4.2).

5 A new bijective proof for the Lebesgue identity

We now turn our attention to the problem of obtaining a representation of the partitions in which even parts are not repeated as a certain class of two-line matrices. As a consequence of our ideas, we provide a new combinatorial proof of the following equality, known as the Lebesgue Identity:

$$\sum_{r=1}^{\infty} \frac{(1+zq)(1+zq^2)\cdots(1+zq^r)}{(1-q)(1-q^2)\cdots(1-q^r)} q^{\binom{r+1}{2}} = \prod_{i=1}^{\infty} (1+zq^{2i})(1+q^i). \tag{5.1}$$

Combinatorial proofs of the Lebesgue Identity have been given by several authors, among them Bessenrodt, Bressoud, Little, Sellers, Alladi, and Gordon (see [6] and [7]). We believe our bijection is simpler. It is based on a new way of representing partitions by two-line matrices, introduced by Santos, Mondek, and Ribeiro in [9].

Let

$$f(z, q) := \sum_{r=1}^{\infty} \frac{(1+zq)(1+zq^2)\cdots(1+zq^r)}{(1-q)(1-q^2)\cdots(1-q^r)} q^{\binom{r+1}{2}} \tag{5.2}$$

be the left-hand side of (5.1). The general term

$$\frac{(1+zq)(1+zq^2)\cdots(1+zq^k)q^{1+2+3+\cdots+k}}{(1-q)(1-q^2)\cdots(1-q^k)}$$

of (5.2) is the generating function of the number of ways of decomposing n as

$$n = (1+j_1)\cdot 1 + (1+j_2)\cdot 2 + \cdots + (1+j_k)\cdot k + d_1 + 2d_2 + \cdots + kd_k,$$

with $d_t \geq 0$ and $j_t \in \{0, 1\}$ or, equivalently, as the sum of the entries of the matrix

$$A = \begin{pmatrix} k + j_1 + \cdots + j_k + d_2 + \cdots + d_k & \cdots & 2 + j_{k-1} + j_k + d_k & 1 + j_k \\ d_1 & \cdots & d_{k-1} & d_k \end{pmatrix}.$$

The exponent of z counts the number of nonvanishing elements among the j_t. Still equivalently, n is the sum of entries of the matrices of the form

$$A = \begin{pmatrix} c_1 & c_2 & \cdots & c_k \\ d_1 & d_2 & \cdots & d_k \end{pmatrix}, \tag{5.3}$$

with non-negative entries and satisfying

$$c_k = i_k \in \{1, 2\}, \tag{5.4}$$

$$c_t = i_t + c_{t+1} + d_{t+1}, \quad \text{with } i_t \in \{1, 2\} \text{ for } t < k, \tag{5.5}$$

$$n = \sum c_t + \sum d_t. \tag{5.6}$$

Denote by $\mathcal{M}_{n,k}$ the set of all matrices of the form (5.3) with non-negative integer coefficients satisfying (5.4), (5.5), and (5.6). Denote by $\mathcal{M}_{n,m,k}$ the subset of all matrices in $\mathcal{M}_{n,k}$ for which $i_t = 2$ for exactly m elements in the first row and let $\mathcal{N}_{n,m} := \bigcup_k \mathcal{M}_{n,m,k}$. The following proposition clearly holds.

Proposition 5.1 *For each fixed k, the coefficient of $z^m q^n$ in the expansion of*

$$\frac{(1 + zq)(1 + zq^2) \cdots (1 + zq^k)}{(1 - q)(1 - q^2) \cdots (1 - q^k)} q^{\binom{k+1}{2}}$$

is the number of matrices in the set $\mathcal{M}_{n,m,k}$. Hence, the coefficient of $z^m q^n$ in the expansion of (5.2) is the number of elements in $\mathcal{N}_{n,m}$.

Denote by $\mathcal{P}_{n,m}$ the set of all partitions of n in which even parts are not repeated and having exactly m even parts. Our main result is the following theorem, from which the Lebesgue Identity will be derived.

Theorem 5.2 *There is a natural bijection between the sets $\mathcal{P}_{n,m}$ and $\mathcal{N}_{n,m}$.*

The construction of the bijection stated in Theorem 5.2 is the object of the next section.

Proposition 5.3 *The coefficient of $z^m q^n$ on the right-hand side of (5.1) is the number of elements of the set $\mathcal{P}_{n,m}$.*

Proof It suffices to rewrite the right-hand side of (5.1) as

$$\prod_{i=1}^{\infty} (1 + zq^{2i})(1 + q^i) = \prod_{i=1}^{\infty} \frac{(1 + zq^{2i})(1 - q^{2i})}{(1 - q^i)} = \prod_{i=1}^{\infty} \frac{(1 + zq^{2i})}{(1 - q^{2i-1})}. \qquad \square$$

We have interpreted the right-hand side and the left-hand side of the Lebesgue Identity (5.1) as generating functions of $\mathcal{N}_{n,m}$ and $\mathcal{P}_{n,m}$, respectively. Hence, to prove the Lebesgue Identity, it suffices to construct a bijection $\psi : \mathcal{N}_{n,m} \longrightarrow \mathcal{P}_{n,m}$. This construction is the object of the next section.

6 Main bijection

In order to prove Theorem 5.2, we now construct a bijection $\psi : \mathcal{N}_{n,m} \longrightarrow \mathcal{P}_{n,m}$. As a matter of fact, we specialize and, for each k, construct a bijection $\psi : \mathcal{M}_{n,m,k} \longrightarrow \mathcal{P}_{n,m,k}$, where $\mathcal{M}_{n,m,k}$ is the set of matrices with k columns defined above and $\mathcal{P}_{n,m,k}$ is a subset of $\mathcal{P}_{n,m}$ to be defined below.

The definition of the set $\mathcal{P}_{n,m,k}$ and the bijection ψ is slightly different, according to the case that the number k of columns is even or odd.

Case 1: k is even. If $k = 2s$, we take $\mathcal{P}_{n,m,k}$ to be the subset of all partitions in $\mathcal{P}_{n,m}$ with exactly s parts greater than or equal to $k + 1 = 2s + 1$. Given any matrix $A \in \mathcal{M}_{n,m,k}$, using (5.4), (5.5), and (5.6) and setting $i_t = 1 + j_t$ $(t = 1, \ldots, k)$, we have

$$n = c_1 + \cdots + c_{2s} + d_1 + \cdots + d_{2s}$$
$$= (j_1 + 2j_2 + \cdots + 2sj_{2s}) + (d_1 + 2d_2 + \cdots + 2sd_{2s}) + (1 + 2 + \cdots + 2s).$$

Since $1 + 2 + \cdots + 2s = s(2s + 1)$, we have

$$n = \left(d_1 + 3d_3 + \cdots + (2s - 1)d_{2s-1}\right) + (2j_2 + 4j_4 + \cdots + 2sj_{2s})$$
$$+ (j_1 + 2s + 1 + 2d_{2s}) + (3j_3 + 2s + 1 + 2d_{2s-2} + 2d_{2s}) + \cdots$$
$$+ \left((2s - 1)j_{2s-1} + 2s + 1 + 2d_2 + \cdots + 2d_{2s}\right).$$

Based on this decomposition, to the $2 \times (2s)$ matrix A we associate the partition π containing

   (i) d_1 parts equal to 1;
  (ii) d_3 parts equal to 3;

     $\vdots$

 (iii) d_{2s-1} parts equal to $2s - 1$;
 (iv) one part equal to $2r$, whenever $j_{2r} = 1, r \in \{1, 2, \ldots, s\}$;
  (v) one part equal to $j_1 + 2s + 1 + 2d_{2s}$;
 (vi) one part equal to $3j_3 + 2s + 1 + 2d_{2s-2} + 2d_{2s}$;

     $\vdots$

(vii) one part equal to $(2s - 1)j_{2s-1} + 2s + 1 + 2d_2 + 2d_4 + \cdots + 2d_{2s}$.

It is easy to see that $\pi \in \mathcal{P}_{n,m,2s} = \mathcal{P}_{n,m,k}$.

The fact that ψ is a bijection from $\mathcal{M}_{n,m,k}$ onto $\mathcal{P}_{n,m,k}$ in Case 1, as well as in Case 2, follows from the following lemma.

Lemma 6.1 *Given k integers such that odd numbers are not repeated, there is a unique way to order them in a certain order* $N_1, N_2, \ldots, N_k$ *and write out*

$$N_1 = 2e_1 + j_1,$$
$$N_2 = 2e_2 + 3j_3,$$
$$N_3 = 2e_3 + 5j_5, \tag{6.1}$$
$$\cdots$$
$$N_k = 2e_k + (2k - 1)j_{2k-1}$$

with

$$j_t \in \{0, 1\}, \tag{6.2}$$
$$e_1 \leq e_2 \leq \cdots \leq e_k. \tag{6.3}$$

To understand the idea of the proof of the above lemma, we provide an example.

Example With $k = 8$, consider the following eight numbers, where odd numbers are not repeated: $(3, 4, 6, 6, 7, 11, 12, 17)$. First, we express the even numbers as

$$4 = 2 \cdot 2,$$
$$6 = 2 \cdot 3,$$
$$6 = 2 \cdot 3, \tag{6.4}$$
$$12 = 2 \cdot 6.$$

Then, we consider the least odd number in the set, which is 3. There is a unique way of expressing it, namely

$$3 = N_1 = 2 \cdot 1 + 1. \tag{6.5}$$

Adding (6.5) to the array (6.4), we obtain

$$3 = 2 \cdot 1 + 1,$$
$$4 = 2 \cdot 2,$$
$$6 = 2 \cdot 3, \tag{6.6}$$
$$6 = 2 \cdot 3,$$
$$12 = 2 \cdot 6.$$

Then, we take the next odd number in the list, i.e., 7. There is only one way of fitting it into the array (6.6), namely at the second place, as N_2, with $j_3 = 1$,

$$3 = 2 \cdot 1 + 1,$$
$$7 = 2 \cdot 2 + 3,$$
$$4 = 2 \cdot 2,$$
$$6 = 2 \cdot 3,$$
$$6 = 2 \cdot 3,$$
$$12 = 2 \cdot 6.$$

(6.7)

The next odd number in the list is 11 and the only way to fit it into the array (6.7) is between 4 and the first 6, i.e., as N_4, with $j_7 = 1$,

$$3 = 2 \cdot 1 + 1,$$
$$7 = 2 \cdot 2 + 3,$$
$$4 = 2 \cdot 2,$$
$$11 = 2 \cdot 2 + 7,$$
$$6 = 2 \cdot 3,$$
$$6 = 2 \cdot 3,$$
$$12 = 2 \cdot 6.$$

(6.8)

Finally, the only way to fit the last odd number 17 into array (6.8) is in the sixth place, as N_6, with $j_{11} = 1$,

$$3 = 2 \cdot 1 + 1,$$
$$7 = 2 \cdot 2 + 3,$$
$$4 = 2 \cdot 2,$$
$$11 = 2 \cdot 2 + 7,$$
$$6 = 2 \cdot 3,$$
$$17 = 2 \cdot 3 + 11,$$
$$6 = 2 \cdot 3,$$
$$12 = 2 \cdot 6.$$

(6.9)

The proof of the above lemma follows the same argument outlined in the example. First, we arrange an array containing the expressions of the even elements in the list. Then, we begin including the odd elements one by one, in increasing order.

Case 2: k is odd. If $k = 2s + 1$, we take $\mathcal{P}_{n,m,k}$ to be the subset of all partitions in $\mathcal{P}_{n,m}$ with at least $s + 1$ parts greater than or equal to $k = 2s + 1$, but at most s of them greater than or equal to $2s + 3$. Given any matrix $A \in \mathcal{M}_{n,m,k}$, we have

$$n = c_1 + \cdots + c_{2s+1} + d_1 + \cdots + d_{2s+1}$$

$$= \left(j_1 + 2j_2 + \cdots + (2s + 1)j_{2s+1}\right) + \left(d_1 + 2d_2 + \cdots + (2s + 1)d_{2s+1}\right)$$

$$+ \left(1 + 2 + \cdots + (2s + 1)\right).$$

Since $1 + 2 + \cdots + (2s + 1) = (s + 1)(2s + 1)$, we have

$$n = \left(d_1 + 3d_3 + \cdots + (2s + 1)d_{2s+1}\right) + (2s + 1 + j_1) + (2j_2 + 4j_4 + \cdots + 2sj_{2s})$$

$$+ (3j_3 + 2s + 1 + 2d_{2s}) + (5j_5 + 2s + 1 + 2d_{2s-2} + 2d_{2s}) + \cdots$$

$$+ \left((2s + 1)j_{2s+1} + 2s + 1 + 2d_2 + \cdots + 2d_{2s}\right).$$

Based on this decomposition, to the $2 \times (2s + 1)$ matrix A we associate the partition π containing

(i) d_1 parts equal to 1;
(ii) d_3 parts equal to 3;

$$\vdots$$

(iii) d_{2s+1} parts equal to $2s + 1$;
(iv) one part equal to $2s + 1 + j_1$;
(v)   one part equal to $2r$, whenever $j_{2r} = 1$, $r \in \{1, 2, \ldots, k\}$;
(vi) one part equal to $3j_3 + 2s + 1 + 2d_{2s}$;
(vii) one part equal to $5j_5 + 2s + 1 + 2d_{2s-2} + 2d_{2s}$;

$$\vdots$$

(viii) one part equal to $(2s + 1)j_{2s+1} + 2s + 1 + 2d_2 + 2d_4 + \cdots + 2d_{2s}$.

It is easy to see that $\pi \in \mathcal{P}_{n,m,2s+1} = \mathcal{P}_{n,m,k}$.

We still have to show that any $\pi \in \mathcal{P}_{n,m}$ belongs to $\mathcal{P}_{n,m,k}$ for some k, i.e., $\mathcal{P}_{n,m} = \bigcup_k \mathcal{P}_{n,m,k}$. Indeed, given $\pi \in \mathcal{P}_{n,m}$, let s be the largest integer such that π has at least s parts greater than or equal to $2s + 1$. If π has exactly s parts greater than or equal to $2s + 1$, then $\pi \in \mathcal{P}_{n,m,2s}$, otherwise $\pi \in \mathcal{P}_{n,m,2s+1}$. This completes the proof of Theorem 5.2.   □

The next example illustrates that our bijective proof of the Lebesgue Identity is different from the ones given in [6] and [7].

Example Consider now the partition $\pi = (22, 21, 19, 18, 15, 10, 9, 9, 7, 4, 2)$, which is the same example as in [7], page 27. There are exactly five parts greater than or equal to 11. Hence, π corresponds to a 2×10 matrix. We first look at parts less than or equal to 10. There is one 7 and two 9s. Therefore, $d_1 = d_3 = d_5 = 0$, $d_7 = 1$, and $d_9 = 2$. There is one of each: 2, 4, and 10. Hence, $j_2 = j_4 = j_{10} = 1$ and $j_6 = j_8 = 0$.

The five parts greater than or equal to 11 are 15, 18, 19, 21, and 22. For these five numbers, the representation given in Lemma 6.1 is:

$$15 = 11 + 2 \cdot 2,$$
$$18 = 11 + 2 \cdot 2 + 3,$$
$$22 = 11 + 2 \cdot 3 + 5,$$
$$19 = 11 + 2 \cdot 4,$$
$$21 = 11 + 2 \cdot 5,$$

and hence,

$$d_{10} = 2,$$
$$d_8 + d_{10} = 2,$$
$$d_6 + d_8 + d_{10} = 3,$$
$$d_4 + d_6 + d_8 + d_{10} = 4,$$
$$d_2 + d_4 + d_6 + d_8 + d_{10} = 5.$$

It follows that $d_2 = 1$, $d_4 = 1$, $d_6 = 1$, $d_8 = 0$, and $d_{10} = 2$.

We now have all the elements to write out the matrix

$$A = \begin{pmatrix} 23 & 21 & 19 & 16 & 14 & 11 & 9 & 8 & 5 & 2 \\ 0 & 1 & 0 & 1 & 0 & 1 & 1 & 0 & 2 & 2 \end{pmatrix}.$$

Using the bijection constructed in Sect. 4, to this 2×10 matrix A we associate a pair (λ, μ) of partitions into distinct parts. We consider $j_t = c_t - c_{t+1} - d_{t+1} - 1$, if $t \leq k - 1$, and $j_k = c_k - 1$. In this concrete example, we have $(j_1, j_2, \ldots, j_{10}) = (0, 1, 1, 1, 1, 0, 0, 0, 0, 1)$. The first element of the pair is the partition λ formed with the non-vanishing elements in $(10 j_{10}, 9 j_9, \ldots, 2 j_2, j_1)$, i.e., $\lambda = (10, 5, 4, 3, 2)$. Consider also the second row of A, $(d_1, d_2, \ldots, d_{10}) = (0, 1, 0, 1, 0, 1, 1, 0, 2, 2)$. The second element of the pair is the partition $\mu = (\mu_1, \mu_2, \ldots, \mu_{10})$ given by

$$\mu_{10} = d_{10} + 1,$$
$$\mu_9 = d_9 + d_{10} + 2,$$
$$\vdots$$
$$\mu_1 = d_1 + \cdots + d_{10} + 10,$$

i.e., $\mu = (18, 17, 15, 14, 12, 11, 9, 7, 6, 3)$. This pair (λ, μ) is different from the one obtained in [7].

References

1. Andrews, G.E.: Enumerative proofs of certains q-identities. Glasg. Math. J. **8**, 33–40 (1967)
2. Andrews, G.E.: Ramanujan's "lost" notebook I, partial θ-functions. Adv. Math. **41**, 137–172 (1981)
3. Andrews, G.E.: Three-quadrant Ferrers graphs. Indian J. Math. **42**, 1–7 (2000)
4. Andrews, G.E., Berndt, B.C.: Ramanujan's Lost Notebook, Part I. Springer, New York (2005)

5. Berndt, B.C.: Number Theory in the Spirit of Ramanujan. AMS, Providence (2006)
6. Little, D.P., Sellers, J.A.: New proofs of identities of Lebesgue and Göllnitz via Tilings. J. Comb. Theory, Ser. A **116**, 223–231 (2009)
7. Pak, I.: Partition bijections, a survey. Ramanujan J. **12**, 5–75 (2006)
8. Santos, J.P.O., Silva, R.: A combinatorial proof for an identity involving partitions with distinct odd parts. South East Asian J. Math. Math. Sci. (accepted)
9. Santos, J.P.O., Mondek, P., Ribeiro, A.C.: New two-line arrays representing partitions. Ann. Comb. (accepted)

Ramanujan J (2010) 23: 297–306
DOI 10.1007/s11139-009-9206-9

Balanced partitions

Sam Vandervelde

Dedicated to George Andrews on the occasion of his 70th birthday

Received: 1 June 2009 / Accepted: 1 October 2009 / Published online: 21 April 2010
© Springer Science+Business Media, LLC 2010

Abstract A famous theorem of Euler asserts that there are as many partitions of n into distinct parts as there are partitions into odd parts. We begin by establishing a less well-known companion result, which states that both of these quantities are equal to the number of partitions of n into even parts along with exactly one triangular part. We then introduce the characteristic of a partition, which is determined in a simple way by the placement of odd parts within the list of all parts. This leads to a refinement of the aforementioned result in the form of a new type of partition identity involving characteristic, distinct parts, even parts, and triangular numbers. Our primary purpose is to present a bijective proof of the central instance of this new type of identity, which concerns balanced partitions—partitions in which odd parts occupy as many even as odd positions within the list of all parts. The bijection is accomplished by means of a construction that converts balanced partitions of $2n$ into unrestricted partitions of n via a pairing of the squares in the Young tableau.

Keywords Integer partition · Distinct parts · Even parts · Triangular number · Characteristic · Bijection

Mathematics Subject Classification (2000) Primary 05A17 · 11P81

1 Introduction

There are as many partitions of n into distinct parts as odd parts; this result still retains its appeal centuries after Euler proved it in 1748 and has formed the basis for generalizations in many directions, some of which are outlined in [1–3], and [4],

S. Vandervelde (✉)
St. Lawrence University, 23 Romoda Drive, Canton, NY 13617, USA
e-mail: svandervelde@stlawu.edu

among others. One natural avenue of inquiry is to ask whether there is a corresponding relationship between partitions into distinct parts and partitions involving even parts. The answer is in the affirmative: there are as many partitions of n into distinct parts as there are partitions of n into even parts along with exactly one triangular part. (A triangular part has size $T_k = \frac{1}{2}k(k+1)$ for some integer k.)

Notwithstanding its elementary nature, this partition identity seems to have been overlooked in recent decades. However, other relationships between partitions into distinct parts and even parts have been found in [2, 5] and [6], for example. It is interesting to note that in each case triangular numbers make an appearance, either overtly or implicitly.

Our aim is not so much to establish this identity (there is a short generating function proof), but to highlight its existence and to show that the search for a bijective proof leads in fruitful directions. In particular, we will define a quantity called the characteristic of a partition and demonstrate how it affords a refinement of this identity, thus leading to a new type of partition identity equating partitions of n into distinct parts having characteristic k with partitions of $n - T_{2k}$ into even parts.

The chief purpose of this paper is to present a bijective proof of the central instance of this type of identity. One formulation of this result states that for all n the number of balanced partitions of $2n$ into distinct parts is the same as the number of unrestricted partitions of n. A balanced partition is one in which the odd parts are equally split between odd positions and even positions when the parts are listed as usual in nonincreasing order. Thus the five balanced partitions of 8 are 8, 7–1, 6–2, 5–3, and 4–3–1, which agrees with the fact that $p(4) = 5$.

2 An initial result

Recall that the kth triangular number is given by $T_k = \frac{1}{2}k(k+1)$. A triangular part of a partition is a part whose size is equal to T_k for some integer k. Thus unlike an even part, a triangular part may have size zero. As usual, the empty partition counts as a partition of 0 into distinct parts or into even parts.

Proposition 1 *For every nonnegative integer n, the number of partitions of n into distinct parts is equal to the number of partitions of n into even parts along with precisely one triangular part.*

To clarify the assertion, consider the cases $n = 9$ and $n = 10$. The partitions of n of each type for these values are listed in Table 1. In the right-hand column of each list, the triangular part is shown in boldface. Note that in the second list the partitions 0–6–4 and 6–4 are counted separately, since the triangular part is distinguished. As predicted, there are an equal number of each type of partition in each list. This fact is quickly established using the Jacobi triple product.

Proof By substituting $x = q^{\frac{1}{2}}$ and $y = q^{\frac{1}{4}}$ in the Jacobi triple product

$$\prod_{m=1}^{\infty} \left(1 - x^{2m}\right)\left(1 + x^{2m-1}y^2\right)\left(1 + x^{2m-1}y^{-2}\right) = \sum_{n=-\infty}^{\infty} x^{n^2} y^{2n}, \tag{1}$$

Table 1 Partitions of 9 and 10 into distinct parts or even parts and a triangular part

distinct	$\chi(\pi)$	Δ + evens	distinct	$\chi(\pi)$	Δ + evens
9	−1	3–6	10	0	10
8–1	1	3–4–2	9–1	0	6–4
7–2	−1	3–2–2–2	8–2	0	6–2–2
6–3	1	1–8	7–3	0	0–10
6–2–1	−1	1–6–2	7–2–1	−2	0–8–2
5–4	−1	1–4–4	6–4	0	0–6–4
5–3–1	−1	1–4–2–2	6–3–1	0	0–6–2–2
4–3–2	1	1–2–2–2–2	5–4–1	−2	0–4–4-2
			5–3–2	0	0–4–2–2–2
			4–3–2–1	2	0–2–2–2–2–2

we obtain the identity

$$\prod_{m=1}^{\infty}\left(1-q^{m}\right)\left(1+q^{m}\right)\left(1+q^{m-1}\right) = \sum_{n=-\infty}^{\infty} q^{\frac{1}{2}n^{2}+\frac{1}{2}n}. \tag{2}$$

Let $T(q) = 1 + q + q^{3} + q^{6} + q^{10} + \cdots$ be the generating function for a single triangular part. Dividing the above equality by 2 and reindexing the $(1+q^{m-1})$ term leads to

$$\prod_{m=1}^{\infty}\left(1-q^{2m}\right)\left(1+q^{m}\right) = T(q), \tag{3}$$

which may be rewritten more productively as

$$(1+q)\left(1+q^{2}\right)\left(1+q^{3}\right)\cdots = T(q)\cdot\frac{1}{1-q^{2}}\cdot\frac{1}{1-q^{4}}\cdot\frac{1}{1-q^{6}}\cdots.$$

The coefficient of q^{n} on the left counts partitions of n into distinct parts, while the coefficient of q^{n} on the right tallies partitions of n into even parts and exactly one triangular part, so we are done. □

Let $p(n)$ denote the number of unrestricted partitions of n and let $p_d(n)$ be the number of partitions of n into distinct parts. Since a partition of $2n$ into even parts is equivalent to a partition of n, there will be $p(\frac{1}{2}(9-1)) = p(4) = 5$ partitions of 9 involving a triangular part of 1 and even parts otherwise. Similarly, there will be $p(\frac{1}{2}(9-3)) = p(3) = 3$ partitions that employ a triangular part of 3, as evidenced by Table 1. Hence we deduce that $p_d(9) = p(4) + p(3)$. In general, this line of reasoning leads to an expression for $p_d(n)$ reminiscent of Euler's Pentagonal Number Theorem. Adopting the standard convention that $p(n) = 0$ for values of n other than nonnegative integers, we have the following

Corollary 1

$$p_d(n) = \sum_{k=0}^{\infty} p\left(\frac{1}{2}(n - T_k)\right). \tag{4}$$

It is interesting to note the similarity of this formula with a relatively recent result obtained by Robbins [6], which states that

$$p_2(n) = \sum_{k=0}^{\infty} p(n - T_k), \tag{5}$$

where $p_2(n)$ is the number of partitions of n in two colors into distinct parts.

3 Characteristic of a partition

We now consider how a bijective proof of Proposition 1 might be obtained. The partitions of n into even parts and a triangular part are naturally grouped by the triangular part used, so we begin by searching for some feature of the partitions of n into distinct parts that gives rise to groups of the same sizes. After some searching, we discover that the characteristic of a partition has the desired property.

Definition 1 Let π be a partition of n, with parts listed in nonincreasing order. Let a_π be the number of odd parts appearing in even positions within the list, and let b_π be the number of odd parts appearing in odd positions. We define the *characteristic* $\chi(\pi)$ of the partition π to be the quantity $a_\pi - b_\pi$. When $\chi(\pi) = 0$, meaning that $a_\pi = b_\pi$, we say that the partition is *balanced*.

This definition applies to any partition of n, not necessarily into distinct parts. Also, we declare that $\chi(\pi) = 0$ for the empty partition.

To illustrate, let π be the partition 7–2–1. Since the odd parts occur in the first and third positions, we have $a_\pi = 0$ and $b_\pi = 2$, so $\chi(\pi) = -2$. Consulting Table 1, we find that among the partitions of 10 into distinct parts, $\chi(\pi) = -2$ also for 5–4–1, while $\chi(\pi) = 2$ only for 4–3–2–1; all other partitions satisfy $\chi(\pi) = 0$. Apparently partitions with $\chi(\pi) = 0, -2$ and 2 should correspond to partitions involving a triangular part of 0, 6 and 10, respectively. Examining the list for $n = 9$ further suggests that partitions with $\chi(\pi) = -1$ or 1 should pair off with partitions having a triangular part of 1 or 3, respectively. In general, we propose the following

Conjecture 1 *Let n be a fixed nonnegative integer. For each integer k, there are as many partitions of n into distinct parts having characteristic k as there are partitions of n into even parts and a single triangular part equal to T_{2k}. That is to say, there are $p(\frac{1}{2}(n - T_{2k}))$ such partitions.*

Therefore, we have found a refinement of the initial result outlined in Proposition 1; summing over all integers k produces Corollary 1. In the subsequent sections, we will provide a bijective proof of this conjecture in the case of balanced partitions.

Fig. 1 Determining diagonal lengths from a Young tableau

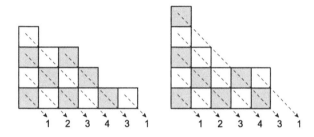

Remark 1 Based on this conjecture, one can determine the generating function for $p_d(n, k)$, the number of partitions of n into distinct parts having characteristic k. Relatively elementary manipulations reveal that

$$\sum_{n=0}^{\infty} \sum_{k=-\infty}^{\infty} p_d(n, k) x^n y^k = P(x^2) \sum_{k=-\infty}^{\infty} y^k x^{T_{2k}}$$

$$= \prod_{m=1}^{\infty} (1 + x^{2m})(1 + x^{4m-1} y)(1 + x^{4m-3} y^{-1}), \quad (6)$$

where $P(x) = \prod_{m=1}^{\infty} (1 - x^m)^{-1}$ is the generating function for unrestricted partitions. The first equality is obtained by considering even and odd values of k separately, while the second follows from a routine application of the Jacobi triple product. Observe that specializing to $y = 1$ neatly gives $P_d(x)$, the generating function for partitions of n into distinct parts. It would be interesting to ascertain the generating function for $p(n, k)$, the number of partitions of n having characteristic k.

The characteristic of a partition of n may also be computed via the lengths of certain diagonals in its Young tableau. Let the lengths of successive diagonals (slanting from upper left to lower right) be denoted by d_1 through d_n, beginning with the lower left corner, as depicted in Fig. 1. Note that squares within the same diagonal need not be adjacent. (For our purposes it will be more natural to order rows of the Young tableau in ascending order, "French style.") Thus the partition 6–4–3–1 has diagonals of length $d_1 = 1$, $d_2 = 2$, $d_3 = 3$, $d_4 = 4$, $d_5 = 3$, $d_6 = 1$, and $d_k = 0$ for $7 \leq k \leq 14$, as shown. Observe that the partition 5–5–2–1–1 yields precisely the same values for d_1 through d_{14}. Thus different partitions may have the same diagonal lengths.

We mention without proof that a sequence $d_1, d_2, \ldots, d_n$ of nonnegative integers are the diagonal lengths for some partition of n if and only if the following three conditions are met:

(i) $d_1 + d_2 + \cdots + d_n = n$,
(ii) $d_1 = 1, d_2 = 2, \ldots, d_m = m$ for some $1 \leq m \leq n$, and
(iii) $d_m \geq d_{m+1} \geq \cdots \geq d_n$.

We will also require the following result on diagonal lengths.

Proposition 2 *Let π be a partition of n having diagonal lengths $d_1, d_2, \ldots, d_n$. Then*

$$\chi(\pi) = \sum_{k=1}^{n} (-1)^k d_k. \tag{7}$$

Furthermore, there is exactly one partition of n into distinct parts having the given diagonal lengths.

Proof Color the Young tableau for π in a chessboard fashion so that the lower left corner is shaded, as done in Fig. 1. Then the odd-numbered diagonals will contain dark squares while the even-numbered diagonals contain light squares, so the sum $\sum(-1)^k d_k$ measures the signed excess of light squares. On the other hand, an even part yields a row of even length, which will contain an equal number of light and dark squares. Meanwhile, an odd part in an even (resp., odd) position corresponds to a row with one extra light (resp., dark) square. Therefore, $\chi(\pi) = a_\pi - b_\pi$ also measures the signed excess of light squares, so these quantities are equal.

Finally, a Young tableau represents a partition into distinct parts if and only if all the squares within each diagonal are adjacent to one another and extend to the lower edge. Otherwise examine the Young tableau at a point along the lowest numbered diagonal where a break occurs to find a pair of equal parts. Hence there exists a unique partition of n into distinct parts having a particular set of diagonal lengths, obtained from a given tableau by "sliding" all blocks within each diagonal down and to the right as far as possible.   □

It is worth noting that one may define an equivalence relation on the set of all partitions of n by declaring that $\pi_1 \sim \pi_2$ whenever π_1 and π_2 have the same diagonal lengths. Then Proposition 2 indicates that there is exactly one partition of n into distinct parts within each equivalence class, hence there are $p_d(n)$ classes in total. In addition, the characteristic of a class is well-defined. (*N.B.* The fact that $\chi(\pi)$ is given by an alternating sum explains our choice of terminology for this quantity.)

4 Describing the bijection

For the remainder of our discussion, we will focus solely on balanced partitions, with the goal of proving that the number of balanced partitions of $2n$ into distinct parts is equal to the number of partitions of n. Consider the Young tableau for any balanced partition of $2n$ having diagonal lengths d_1 to d_{2n}, shaded as in Fig. 1. We perform the following algorithm, which has the effect of pairing light and dark squares in the process of creating a partition of n.

(a) Keep the dark square from the first diagonal in its original position.
(b) For each subsequent diagonal, use as many squares as necessary from that diagonal to cover all unpaired squares (if any) left over from the previous step.
(c) Place the remaining squares in the rectangular block reserved for that particular diagonal as indicated in Fig. 2, starting at the left-hand (or bottom) edge and filling in to the right (or up).

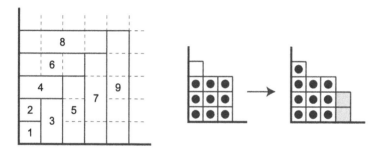

Fig. 2 Blocks used for creating a partition of n, numbered according to the diagonal they serve

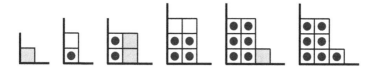

Fig. 3 Applying the bijective algorithm to a partition with diagonal lengths 1, 2, 3, 4, 3, and 1

(d) Repeat steps (b) and (c) until all diagonals have been incorporated and every square has been paired with a square of the opposite color.

One step of this algorithm is illustrated in Fig. 2 for a certain partition with $d_7 = 3$. Squares along the seventh diagonal are dark; we use one of them to cover the currently unpaired white square in the fourth row, then begin filling the block in the fourth column with the remaining two dark squares. (The solid circles represent squares that have already been paired.) The entire process is illustrated in Fig. 3 for the partitions in Fig. 1, ultimately yielding the partition 3–2–2.

Before continuing, the reader is encouraged to perform this algorithm for each of the balanced partitions of 10 into distinct parts. (They are 10, 9–1, 8–2, 7–3, 6–4, 6–3–1, and 5–3–2.) This can easily be done with five red and five black cards from a standard deck to represent the five light and five dark squares. It is quite marvelous to see the Young tableaux for each partition of 5 appear in turn. The intuition gained from such an exercise will also greatly clarify the subsequent arguments.

Proposition 3 *Let π be a balanced partition of $2n$. Then the algorithm described above yields a valid Young tableau for a partition of n.*

Proof Suppose π has diagonal lengths $d_1 = 1$, $d_2 = 2$, ..., $d_m = m$ followed by lengths satisfying $d_m \geq d_{m+1} \geq \cdots \geq d_{2n}$. (Here m is a constant depending on π.) The first m steps of the algorithm proceed in an orderly fashion: the squares within each diagonal cover all unpaired squares from the previous step and then exactly fill out their designated block shown in Fig. 2. During this stage the Durfee square of the new partition is filled out.

We will show that from this point on each diagonal is long enough to cover all the unpaired squares left over from the previous step, but not so long as to subsequently fill the allotted space in the block reserved for that diagonal. Assume, for the sake of argument, that we are handling diagonal $2k$, which has light squares, for some $2k > m$. Then there should be at least as many light as dark squares in the first $2k$ diagonals, but the excess should be less than k. In other words, we must have

$$0 \le (d_{2k} + d_{2k-2} + \cdots + d_2) - (d_{2k-1} + d_{2k-3} + \cdots + d_1) < k.$$

But $d_{2j} - d_{2j-1} = 1$ when $2 \le 2j \le m$ and $d_{2j} - d_{2j-1} \le 0$ for $2j > m$, which establishes the right-hand inequality. Furthermore, using the fact that $\chi(\pi) = 0$ we may rewrite the left-hand inequality as $(d_{2n} + \cdots + d_{2k+2}) - (d_{2n-1} + \cdots + d_{2k+1}) \le 0$, which follows immediately from the fact that $d_{2j} - d_{2j-1} \le 0$ whenever $2j > m$.

For diagonal $2k + 1 > m$ consisting of dark squares, we must instead show that

$$0 \le (d_{2k+1} + d_{2k-1} + \cdots + d_1) - (d_{2k} + d_{2k-2} + \cdots + d_2) < k + 1.$$

Pairing terms and using $d_1 = 1$ gives the right-hand inequality in the same manner as above. We may use $\chi(\pi) = 0$ to rewrite the left-hand inequality as before, then pair up terms and note that $d_{2n} \ge 0$ to finish the odd case.

To complete the proof we must show that the portions of the Young tableau to the right of and above the Durfee square form a nonincreasing sequence of columns and rows. The height of the $(k+1)$st column is equal to the excess of dark squares in the first $2k + 1$ diagonals, thus is given by

$$(d_{2k+1} + d_{2k-1} + \cdots + d_1) - (d_{2k} + d_{2k-2} + \cdots + d_2).$$

Hence the difference in height between the kth and $(k+1)$st columns is $d_{2k} - d_{2k+1}$, which is nonnegative since $2k + 1 > m$, i.e., we are to the right of the Durfee square. The same reasoning shows that the rows above the Durfee square are also nonincreasing, thus completing the proof.   □

5 Proof of the bijection

We now show that the construction just described is, in fact, a bijection, which will prove our main result.

Theorem 1 *For each nonnegative integer n, the number of balanced partitions of $2n$ into distinct parts is equal to the number of partitions of n.*

Proof When $n = 0$ there is one partition of each type, so assume that $n \ge 1$. There are $p_d(2n)$ equivalence classes of partitions of $2n$ when they are grouped according to diagonal lengths, since there is exactly one partition into distinct parts within each class, by Proposition 2. The above construction maps each class to a partition of n, so we must establish that this map is injective and surjective to prove the theorem.

Suppose that partitions π_1 and π_2 belong to distinct classes, and let their diagonal lengths differ for the first time at diagonal k. Then, clearly, the construction will result in a different number of filled squares appearing in block k in Fig. 2. Since no later

Fig. 4 Proving that the
construction is surjective in the
case of a 6–5–4–2–2 partition

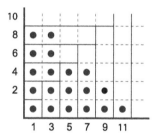

step in the construction affects the number of squares in that block, the resulting partitions of n will be distinct, hence the map is injective.

We now show that every partition of n arises via our construction. Given a Ferrer's diagram for a partition of n, overlay the blocks used in the algorithm, as illustrated in Fig. 4 for the partition 6–5–4–2–2. Set $d_1 = 1$, and for each $1 < j \le 2n$ define d_j to be the total number of dots contained within blocks j and $j - 1$. We claim that these values constitute a valid set of diagonal lengths for a partition of $2n$. For instance, the nonzero values of the d_j for the partition illustrated in Fig. 4 are d_1 through d_{12}, equal to 1, 2, 3, 4, 5, 5, 5, 5, 4, 2, 1, 1. These correspond to the balanced partition 12–9–7–6–4 of 38.

Note that all n dots are contained within the $n \times (n + 1)$ rectangle consisting of blocks 1 to $2n$. Also, every dot is counted exactly twice when assigning values for the diagonal lengths (since no dots reach block $2n$), so $d_1 + d_2 + \cdots + d_{2n} = 2n$. Next let m be the smallest value for which block m is completely filled but block $m + 1$ is not. Then, clearly, we have $d_1 = 1, d_2 = 2, \ldots, d_m = m$. But as soon as the dots fail to completely fill one of the blocks—as is the case for block 6 in Fig. 4—the dots in all subsequent blocks may not extend beyond those in the previous adjacent block, since this would imply that some row or column of the Ferrer's diagram had a gap. Note also that block $m + 1$ has no more dots than block $m - 1$, since the latter block is filled but the former is not. By definition, $d_k - d_{k+1}$ is equal to the difference between the number of dots in (adjacent) blocks $k - 1$ and $k + 1$, which we have just argued is zero or positive when $k \ge m$. Therefore, we conclude that $d_m \ge d_{m+1} \ge \cdots \ge d_{2n}$.

In summary, we have shown that the values for d_1 through d_{2n} represent the diagonal lengths for some partition of $2n$. Hence the map defined by our construction is surjective, and therefore bijective, which completes the proof. $\square$

Corollary 2 *The number of ordered $2n$-tuples $(d_1, d_2, \ldots, d_{2n})$ of nonnegative integers satisfying the following conditions is equal to $p(n)$.*

(i) $d_1 + d_2 + \cdots + d_{2n} = 2n$,
(ii) $d_1 + d_3 + \cdots + d_{2n-1} = d_2 + d_4 + \cdots + d_{2n}$,
(iii) $d_1 = 1, d_2 = 2, \ldots, d_m = m$ *for some* $1 \le m \le 2n$,
(iv) $d_m \ge d_{m+1} \ge \cdots \ge d_{2n}$.

Proof Such ordered $2n$-tuples comprise all the possible diagonal lengths for balanced partitions of $2n$, which are in one-to-one correspondence with partitions of n, by Theorem 5. $\square$

Remark 2 We made a choice to orient even-numbered blocks horizontally and odd-numbered blocks vertically when setting up our algorithm. However, the mechanics of the proofs would have proceeded just as smoothly if Fig. 2 were reflected over the line $y = x$. In this case, the algorithm would produce a partition conjugate to the one it currently produces, as one might imagine. We chose the placement of blocks described above because this version of the algorithm has the additional property that it converts a balanced partition of $2n$ with all even parts to the partition of n having parts half as large, as the reader may confirm.

Acknowledgements The author would like to thank the organizers of the Combinatory Analysis 2008 conference for their invitation to give a talk on the material presented in this paper. The author is also grateful to the referee for noticing a careless mistake and making several improvements to the paper.

References

1. Alder, H.L.: Partition identities—from Euler to the present. Am. Math. Mon. **76**, 733–746 (1969)
2. Andrews, G.E.: On generalizations of Euler's partition theorem. Mich. Math. J. **13**, 491–498 (1966)
3. Andrews, G.E.: An algebraic identity of F. H. Jackson and its implications for partitions. Discrete Math. **232**, 77–83 (2001)
4. Bousquet-Mélou, M., Eriksson, K.: Lecture hall partitions. Ramanujan J. **1**, 101–111 (1997)
5. Hirschhorn, M.D.: Sylvester's partition theorem, and a related result. Mich. Math. J. **21**, 133–136 (1974)
6. Robbins, C.: On two-color partitions. Util. Math. **65**, 193–200 (2004)

Ramanujan J (2010) 23: 307–313
DOI 10.1007/s11139-009-9189-6

Partitions with rounded occurrences and attached parts

Jeremy Lovejoy

Dedicated to George Andrews, with admiration, on the occasion of his 70th birthday

Received: 3 June 2009 / Accepted: 30 June 2009 / Published online: 16 April 2010
© Springer Science+Business Media, LLC 2010

Abstract We introduce the number of (k, i)-rounded occurrences of a part in a partition and use q-difference equations to interpret a certain q-series $S_{k,i}(a; x; q)$ as the generating function for partitions with bounded (k, i)-rounded occurrences and attached parts. When $a = 0$ these partitions are the same as those studied by Bressoud in his extension of the Rogers-Ramanujan-Gordon identities to even moduli. When $a = 1/q$ we obtain a new family of partition identities.

Keywords Partitions · q-difference equations · Overpartitions · Bressoud's theorem · Rogers-Ramanujan identities

Mathematics Subject Classification (2000) 11P81 · 05A17

1 Introduction and statement of results

In 1968, greatly generalizing work of Rogers [25] and Selberg [26], Andrews defined a family of basic hypergeometric series $J_{\lambda,k,i}(a_1, a_2, \ldots, a_\lambda; x; q)$ and established q-difference equations involving them [5]. This work became one of the foundations of modern partition theory. Andrews had already seen how to use some of these q-difference equations to prove families of partition identities [1–4], including Gordon's combinatorial generalization of the Rogers-Ramanujan identities, and over the next decade many further partition identities [6, 8, 9, 14, 16] would be deduced from the $J_{\lambda,k,i}(a_1, a_2, \ldots, a_\lambda; x; q)$ and their q-difference equations.

With the focus on analytic identities, motivated in large part by the burgeoning applications in statistical mechanics [10] and the advent of the powerful Bailey pair method [11], the 80's and 90's saw the study of q-difference equations fall out of

J. Lovejoy (✉)
CNRS, LIAFA, Université Denis Diderot - Paris 7, Case 7014, 75205 Paris Cedex 13, France
e-mail: lovejoy@liafa.jussieu.fr

favor. Over the last decade or so, however, a string of papers have shown that there is still much to be discovered in Andrews' $J_{\lambda,k,i}(a_1, a_2, \ldots, a_\lambda; x; q)$ [12, 13, 18, 20–24]. The present work is yet another contribution to this list.

We study the series $S_{k,i}(a; x; q)$, defined for $k \geq 2$ and $1 \leq i \leq k$ using the usual q-series notation [19] by

$$S_{k,i}(a; x; q) := \frac{1}{(xq)_\infty} \sum_{n \geq 0} \frac{a^n x^{(k-1)n} q^{(k-1)n^2+(k-i+1)n}(x^2 q^2, 1/a; q^2)_n}{(q^2, ax^2 q^2; q^2)_n}$$

$$\times \left(1 + \frac{ax^i q^{(2n+1)i-2n}(1 - q^{2n}/a)}{(1 - ax^2 q^{2n+2})}\right). \tag{1.1}$$

In terms of Andrews' series, we have

$$S_{k,i}(a; x; q) := \frac{(-xq)_\infty}{(ax^2 q^2; q^2)_\infty} J_{1, \frac{k-1}{2}, \frac{i}{2}}(1/a; x^2; q^2). \tag{1.2}$$

We will describe the coefficient of $a^t x^m q^n$ of $S_{k,i}(a; x; q)$ in terms of partition pairs, using the number of (k, i)-*rounded occurrences* of a part j in a partition λ.

Definition 1.1 Denote by $f_j(\lambda)$ the number of occurrences of j in λ. The number of (k, i)-rounded occurrences of a part j in a partition λ, denoted $f_j^{(k,i)}(\lambda)$, is defined using the usual characteristic function by

$$f_{2j}^{(k,i)}(\lambda) := f_{2j}(\lambda) + \chi(f_{2j} \not\equiv (k - i) \,(\text{mod}\,2)), \tag{1.3}$$

$$f_{2j+1}^{(k,i)}(\lambda) := f_{2j+1}(\lambda) + \chi(f_{2j+1} \not\equiv (i - 1) \,(\text{mod}\,2)). \tag{1.4}$$

Notice that with this definition we require nothing about the parity of the number of occurrences of a part, only that the number of occurrences be *regarded* as having a certain parity by rounding up, if necessary. To illustrate the definition, consider the partition $\lambda = (6, 6, 6, 4, 4, 3, 3, 3, 3, 1)$. Then we have, for example, $f_1^{(4,3)} = 2$, $f_2^{(4,2)} = 0$, $f_3^{(5,2)} = 5$, $f_4^{(4,4)} = 2$, $f_5^{(4,4)} = 1$, and $f_6^{(5,3)} = 4$.

We now define the partitions pairs of interest.

Definition 1.2 Let $s_{k,i}(n)$ denote the number of partition pairs (λ, μ) of n such that:

(i) $f_1^{(k,i)}(\lambda) \leq i - 1$,

(ii) if $f_1^{(k,i)}(\lambda) = i - 1$ then 1 may occur an even number of times in μ,

(iii) if $i = 1$ then 1 may occur unrestricted in μ,

(iv) for each $j \geq 1$ we have $f_j^{(k,i)}(\lambda) + f_{j+1}^{(k,i)}(\lambda) \leq k - 1$,

(v) for each $j \geq 1$, if $f_j^{(k,i)}(\lambda) + f_{j+1}^{(k,i)}(\lambda) = k - 1$, then $j + 1$ may occur an even number of times in μ,

(vi) for each $j \geq 1$, if $f_j^{(k,i)}(\lambda) = k - 1$, then $j + 1$ may occur unrestricted in μ.

We are now ready to state the main theorem.

Theorem 1.3 *Let $s_{k,i}(t, m, n)$ denote the number of partition pairs counted by $s_{k,i}(n)$ such that $m = \sum_j (f_j(\lambda) + f_j(\mu))$ and $t = \sum_j \lceil \frac{f_j(\mu)}{2} \rceil$. Then*

$$\sum_{t,m,n\geq 0} s_{k,i}(t, m, n) a^t x^m q^n = S_{k,i}(a; x; q). \qquad (1.5)$$

Theorem 1.3 shows that special cases of the functions $S_{k,i}(a; x; q)$ are generating functions for some well-known partitions. For example, a few moments' consideration (or an appeal to (1.1)) reveals that ordinary partitions are generated by $S_{k,i}(1; x; q)$ (for any k and i). It is also not hard to see that partitions into distinct parts are generated by $S_{3,2}(0; 1; q)$. More generally, the partitions generated by $S_{k,i}(0; x; q)$ may be identified with those studied by Bressoud in his extension to even moduli of Gordon's generalization of the Rogers-Ramanujan identities (i.e., the partitions counted by $b_{k,i}(m, n)$ in [14]). Setting $a = 0$ and $x = 1$ in (1.1) and appealing to the triple product identity [19, p. 239, Eq. (II.28)],

$$\sum_{n\in\mathbb{Z}} z^n q^{n^2} = (-q/z, -zq, q^2; q^2)_\infty, \qquad (1.6)$$

we then recover Bressoud's result [14, Theorem, $j = 0$] in the following form:

Corollary 1.4 (Bressoud [14]) *For $k \geq 2$ and $1 \leq i < k$, let $B_{k,i}(n)$ denote the number of partitions λ of n such that:*

(i) $f_1^{(k,i)}(\lambda) \leq i - 1$,

(ii) *for each $j \geq 1$ we have $f_j^{(k,i)}(\lambda) + f_{j+1}^{(k,i)}(\lambda) \leq k - 1$.*

Let $A_{k,i}(n)$ denote the number of partitions of n into parts not congruent to 0 or $\pm i$ modulo $2k$. Then $A_{k,i}(n) = B_{k,i}(n)$.

Another interesting consequence of Theorem 1.3 arises when we set $a = 1/q$. It is convenient to state this result in terms of overpartitions, which are partitions in which the first occurrence of a part may be overlined.

Corollary 1.5 *For $k \geq 2$ and $1 < i < k$, let $\mathcal{B}_{k,i}(n)$ denote the number of overpartition pairs (λ, μ) of n such that:*

(i) *λ is an ordinary partition counted by $B_{k,i}(n)$ (see Corollary 1.4),*

(ii) *if $f_1^{(k,i)}(\lambda) = i - 1$, then 1 may occur (non-overlined and unrestricted) in μ,*

(iii) *for $j \geq 1$, if $f_j^{(k,i)}(\lambda) + f_{j+1}^{(k,i)}(\lambda) = k - 1$, then $2j + 1$ may occur (non-overlined and unrestricted) in μ,*

(iv) *for $j \geq 1$ if $f_j^{(k,i)}(\lambda) = k - 1$ then $\bar{j}$ may appear in μ.*

Let $\mathcal{A}_{k,i}(n)$ denote the number of overpartitions of n where non-overlined parts are not divisible by $2k - 2$ and overlined parts are $\pm(i-1) \pmod{2k-2}$. Then $\mathcal{A}_{k,i}(n) = \mathcal{B}_{k,i}(n)$.

Despite the requirement that $1 < i < k$ above, there is still an identity when $i = 1$ or k. Indeed, the proof of Corollary 1.5 presented in Sect. 2 applies equally well when $i = 1$ or k. The definition of $\mathcal{B}_{k,i}(n)$ is still valid (with a suitable modification for $i = 1$ arising from condition (iii) in Definition 1.2), and the generating functions for $\mathcal{A}_{k,1}(n)$ and $\mathcal{A}_{k,k}(n)$ are

$$\mathcal{A}_{k,1}(n) = \frac{2(-q^{2k-2}; q^{2k-2})_\infty^2 (q^{2k-2}; q^{2k-2})_\infty}{(q)_\infty}$$

and

$$\mathcal{A}_{k,k}(n) = \frac{(-q^{k-1}; q^{2k-2})_\infty^2 (q^{2k-2}; q^{2k-2})_\infty}{(q)_\infty}.$$

We let the reader interpret these products as he pleases.

2 Proofs of Theorem 1.3 and Corollary 1.5

Using (1.2) and [5, Theorem 1] one may compute that

$$S_{k,1}(a; x; q) = \frac{(1 + axq)}{(1 - ax^2q^2)} S_{k,k}(a; xq; q), \tag{2.1}$$

$$S_{k,2}(a; x; q) = \frac{(1 + xq)}{(1 - ax^2q^2)} S_{k,k-1}(a; xq; q), \tag{2.2}$$

and for $3 \le i \le k$,

$$S_{k,i}(a; x; q) - S_{k,i-2}(a; x; q) = \frac{(xq)^{i-2}(1 + xq)}{(1 - ax^2q^2)} S_{k,k-i+1}(a; xq; q)$$

$$- \frac{a(xq)^{i-2}(1 + xq)}{(1 - ax^2q^2)} S_{k,k-i+3}(a; xq; q). \tag{2.3}$$

The final q-difference equation is not terribly useful combinatorially. However, there is another q-difference equation which may be easily deduced from (2.1), (2.2), and (2.3) using induction ((2.3) providing the induction step). This method of eliminating the minus sign is inspired by [13].

Lemma 2.1 *If $i \ge 2$ is even then*

$$S_{k,i}(a; x; q) = \frac{(xq)^{i-2}(1 + xq)}{(1 - ax^2q^2)} S_{k,k-i+1}(a; xq; q)$$

$$+ \sum_{v=1}^{(i-2)/2} (xq)^{2v-2}(1 + xq) S_{k,k-2v+1}(a; xq; q), \tag{2.4}$$

and if $i \geq 3$ is odd then

$$S_{k,i}(a; x; q) = S_{k,k}(a; xq; q) + \frac{(xq)^{i-2}(1+xq)}{(1-ax^2q^2)} S_{k,k-i+1}(a; xq; q)$$

$$+ \sum_{v=1}^{(i-3)/2} (xq)^{2v-1}(1+xq) S_{k,k-2v}(a; xq; q). \tag{2.5}$$

Proof of Theorem 1.3 Notice that together with the initial condition $S_{k,i}(a; 0; q) = 1$, the q-difference equations (2.1), (2.4), and (2.5) uniquely define the functions $S_{k,i}(a; x; q)$. To prove Theorem 1.3 then, we define

$$\widehat{S}_{k,i}(a; x; q) := \sum_{t,m,n \geq 0} s_{k,i}(t, m, n) a^t x^m q^n$$

and show that the $\widehat{S}_{k,i}(a; x; q)$ satisfy the same defining conditions. That $\widehat{S}_{k,i}(a; 0; q) = 1$ follows from the fact that the only partition without any parts whatsoever is the empty partition of 0.

We now turn to (2.1). Let (λ, μ) be a partition pair counted by $\widehat{S}_{k,1}(a; x; q)$. By definition, we have $f_1(\lambda) = 0$, $f_2^{(k,1)}(\lambda) \leq k - 1$, and $f_1(\mu)$ is unrestricted. Removing the 1's and subtracting one from each part ≥ 2, we see that

$$\widehat{S}_{k,1}(a; x; q) = \frac{(1+axq)}{(1-ax^2q^2)} \widehat{S}_{k,k}(a; xq; q).$$

(Notice that for $(k, i) = (k, 1)$ and (k, k), the residue classes modulo 2 of $(k - i)$ and $(i - 1)$ are interchanged, so that subtracting one from each part is consistent with the definition of the number of (k, i)-rounded occurrences in Definition 1.1 and the conditions on the $s_{k,i}(t, m, n)$ in Theorem 1.3. This will be the case throughout the proof, though we shall not mention it again.)

Next we treat (2.4). Suppose that (λ, μ) is a partition pair counted by $\widehat{S}_{k,i}(a; x; q)$, where $i \geq 2$ is even. We have $0 \leq f_1(\lambda) \leq i - 1$. For each v with $1 \leq v \leq i/2$, if $f_1(\lambda) = 2v - 1$ or $2v - 2$ then $f_1^{(k,i)}(\lambda) = 2v - 1$. In the case $v = i/2$, we have $f_2^{(k,i)}(\lambda) \leq k - i$ and $f_1(\mu)$ is even. Removing the 1's and subtracting one from each remaining part we see that these pairs are generated by $((xq)^{i-2} + (xq)^{i-1})/(1 - ax^2q^2)\widehat{S}_{k,k-i+1}(a; xq; q)$. Now for $1 \leq v \leq (i - 2)/2$, we have $f_1(\mu) = 0$ and $f_2^{(k,i)}(\lambda) \leq k - 2v$. Again removing the 1's and subtracting one from each part, these pairs are generated by $(xq)^{2v-2}(1 + xq)\widehat{S}_{k,k-2v+1}(a; xq; q)$. This gives (2.4).

To prove (2.5), suppose that (λ, μ) is a partition pair counted by $\widehat{S}_{k,i}(a; x; q)$, where $i \geq 3$ is odd. For each v with $1 \leq v \leq (i - 1)/2$, if $f_1(\lambda) = 2v$ or $2v - 1$, then $f_1^{(k,i)}(\lambda) = 2v$. The argument now proceeds as above, except that we have left out the case $f_1(\lambda) = 0$ because i is odd. This accounts for the extra term $\widehat{S}_{k,k}(a; xq; q)$. This concludes the proof of Theorem 1.3. $\qquad\square$

We now turn to Corollary 1.5. First, setting $a = 1/q$ and $x = 1$ on the right-hand side of (1.1), we have

$$S_{k,i}(1/q; 1; q) = \frac{1}{(q)_\infty} \sum_{n \geq 0} q^{(k-1)n^2 + (k-i)n}(1 + q^{(2n+1)(i-1)})$$

$$= \frac{1}{(q)_\infty} \sum_{n \in \mathbb{Z}} q^{(k-1)n^2 + (k-i)n}$$

$$= \frac{(-q^{i-1}, -q^{2k-i+1}, q^{2k-2}; q^{2k-2})_\infty}{(q)_\infty}$$

$$= \sum_{n \geq 0} \mathcal{A}_{k,i}(n)q^n,$$

where the penultimate line follows from the Jacobi triple product identity (1.6). On the other hand, if we let $a = 1/q$ and $x = 1$ in Theorem 1.3 and consider the effect on partition pairs counted by $s_{k,i}(t, m, n)$, then parts $j + 1$ occurring an even number of times in μ may be regarded as repeatable parts of the form $2j + 1$, while the eventual leftover occurrence of $j + 1$ becomes $\overline{j}$. This gives the pairs counted by $\mathcal{B}_{k,i}(n)$ and completes the proof of Corollary 1.5.   □

3 Conclusion

In addition to Andrews' $J_{\lambda,k,i}(a_1, a_2, \ldots, a_\lambda; x; q)$, there are several other families of q-series whose q-difference equations are worth exploring. We indicate three of these here. First, Andrews has developed q-difference equations for some series $K_{\lambda,k,i}(a_1, a_2, \ldots, a_\lambda; x; q)$ [7, Section 3] which may be regarded as bilateral series analogues of the $J_{\lambda,k,i}(a_1, a_2, \ldots, a_\lambda; x; q)$. Second, Bressoud's $F_{\lambda,k,i}(c_1, c_2, a_1, a_2, \ldots, a_\lambda; x; q)$ [15] reduce to Andrews' $J_{\lambda,k,i}(a_1, a_2, \ldots, a_\lambda; x; q)$ when $c_1, c_2 \to \infty$ and $x = xq$. When $c_1 \to \infty$ and $c_2 = -q$, q-difference equations and their combinatorial implications have been worked out for $\lambda = 1$ in [17] and for $\lambda = 2$ in [24]. Surely many more instances of Bressoud's series satisfy meaningful q-difference equations. Finally, there are nice q-difference equations for a family of series containing both $F_{1,k,i}(-q, \infty, a_1; xq; q)$ and $F_{1,k,i}(\infty, \infty, a_1; xq; q)$ presented in [17, Section 6].

Acknowledgement   The author is indebted to Ae Ja Yee for her reading of an earlier version of this paper.

References

1. Andrews, G.E.: An analytic proof of the Rogers-Ramanujan-Gordon identities. Am. J. Math. **88**, 844–846 (1966)
2. Andrews, G.E.: Some new partition theorems. J. Combin. Theory **2**, 431–436 (1967)
3. Andrews, G.E.: Partition theorems related to the Rogers-Ramanujan identities. J. Combin. Theory **2**, 422–430 (1967)

4. Andrews, G.E.: A generalization of the Göllnitz-Gordon partition identities. Proc. Am. Math. Soc. **8**, 945–952 (1967)
5. Andrews, G.E.: On q-difference equations for certain well-poised basic hypergeometric series. Quart. J. Math. **19**, 433–447 (1968)
6. Andrews, G.E.: A generalization of the classical partition theorems. Trans. Am. Math. Soc. **145**, 205–221 (1968)
7. Andrews, G.E.: Applications of basic hypergeometric series. SIAM Rev. **16**, 441–484 (1974)
8. Andrews, G.E.: On the General Rogers-Ramanujan Theorem. Mem. Am. Math. Soc., vol. 152. Am. Math. Soc., Providence (1974). 86 pp.
9. Andrews, G.E.: On the Alder polynomials and a new generalization of the Rogers-Ramanujan identities. Trans. Am. Math. Soc. **204**, 40–64 (1975)
10. Andrews, G.E.: The hard-hexagon model and the Rogers-Ramanujan type identities. Proc. Nat. Acad. Sci. **78**, 5290–5292 (1981)
11. Andrews, G.E.: Multiple series Rogers-Ramanujan type identities. Pacific J. Math. **114**, 267–283 (1984)
12. Andrews, G.E.: Parity in partition identities, Ramanujan J., to appear
13. Andrews, G.E., Santos, J.P.O.: Rogers-Ramanujan type identities for partitions with attached odd parts. Ramanujan J. **1**, 91–99 (1997)
14. Bressoud, D.M.: A generalization of the Rogers-Ramanujan identities for all moduli. J. Combin. Theory **27**, 64–68 (1979)
15. Bressoud, D.M.: Analytic and combinatorial generalizations of the Rogers-Ramanujan identities. Mem. Am. Math. Soc. **227**, 54 (1980)
16. Connor, W.G.: Partition theorems related to some identities of Rogers and Watson. Trans. Am. Math. Soc. **214**, 95–111 (1975)
17. Corteel, S., Lovejoy, J., Mallet, O.: An extension to overpartitions of the Rogers-Ramanujan identities for even moduli. J. Number Theory **128**, 1602–1621 (2008)
18. Corteel, S., Mallet, O.: Overpartitions, lattice paths, and the Rogers-Ramanujan identities. J. Combin. Theory Ser. A **114**, 1407–1437 (2007)
19. Gasper, G., Rahman, M.: Basic Hypergeometric Series. Cambridge University Press, Cambridge (1990)
20. Lovejoy, J.: Gordon's theorem for overpartitions. J. Combin. Theory Ser. A **103**, 393–401 (2003)
21. Lovejoy, J.: Overpartition theorems of the Rogers-Ramanujan type. J. London Math. Soc. **69**, 562–574 (2004)
22. Lovejoy, J.: Overpartition pairs. Ann. Inst. Fourier (Grenoble) **56**, 781–794 (2006)
23. Lovejoy, J.: Partitions and overpartitions with attached parts. Arch. Math. (Basel) **88**, 316–322 (2007)
24. Lovejoy, J., Mallet, O.: Overpartition pairs and two classes of basic hypergeometric series. Adv. Math. **217**, 386–418 (2008)
25. Rogers, L.J.: Proof of certain identities in combinatory analysis. Proc. Cambridge Philos. Soc. **19**, 211–214 (1919)
26. Selberg, A.: Über einige arithmetische Identitäten. Avhl. Norske Vid. **8**, 23 (1936)

Ramanujan J (2010) 23: 315–333
DOI 10.1007/s11139-010-9282-x

Shifted versions of the Bailey and Well-Poised Bailey lemmas

Frédéric Jouhet

Dedicated to George Andrews for his 70th birthday

Received: 9 June 2009 / Accepted: 26 October 2010 / Published online: 12 November 2010
© Springer Science+Business Media, LLC 2010

Abstract The Bailey lemma is a famous tool to prove Rogers–Ramanujan type identities. We use shifted versions of the Bailey lemma to derive m-versions of multisum Rogers–Ramanujan type identities. We also apply this method to the Well-Poised Bailey lemma and obtain a new extension of the Rogers–Ramanujan identities.

Keywords Bailey lemma · WP-Bailey lemma · q-Series · Rogers–Ramanujan identities

Mathematics Subject Classification (2000) 33D15

1 Introduction

The Rogers–Ramanujan identities

$$\sum_{k=0}^{\infty} \frac{q^{k^2}}{(1-q)\cdots(1-q^k)} = \prod_{n\geq 0} \frac{1}{(1-q^{5n+1})(1-q^{5n+4})}, \qquad (1.1)$$

$$\sum_{k=0}^{\infty} \frac{q^{k^2+k}}{(1-q)\cdots(1-q^k)} = \prod_{n\geq 0} \frac{1}{(1-q^{5n+2})(1-q^{5n+3})} \qquad (1.2)$$

are among the most famous q-series identities in partition theory and combinatorics. Since their discovery, they have been proved and generalized in various ways (see [4, 9, 15] and the references cited there). A classical approach to get this kind of identities

F. Jouhet (✉)
UMR 5208 Institut Camille Jordan, Université de Lyon, Université Lyon I, CNRS, Bâtiment du
Doyen Jean Braconnier, 43, bd du 11 Novembre 1918, 69622 Villeurbanne Cedex, France
e-mail: jouhet@math.univ-lyon1.fr
url: http://math.univ-lyon1.fr/~jouhet

is the Bailey lemma, originally proved by Bailey [8] and later strongly highlighted by Andrews [3–5]. The goal of this paper is to use bilateral extensions of this tool to derive new generalizations of (1.1) and (1.2) as well as other famous identities of the same kind.

First, recall some standard notations for q-series which can be found in [16]. Let q be a fixed complex parameter (the "base") with $0 < |q| < 1$. The q-shifted factorial is defined for any complex parameter a by

$$(a)_\infty \equiv (a; q)_\infty := \prod_{j \geq 0}(1 - aq^j) \quad \text{and} \quad (a)_k \equiv (a; q)_k := \frac{(a; q)_\infty}{(aq^k; q)_\infty},$$

where k is any integer. Since the same base q is used throughout this paper, it may be readily omitted (in notation, writing $(a)_k$ instead of $(a; q)_k$, etc.) which will not lead to any confusion. For brevity, write

$$(a_1, \ldots, a_m)_k := (a_1)_k \cdots (a_m)_k,$$

where k is an integer or infinity. The q-binomial coefficient is defined as follows:

$$\begin{bmatrix} n \\ k \end{bmatrix}_q := \frac{(q)_n}{(q)_k(q)_{n-k}},$$

and we assume that $\begin{bmatrix} n \\ k \end{bmatrix}_q = 0$ if $k < 0$ or $k > n$. Further, recall the basic hypergeometric series

$$_s\phi_{s-1}\begin{bmatrix} a_1, \ldots, a_s \\ b_1, \ldots, b_{s-1} \end{bmatrix} := \sum_{k=0}^{\infty} \frac{(a_1, \ldots, a_s)_k}{(q, b_1, \ldots, b_{s-1})_k} z^k,$$

and the bilateral basic hypergeometric series

$$_s\psi_s\begin{bmatrix} a_1, \ldots, a_s \\ b_1, \ldots, b_s \end{bmatrix} := \sum_{k=-\infty}^{\infty} \frac{(a_1, \ldots, a_s)_k}{(b_1, \ldots, b_s)_k} z^k.$$

The set of nonnegative (resp., positive) integers will be denoted by $\mathbb{N}$ (resp., $\mathbb{N}^*$). We will use throughout this paper the following results on q-series, which are the finite q-binomial [16, Appendix, (II.4)], q-Pfaff–Saalschütz [16, Appendix, (II.12)] and Jacobi triple product [16, Appendix, (II.28)] identities, respectively:

$$_1\phi_0\begin{bmatrix} q^{-n} \\ - \end{bmatrix} = (zq^{-n})_n \quad \text{for } n \in \mathbb{N}, \tag{1.3}$$

$$_3\phi_2\begin{bmatrix} a, b, q^{-n} \\ c, abq^{1-n}/c \end{bmatrix} = \frac{(c/a, c/b)_n}{(c, c/ab)_n} \quad \text{for } n \in \mathbb{N}, \tag{1.4}$$

$$\sum_{n \in \mathbb{Z}} (-1)^n z^n q^{\binom{n}{2}} = (q, z, q/z)_\infty. \tag{1.5}$$

Recall [5] that a Bailey pair $(\alpha_n(a,q), \beta_n(a,q))$ related to a and q is defined by the relation:

$$\beta_n(a,q) = \sum_{r=0}^{n} \frac{\alpha_r(a,q)}{(q)_{n-r}(aq)_{n+r}} \qquad \forall n \in \mathbb{N}. \qquad (1.6)$$

The Bailey lemma describes how, from a Bailey pair, one can produce infinitely many of them:

Theorem 1.1 (Bailey lemma) *If $(\alpha_n(a,q), \beta_n(a,q))$ is a Bailey pair related to a and q, then so is $(\alpha'_n(a,q), \beta'_n(a,q))$, where*

$$\alpha'_n(a,q) = \frac{(\rho_1,\rho_2)_n (aq/\rho_1\rho_2)^n}{(aq/\rho_1, aq/\rho_2)_n} \alpha_n(a,q)$$

and

$$\beta'_n(a,q) = \sum_{j\geq 0} \frac{(\rho_1,\rho_2)_j (aq/\rho_1\rho_2)_{n-j}(aq/\rho_1\rho_2)^j}{(q)_{n-j}(aq/\rho_1, aq/\rho_2)_n} \beta_j(a,q).$$

In [5], the following unit Bailey pair is considered:

$$\alpha_n = \frac{(-1)^n q^{n(n-1)/2}(a)_n(1-aq^{2n})}{(1-a)(q)_n}, \qquad \beta_n = \delta_{n,0}, \qquad (1.7)$$

and two iterations of Theorem 1.1 applied to (1.7) prove Watson's transformation [16, Appendix, (III.18)], which is a six parameter finite extension of (1.1) and (1.2).

Now we want to point out that in the definition of a Bailey pair (1.6), the condition that the sum on the right-hand side must vanish for $r > n$ is "natural" in the sense that $1/(q)_{n-r} = 0$ for $n - r < 0$. However, the fact that this sum starts at $r = 0$ cannot be omitted, therefore the definition of a Bailey pair would be slightly different if the sum could start from $-\infty$ up to n. As noticed in [11], one can define for all $n \in \mathbb{Z}$ a *bilateral Bailey pair* $(\alpha_n(a,q), \beta_n(a,q))$ related to a and q by the relation:

$$\beta_n(a,q) = \sum_{r\leq n} \frac{\alpha_r(a,q)}{(q)_{n-r}(aq)_{n+r}} \qquad \forall n \in \mathbb{Z}. \qquad (1.8)$$

It is, of course, possible to find bilateral Bailey pairs with general a, but it seems difficult to express $\beta_n(a,q)$ in a nice (closed) form. However, we will see in the remainder of this paper that it becomes easier in the special case $a = q^m$, $m \in \mathbb{N}$, the reason being that the sum on the right-hand side of (1.8) will run from $-m - n$ to n, and therefore will be finite. We found in this case more appropriate to call such a bilateral Bailey pair $(\alpha_n(q^m,q), \beta_n(q^m,q))$ a *shifted Bailey pair*. Note that among other things, bilateral Bailey pairs have also been considered in the survey paper [20].

In [11], the Bailey lemma is extended in the following way:

Theorem 1.2 (Bilateral Bailey lemma) *If $(\alpha_n(a, q), \beta_n(a, q))$ is a bilateral Bailey pair related to a and q, then so is $(\alpha'_n(a, q), \beta'_n(a, q))$, where*

$$\alpha'_n(a, q) = \frac{(\rho_1, \rho_2)_n (aq/\rho_1\rho_2)^n}{(aq/\rho_1, aq/\rho_2)_n} \alpha_n(a, q)$$

and

$$\beta'_n(a, q) = \sum_{j \leq n} \frac{(\rho_1, \rho_2)_j (aq/\rho_1\rho_2)_{n-j} (aq/\rho_1\rho_2)^j}{(q)_{n-j}(aq/\rho_1, aq/\rho_2)_n} \beta_j(a, q),$$

subject to convergence conditions on the sequences $\alpha_n(a, q)$ and $\beta_n(a, q)$, which make the relevant infinite series absolutely convergent.

Remark 1.3 In Theorem 1.1, no problem occurs with changing summations as the sum in (1.6) is finite. This is not true any more in the bilateral version, therefore one needs to add absolute convergence conditions to change summations before using the q-Pfaff–Saalschütz identity. Note also that these convergence conditions are not needed in the particular case $a = q^m$, $m \in \mathbb{N}$, of shifted Bailey pairs, which will be often used throughout the paper.

There is an extension of the Bailey lemma, the Well-Poised (or WP-) Bailey lemma [6], which also has a bilateral version. Indeed, define for all $n \in \mathbb{Z}$ a *WP-bilateral Bailey pair* $(\alpha_n(a, \alpha), \beta_n(a, \alpha))$ related to a and α by the relation:

$$\beta_n(a, \alpha) = \sum_{r \leq n} \frac{(\alpha/a)_{n-r}(\alpha)_{n+r}}{(q)_{n-r}(aq)_{n+r}} \alpha_r(a, \alpha) \quad \forall n \in \mathbb{Z}. \tag{1.9}$$

The results of [6] can also be extended to the bilateral case. We omit the proof, as it is exactly the same as in [6]: it requires Jackson's $_8\phi_7$ finite summation [16, Appendix, (II. 22)], and the q-Pfaff–Saalschütz identity (1.4).

Theorem 1.4 (WP-bilateral Bailey lemma) *If $(\alpha_n(a, \alpha), \beta_n(a, \alpha))$ is a WP-bilateral Bailey pair related to a and α, then so are $(\alpha'_n(a, \alpha), \beta'_n(a, \alpha))$ and $(\widetilde{\alpha}_n(a, \alpha), \widetilde{\beta}_n(a, \alpha))$, where*

$$\alpha'_n(a, \alpha) = \frac{(\rho_1, \rho_2)_n}{(aq/\rho_1, aq/\rho_2)_n} (\alpha/c)^n \alpha_n(a, c),$$

$$\beta'_n(a, \alpha) = \frac{(\alpha\rho_1/a, \alpha\rho_2/a)_n}{(aq/\rho_1, aq/\rho_2)_n} \sum_{j \leq n} \frac{(\rho_1, \rho_2)_j}{(\alpha\rho_1/a, \alpha\rho_2/a)_j}$$

$$\times \frac{1 - cq^{2j}}{1 - c} \frac{(\alpha/c)_{n-j}(\alpha)_{n+j}}{(q)_{n-j}(qc)_{n+j}} (\alpha/c)^j \beta_j(a, c),$$

with $c = \alpha\rho_1\rho_2/aq$, and

$$\widetilde{\alpha}_n(a, \alpha) = \frac{(qa^2/\alpha)_{2n}}{(\alpha)_{2n}} (\alpha^2/qa^2)^n \alpha_n(a, qa^2/\alpha),$$

$$\tilde{\beta}_n(a,\alpha) = \sum_{j \le n} \frac{(\alpha^2/qa^2)_{n-j}}{(q)_{n-j}} (\alpha^2/qa^2)^j \beta_j(a, qa^2/\alpha),$$

subject to convergence conditions on the sequences α_n and β_n, which make the relevant infinite series absolutely convergent.

Note that if $\alpha = 0$, then the first instance of the previous theorem reduces to Theorem 1.2. As before, we will often avoid convergence conditions by setting $a = q^m$, $m \in \mathbb{N}$, and such WP-bilateral Bailey pairs will be called *WP-shifted Bailey pairs*.

Remark 1.5 One can see that any shifted Bailey pair (resp., WP-shifted Bailey pair) is equivalent to a classical Bailey pair (resp., WP-Bailey pair) related to $a = 1$ or $a = q$, according to the parity of m. Thus, the concept of (WP-) shifted Bailey pairs is nothing else but an appropriate and useful way of writing some (WP-) Bailey pairs, yielding surprising (known and new) identities.

We also want to point out that Schlosser proved a very general bilateral well-poised Bailey lemma, based on a matrix inversion [21], and which is different from Theorem 1.4.

This paper is organized as follows. In Sect. 2, we give a shifted Bailey pair, which is used to prove in an elementary way m-versions of multisum Rogers–Ramanujan type identities. We will also point out some interesting special cases, including the m-versions of the Rogers–Ramanujan identities from [15]. In Sect. 3, we give some results concerning bilateral versions of the change of base in Bailey pairs from [9], yielding m-versions of other multisum Rogers–Ramanujan type identities. In Sect. 4, we first give a "unit" WP-bilateral Bailey pair which, by applying Theorem 1.4, yields a bilateral transformation generalizing both Ramanujan's $_1\psi_1$ and Bailey's $_6\psi_6$ summation formulae. We also find a WP-shifted Bailey pair, which yields a striking extension of the Rogers–Ramanujan identities, generalizing some other results of [15]. Finally, in the last section, we will give a few concluding remarks.

2 A shifted Bailey pair and applications

The following result gives a shifted Bailey pair, i.e., a bilateral Bailey pair related to $a = q^m$, $m \in \mathbb{N}$, which was already mentioned, but in another form, in [7], where the authors generalize this Bailey pair to the A_2 case.

Proposition 2.1 *For $m \in \mathbb{N}$, $(\alpha_n(q^m, q), \beta_n(q^m, q))$ is a shifted Bailey pair, where*

$$\alpha_n(q^m, q) = (-1)^n q^{\binom{n}{2}}$$

and

$$\beta_n(q^m, q) = (q)_m (-1)^n q^{\binom{n}{2}} \begin{bmatrix} m+n \\ m+2n \end{bmatrix}_q.$$

Proof We have by definition

$$\beta_n(q^m, q) = \sum_{k \leq n} \frac{(-1)^k q^{\binom{k}{2}}}{(q)_{n-k}(q^{1+m})_{n+k}}$$

so, as $1/(q^{1+m})_{n+k} = 0$ if $n + k + m < 0$, we can see that $\beta_n = 0$ unless $2n + m \geq 0$. In that case, one has

$$\beta_n(q^m, q) = \sum_{k \geq 0} \frac{(-1)^{n-k} q^{\binom{n-k}{2}}}{(q)_k (q^{1+m})_{2n-k}} = \frac{(-1)^n q^{\binom{n}{2}}}{(q^{1+m})_{2n}} \sum_{k \geq 0} \frac{(q^{-2n-m})_k}{(q)_k} \left(q^{m+n+1}\right)^k.$$

As $2n + m \geq 0$, we can apply (1.3) to the sum over k. We get

$$\beta_n(q^m, q) = (-1)^n q^{\binom{n}{2}} \frac{(q)_m}{(q)_{m+2n}} \left(q^{-n+1}\right)_{m+2n} = (-1)^n q^{\binom{n}{2}} (q)_m \frac{(q)_{m+n}}{(q)_{m+2n}(q)_{-n}},$$

which is the desired result.   □

Remark 2.2 The special cases $m = 0$ and 1 in Proposition 2.1 correspond to the unit Bailey pair (1.7) with $a = 1$ and q. These two values of the parameter a are in all classical uses of the Bailey lemma the only ones for which Jacobi triple product identity (1.5) can be used to get interesting Rogers–Ramanujan type identities. The clue in the present shifted case is that (1.5) can be used for all $a = q^m$, $m \in \mathbb{N}$.

We will need two instances of the bilateral Bailey lemma, which are given by specializing $\rho_1, \rho_2 \to \infty$ and $\rho_1 = \sqrt{aq}$, $\rho_2 \to \infty$ in Theorem 1.2, respectively:

$$\alpha'_n(a, q) = q^{n^2} a^n \alpha_n(a, q), \qquad \beta'_n(a, q) = \sum_{j \leq n} \frac{q^{j^2} a^j}{(q)_{n-j}} \beta_j(a, q), \qquad (2.1)$$

and

$$\alpha'_n(a, q) = q^{n^2/2} a^{n/2} \alpha_n(a, q), \qquad \beta'_n(a, q) = \sum_{j \leq n} \frac{q^{j^2/2} a^{j/2}}{(q)_{n-j}} \frac{(-\sqrt{aq})_j}{(-\sqrt{aq})_n} \beta_j(a, q).$$

$$(2.2)$$

Now we can state a first consequence of Proposition 2.1.

Theorem 2.3 *For all $k \in \mathbb{N}^*$ and $m \in \mathbb{N}$ we have*

$$\sum_{-\lfloor m/2 \rfloor \leq n_k \leq n_{k-1} \leq \cdots \leq n_1} \frac{q^{n_1^2 + \cdots + n_k^2 + m(n_1 + \cdots + n_k)}}{(q)_{n_1 - n_2} \cdots (q)_{n_{k-1} - n_k}} (-1)^{n_k} q^{\binom{n_k}{2}} \begin{bmatrix} m + n_k \\ m + 2n_k \end{bmatrix}_q$$

$$= \frac{(q^{2k+1}, q^{k(m+1)}, q^{k(1-m)+1}; q^{2k+1})_\infty}{(q)_\infty}, \qquad (2.3)$$

and

$$\sum_{-\lfloor m/2 \rfloor \leq n_k \leq n_{k-1} \leq \cdots \leq n_1} \frac{q^{n_1^2/2 + n_2^2 + \cdots + n_k^2 + m(n_1/2 + n_2 + \cdots + n_k)}(-q^{(m+1)/2})_{n_1}}{(q)_{n_1 - n_2} \cdots (q)_{n_{k-1} - n_k}}$$

$$\times (-1)^{n_k} q^{\binom{n_k}{2}} \begin{bmatrix} m + n_k \\ m + 2n_k \end{bmatrix}_q$$

$$= \frac{(-q^{(m+1)/2})_\infty}{(q)_\infty} \left(q^{2k}, q^{(k-1/2)(m+1)}, q^{k(1-m)+(m+1)/2}; q^{2k} \right)_\infty. \tag{2.4}$$

Proof We apply k times the instance (2.1) of the bilateral Bailey lemma with $a = q^m$ to our shifted Bailey pair $(\alpha_n(q^m, q), \beta_n(q^m, q))$ of Proposition 2.1, so we get a shifted Bailey pair $(\alpha_n^{(k)}(q^m, q), \beta_n^{(k)}(q^m, q))$, where

$$\alpha_n^{(k)}(q^m, q) = q^{kn^2 + kmn} \alpha_n(q^m, q)$$

and

$$\beta_n^{(k)}(q^m, q) = \sum_{n_k \leq n_{k-1} \leq \cdots \leq n_1 \leq n} \frac{q^{n_1^2 + \cdots + n_k^2 + m(n_1 + \cdots + n_k)}}{(q)_{n - n_1}(q)_{n_1 - n_2} \cdots (q)_{n_{k-1} - n_k}} \beta_{n_k}(q^m, q).$$

Invoking Tannery's Theorem [12] to interchange a limit and a summation, (2.3) follows by letting $n \to +\infty$ in the relation

$$\beta_n^{(k)}(q^m, q) = \sum_{j \leq n} \frac{\alpha_j^{(k)}(q^m, q)}{(q)_{n-j}(q^{1+m})_{n+j}}$$

and finally using (1.5) to factorize the right-hand side.

For (2.4), we apply $k - 1$ times the instance (2.1) of the bilateral Bailey lemma with $a = q^m$ to our shifted Bailey pair $(\alpha_n(q^m, q), \beta_n(q^m, q))$, and then once the instance (2.2), so we get a bilateral Bailey pair $(\alpha_n^{(k)}(q^m, q), \beta_n^{(k)}(q^m, q))$, where

$$\alpha_n^{(k)}(q^m, q) = q^{(k-1/2)n^2 + (k-1/2)mn} \alpha_n(q^m, q)$$

and

$$\beta_n^{(k)}(q^m, q) = \sum_{n_k \leq n_{k-1} \leq \cdots \leq n_1 \leq n} \frac{q^{n_1^2/2 + n_2^2 + \cdots + n_k^2 + m(n_1/2 + n_2 + \cdots + n_k)}}{(q)_{n - n_1}(q)_{n_1 - n_2} \cdots (q)_{n_{k-1} - n_k}}$$

$$\times \frac{(-q^{(m+1)/2})_{n_1}}{(-q^{(m+1)/2})_n} \beta_{n_k}(q^m, q).$$

The result follows as before by letting $n \to +\infty$ in the relation

$$\beta_n^{(k)} = \sum_{j \leq n} \frac{\alpha_j^{(k)}}{(q)_{n-j}(q^{1+m})_{n+j}}$$

and finally using (1.5) to factorize the right-hand side. $\qquad \square$

Remark 2.4 Identity (2.3) (resp., (2.4)) is an m-version of the Andrews–Gordon identities (resp., the generalized Göllnitz–Gordon identities) which are obtained by setting $m = 0$ and 1 in (2.3) (resp., $m = 0$ and 2 in (2.4)). However, we do not get here m-versions of the *full* Andrews–Gordon or Göllnitz–Gordon identities (see, for instance, [2, p. 111] and [9]).

In the case $k = 1$, we derive the following interesting identities.

Corollary 2.5 *For all $m \in \mathbb{N}$, we have:*

$$\sum_{j=0}^{\lfloor m/2 \rfloor} (-1)^j q^{\binom{j}{2}} \begin{bmatrix} m - j \\ j \end{bmatrix}_q = \begin{cases} (-1)^{\lfloor m/3 \rfloor} q^{m(m-1)/6} & \text{if } m \not\equiv 2 \ (\text{mod } 3), \\ 0 & \text{if } m \equiv 2 \ (\text{mod } 3), \end{cases} \tag{2.5}$$

$$\sum_{j=0}^{m} (-1)^j q^{2\binom{j}{2}} \begin{bmatrix} 2m - j \\ j \end{bmatrix}_{q^2} (-q; q^2)_{m-j} = (-1)^{\lfloor m/2 \rfloor} q^{m(3m-1)/2}, \tag{2.6}$$

$$\sum_{j=0}^{m} (-1)^j q^{\binom{j}{2}} \begin{bmatrix} 2m + 1 - j \\ j \end{bmatrix}_q (-q)_{m-j}$$
$$= \begin{cases} (-1)^{\lfloor m/2 \rfloor} q^{m(3m+2)/4} & \text{if } m \text{ even,} \\ 0 & \text{if } m \text{ odd.} \end{cases} \tag{2.7}$$

Proof In (2.3), take $k = 1$ and replace the single index of summation by $-j$ to get:

$$\sum_{j=0}^{\lfloor m/2 \rfloor} (-1)^j q^{3\binom{j}{2}-(m-2)j} \begin{bmatrix} m - j \\ j \end{bmatrix}_q = \frac{(q^3, q^{m+1}, q^{2-m}; q^3)_\infty}{(q)_\infty}.$$

Write the right-hand side as

$$\frac{(q^{m+1}, q^{2-m}; q^3)_\infty}{(q, q^2; q^3)_\infty} = \begin{cases} (-1)^{\lfloor m/3 \rfloor} q^{-m(m-1)/6} & \text{if } m \not\equiv 2 \ (\text{mod } 3), \\ 0 & \text{if } m \equiv 2 \ (\text{mod } 3), \end{cases}$$

then replace q by q^{-1} and use $\begin{bmatrix} n \\ k \end{bmatrix}_{q^{-1}} = q^{k(k-n)} \begin{bmatrix} n \\ k \end{bmatrix}_q$ to get (2.5).

Now (2.4) with $k = 1$ and q replaced by q^2 can be rewritten:

$$\sum_{j=0}^{\lfloor m/2 \rfloor} (-1)^j q^{4\binom{j}{2}-(m-3)j} \begin{bmatrix} m - j \\ j \end{bmatrix}_{q^2} \frac{(-q^{m+1}; q^2)_{-j}}{(-q^{m+1}; q^2)_\infty} = \frac{(q^4, q^{m+1}, q^{3-m}; q^4)_\infty}{(q^2; q^2)_\infty}. \tag{2.8}$$

Next the m even and odd cases have to be considered separately to simplify (2.8). Replace first m by $2m$, multiply both sides by $(-q; q^2)_\infty$ and write the right-hand side as:

$$\frac{(q^{2m+1}, q^{3-2m}; q^4)_\infty}{(q, q^3; q^4)_\infty} = (-1)^{\lfloor m/2 \rfloor} q^{-m(m-1)/2},$$

where the equality is obtained by considering the parity of m. Replacing q by q^{-1} yields (2.6) after a few simplifications.

Finally, if we replace m by $2m+1$ in (2.8), multiply both sides by $(-q^2;q^2)_\infty$ and write the right-hand side as:

$$\frac{(q^{2m+2},q^{2-2m};q^4)_\infty}{(q^2,q^2;q^4)_\infty} = \begin{cases} (-1)^{\lfloor m/2 \rfloor} q^{-m^2/2} & \text{if } m \text{ even,} \\ 0 & \text{if } m \text{ odd,} \end{cases}$$

then we obtain (2.7) after replacing q by $q^{-1/2}$ and simplifying. $\qquad\square$

Remark 2.6 Identity (2.5) is a well-known polynomial analogue of Euler's pentagonal number theorem which has been generalized to a multivariable version by Guo and Zeng in [17], and extensively studied in the framework of q-Fibonacci polynomials by Cigler in [13]. In [23, Corollary 4.13], Warnaar generalizes (2.5) to a cubic summation formula for elliptic hypergeometric series. Identity (2.6) is a hidden special case of the terminating q-analogue of Whipple's $_3F_2$ sum [16, Appendix, (II.19)]. Finally, (2.7) is a special case of an identity obtained by Gessel and Stanton through q-Lagrange inversion, generalized to the elliptic case by Warnaar in [23, Corollary 4.11].

Now we study further the case $k = 2$ of (2.3) and (2.4).

Corollary 2.7 *For all $m \in \mathbb{N}$, we have:*

$$\sum_{j \geq 0} (-1)^j q^{5\binom{j}{2}-(2m-3)j} \begin{bmatrix} m-j \\ j \end{bmatrix}_q \sum_{k \geq 0} \frac{q^{k^2+(m-2j)k}}{(q)_k}$$

$$= \frac{(q^5, q^{2m+2}, q^{3-2m}; q^5)_\infty}{(q)_\infty}, \tag{2.9}$$

and

$$\sum_{j \geq 0} (-1)^j q^{8\binom{j}{2}-(3m-4)j} \begin{bmatrix} m-j \\ j \end{bmatrix}_{q^2} \sum_{k \geq 0} q^{k^2+(m-2j)k} \frac{(-q^{m+1};q^2)_k}{(q^2;q^2)_k}$$

$$= \frac{(-q^{m+1};q^2)_\infty}{(q^2;q^2)_\infty} (q^8, q^{3m+3}, q^{5-3m}; q^8)_\infty. \tag{2.10}$$

Proof Take $k = 2$ in (2.3), then the left-hand side, after a few rearrangements, is equal to:

$$\sum_{j \geq 0} (-1)^j q^{5\binom{j}{2}-(2m-3)j} \begin{bmatrix} m-j \\ j \end{bmatrix}_q \sum_{k \geq 0} \frac{q^{k^2+(m-2j)k}}{(q)_k},$$

and this yields (2.9). For (2.10), let $k = 2$ in (2.4), simplify as before and then replace q by q^2. $\qquad\square$

Identity (2.9) is an m-version of the Rogers–Ramanujan identities, which was discovered by Garrett, Ismail and Stanton in [15, Theorem 3.1] using the theory of q-orthogonal polynomials and integral evaluation. The authors derived with the same method the following identity:

$$\sum_{n\geq 0}\frac{q^{n^2+nm}}{(q)_n}=\frac{1}{(q)_\infty}\sum_{k=0}^{m}\begin{bmatrix}m\\k\end{bmatrix}_q q^{2k(k-m)}\left(q^5,q^{3+4k-2m},q^{2-4k+2m};q^5\right)_\infty. \quad (2.11)$$

Note that (2.11) is a famous m-version of the Rogers–Ramanujan identities, which is the inverse of (2.9). Other identities related to (2.11) are proved in [22]. To our knowledge, the m-version (2.10) of the Göllnitz–Gordon identities seems to be new. In view of (2.9) and (2.11), it is possible to invert (2.3) through the classical Bailey inversion (see, for instance, [5]). This is done in the following theorem, which is a k-generalization of (2.9). Unfortunately, it seems not possible to get in the same way nice inversions of (2.4) or (2.10).

Theorem 2.8 *For all $k \in \mathbb{N}^*$ and $m \in \mathbb{N}$ we have*

$$\sum_{0\leq n_{k-1}\leq\cdots\leq n_1}\frac{q^{n_1^2+\cdots+n_{k-1}^2+m(n_1+\cdots+n_{k-1})}}{(q)_{n_1-n_2}\cdots(q)_{n_{k-2}-n_{k-1}}(q)_{n_{k-1}}}$$

$$=\sum_{j=0}^{m}\begin{bmatrix}m\\j\end{bmatrix}_q q^{kj(j-m)}\times\frac{(q^{2k+1},q^{k(m-2j+1)},q^{k(1-m+2j)+1};q^{2k+1})_\infty}{(q)_\infty}. \quad (2.12)$$

Proof We will only consider the even case where m is replaced by $2m$, the process is the same in the odd case. Shift $n_i \to n_i - n_k$ for $1 \leq i \leq k-1$, set $j = -n_k$ and finally replace j by $m - j$ to get

$$a_m=\frac{1-q^{2m+1}}{1-q}\sum_{j=0}^{m}(-1)^{m-j}q^{\binom{m-j}{2}}\frac{(q)_{m+j}}{(q)_{m-j}}b_j,$$

where

$$a_m:=q^{km^2-m}\frac{1-q^{2m+1}}{1-q}\frac{(q^{2k+1},q^{k(2m+1)},q^{k(1-2m)+1};q^{2k+1})_\infty}{(q)_\infty},$$

and

$$b_m:=\frac{q^{km^2-m}}{(q)_{2m}}\sum_{0\leq n_{k-1}\leq\cdots\leq n_1}\frac{q^{n_1^2+\cdots+n_{k-1}^2+2m(n_1+\cdots+n_{k-1})}}{(q)_{n_1-n_2}\cdots(q)_{n_{k-2}-n_{k-1}}(q)_{n_{k-1}}}.$$

The classical Bailey inversion [5] gives $b_m=\sum_{j=0}^{m}\frac{a_j}{(q)_{m-j}(q^2)_{m+j}}$, which can be rewritten as

$$b_m=\frac{q^{km^2-m}}{(q)_{2m}}\sum_{j=0}^{m}\frac{q^j-q^{2m-j+1}}{1-q^{2m-j+1}}\begin{bmatrix}2m\\j\end{bmatrix}_q q^{kj^2-2kjm}$$

$$\times \frac{(q^{2k+1}, q^{k(2m-2j+1)}, q^{k(1-2m+2j)+1}; q^{2k+1})_\infty}{(q)_\infty}.$$

Writing $\frac{q^j - q^{2m-j+1}}{1 - q^{2m-j+1}} = 1 + \frac{q^j - 1}{1 - q^{2m-j+1}}$, splitting the sum over j into two parts, and replacing j by $2m + 1 - j$ in the second sum, the resulting identity is (2.12). $\qquad\square$

In [14], Garrett obtained m-versions of the full Andrews–Gordon identities, thus generalizing (2.12). Besides, Berkovich and Paule proved with another method in [10, (3.21)] a *negative* m-version of the full Andrews–Gordon identities. Warnaar also obtained other identities of the same kind in [24], by using a different approach from ours (although related to the Bailey lemma). It could be interesting to derive all these results of [10, 14, 24] using our approach. A bilateral version of the famous Bailey lattice [1] would probably be needed, and we will come back to these questions in a forthcoming paper. Before ending this section, we note that (2.3) is, in fact, closely related to the full Andrews–Gordon identities. Indeed, replace m by $2m$, and then shift $n_i \to n_i - m$ for $1 \le n_i \le k$ in the left-hand side of (2.3). Using

$$\left(q^{k(2m+1)}, q^{k(1-2m)+1}; q^{2k+1}\right)_\infty = (-1)^m q^{-km^2 + \binom{m+1}{2}} \left(q^{k+m+1}, q^{k-m}; q^{2k+1}\right)_\infty$$

yields

$$\sum_{0 \le n_k \le n_{k-1} \le \cdots \le n_1} \frac{q^{n_1^2 + \cdots + n_k^2}}{(q)_{n_1 - n_2} \cdots (q)_{n_{k-1} - n_k}} (-1)^{n_k} q^{\binom{n_k}{2} - mn_k} \begin{bmatrix} m + n_k \\ 2n_k \end{bmatrix}_q$$

$$= \frac{(q^{2k+1}, q^{k+m+1}, q^{k-m}; q^{2k+1})_\infty}{(q)_\infty}. \tag{2.13}$$

Notice that the right-hand side of (2.13) is the same as in the full Andrews–Gordon identities, thus identifying the right-hand sides yields for all $k \in \mathbb{N}^*$ and $m \in \{1, \ldots, k-1\}$:

$$\sum_{0 \le n_k \le n_{k-1} \le \cdots \le n_1} \frac{q^{n_1^2 + \cdots + n_k^2}}{(q)_{n_1 - n_2} \cdots (q)_{n_{k-1} - n_k}} (-1)^{n_k} q^{\binom{n_k}{2} - mn_k} \begin{bmatrix} m + n_k \\ 2n_k \end{bmatrix}_q$$

$$= \sum_{0 \le n_{k-1} \le \cdots \le n_1} \frac{q^{n_1^2 + \cdots + n_{k-1}^2 + n_{k-m} + \cdots + n_{k-1}}}{(q)_{n_1 - n_2} \cdots (q)_{n_{k-1}}}. \tag{2.14}$$

Proving directly (2.14) (i.e., without appealing to the full Andrews–Gordon identities) does not seem to be obvious.

The same link can be done between (2.4) and the full Göllnitz–Gordon identities.

3 Change of base

In [9], many multisums of Rogers–Ramanujan type are proved as consequences of change of base in Bailey pairs. In the same vein as Sect. 2, many results concerning

Bailey pairs in [9] have bilateral versions. Here we will only highlight the following bilateral version of [9, Theorem 2.1].

Theorem 3.1 *If $(\alpha_n(a, q), \beta_n(a, q))$ is a bilateral Bailey pair related to a and q, then so is $(\alpha'_n(a, q), \beta'_n(a, q))$, where*

$$\alpha'_n(a, q) = \frac{(-b)_n}{(-aq/b)_n} b^{-n} q^{-\binom{n}{2}} \alpha_n(a^2, q^2)$$

and

$$\beta'_n(a, q) = \sum_{k \leq n} \frac{(-aq)_{2k}(b^2; q^2)_k (q^{-k}/b, bq^{k+1})_{n-k}}{(b, -aq/b)_n (q^2; q^2)_{n-k}} b^{-k} q^{-\binom{k}{2}} \beta_k(a^2, q^2),$$

provided the relevant series are absolutely convergent.

Proof As in [9], we only need to use the definition (1.8) of a bilateral Bailey pair, interchange summations and apply Singh's quadratic transformation [16, Appendix, (III.21)] summed with q-Pfaff–Saalschütz (1.4). □

The following result gives an m-version of Bressoud's identities for even moduli:

Theorem 3.2 *For all integers $m \in \mathbb{N}$ and $k \geq 1$ we have*

$$\sum_{-\lfloor m/2 \rfloor \leq n_k \leq n_{k-1} \leq \cdots \leq n_1} \frac{q^{n_1^2 + \cdots + n_k^2 + m(n_1 + \cdots + n_{k-1}) + n_{k-1} - 2n_k} (-q)_{2n_k + m}}{(q)_{n_1 - n_2} \cdots (q)_{n_{k-2} - n_{k-1}} (q^2; q^2)_{n_{k-1} - n_k}}$$

$$\times (-1)^{n_k} \begin{bmatrix} m + n_k \\ m + 2n_k \end{bmatrix}_{q^2}$$

$$= \frac{(q^{2k}, q^{(k-1)(m+1)}, q^{(k-1)(1-m)+2}; q^{2k})_\infty}{(q)_\infty}. \tag{3.1}$$

Proof Let $b \to \infty$ in Theorem 3.1:

$$\alpha'_n(a, q) = \alpha_n(a^2, q^2), \qquad \beta'_n(a, q) = \sum_{j \leq n} \frac{(-aq)_{2j}}{(q^2; q^2)_{n-j}} q^{n-j} \beta_j(a^2, q^2). \tag{3.2}$$

Apply (3.2) to the shifted Bailey pair from Proposition 2.1. This gives a new shifted Bailey pair $(\alpha'_n(q^m, q), \beta'_n(q^m, q))$, where

$$\alpha'_n(q^m, q) = \alpha_n(q^{2m}, q^2) = (-1)^n q^{2\binom{n}{2}}$$

and

$$\beta'_n(q^m, q) = \sum_{j \leq n} \frac{(-q^{1+m})_{2j}}{(q^2; q^2)_{n-j}} q^{n-j} \beta_j(q^{2m}, q^2)$$

$$= (q^2; q^2)_m \sum_{j \leq n} \frac{(-q^{1+m})_{2j}}{(q^2; q^2)_{n-j}} q^{n-j} (-1)^j q^{2\binom{j}{2}} \begin{bmatrix} m+j \\ m+2j \end{bmatrix}_{q^2}.$$

Next apply $k-1$ times the instance (2.1) of the bilateral Bailey lemma to the new pair $(\alpha'_n(q^m, q), \beta'_n(q^m, q))$, this gives a shifted Bailey pair $(\alpha_n^{(k)}(q^m, q), \beta_n^{(k)}(q^m, q))$, where

$$\alpha_n^{(k)}(q^m, q) = q^{(k-1)n^2 + (k-1)mn} (-1)^n q^{2\binom{n}{2}}$$

and

$$\beta_n^{(k)}(q^m, q) = \sum_{n_k \leq n_{k-1} \leq \cdots \leq n_1 \leq n} \frac{q^{n_1^2 + \cdots + n_k^2 + m(n_1 + \cdots + n_{k-1}) + n_{k-1} - 2n_k} (-q^{1+m})_{2n_k}}{(q)_{n-n_1} (q)_{n_1 - n_2} \cdots (q)_{n_{k-2} - n_{k-1}} (q^2; q^2)_{n_{k-1} - n_k}}$$

$$\times (q^2; q^2)_m (-1)^{n_k} \begin{bmatrix} m + n_k \\ m + 2n_k \end{bmatrix}_{q^2}.$$

Writing $(-q^{1+m})_{2n_k} = \frac{(-q)_{2n_k+m}}{(-q)_m}$, the result then follows by letting $n \to +\infty$ and invoking Tannery's theorem [12] in the relation

$$\beta_n^{(k)} = \sum_{j \leq n} \frac{\alpha_j^{(k)}}{(q)_{n-j} (q^{1+m})_{n+j}}$$

and finally using (1.5) to factorize the right-hand side. □

Remark 3.3 The case $k = 1$ of (3.1) is trivial, while the case $k = 2$, after a few simplifications, appears to yield exactly identities (2.6) and (2.7).

As in Theorem 2.8, if we invert (3.1) by using the classical Bailey inversion, then we obtain the following result.

Theorem 3.4 *For all $k \in \mathbb{N}^*$ and $m \in \mathbb{N}$ we have*

$$\sum_{0 \leq n_{k-1} \leq \cdots \leq n_1} \frac{q^{n_1^2 + \cdots + n_{k-1}^2 + m(n_1 + \cdots + n_{k-1}) + n_{k-1}}}{(q)_{n_1 - n_2} \cdots (q)_{n_{k-2} - n_{k-1}} (q^2; q^2)_{n_{k-1}}}$$

$$= \sum_{j=0}^m \begin{bmatrix} m \\ j \end{bmatrix}_{q^2} \frac{q^{(k-1)j(j-m)}}{(-q)_m} \frac{(q^{2k}, q^{(k-1)(m-2j+1)}, q^{(k-1)(1-m+2j)+2}; q^{2k})_\infty}{(q)_\infty}.$$

$$(3.3)$$

Note that (3.3) is a special case of the results in [14], where m-versions of the *full* Bressoud identities for even moduli (see, for instance, [9]) are obtained. As at the end of the previous section, by replacing m by $2m$ and simplifying, it is possible to see that (3.1) is related to the full Bressoud identities.

4 Applications of the bilateral WP-Bailey lemma

As in [25], it is possible to invert the relation (1.9) by using a matrix inversion, which gives:

$$\alpha_n(a,\alpha,q) = \frac{1-aq^{2n}}{1-a} \sum_{r\le n} \frac{(a)_{n+r}}{(q)_{n-r}} \frac{(a/\alpha)_{n-r}}{(\alpha q)_{n+r}} \frac{1-\alpha q^{2r}}{1-\alpha} \left(\frac{\alpha}{a}\right)^{n-r} \beta_r(a,\alpha,q), \quad (4.1)$$

for all $n \in \mathbb{Z}$. Set $m \in \mathbb{N}$, then the following form a WP-bilateral Bailey pair:

$$\begin{cases} \alpha_n(a,\alpha,q) = \dfrac{1-aq^{2n}}{1-a} \dfrac{(a)_{n-m}}{(q)_{n+m}} \dfrac{(a/\alpha)_{n+m}}{(\alpha q)_{n-m}} \dfrac{1-\alpha q^{-2m}}{1-\alpha} \left(\dfrac{\alpha}{a}\right)^{n+m}, \\ \beta_n(a,\alpha,q) = \delta_{n+m,0}. \end{cases} \quad (4.2)$$

In the case $m = 0$, we recover the unit WP-Bailey pair from [6]. Recall that two iterations of the first instance of the WP-Bailey lemma applied to the unit WP-Bailey pair gives Bailey's transformation between two terminating very-well poised $_{10}\phi_9$ [16, Appendix, (III.28)]. We will show that two iterations of the WP-bilateral Bailey lemma to (4.2) yields an extension of both Ramanujan's $_1\psi_1$ summation [16, Appendix, (II.29)], and Bailey's $_6\psi_6$ summation [16, Appendix, (II.33)] formulae. This is stated in the following result.

Proposition 4.1 *For $m \in \mathbb{N}$, $|\alpha/a| < 1$ and $|aq/\mu_1\mu_2| < 1$, we have:*

$$_8\psi_8\left[\begin{matrix} q\sqrt{a}, -q\sqrt{a}, \rho_1, \rho_2, \mu_1, \mu_2, aq^{-m}, a^3q^{2+m}/\alpha\rho_1\rho_2\mu_1\mu_2 \\ \sqrt{a}, -\sqrt{a}, aq/\rho_1, aq/\rho_2, aq/\mu_1, aq/\mu_2, q^{1+m}, \alpha\rho_1\rho_2\mu_1\mu_2 q^{-m}/qa^2 \end{matrix}; q, \frac{\alpha}{a}\right]$$

$$= \frac{(aq, \lambda q/\mu_1, \lambda q/\mu_2, aq/\mu_1\mu_2)_\infty}{(\alpha/a, aq/\mu_1, aq/\mu_2, \lambda q)_\infty} \times \frac{(q/a, aq/\lambda\rho_1, aq/\lambda\rho_2, aq/\rho_1\rho_2)_m}{(q/\rho_1, q/\rho_2, q/\lambda, qa^2/\lambda\rho_1\rho_2)_m}$$

$$\times {}_8\psi_8\left[\begin{matrix} q\sqrt{\lambda}, -q\sqrt{\lambda}, \mu_1, \mu_2, \lambda\rho_1/a, \lambda\rho_2/a, \lambda q^{-m}, aq^{1+m}/\rho_1\rho_2 \\ \sqrt{\lambda}, -\sqrt{\lambda}, \lambda q/\mu_1, \lambda q/\mu_2, aq/\rho_1, aq/\rho_2, q^{1+m}, \lambda\rho_1\rho_2 q^{-m}/a \end{matrix}; q, \frac{\alpha}{\lambda}\right],$$

$$(4.3)$$

where $\lambda := \alpha\mu_1\mu_2/aq$.

Proof Apply twice the first instance of the WP-bilateral Bailey lemma to (4.2), this gives a WP-bilateral Bailey pair with four new parameters ρ_1, ρ_2, μ_1 and μ_2. Replacing it in (1.9) yields (4.3) after letting $n \to +\infty$ under the necessary conditions $|\alpha/a| < 1$ and $|aq/\mu_1\mu_2| < 1$ to use Tannery's theorem [12], and simplifying. □

Now in (4.3), set $\alpha = 0$, $\mu_1 = b$, $\mu_2 = aq/bz$, $\rho_1 = aq/c$, $\rho_2 = bz$, and finally $a = b$ and $m \to +\infty$. This gives after using on the left-hand side a limit case of the terminating very-well poised $_6\phi_5$ summation formula [16, Appendix, (II-20)]:

$$\frac{(q, c/b, bz, q/bz)_\infty}{(c, q/b, z, c/bz)_\infty} = {}_1\psi_1\left[\begin{matrix} b \\ c \end{matrix}; q, z\right], \quad (4.4)$$

which is Ramanujan's $_1\psi_1$ summation formula, valid for $|q| < 1$ and $|c/b| < |z| < 1$.

Next, setting in (4.3) $\rho_1 = b$, $\rho_2 = c$, $\mu_2 = e$, and $\mu_1 = d = aq/\alpha$ (which gives $\lambda = e$) yields after letting $m \to +\infty$ and using on the right-hand side the same limit case of the terminating very-well poised $_6\phi_5$ summation formula [16, Appendix, (II-20)]:

$$_6\psi_6 \left[\begin{matrix} q\sqrt{a}, -q\sqrt{a}, b, c, d, e \\ \sqrt{a}, -\sqrt{a}, aq/b, aq/c, aq/d, aq/e \end{matrix}; q, \frac{qa^2}{bcde} \right]$$

$$= \frac{(q, aq, q/a, aq/bc, aq/bd, aq/be, aq/cd, aq/ce, aq/de)_\infty}{(q/b, q/c, q/d, q/e, aq/b, aq/c, aq/d, aq/e, a^2q/bcde)_\infty}, \tag{4.5}$$

which is Bailey's $_6\psi_6$ summation, valid for $|a^2q/bcde| < 1$.

Remark 4.2 Note that by shifting the index of summation $k \to k - m$ on both sides of (4.3), one recovers the instance $n \to +\infty$ of Bailey's $_{10}\phi_9$ transformation formula [16, Appendix, (III.28)], which corresponds to [16, Appendix, (III.23)]. Now if $m \to +\infty$ in (4.3), the left-hand side is independent of α, and we recover a $_6\psi_6$ transformation formula from [18], which can be iterated to yield directly (4.5), without appealing to the $_6\phi_5$ summation formula.

In what follows, we give a new WP-shifted Bailey pair (recall that this means setting $a = q^m$, $m \in \mathbb{N}$ to avoid convergence problems), which generalizes the shifted Bailey pair from Sect. 2:

Proposition 4.3 *For $m \in \mathbb{N}$, $(\alpha_n(q^m, \alpha), \beta_n(q^m, \alpha))$ is a WP-shifted Bailey pair related to $a = q^m$ and α, where*

$$\alpha_n(q^m, \alpha) = \frac{(q^m/\alpha)_n}{(\alpha q^{-m})_n}(\alpha q^{-m})^n$$

and

$$\beta_n(q^m, \alpha) = \frac{(q)_m(q/\alpha)_{m-n}(\alpha^2 q^{-2m})_{m+2n}}{(q/\alpha, \alpha q^{-m})_m(\alpha q^{1-m})_{m+n}} \begin{bmatrix} m+n \\ m+2n \end{bmatrix}_q (q^m/\alpha)^n.$$

Proof We have by definition

$$\beta_n(q^m, \alpha) = \sum_{r \leq n} \frac{(\alpha q^{-m})_{n-r}(\alpha)_{n+r}}{(q)_{n-r}(q^{1+m})_{n+r}} \alpha_r(q^m, \alpha)$$

so, as $1/(q^{1+m})_{n+r} = 0$ if $n + r + m < 0$, we can see that $\beta_n = 0$ unless $2n + m \geq 0$. In that case, one has

$$\beta_n(q^m, \alpha) = \sum_{k \geq 0} \frac{(\alpha q^{-m})_k(\alpha)_{2n-k}(q^m/\alpha)_{n-k}}{(q)_k(q^{1+m})_{2n-k}(\alpha q^{-m})_{n-k}}(\alpha q^{-m})^{n-k}$$

$$= \frac{(\alpha)_{2n}(q^m/\alpha)_n}{(q^{1+m})_{2n}(\alpha q^{-m})_n}(\alpha q^{-m})^n \sum_{k \geq 0} \frac{(\alpha q^{-m}, q^{-2n-m}, q^{1-n+m}/\alpha)_k}{(q, q^{1-2n}/\alpha, \alpha q^{1-n-m})_k} q^k.$$

As $2n + m \geq 0$, the last sum can be evaluated by the q-Pfaff–Saalschütz formula (1.4). We then get

$$\beta_n(q^m, \alpha) = \frac{(\alpha)_{2n}(q^m/\alpha)_n}{(q^{1+m})_{2n}(\alpha q^{-m})_n} \left(\alpha q^{-m}\right)^n \frac{(\alpha^2 q^{-2m}, q^{1-n})_{m+2n}}{(\alpha q^{1-n-m}, \alpha q^{-m})_{m+2n}}$$

which is the desired result.   □

Remark 4.4 When $\alpha \to 0$ in Proposition 4.3, the WP-shifted Bailey pair becomes the shifted Bailey pair of Proposition 2.1.

As an application, we prove the following new transformation, which generalizes a result of Garrett, Ismail and Stanton [15, (6.3)].

Theorem 4.5 *For all non-negative integer m and real parameters β, γ, ρ such that $|q/\beta^2| < 1$, we have*

$$\frac{(\beta)_m}{(q/\beta)_m} \sum_{n \in \mathbb{Z}} \frac{(1/\gamma, \rho, \gamma q^{1+m}/\beta\rho)_n}{(\gamma, q^{1+m}/\rho, \beta\rho/\gamma)_n} \frac{(\beta q^m)_{2n}}{(q^{1+m}/\beta)_{2n}} (q/\beta)^n$$

$$= \frac{(q, q/\beta^2)_\infty}{(q/\beta, q/\beta)_\infty} \sum_{s=0}^{\lfloor m/2 \rfloor} (\beta^3/q)^s \frac{(q/\gamma, \gamma q/\beta\rho, \rho q^{-m})_s}{(q, q/\rho, \beta\rho q^{-m}/\gamma)_s} \frac{1 - \gamma q^{m-2s}}{1 - \gamma}$$

$$\times \frac{(\beta, \gamma^2)_{m-2s}}{(q, \gamma q)_{m-2s}} \frac{(q)_{m-s}}{(\gamma q)_{m-s}} {}_4\phi_3 \left[\begin{matrix} \beta/\gamma, \beta q^{m-2s}, \rho\beta q^{-s}, \gamma q^{1+m-s}/\rho \\ \gamma q^{1+m-2s}, \beta\rho q^{-s}/\gamma, q^{1+m-s}/\rho \end{matrix} ; q, q/\beta^2 \right].$$

(4.6)

Proof Apply the first instance of Theorem 1.4 to Proposition 4.3 to get the WP-bilateral Bailey pair:

$$\alpha'_n(q^m, \alpha) = \frac{(\rho_1, \rho_2)_n}{(q^{1+m}/\rho_1, q^{1+m}/\rho_2)_n} (\alpha/c)^n \alpha_n(q^m, c)$$

$$= \frac{(\rho_1, \rho_2)_n}{(q^{1+m}/\rho_1, q^{1+m}/\rho_2)_n} \left(\alpha q^{-m}\right)^n \frac{(q^m/c)_n}{(cq^{-m})_n},$$

and

$$\beta'_n(q^m, \alpha) = \frac{(\alpha\rho_1/q^m, \alpha\rho_2/q^m)_n}{(q^{1+m}/\rho_1, q^{1+m}/\rho_2)_n} \sum_{j \leq n} \frac{(\rho_1, \rho_2)_j}{(\alpha\rho_1/q^m, \alpha\rho_2/q^m)_j}$$

$$\times \frac{1 - cq^{2j}}{1 - c} \frac{(\alpha/c)_{n-j}(\alpha)_{n+j}}{(q)_{n-j}(qc)_{n+j}} (\alpha/c)^j \beta_j(q^m, c),$$

with $c = \alpha\rho_1\rho_2/q^{1+m}$.

Now use the second instance of Theorem 1.4 to derive

$$\tilde{\alpha}'_n(q^m, \alpha) = \frac{(q^{1+2m}/\alpha)_{2n}}{(\alpha)_{2n}} \frac{(\rho_1, \rho_2, \alpha/\rho_1\rho_2)_n}{(q^{1+m}/\rho_1, q^{1+m}/\rho_2, \rho_1\rho_2/\alpha)_n} \left(\alpha q^{-m}\right)^n$$

and

$$
\tilde{\beta}'_n\left(q^m, \alpha\right) = \sum_{j \le n} \frac{(\alpha^2/q^{1+2m})_{n-j}}{(q)_{n-j}} (\alpha^2/q^{1+2m})^j \beta'_j\left(q^m, q^{1+2m}/\alpha\right).
$$

Under the convergence condition $|\alpha^2/q^{1+2m}| < 1|$, let $n \to +\infty$ in the relation

$$
\tilde{\beta}'_n\left(q^m, \alpha\right) = \sum_{r \le n} \frac{(\alpha q^{-m})_{n-r}(\alpha)_{n+r}}{(q)_{n-r}(q^{1+m})_{n+r}} \tilde{\alpha}'_r\left(q^m, \alpha\right).
$$

This yields

$$
\sum_{n \in \mathbb{Z}} \frac{(q^{1+2m}/\alpha)_{2n}}{(\alpha)_{2n}} \frac{(\rho_1, \rho_2, \alpha/\rho_1\rho_2)_n}{(q^{1+m}/\rho_1, q^{1+m}/\rho_2, \rho_1\rho_2/\alpha)_n} (\alpha q^{-m})^n
$$

$$
= \frac{(q^{1+m}, \alpha^2/q^{1+2m})_\infty}{(\alpha q^{-m}, \alpha)_\infty} \sum_{-\lfloor m/2 \rfloor \le s \le j} f(s, j)
$$

$$
= \frac{(q^{1+m}, \alpha^2/q^{1+2m})_\infty}{(\alpha q^{-m}, \alpha)_\infty} \sum_{s=0}^{\lfloor m/2 \rfloor} \sum_{j \ge 0} f(-s, j - s), \tag{4.7}
$$

where

$$
f(s, j) := \left(\frac{\alpha^2}{q^{1+2m}}\right)^j \left(\frac{\alpha q^{1+m}}{\rho_1^2 \rho_2^2}\right)^s \times \frac{1 - \rho_1\rho_2 q^{m+2s}/\alpha}{1 - \rho_1\rho_2 q^m/\alpha} \begin{bmatrix} m + s \\ m + 2s \end{bmatrix}_q
$$

$$
\times \frac{(\rho_1 q^{1+m}/\alpha, \rho_2 q^{1+m}/\alpha)_j}{(q^{1+m}/\rho_1, q^{1+m}/\rho_2)_j} \times \frac{(\rho_1, \rho_2)_s}{(\rho_1 q^{1+m}/\alpha, \rho_2 q^{1+m}/\alpha)_s}
$$

$$
\times \frac{(q^{1+m}/\rho_1\rho_2)_{j-s}(q^{1+2m}/\alpha)_{j+s}}{(q)_{j-s}(\rho_1\rho_2 q^{1+m}/\alpha)_{s+j}} \times \frac{(q)_m(\alpha q^{1-m}/\rho_1\rho_2)_{m-s}}{(\alpha q^{1-m}/\rho_1\rho_2, \rho_1\rho_2/\alpha)_m}
$$

$$
\times \frac{(\rho_1^2 \rho_2^2/\alpha^2)_{m+2s}}{(q\rho_1\rho_2/\alpha)_{m+s}}.
$$

Setting $\beta = q^{1+m}/\alpha$, $\gamma = \rho_1\rho_2/\alpha = \beta\rho_1\rho_2/q^{1+m}$, $\rho_1 = \rho$, and finally rearranging the right-hand side of (4.7), we get (4.6). □

5 Concluding remarks

We proved through extensions of the classical Bailey lemma many results from [15]. It could be a challenging problem to find a proof of the quintic formula [15, Theorem 7.1] using our approach.

We also want to point out that orthogonality relations and connection coefficient formulas for the q-Hermite and q-ultraspherical polynomials are at the heart of the proofs in [15]. Recall that the q-ultraspherical polynomials have the explicit representation:

$$C_n(\cos\theta; \beta|q) = \sum_{k=0}^{n} \frac{(\beta)_k(\beta)_{n-k}}{(q)_k(q)_{n-k}} e^{-i(n-2k)\theta}. \tag{5.1}$$

One can see that (5.1) is equivalent to saying that $(\alpha_n(q^m, \beta q^m), \beta_n(q^m, \beta q^m))$ is a WP-shifted Bailey pair, where

$$\alpha_n(q^m, \beta q^m) = e^{2in\theta} \quad \text{and} \quad \beta_n(q^m, \beta q^m) = e^{-im\theta} \frac{(q)_m}{(\beta)_m} C_{2n+m}(\cos\theta; \beta|q).$$

Applying the first instance of Theorem 1.4 with $\rho_2 \to +\infty$, $\rho_1 \to 0$ and $c = \beta\rho_1\rho_2/q < \infty$ yields the connection coefficient formula for C_n:

$$C_n(\cos\theta; c|q) = \sum_{k=0}^{\lfloor n/2 \rfloor} \frac{(c/\beta)_k(c)_{n-k}}{(q)_k(q\beta)_{n-k}} \beta^k \frac{1 - \beta q^{n-2k}}{1 - \beta} C_{n-2k}(\cos\theta; \beta|q).$$

A natural question is then to ask whether our method applied to the other Bailey lemmas (classical, Well-Poised or elliptic from [25]) could prove or highlight properties for more general q-orthogonal polynomials.

Besides, apart from the one mentioned in Sect. 3, there are many other changes of base in [9], which have a bilateral version. Moreover, there should be bilateral versions of the other WP Bailey lemmas proved by Warnaar [25] or Mc Laughlin and Zimmer [19]. It could be interesting to derive applications of our method from all of these.

Acknowledgements   We would like to thank Mourad Ismail for interesting discussions during his visit in Lyon. We also thank very much Michael Schlosser and Ole Warnaar for very useful comments on an earlier version of this paper.

References

1. Agarwal, A., Andrews, G.E., Bressoud, D.: The Bailey lattice. J. Indian Math. Soc. **51**, 57–73 (1987)
2. Andrews, G.E.: The Theory of Partitions. Encyclopedia of Mathematics and Its Applications, vol. 2. Addison-Wesley, Reading (1976)
3. Andrews, G.E.: Multiple series Rogers–Ramanujan type identities. Pac. J. Math. **114**, 267–283 (1984)
4. Andrews, G.E.: q-Series: Their Development and Application in Analysis, Number Theory, Combinatorics, Physics and Computer Algebra. CBMS Regional Conference Series in Mathematics, vol. 66. AMS, Providence (1986)
5. Andrews, G.E., Askey, R., Roy, M.: Special Functions. Encyclopedia of Mathematics and Its Applications, vol. 71. Cambridge University Press, Cambridge (1999)
6. Andrews, G.E., Berkovich, A.: The WP-Bailey tree and its implications. J. Lond. Math. Soc. **66**, 529–549 (2002)
7. Andrews, G.E., Schilling, A., Warnaar, S.O.: An A_2 Bailey lemma and Rogers–Ramanujan-type identities. J. Am. Math. Soc. **12**(3), 677–702 (1999)
8. Bailey, W.N.: Identities of the Rogers–Ramanujan type. Proc. Lond. Math. Soc. (2) **50**, 1–10 (1949)
9. Bressoud, D., Ismail, M., Stanton, D.: Change of base in Bailey pairs. Ramanujan J. **4**, 435–453 (2000)
10. Berkovich, A., Paule, P.: Variants of the Andrews–Gordon identities. Ramanujan J. **5**, 391–404 (2001)

11. Berkovich, A., McCoy, B.M., Schilling, A.: $N = 2$ supersymmetry and Bailey pairs. Physica A **228**, 33–62 (1996)
12. Bromwich, T.J.l'A: An Introduction to the Theory of Infinite Series, 2nd edn. Macmillan, London (1949)
13. Cigler, J.: A new class of q-Fibonnacci polynomials. Electron. J. Comb. **10**, #R19 (2003)
14. Garrett, K.: New generalizations of Rogers–Ramanujan type multisum identities. Preprint
15. Garrett, K., Ismail, M.E.H., Stanton, D.: Variants of the Rogers–Ramanujan identities. Adv. Appl. Math. **23**, 274–299 (1999)
16. Gasper, G., Rahman, M.: Basic Hypergeometric Series, 2nd edn. Encyclopedia of Mathematics and Its Applications, vol. 96. Cambridge University Press, Cambridge (2004)
17. Guo, V.J.W., Zeng, J.: Multiple extensions of a finite Euler's pentagonal number theorem and the Lucas formulas. Discrete Math. **308**, 4069–4078 (2008)
18. Jouhet, F., Schlosser, M.: Another proof of Bailey's $_6\psi_6$ summation. Aequ. Math. **70**(1–2), 43–50 (2005)
19. McLaughlin, J., Zimmer, P.: General WP-Bailey Chains. Preprint. http://math.wcupa.edu/~mclaughlin/newbailey05nov2008.pdf
20. Paule, P.: The concept of Bailey chains. Sémin. Lothar. Comb. B **18f**, 24 pp. (1987)
21. Schlosser, M.: Private communication
22. Warnaar, S.O.: Partial-sum analogues of the Rogers–Ramanujan identities. J. Comb. Theory A **99**, 143–161 (2002)
23. Warnaar, S.O.: Summation and transformation formulas for elliptic hypergeometric series. Constr. Approx. **18**, 479–502 (2002)
24. Warnaar, S.O.: Partial theta functions. I. Beyond the lost notebook. Proc. Lond. Math. Soc. **87**, 363–395 (2003)
25. Warnaar, S.O.: Extensions of the well-poised and elliptic well-poised Bailey lemma. Indag. Math., New Ser. **14**, 571–588 (2003)

Ramanujan J (2010) 23: 335–339
DOI 10.1007/s11139-010-9243-4

Column-to-row operations on partitions: Garden of Eden partitions

Brian Hopkins · Louis Kolitsch

Dedicated to George Andrews on the occasion of his 70th birthday

Received: 12 June 2009 / Accepted: 6 May 2010 / Published online: 31 August 2010
© Springer Science+Business Media, LLC 2010

Abstract Conjugation and the Bulgarian solitaire move are the extreme cases of general column-to-row operations on integer partitions. Each operation generates a state diagram on the partitions of n. Garden of Eden states are those with no preimage under the operation in question. In this note, we determine the number of Garden of Eden partitions for all n and column-to-row operations.

Keywords Partitions · Bulgarian solitaire

Mathematics Subject Classification (2000) Primary 05A17 · 37E15

1 Introduction

Let $P(n)$ denote the set of integer partitions of n. Write a partition $\lambda \in P(n)$ as $(\lambda_1, \ldots, \lambda_{\ell(\lambda)})$ where $\ell(\lambda)$ denotes the partition's length, its number of parts. The conjugate partition λ' is defined as $\lambda' = (\lambda'_1, \ldots, \lambda'_s)$ where λ'_i is the number of parts $\{\lambda_i\}$ greater than or equal to i. The sequence of operations D^k are defined in [4] as

$$D^k(\lambda) = (\lambda'_1, \ldots, \lambda'_k, \lambda_1 - k, \ldots, \lambda_{\ell(\lambda)} - k) \in P(n),$$

where any nonpositive numbers are removed and the parts may not be in the standard nonincreasing order. In terms of the Ferrers diagram of the partition, D^k takes the leftmost k columns and makes them rows. The $k = 1$ case corresponds to the

B. Hopkins (✉)
Department of Mathematics, Saint Peter's College, Jersey City, NJ 07306, USA
e-mail: bhopkins@spc.edu

L. Kolitsch
Department of Mathematics, University of Tennessee at Martin, Martin, TN 38238, USA
e-mail: lkolitsc@utm.edu

Fig. 1 The D^2 operation on $P(5)$, written with superscripts for parts repeated three or more times

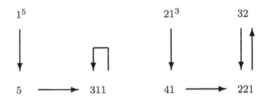

operation introduced to the research literature by Brandt [2] and popularized as Bulgarian solitaire by Gardner [3]. For $k \geq n - 1$, the D^k operations are equivalent to conjugation (see [4], Lemma 1).

Each D^k operation produces a state diagram on $P(n)$. For example, the effect of D^2 on $P(5)$ is shown in Fig. 1. Among the partitions distinguished by the operation, those without a preimage are called Garden of Eden partitions (GE-partitions hereafter). Let $GE(n, k)$ denote the GE-partitions in $P(n)$ under the operation D^k. From Fig. 1, we see that $GE(5, 2) = \{(1, 1, 1, 1, 1), (2, 1, 1, 1)\}$. Write $ge(n, k)$ for the cardinality of $GE(n, k)$; we have $ge(5, 2) = 2$. Similarly, write $p(n)$ for the number of partitions of n.

Previous results include formulas for $ge(n, 1)$ [5] and $ge(n, k)$ for $\frac{n-1}{2} \leq k$ [4]. The result of this note establishes a formula for arbitrary $ge(n, k)$; the previous results are shown to follow as a corollary.

2 Garden of Eden partitions

We repeat the characterization of GE-partitions from [4] for completeness.

Lemma 1 *A partition* $\lambda = (\lambda_1, \ldots, \lambda_{\ell(\lambda)}) \in P(n)$ *is in* $GE(n, k)$ *exactly when* $\lambda_k - \ell(\lambda) \leq -1 - k$.

Proof In terms of the Ferrers diagram, λ has a preimage under D^k for every set of k rows each greater than or equal to $\ell(\lambda) - k$, i.e., long enough to be moved to become columns to the left side of the remaining dots. This fails when $\lambda_k < \ell(\lambda) - k$. $\square$

We use standard q-series notation, with $(q; q)_m = (1 - q)(1 - q^2) \cdots (1 - q^m)$ and the q-binomial coefficient $\begin{bmatrix} n \\ m \end{bmatrix}$. We can now state and prove our theorem.

Theorem 1 *The values of* $ge(n, k)$ *are given by*

$$\sum_{n \geq 0} ge(n, k)q^n = \frac{1}{(q; q)_\infty} \sum_{j=1}^{\infty} (-1)^{j+1} q^{\frac{3j^2 + (2k+1)j}{2}} \begin{bmatrix} k + j - 1 \\ k - 1 \end{bmatrix}.$$

Proof Consider a partition of n that does have a preimage under D^k with the additional condition that $\lambda_k = s$. From the lemma, we have $\ell(\lambda) \leq k + s$. The Ferrers diagram of such a partition can be broken into three parts: a $k \times s$ array of dots, a section to the right of the array with up to $k - 1$ rows, and a section under the array that must fit inside an $s \times s$ box; see Fig. 2.

Fig. 2 Partition with $\lambda_k = s$ that has a preimage under D^k

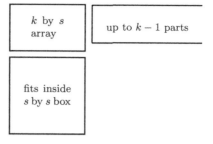

These partitions with preimages under D^k are therefore counted by

$$\frac{1}{(q;q)_{k-1}} \sum_{s=0}^{\infty} q^{sk} \begin{bmatrix} 2s \\ s \end{bmatrix}$$

using Sylverster's interpretation of q-binomial coefficients counting partitions inside a box, Theorem 3.1 in [1].

The GE-partitions we wish to count are the partitions of n left after the partitions with preimages have been removed. This establishes the following equation.

$$\sum_{n=0}^{\infty} ge(n,k)q^n = \left(\sum_{n=0}^{\infty} p(n)q^n \right) - \frac{1}{(q;q)_{k-1}} \sum_{s=0}^{\infty} q^{sk} \begin{bmatrix} 2s \\ s \end{bmatrix}. \tag{1}$$

Now we focus on part of the second term in the difference.

$$\sum_{s=0}^{\infty} q^{sk} \begin{bmatrix} 2s \\ s \end{bmatrix} = \sum_{s=0}^{\infty} q^{sk} \frac{(q;q)_{2s}}{(q;q)_s^2}$$

$$= \sum_{s=0}^{\infty} \frac{q^{sk}}{(q;q)_s} (q^{s+1};q)_s$$

$$= \sum_{s=0}^{\infty} \frac{q^{sk}}{(q;q)_s} \sum_{j=0}^{s} \begin{bmatrix} s \\ j \end{bmatrix} (-1)^j q^{sj} q^{\frac{j^2+j}{2}} \tag{2}$$

$$= \sum_{j=0}^{\infty} \sum_{s=j}^{\infty} \frac{q^{(j+k)s}}{(q;q)_s} \begin{bmatrix} s \\ j \end{bmatrix} (-1)^j q^{\frac{j^2+j}{2}}$$

$$= \sum_{j=0}^{\infty} \sum_{s=0}^{\infty} \frac{q^{(k+j)(s+j)}}{(q;q)_{s+j}} \begin{bmatrix} s+j \\ j \end{bmatrix} (-1)^j q^{\frac{j^2+j}{2}}$$

$$= \sum_{j=0}^{\infty} \frac{(-1)^j q^{\frac{3j^2+j}{2}+kj}}{(q;q)_j} \sum_{s=0}^{\infty} \frac{q^{(k+j)s}}{(q;q)_s}$$

$$= \sum_{j=0}^{\infty} \frac{(-1)^j q^{\frac{3j^2+(2k+1)j}{2}}}{(q;q)_j} \frac{(q;q)_{k+j-1}}{(q;q)_\infty} \tag{3}$$

$$= \frac{(q;q)_{k-1}}{(q;q)_\infty} \sum_{j=0}^{\infty} (-1)^j q^{\frac{3j^2+(2k+1)j}{2}} \begin{bmatrix} k+j-1 \\ j \end{bmatrix}.$$

Line (2) follows from identity 3.3.6 in [1] and line (3) is an application of identity 2.2.1 in [1]. Substituting this into Eq. (1) and using Euler's result $\frac{1}{(q;q)_\infty} = \sum p(n)q^n$ completes the proof.

$$\sum_{n=0}^{\infty} ge(n,k)q^n = \frac{1}{(q;q)_\infty} - \frac{1}{(q;q)_\infty} \sum_{j=0}^{\infty} (-1)^j q^{\frac{3j^2+(2k+1)j}{2}} \begin{bmatrix} k+j-1 \\ j \end{bmatrix}$$

$$= \frac{1}{(q;q)_\infty} \sum_{j=1}^{\infty} (-1)^{j+1} q^{\frac{3j^2+(2k+1)j}{2}} \begin{bmatrix} k+j-1 \\ k-1 \end{bmatrix}. \qquad \square$$

We now show that previous results about counting GE-partitions follow from the theorem.

Corollary 1

(a) $\quad ge(n,1) = \sum_{j=1}^{\infty} (-1)^{j+1} p\left(n - \frac{3j^2+3j}{2}\right).$

(b) $\quad ge(n,2) = \sum_{j=1}^{\infty} \sum_{\ell=0}^{j} (-1)^{j+1} p\left(n - \frac{3j^2+5j}{2} - \ell\right).$

(c) $\quad$ If $\frac{n-1}{2} \le k \le n-2$, then $ge(n,k) = \sum_{i=0}^{n-k-2} p(i).$

(d) $\quad ge(n, n-1) = ge(n,n) = 0.$

Proof (a) When $k = 1$, the q-binomial coefficient in the theorem equation reduces to 1. Writing the expression in terms of $p(n)$ follows from comparing coefficients. This formula for $ge(n,1)$ is proven in [5] in two ways: a (different) generating function argument, and a combinatorial proof.

(b) When $k = 2$, the q-binomial coefficient reduces to $\sum_{\ell=0}^{j} q^\ell$. This confirms a conjecture of [4]. Simplified formulas for other small values of k can be derived from the theorem.

(c) When $\frac{n-1}{2} \le k \le n-2$, only the $j = 1$ term on the right hand side gives an exponent of q that is not too large, leaving a sum of partition numbers. This result is proven in [4] combinatorially.

(d) When $k = n - 1$ or more, all coefficients on the right hand side of the theorem equation are too large to match coefficients. This corresponds to the fact that every partition has a preimage under conjugation. □

We conclude with a note on combinatorial proofs. For $3 \leq k < \frac{n-1}{2}$, the formulas for $ge(n, k)$ include $p(i)$ terms with coefficients greater than 1, which could make a combinatorial proof more challenging. However, the $k = 2$ case, part (b) of the corollary, has all $p(i)$ terms with coefficient 1, like the $k = 1$ and $\frac{n-1}{2} \leq k$ cases that have combinatorial proofs as mentioned above. Therefore we encourage work on a combinatorial proof of the $ge(n, 2)$ formula, if not the full $ge(n, k)$ result.

Acknowledgements   These results were presented at the conference Combinatory Analysis 2008: Partitions, q-series, and Applications, held December 2008 at The Pennsylvania State University in honor of George Andrews' 70th birthday. We are grateful to the conference organizers, Krishnaswami Alladi, Peter Paule, James Sellers, and Ae Ja Yee, for making this gathering possible, and of course to the honoree himself for his leadership and encouragement.

References

1. Andrews, G.: The Theory of Partitions. Cambridge University Press, Cambridge (1984)
2. Brandt, J.: Cycles of partitions. Proc. Am. Math. Soc. **85**, 483–486 (1982)
3. Gardner, M.: Bulgarian solitaire and other seemingly endless tasks. Sci. Am. **249**, 12–21 (1983)
4. Hopkins, B.: Column-to-row operations on partitions: the envelopes. In: Combinatorial Number Theory, Proceedings in Mathematics, pp. 65–76. de Gruyter, Berlin (2009). Also available in Integers 9 Supplement, Article 6 (2009)
5. Hopkins, B., Sellers, J.: Exact enumeration of Garden of Eden partitions. Integers **7**(2), A19 (2007)

Ramanujan J (2010) 23: 341–353
DOI 10.1007/s11139-009-9195-8

Some Fine combinatorics

David P. Little

Dedicated to George Andrews on the occasion of his 70th birthday

Received: 30 June 2009 / Accepted: 12 August 2009 / Published online: 31 August 2010
© Springer Science+Business Media, LLC 2010

Abstract In 1988, N.J. Fine published a monograph entitled *Basic Hypergeometric Series and Applications* in which he proved a number of results concerning the series $F(a, b; t : q)$. In this paper, we present a new combinatorial interpretation for the series $F(a, b; t : q)$ and use Fine's work as a guide for proving the Rogers–Fine identity and many of its properties in this setting.

Keywords Rogers–Fine identity · Basic hypergeometric series · q-difference equations

Mathematics Subject Classification (2000) Primary 05A15

1 Introduction

N.J. Fine [2] presented an introduction to the theory of basic hypergeometric series by giving an extensive study of the series

$$F(a, b; t : q) = \sum_{n=0}^{\infty} \frac{(aq)_n t^n}{(bq)_n},$$

where $(z)_n = (1 - z)(1 - zq) \cdots (1 - zq^{n-1})$. By deriving numerous functional equations satisfied by $F(a, b; t : q)$, one of the many results Fine proved is the so-called

---

D.P. Little (✉)
Department of Mathematics, Penn State University, University Park, PA 16802, USA
e-mail: dlittle@math.psu.edu

Rogers–Fine identity [3]

$$F(a, b; t : q) = \sum_{n=0}^{\infty} \frac{(aq)_n(atq/b)_n(1 - atq^{2n+1})b^n t^n q^{n^2}}{(bq)_n(t)_{n+1}} \tag{1}$$

along with numerous well-known identities that appear as special cases of (1). It is our goal to provide a new combinatorial context for $F(a, b; t : q)$ that easily explains the Rogers–Fine identity, including many of its properties and applications.

To this end, we consider weighted permutations of a multiset that, for illustrative purposes, we present as tilings of a single row of finite length using black, gray, and white squares. Given a tile t, we define its weight $w(t)$ as

$$w(t) = \begin{cases} aq^i & \text{if } t \text{ is a } \blacksquare \text{ with } i \; \square \text{ or } \square \text{ to its left,} \\ bq^i & \text{if } t \text{ is a } \square \text{ with } i \; \square \text{ or } \square \text{ to its left,} \\ c & \text{if } t \text{ is a } \square. \end{cases}$$

The weight of a tiling T is given by

$$w_T(a, b, c; q) = \prod_{t \in T} w(t).$$

For example, if the tiling T is given by

then the weight of T is

$$\begin{aligned} w_T(a, b, c; q) &= c \cdot aq \cdot c \cdot bq^2 \cdot c \cdot c \cdot c \cdot bq^6 \cdot aq^7 \cdot c \cdot c \cdot bq^9 \cdot c \cdot aq^{11} \\ &= a^3 b^3 c^8 q^{36}. \end{aligned}$$

Clearly the power of a in $w_T(a, b, c; q)$ keeps track of the number of black squares in T, the power of b keeps track of the number of gray squares, and the power of c keeps track of the number of white squares. The power of q is very similar to the usual inversion statistic of a permutation, except that it also includes the number of pairs of gray squares.

For the purposes of this paper, we will restrict ourselves to considering tilings that do not end with a white square. In particular, define

$$G(a, b, c; q) = \sum_T w_T(a, b, c; q),$$

where the sum is over all tilings of finite length that end with a black or gray square, including the empty tiling, which has weight 1. Note that we are considering all tilings that do not end with a white square and not restricting ourselves to tilings of a certain length.

It should be clear that $aG(a, b, c; q)$ is the generating function for tilings that start with a black square and do not end with a white square and that $c^j G(a, b, c; q)$ is the generating function for tilings that end with exactly j consecutive white squares. Consequently,

$$(1 - a)G(a, b, c; q)$$

can be interpreted as the generating function for tilings that do not start with a black square and do not end with a white square, and

$$\frac{G(a, b, c; q)}{1 - c}$$

is the generating function for all tilings without restriction. While we could have chosen either of these types of tilings as our focal point, as it turns out, tilings that do not end with a white square will be in keeping with the identities as presented in [2].

2 Proof of the Rogers–Fine identity

We begin by making the substitutions $a \to -\frac{b}{aq}$, $b \to c$ and $t \to a$ in (1), which results in

$$\sum_{n=0}^{\infty} \frac{(-b/a)_n a^n}{(cq)_n} = \sum_{n=0}^{\infty} \frac{(-b/a)_n (-b/c)_n a^n c^n q^{n^2} (1 + bq^{2n})}{(a)_{n+1}(cq)_n}.$$

We will show that both sides of this transformed identity are equal to $G(a, b, c; q)$.

Consider all tilings that do not end with a white square and have exactly n black or gray squares. Each such tiling can be broken up into n segments consisting of consecutive white squares followed by a single black or gray square. In other words, each black or gray square marks the end of a segment. The example from the previous section would be broken up in the following manner:

Suppose that the ith segment from left to right consists of exactly $j \geq 0$ white squares followed by a single tile t, which is a black or gray square. Note that each white square in this segment has exactly $i - 1$ black or gray squares to its left and $n + 1 - i$ black or gray squares to its right. Therefore, each white square contributes a factor of cq^{n+1-i} to the weight of the tiling since each of the last $n + 1 - i$ black or gray squares will count the white square as being to its left. Tile t contributes a factor of a if it is a black square or a factor of bq^{n-i} if it is a gray square since each of the last $n - i$ black or gray squares will count tile t as being to its left. Allowing for all possible values of j and t, the ith segment of tiles accounts for a weight of

$$\left(a + bq^{n-i}\right) \sum_{j=0}^{\infty} c^j q^{j(n+1-i)} = \frac{a + bq^{n-i}}{1 - cq^{n+1-i}}.$$

Therefore, the generating function for all tilings that have exactly n black or gray squares is given by

$$\prod_{i=1}^{n} \frac{a + bq^{n-i}}{1 - cq^{n+1-i}} = \frac{(-b/a)_n a^n}{(cq)_n}.$$

Summing over all $n \geq 0$ accounts for all tilings that do not end with a white square and yields

$$G(a, b, c; q) = \sum_{n=0}^{\infty} \frac{(-b/a)_n a^n}{(cq)_n}.$$

To complete the proof of the Rogers–Fine identity, we need to construct the same set of tilings in a different manner. To this end, we define the *weighted center* of a tiling to be the place on the board where the number of gray or white squares to its left is the same as the number of black or gray squares to its right. To justify the existence of such a position, the weighted center of a tiling can be found in the following manner. Initially place one marker ($\blacktriangle$) to the left of the first tile and another marker to the right of the last tile. Then alternate applying the following two steps, until a weighted center is found.

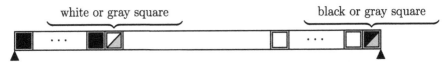

Step 1: If the left marker has a black square immediately to its right, then repeatedly move the left marker one tile to its right until this is no longer the case. At the same time, if the right marker has a white square immediately to its left, then repeatedly move the right marker one tile to its left until this is no longer the case.

Step 2: If the two markers are now in the same location, then that location is the weighted center. If there is exactly one tile separating the two markers, then that tile must be a gray square, and the middle of that gray square is the weighted center. Otherwise, move the left marker one tile to its right, move the right marker one tile to its left, and repeat Step 1.

Note that after either step, the number of white or gray squares to the left of the left marker is always equal to the number of black or gray squares to the right of the right marker. Thus when the process terminates, we have located a position on the board that satisfies the definition of a weighted center. And each time the process repeats, the markers are closer together than when they started, and therefore the process must terminate.

Now suppose that there are two places on the board that satisfy the definition of a weighted center. Let the two positions be indicated by c_1 and c_2 with c_1 being to the left of c_2. If there are x_i white or gray squares strictly to the left of c_i for $i = 1, 2$, then $x_1 \le x_2$ since c_2 is to the right of c_1. However, we must also have $x_1 \ge x_2$ since x_i is also the number of black or gray squares to the right of c_i and c_1 is to the left of c_2. Thus $x_1 = x_2$.

Next, let tile t_1 be the x_1th white or gray square from the left, and let tile t_2 be the x_1th black or gray square from the right. Since the weighted center must be between these two tiles, the only tiles separating t_1 and t_2 could be a sequence of black squares, followed by at most one gray square, followed by a sequence of white squares.

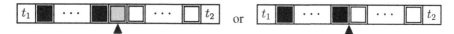

However, the weighted center for this portion of the tiling is unique (indicated by a single marker in each of the above diagrams), and thus, c_1 and c_2 must mark the same location on the board. In other words, every tiling must have a *unique* weighted center.

We say that a tiling is of *degree n* if there are exactly n gray or white squares strictly to the left (or equivalently, n black or gray squares strictly to the right) of its weighted center. If the weighted center falls on a gray square, that gray square is not counted as part of the degree of the tiling. For example, the following tiling is of degree 4, and the weighted center is indicated by a single marker.

Notice that inserting any number of black squares before the weighted center and/or white squares after the weighted center does not change the degree of the tiling.

Furthermore, if the weighted center is between two tiles, then inserting a gray square at the weighted center does not change the degree of the tiling; however, the weighted center of the resulting tiling would coincide with the inserted tile.

Now back to our proof. Consider all tilings of degree n. To construct such a tiling, first place n gray and/or white squares in positions 1 through n. A white square placed in position j contributes a factor of c to the weight of the tiling. A gray square in position j contributes a factor of bq^{j-1} since there are exactly $j - 1$ gray or white squares to its left. In other words, the factor of $c + bq^{j-1}$ represents the choice of initially placing a white or gray square, respectively, in position j for $1 \le j \le n$.

Therefore, allowing for all possible arrangements of gray and white squares in the first n positions contributes

$$\prod_{j=1}^{n}\left(c + bq^{j-1}\right) = (-b/c)_n c^n$$

to the weight of all tilings of degree n.

Next, place n black and/or gray squares in positions $n + 1$ through $2n$. A black square placed in position $n + j$ contributes a factor of aq^n to the weight of the tiling. A gray square placed in position $n + j$ contributes a factor of bq^{2n-j} since the n squares in positions 1 through n are to its left and each of the last $n - j$ black or gray squares are to its right. In other words, the factor of $aq^n + bq^{2n-j}$ represents the choice of initially placing a black or gray square, respectively, in position $n + j$ for $1 \leq j \leq n$. Therefore, allowing for all possible arrangements of black and gray squares contributes

$$\prod_{j=1}^{n}\left(aq^n + bq^{2n-j}\right) = (-b/a)_n a^n q^{n^2}$$

to the weight of all tilings of degree n.

Now decide whether or not the weighted center coincides with the position of a gray square. Or equivalently, whether or not to insert a gray square immediately after the nth square. If you do not insert a gray square, then the weight of the tiling is unchanged. However, if a gray square is inserted, the weight of the tiling changes, but the degree does not. In particular, the weight of the gray square is bq^n, and its presence increases the weight of each of the n black or gray squares that appear to its right by a factor of q. Therefore, if there is a gray square at the weighted center, then it increases the weight of the tiling by a factor of bq^{2n}. Thus the factor

$$\left(1 + bq^{2n}\right)$$

represents the choice of having a gray square at the weighted center or not.

At this point, we have guaranteed that the tiling is of degree n, but we have not yet constructed all such tilings. Recall that inserting any number of black squares before the weighted center and/or any number of white squares after the weighted center does not change the degree of the tiling.

Suppose that exactly $j \geq 0$ consecutive black squares are inserted immediately *after* the ith gray or white square for $i = 0, 1, 2, \ldots, n$, where the $i = 0$ case means that the black squares are placed *before* the first gray or white square. Each of the j black squares contributes a factor of aq^i to the weight of the tiling since there are exactly i gray or white squares to its left. Allowing for all possible values of j, the consecutive black squares immediately after the ith gray or white square contributes

$$\sum_{j=0}^{n} a^j q^{ji} = \frac{1}{1 - aq^i}$$

to the weight of all tilings of degree n.

Now suppose that exactly $j \geq 0$ consecutive white squares are inserted immediately before the ith black or gray square to the right of the weighted center for $i = 1, 2, \ldots, n$. Each of the j white squares contributes a factor of cq^{n+1-i} to the weight of the tiling since there are exactly $n + 1 - i$ black or gray squares to its right. Allowing for all possible values of j, the consecutive white squares immediately before the ith black or gray square after the weighted center contributes

$$\sum_{j=0}^{n} c^j q^{j(n+1-i)} = \frac{1}{1 - cq^{n+1-i}}$$

to the weight of all tilings of degree n. Therefore, the generating function for all tilings of degree n is given by

$$\frac{(-b/a)_n (-b/c)_n a^n c^n q^{n^2} (1 + bq^{2n})}{(1-a)\cdots(1-aq^n)(1-cq)\cdots(1-cq^n)} = \frac{(-b/a)_n (-b/c)_n a^n c^n q^{n^2} (1 + bq^{2n})}{(a)_{n+1}(cq)_n}.$$

Summing over all $n \geq 0$ again accounts for all tilings that do not end with a white square and yields

$$G(a, b, c; q) = \sum_{n=0}^{\infty} \frac{(-b/a)_n (-b/c)_n a^n c^n q^{n^2} (1 + bq^{2n})}{(a)_{n+1}(cq)_n},$$

which completes the proof of the Rogers–Fine identity.

3 Functional equations

In Sects. 2 and 4 of [2], Fine presented a number of functional equations satisfied by $F(a, b; t : q)$, one of which was used to prove the Rogers–Fine identity. For the sake of completeness, we present here many of these functional equations in the context of the series $G(a, b, c; q)$.

For each of the functional equations presented in this section, we will make use of the following trivial observation regarding the substitution $a \to aq$. If tiling T contains exactly n black squares, then

$$w_T(aq, b, c; q) = q^n w_T(a, b, c; q).$$

In other words, to increase the weight of the tiling by a factor of q for each black square, simply replace a with aq. In a similar manner, the substitutions $b \to bq$ and $c \to cq$ can be used to increase the weight of a tiling by the number of gray and white squares, respectively. With that in mind, we are now ready to prove the following identities.

Theorem 1

$$G(a, b, c; q) = 1 + \frac{a+b}{1-cq} G(a, bq, cq; q).$$

Proof Since the weight of the empty tiling is 1, $G(a, b, c; q) - 1$ is the generating function for all weighted tilings that have at least one black or gray square. Any such tiling, T, can be uniquely decomposed into a (possibly empty) tiling, T', followed by a sequence of $j \geq 0$ consecutive white squares, followed by a single black or gray square, t.

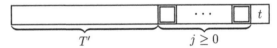

The power of q in $w(t)$ is j plus the number of gray or white squares in T', which is determined by replacing b with bq and c with cq in $w_{T'}(a, b, c; q)$. In other words, the weight of T is given by

$$w_T(a, b, c; q) = \begin{cases} ac^j q^j w_{T'}(a, bq, cq; q) & \text{if } t \text{ is a } \blacksquare, \\ bc^j q^j w_{T'}(a, bq, cq; q) & \text{if } t \text{ is a } \square. \end{cases}$$

Summing over all possible tilings T' and nonnegative integers j yields

$$G(a, b, c; q) - 1 = \sum_{j=0}^{\infty} \sum_{T'} ac^j q^j w_{T'}(a, bq, cq; q) + bc^j q^j w_{T'}(a, bq, cq; q)$$

$$= (a + b) \sum_{j \geq 0} c^j q^j \sum_{T'} w_{T'}(a, bq, cq; q)$$

$$= \frac{a + b}{1 - cq} G(a, bq, cq; q),$$

as desired. $\square$

Theorem 2

$$G(a, b, c; q) = \frac{1 - c}{1 - a} + \frac{b + c}{1 - a} G(aq, bq, c; q).$$

Proof We begin our proof by pointing out that any tiling T falls into exactly one of the following three categories based on the first nonblack tile, if any, to appear in T.

If all of the tiles in T are black squares, then suppose that T consists of exactly $j \geq 0$ consecutive black squares and no other tiles. Since each square contributes a factor of a to the weight of T, the generating function for all such tilings, including the empty tiling, is given by

$$\sum_{j=0}^{\infty} a^j = \frac{1}{1 - a}.$$

If the first nonblack tile is a white square, then T can be decomposed into $j \geq 0$ consecutive black squares followed by a single white square followed by a nonempty tiling, T', since T cannot end with a white square.

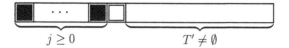

$$j \geq 0 \qquad\qquad T' \neq \emptyset$$

The first white square contributes a factor of c and one factor of q for each black or gray square in T'. These additional factors of q can be obtained by replacing a with aq and b with bq in $w_{T'}(a, b, c; q)$. Therefore, the generating function for all such T is given by

$$\sum_{j=0}^{\infty} \sum_{T' \neq \emptyset} ca^j w_{T'}(aq, bq, c; q) = \frac{c}{1-a}\big[G(aq, bq, c; q) - 1\big].$$

Otherwise, the first nonblack tile is a gray square, in which case T can be decomposed into $j \geq 0$ consecutive black squares followed by a single gray square followed by any tiling, T'.

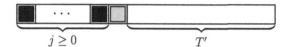

$$j \geq 0 \qquad\qquad T'$$

The first gray square contributes a factor of b and one factor of q for each black or gray square in T'. Therefore, the generating function for all such T is given by

$$\sum_{j=0}^{\infty} \sum_{T'} ba^j w_{T'}(aq, bq, c; q) = \frac{b}{1-a} G(aq, bq, c; q)$$

and

$$G(a, b, c; q) = \frac{1}{1-a} + \frac{c}{1-a}\big[G(aq, bq, c; q) - 1\big] + \frac{b}{1-a}G(aq, bq, c; q)$$

$$= \frac{1-c}{1-a} + \frac{b+c}{1-a}G(aq, bq, c; q),$$

as claimed. $\square$

Theorem 3

$$G(a, b, c; q) = \frac{1+b}{1-a} + \frac{(a+b)(b+c)q}{(1-a)(1-cq)}G(aq, bq^2, cq; q).$$

Proof For this proof, we will break up tilings into two categories, based on whether or not the tiling is of degree 0.

Consider all tilings of degree 0. Each such tiling consists of exactly $j \geq 0$ consecutive black squares and no other tiles or a sequence of exactly $j \geq 0$ consecutive black squares followed by a single gray square. The generating function for all such tilings is given by

$$\sum_{j=0}^{\infty} a^j + a^j b = \frac{1+b}{1-a}.$$

Now consider all tilings of degree at least 1. Any such tiling consists of $j \geq 0$ consecutive black squares, followed by a single gray or white square, t_1, followed by a tiling T' of degree at least 0 (i.e., any tiling), followed by $k \geq 0$ consecutive white squares, followed by a single black or gray square, t_2.

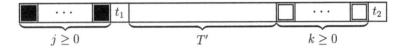

The tile t_1 contributes a factor of q to the weight of T for each black or gray square in T'. The tile t_2 contributes a factor of q for each gray or white square in T'. In other words, these additional factors of q can be obtained by replacing a with aq, b with bq^2, and c with cq in $w_{T'}(a,b,c;q)$. Therefore, the generating function for all such T is given by

$$(b+c)(a+b)q \sum_{j=0}^{\infty} \sum_{T'} \sum_{k=0}^{\infty} a^j w_{T'}(aq, bq^2, cq,)c^k q^k$$

$$= \frac{(b+c)(a+b)q}{(1-a)(1-cq)} G(aq, bq^2, cq; q).$$

The factor of $(b+c)$ accounts for the possibility of t_1 being a gray or white square, the factor of $(a+b)$ accounts for the possibility of t_2 being a black or gray square, and the factor of q accounts for the fact that t_1 is to the left of t_2. Adding the above two cases completes the proof.   □

4 Specializations

In Sect. 6 of [2], Fine considered several applications of iterations of the substitution $t \rightarrow tq$. In particular, Fine was led to identities involving $F(a,0;t:q)$ and $F(a,1;t:q)$. We present here the corresponding equations for $G(a,b,0;q)$ and $G(a,b,1;q)$.

In the case $c = 0$ (i.e., the tiling has no white squares) we can count tilings based on the number of gray squares. More specifically, consider all tilings that have exactly n gray squares. Each such tiling starts with $j \geq 0$ consecutive black squares followed by n segments, where each segment consists of a single gray square followed by any number of consecutive black squares.

If the ith segment contains $k \geq 0$ black squares, then the weight of the ith gray square is bq^{i-1} and the weight of each of the black squares in the ith segment is aq^i. Allowing for all possible values of k, the weight of the ith segment contributes

$$bq^{i-1} \sum_{k=0}^{\infty} a^k q^{ik} = \frac{bq^{i-1}}{1 - aq^i}$$

to the generating function $G(a, b, 0; q)$. Therefore, the generating function for tilings with no white squares and exactly n gray squares is given by

$$\sum_{j=0}^{\infty} a^j \prod_{i=1}^{n} \frac{bq^{i-1}}{1 - aq^i} = \frac{b^n q^{(n^2 - n)/2}}{(1 - a)(aq)_n}$$

$$= \frac{b^n q^{(n^2 - n)/2}}{(a)_{n+1}},$$

and thus

$$\sum_{n=0}^{\infty} (-b/a)_n a^n = G(a, b, 0; q) = \sum_{n=0}^{\infty} \frac{b^n q^{(n^2 - n)/2}}{(a)_{n+1}}.$$

In the case $c = 1$, Theorem 2 yields

$$G(a, b, 1; q) = \frac{1 + b}{1 - a} G(aq, bq, 1; q).$$

Applying this functional equation n times yields

$$G(a, b, 1; q) = \frac{(1 + b)}{(1 - a)} \frac{(1 + bq)}{(1 - aq)} \frac{(1 + bq^2)}{(1 - aq^2)} \cdots \frac{(1 + bq^{n-1})}{(1 - aq^{n-1})} G(aq^n, bq^n, 1; q)$$

$$= \frac{(-b)_n}{(a)_n} G(aq^n, bq^n, 1; q).$$

If we assume that $|q| < 1$, taking the limit as n tends to infinity produces

$$G(a, b, 1; q) = \frac{(-b)_\infty}{(a)_\infty} G(0, 0, 1; q)$$

$$= \frac{(-b)_\infty}{(a)_\infty}$$

since $G(0, 0, 1; q) = 1$. This completes a proof of Cauchy's q-analog of the binomial theorem [1] since

$$\sum_{n=0}^{\infty} \frac{(-b/a)_n a^n}{(q)_n} = G(a, b, 1; q) = \prod_{i=0}^{\infty} \frac{1 + bq^i}{1 - aq^i}.$$

Cauchy's result may also be obtained directly by pointing out that every tiling can be broken up into segments consisting of any number of consecutive black squares followed by a single gray or white square. In other words, every gray or white square marks the end of a segment.

Note that this means we now have to allow for tilings to end with a white square. In particular, since the weight of a white square is 1, we can think of our tilings as being an infinitely long sequence of tiles with only a finite number of black and/or gray squares.

5 Symmetry of $(1-a)G(a, b, c; q)$

The final result in Sect. 6 of [2] is that $(1 - t)F(a, b; t : q)$ is invariant after the substitution $a \to at/b$, $b \to t$, and $t \to b$. In this final section, we prove the same result in terms of the function $(1 - a)G(a, b, c; q)$. To this end, recall that

$$(1-a)G(a, b, c; q)$$

can be interpreted as the generating function for tilings that do not start with a black square and as usual, do not end with a white square. We will show that this series is symmetric in the variables a and c. In other words,

$$(1 - a)G(a, b, c; q) = (1 - c)G(c, b, a; q) \qquad (2)$$

or equivalently,

$$G(a, b, c; q) = \frac{1 - c}{1 - a} G(c, b, a; q).$$

We begin by noting that $(1 - c)G(c, b, a; q)$ can be interpreted as the generating function for tilings that do not start with a black square and do not end with a white square, where the weight of each tiling is computed using the new weight function, $\tilde{w}(t)$, defined by

$$\tilde{w}(t) = \begin{cases} cq^i & \text{if } t \text{ is a } \blacksquare \text{ with } i \ \square \text{ or } \square \text{ to its left,} \\ bq^i & \text{if } t \text{ is a } \square \text{ with } i \ \square \text{ or } \square \text{ to its left,} \\ a & \text{if } t \text{ is a } \square \end{cases}$$

and $\tilde{w}_T(a, b, c; q) = \prod_{t \in T} \tilde{w}(t)$. To prove (2), it suffices to describe an involution, φ, on the set of tilings such that

$$w_T(a, b, c; q) = \tilde{w}_{\varphi(T)}(a, b, c; q).$$

Given a tiling T, $\varphi(T)$ is obtained by reversing the order of the tiles in T and then replacing each black square with a white square, and vice versa. For example, if T is given by

then $\varphi(T)$ is

Clearly this operation is an involution on the set of tilings that do not start with a black square and do not end with a white square. Furthermore, the number of black squares in T is equal to the number of white squares in $\varphi(T)$, the number of gray squares in T is equal to the number of gray squares in $\varphi(T)$, and the number of white squares in T is equal to the number of black squares in $\varphi(T)$. This implies that the power of a, b, and c, respectively, in $w_T(a, b, c; q)$ is the same as the power of a, b, and c in $\tilde{w}_{\varphi(T)}(a, b, c; q)$.

It remains to show that $w_T(a, b, c; q)$ and $\tilde{w}_{\varphi(T)}(a, b, c; q)$ have the same power of q. However, this is a simple consequence of the following observations:

- T has a ☐ in position i and a ■ in position j if and only if
  $\varphi(T)$ has a ☐ in position $n + 1 - j$ and a ■ in position $n + 1 - i$.
- T has a ▦ in position i and a ■ in position j if and only if
  $\varphi(T)$ has a ☐ in position $n + 1 - j$ and a ▦ in position $n + 1 - i$.
- T has a ☐ in position i and a ▦ in position j if and only if
  $\varphi(T)$ has a ▦ in position $n + 1 - j$ and a ■ in position $n + 1 - i$.
- T has a ▦ in position i and a ▦ in position j if and only if
  $\varphi(T)$ has a ▦ in position $n + 1 - j$ and a ▦ in position $n + 1 - i$,

where n is the total number of tiles in T. In other words, every factor of q that arises from the weight of a black or gray square in T corresponds to a factor of q from the weight of a black or gray square in $\varphi(T)$, and vice versa. For example, the weight of $\varphi(T)$ is given by

$$\tilde{w}_{\varphi(T)}(a, b, c; q) = a \cdot cq \cdot bq \cdot cq^2 \cdot cq^2 \cdot a \cdot bq^3 \cdot cq^4 \cdot cq^4 \cdot cq^4 \cdot bq^4 \cdot cq^5 \cdot a \cdot cq^6$$

$$= a^3 b^3 c^8 q^{36}$$

$$= w_T(a, b, c; q).$$

References

1. Andrews, G.E.: The Theory of Partitions. Cambridge Mathematical Library. Cambridge University Press, Cambridge (1998). Reprint of the 1976 original
2. Fine, N.J.: Basic Hypergeometric Series and Applications. Mathematical Surveys and Monographs, vol. 27. American Mathematical Society, Providence (1988). With a foreword by George E. Andrews
3. Rogers, L.J.: On two theorems of combinatory analysis and some allied identities. Proc. Lond. Math. Soc. **16**(2), 315–336 (1917)

Ramanujan J (2010) 23: 355–369
DOI 10.1007/s11139-010-9232-7

Symmetrically constrained compositions

**Matthias Beck · Ira M. Gessel · Sunyoung Lee ·
Carla D. Savage**

Dedicated to George Andrews on the occasion of his seventieth birthday

Received: 30 June 2009 / Accepted: 25 March 2010 / Published online: 22 April 2010
© Springer Science+Business Media, LLC 2010

Abstract Given integers $a_1, a_2, \ldots, a_n$, with $a_1 + a_2 + \cdots + a_n \geq 1$, a symmetrically constrained composition $\lambda_1 + \lambda_2 + \cdots + \lambda_n = M$ of M into n nonnegative parts is one that satisfies each of the $n!$ constraints $\{\sum_{i=1}^{n} a_i \lambda_{\pi(i)} \geq 0 : \pi \in S_n\}$. We show how to compute the generating function of these compositions, combining methods from partition theory, permutation statistics, and lattice-point enumeration.

Keywords Symmetrically constrained composition · Partition analysis ·
Permutation statistics · Generating function · Lattice-point enumeration

Mathematics Subject Classification (2000) Primary 05A17 · 05A15 · 11P21

This work was supported in part by the National Science Foundation Grants INT-0230800,
DMS-0300034, and DMS-0810105 and the National Security Agency Grant H98230-08-1-0072.

M. Beck
Department of Mathematics, San Francisco State University, San Francisco, CA 94132, USA
e-mail: beck@math.sfsu.edu

I.M. Gessel
Department of Mathematics, Brandeis University, Waltham, MA 02454-9110, USA
e-mail: gessel@brandeis.edu

S. Lee · C.D. Savage (✉)
Department of Computer Science, North Carolina State University, Raleigh, NC 27695-8206, USA
e-mail: savage@ncsu.edu

S. Lee
e-mail: sunyoung92@gmail.com

Present address:
S. Lee
Department of Interaction Science, Sungkyunkwan University, Seoul 110-745, South Korea

1 Introduction

1.1 Constrained compositions

This work was inspired by the "constrained compositions" introduced by Andrews, Paule, and Riese in [4]. We consider the problem of enumerating *symmetrically constrained compositions*, that is, compositions of an integer M into n nonnegative parts

$$M = \lambda_1 + \lambda_2 + \cdots + \lambda_n = |\lambda|,$$

where the sequence $(\lambda_1, \lambda_2, \ldots, \lambda_n)$ is constrained to satisfy a symmetric system of linear inequalities. For example, the compositions $\lambda_1 + \lambda_2 + \lambda_3$ of M satisfying

$$\lambda_{\pi(1)} + \lambda_{\pi(2)} \geq \lambda_{\pi(3)} \tag{1}$$

for every permutation $\pi \in S_3$, are known as *integer-sided triangles* of perimeter M [1, 2, 11, 14]. The number Δ_M of *incongruent* triangles of perimeter M is given by

$$\sum_{M \geq 0} \Delta_M q^M = \sum_{\substack{\lambda_1 \geq \lambda_2 \geq \lambda_3 \geq 0 \\ \lambda_2 + \lambda_3 \geq \lambda_1}} q^{|\lambda|} = \frac{1}{(1-q^2)(1-q^3)(1-q^4)}.$$

(Note that, in contrast to [1, 2, 11, 14], we allow $\lambda_i = 0$.)

However, $3 + 2 + 1$ and $2 + 3 + 1$ are different *compositions* (i.e., ordered partitions) of 6 and counting the number Δ_M^* of *ordered* solutions to (1) gives

$$\sum_{M \geq 0} \Delta_M^* q^M = \sum_{\substack{\lambda_1 + \lambda_2 \geq \lambda_3 \\ \lambda_1 + \lambda_3 \geq \lambda_2 \\ \lambda_2 + \lambda_3 \geq \lambda_1}} q^{|\lambda|} = \frac{1 + 2q^2 + 2q^4 + q^6}{(1-q^2)(1-q^3)(1-q^4)} = \frac{1+q^3}{(1-q^2)^3}. \tag{2}$$

One could generalize this example in several ways. For example, moving to n dimensions, one could ask for the integer sequences $(\lambda_1, \lambda_2, \ldots, \lambda_n)$ satisfying

$$\lambda_{\pi(1)} + \lambda_{\pi(2)} + \cdots + \lambda_{\pi(n-1)} \geq \lambda_{\pi(n)}$$

for all $n!$ permutations $\pi \in S_n$. Another generalization would be to study, given positive integers k, ℓ, m, the integer sequences $(\lambda_1, \lambda_2, \ldots, \lambda_n)$ satisfying

$$k\lambda_{\pi(1)} + \ell\lambda_{\pi(2)} \geq m\lambda_{\pi(3)} \quad \text{for all } \pi \in S_n.$$

Another example, considered in [4], was inspired by a Putnam exam problem [12, Problem B3]: Enumerate all compositions of $M = \lambda_1 + \lambda_2$ into two parts satisfying

$$2\lambda_1 \geq \lambda_2 \quad \text{and} \quad 2\lambda_2 \geq \lambda_1.$$

It is shown that

$$\sum_{\substack{2\lambda_1 \geq \lambda_2 \\ 2\lambda_2 \geq \lambda_1}} x^{\lambda_1} y^{\lambda_2} = \frac{1 + xy + x^2 y^2}{(1 - xy^2)(1 - x^2 y)}, \tag{3}$$

giving a complete parametrization of all solutions.

In [4], Andrews, Paule, and Riese demonstrate the suitability of the Omega package [3] for experimenting with problems of this sort and the power of MacMahon's partition analysis [2] to prove some elegant generalizations.

The goal of this paper is (1) to formulate a generalization of the symmetrically constrained compositions enumeration problem; (2) to show how this problem is connected to permutation statistics; (3) to show that the permutation statistics approach gives, for many cases, an effective computation method and, for certain cases, a way to derive compact formulas; and (4) to show that the insight provided by the geometry of lattice-point enumeration aids in the handling of the most general case.

1.2 The symmetrically constrained compositions enumeration problem

Let $\mathbb{Z}_{\geq 0}$ denote the set of nonnegative integers. Fix integers $a_1, a_2, \ldots, a_n$ (which may be negative). We are interested in enumerating compositions $\lambda = (\lambda_1, \lambda_2, \ldots, \lambda_n) \in \mathbb{Z}_{\geq 0}^n$ that satisfy the $n!$ homogeneous linear constraints

$$a_1 \lambda_{\pi(1)} + a_2 \lambda_{\pi(2)} + \cdots + a_n \lambda_{\pi(n)} \geq 0 \quad \text{for all } \pi \in S_n.$$

Specifically, we are interested in computing the generating functions

$$F(z_1, z_2, \ldots, z_n) := \sum_{\lambda} z_1^{\lambda_1} z_2^{\lambda_2} \cdots z_n^{\lambda_n}$$

and

$$F(q) := F(q, q, \ldots, q) = \sum_{\lambda} q^{\lambda_1 + \lambda_2 + \cdots + \lambda_n},$$

by exploiting the symmetry of the constraints. Note that because of the symmetry, there is no loss of generality in assuming that

$$a_1 \leq a_2 \leq \cdots \leq a_n,$$

which we will do from now on.

In Sect. 2, we show how to solve the enumeration problem when $\sum_{i=1}^{n} a_i = 1$. In certain special cases, we show that permutation statistics can be used to derive elegant formulas. We note that even this simple case is difficult for general purpose software like the Omega Package [3], Xin's improvement of Omega [15], and LattE macchiato [10, 13], designed to enumerate solutions to linear Diophantine equations and inequalities. In Sect. 3 we solve the general problem. We close this section with some notation and background on permutation statistics.

1.3 Permutation statistics

Throughout the paper, the following notation is used: $[n]_q = (1 - q^n)/(1 - q)$;
$[n]_q! = \prod_{i=1}^{n} [i]_q$; and $(a; q)_n = \prod_{i=0}^{n-1} (1 - aq^i)$.

For a permutation $\pi = \pi(1)\pi(2) \cdots \pi(n)$ of $[n] := \{1, 2, \ldots, n\}$, the *descent set of* π is

$$D_\pi = \{ j : \pi(j) > \pi(j+1) \}.$$

The statistic $\text{des}(\pi) = |D(\pi)|$ is the number of descents of π and the *major index* of π is the sum of the descent positions: $\text{maj}(\pi) = \sum_{i \in D(\pi)} i$. It is well known that

$$\sum_{\pi \in S_n} q^{\text{maj}(\pi)} = \prod_{i=1}^{n} \frac{1 - q^i}{1 - q} = [n]_q! \tag{4}$$

(see, e.g., [14]). The joint distribution of $\text{des}(\pi)$ and $\text{maj}(\pi)$ over the set S_n of all permutations of $[n]$ is given by Carlitz's q-Eulerian polynomial [6, 7]:

$$C_n(x, q) = \sum_{\pi \in S_n} x^{\text{des}(\pi)} q^{\text{maj}(\pi)} = \prod_{i=0}^{n} (1 - xq^i) \sum_{j=1}^{\infty} [j]_q^n x^{j-1}.$$

Applying the definition of $[j]_q$ and the binomial theorem, we can rewrite this as

$$C_n(x, q) = \frac{(x; q)_{n+1}}{(1 - q)^n} \sum_{i=0}^{n} \binom{n}{i} \frac{(-q)^i}{1 - q^i x}. \tag{5}$$

So, for example,

$$C_1(x, q) = 1,$$
$$C_2(x, q) = 1 + xq, \tag{6}$$
$$C_3(x, q) = 1 + 2xq + 2xq^2 + x^2 q^3.$$

If we take the limit as $x \to q^{-n}$ in (5) all terms except $i = n$ in the sum are canceled by $(q^{-n}; q)_{n+1} = 0$ in the numerator, so

$$C_n(q^{-n}, q) = \frac{(-q)^n}{(1 - q)^n} \lim_{x \to q^{-n}} \frac{(x; q)_{n+1}}{1 - q^n x} = \frac{(-q)^n}{(1 - q)^n} \lim_{x \to q^{-n}} (x; q)_n$$

$$= \frac{(-q)^n}{(1 - q)^n} (q^{-n}; q)_n. \tag{7}$$

Finally, for $i \leq n - 1$, let $S_n^{(i)}$ be the set of permutations of $[n]$ that have no descent in positions $\{n - i, n - i + 1, \ldots, n - 1\}$. Let

$$C_n^{(i)}(x, q) := \sum_{\pi \in S_n^{(i)}} x^{\text{des}(\pi)} q^{\text{maj}(\pi)}.$$

In [8], it is shown that

$$C_n^{(i)}(x,q) = \frac{C_n(x,q)}{(xq^{n-i};q)_i} - \sum_{k=1}^{i} \binom{n}{k} xq^{n-k} \frac{C_{n-k}(x,q)}{(xq^{n-i};q)_{i-k+1}}$$

so, in particular,

$$C_n^{(1)}(x,q) = \frac{C_n(x,q) - nxq^{n-1}C_{n-1}(x,q)}{1 - xq^{n-1}}. \tag{8}$$

2 Symmetrically constrained compositions when $\sum a_i = 1$

2.1 The main theorem

Theorem 1 *Given integers $a_1 \le a_2 \le \cdots \le a_n$ satisfying $\sum_{i=1}^{n} a_i = 1$, the generating function for those $\lambda \in \mathbb{Z}_{\ge 0}^n$ satisfying*

$$\sum_{j=1}^{n} a_j \lambda_{\pi(j)} \ge 0 \quad \text{for all } \pi \in S_n$$

is

$$F(z_1, z_2, \ldots, z_n) = \sum_{\pi \in S_n} \frac{\prod_{j \in D_\pi} (z_1^{b_{1,j}} z_2^{b_{2,j}} \cdots z_n^{b_{n,j}})}{\prod_{j=1}^{n} (1 - z_1^{b_{1,j}} z_2^{b_{2,j}} \cdots z_n^{b_{n,j}})},$$

where

$$b_{i,j} = \begin{cases} 1 & \text{if } j = n, \\ -(a_1 + \cdots + a_j) & \text{if } n \ge i > j \ge 1, \\ 1 - (a_1 + \cdots + a_j) & \text{if } 1 \le i \le j < n. \end{cases}$$

In particular, setting $z_1 = \cdots = z_n = q$ yields

$$F(q) = \frac{\sum_{\pi \in S_n} \prod_{j \in D_\pi} q^{j - n \sum_{i=1}^{j} a_i}}{(1 - q^n) \prod_{j=1}^{n-1} (1 - q^{j - n \sum_{i=1}^{j} a_i})}.$$

Proof To simplify notation, let

$$F(z) = F(z_1, z_2, \ldots, z_n).$$

For $b \in \mathbb{Z}^n$, let

$$z^b = z_1^{b_1} z_2^{b_2} \cdots z_n^{b_n}$$

and for $\pi \in S_n$, let

$$z_\pi = (z_{\pi(1)}, z_{\pi(2)}, \ldots, z_{\pi(n)}).$$

With

$$L := \left\{ \lambda \in \mathbb{Z}_{\geq 0}^n : \sum_{j=1}^n a_j \lambda_{\pi(j)} \geq 0 \text{ for all } \pi \in S_n \right\}$$

we have

$$F(z) = \sum_{\lambda \in L} z^\lambda.$$

Now we use the standard method of partitioning the elements of L into classes L_π indexed by permutations $\pi \in S_n$:

$$L_\pi = \left\{ \lambda \in \mathbb{Z}^n : \lambda_{\pi(1)} \geq \lambda_{\pi(2)} \geq \cdots \geq \lambda_{\pi(n)}, \right.$$

$$\left. \sum_{i=1}^n a_i \lambda_{\sigma(i)} \geq 0 \text{ for all } \sigma \in S_n, \text{ and } \lambda_{\pi(i)} > \lambda_{\pi(i+1)} \text{ if } i \in D_\pi \right\}.$$

Since the last condition guarantees that no λ is in more than one class, L is the disjoint union

$$L = \bigcup_{\pi \in S_n} L_\pi.$$

Our goal now simplifies to computing

$$F_\pi(z) := \sum_{\lambda \in L_\pi} z^\lambda,$$

because $F(z) = \sum_{\pi \in S_n} F_\pi(z)$. In L_π, since

$$\lambda_{\pi(1)} \geq \lambda_{\pi(2)} \geq \cdots \geq \lambda_{\pi(n)}$$

and since, by our assumption, $a_1 \leq a_2 \leq \cdots \leq a_n$, the $n!$ constraints

$$a_1 \lambda_{\sigma(1)} + a_2 \lambda_{\sigma(2)} + \cdots + a_n \lambda_{\sigma(n)} \geq 0 \quad \text{for all } \sigma \in S_n$$

are all implied by the single constraint

$$a_1 \lambda_{\pi(1)} + a_2 \lambda_{\pi(2)} + \cdots + a_n \lambda_{\pi(n)} \geq 0,$$

so that we get the more compact description

$$L_\pi = \left\{ \lambda \in \mathbb{Z}^n : \begin{array}{l} \lambda_{\pi(1)} \geq \lambda_{\pi(2)} \geq \cdots \geq \lambda_{\pi(n)} \geq 0 \text{ and } \lambda_{\pi(j)} > \lambda_{\pi(j+1)} \text{ if } j \in D_\pi \\ a_1 \lambda_{\pi(1)} + a_2 \lambda_{\pi(2)} + \cdots + a_n \lambda_{\pi(n)} \geq 0 \end{array} \right\}.$$

But this means that all L_π look similar, except for the strict inequalities determined by D_π. More precisely, if we let

$$\tilde{L}_\pi := \{ \lambda \in L_{\mathrm{Id}} : \lambda_j > \lambda_{j+1} \text{ if } j \in D_\pi \}$$

and $G_\pi(z) := \sum_{\lambda \in \tilde{L}_\pi} z^\lambda$, then

$$F_\pi(z) = G_\pi(z_\pi).$$

So it remains to find $G_\pi(z)$, the generating function for

$$\tilde{L}_\pi = \left\{ \lambda \in \mathbb{Z}^n : \begin{array}{l} \lambda_1 \geq \lambda_2 \geq \cdots \geq \lambda_n \geq 0 \text{ and } \lambda_j > \lambda_{j+1} \text{ if } j \in D_\pi \\ a_1\lambda_1 + a_2\lambda_2 + \cdots + a_n\lambda_n \geq 0 \end{array} \right\},$$

for a given $\pi \in S_n$.

The constraints of $\tilde{L}_\pi$ are given by the system

$$
\begin{bmatrix}
1 & -1 & & & \\
& 1 & -1 & & \\
& & \ddots & & \\
& & & 1 & -1 \\
a_1 & a_2 & a_3 & \cdots & a_{n-1} & a_n
\end{bmatrix}
\lambda \geq
\begin{bmatrix}
e_1 \\
e_2 \\
\vdots \\
e_{n-1} \\
e_n
\end{bmatrix},
\tag{9}
$$

where

$$
e_j = \begin{cases}
0 & \text{if } j \notin D_\pi, \\
1 & \text{if } j \in D_\pi.
\end{cases}
$$

We make use of the following lemma, a well known result in lattice-point enumeration. This version was formulated in [9] for easy application to partition and composition enumeration problems.

Lemma 1 *Let $C = [c_{i,j}]$ be an $n \times n$ matrix of integers such that $C^{-1} = B = [b_{i,j}]$ exists and $b_{i,j}$ are all nonnegative integers. Let $e_1, \ldots, e_n$ be nonnegative integer constants. For each $1 \leq i \leq n$, let c_i be the constraint*

$$c_{i,1}\lambda_1 + c_{i,2}\lambda_2 + \cdots + c_{i,n}\lambda_n \geq e_i.$$

Let S_C be the set of nonnegative integer sequences $\lambda = (\lambda_1, \lambda_2, \ldots, \lambda_n)$ satisfying the constraints c_i for all i, $1 \leq i \leq n$. Then the generating function for S_C is:

$$F_C(x_1, x_2, \ldots, x_n) = \sum_{\lambda \in S_C} x_1^{\lambda_1} x_2^{\lambda_2} \cdots x_n^{\lambda_n} = \frac{\prod_{j=1}^n (x_1^{b_{1,j}} x_2^{b_{2,j}} \cdots x_n^{b_{n,j}})^{e_j}}{\prod_{j=1}^n (1 - x_1^{b_{1,j}} x_2^{b_{2,j}} \cdots x_n^{b_{n,j}})}.$$

Now let C be the matrix on the left side of (9). Then $\det(C) = a_1 + \cdots + a_n = 1$, so C is invertible and $B = C^{-1}$ has all integer entries:

$$
b_{i,j} = \begin{cases}
1 & \text{if } j = n, \\
-(a_1 + \cdots + a_j) & \text{if } n \geq i > j \geq 1, \\
1 - (a_1 + \cdots + a_j) & \text{if } 1 \leq i \leq j < n.
\end{cases}
$$

If, in addition, $a_1 + \cdots + a_j \leq 0$ for $1 \leq j \leq n-1$, the integer entries of $B = C^{-1}$ are all nonnegative and Lemma 1 gives the generating function $G_\pi(z)$ and the theorem follows.

To complete the proof, we show that if $a_1 + \cdots + a_n = 1$ and $a_1 \leq a_2 \leq \cdots \leq a_n$, then for $1 \leq j \leq n-1$ we have $a_1 + \cdots + a_j \leq 0$.

Let j be the smallest index satisfying $1 \leq j \leq n-1$ and $a_1 + \cdots + a_j \leq 0$, but $a_1 + \cdots + a_{j+1} > 0$. Then $a_{j+1} > -(a_1 + \cdots + a_j) \geq 0$. Thus

$$1 \leq a_{j+1} \leq \cdots \leq a_n,$$

so

$$1 = a_1 + \cdots + a_n \geq a_1 + \cdots + a_{j+1} + n - j - 1 > n - j - 1.$$

So $j = n-1$ and therefore $a_1 + \cdots + a_j \leq 0$ for $1 \leq j \leq n-1$. $\qquad\square$

In Sect. 2.3 we derive an algorithm based on Theorem 1 for efficient computation of $F(q)$, given the a_i. In the next section, we give examples of how to combine Theorem 1 with results on permutation statistics to derive formulas for $F(q)$ in special cases.

2.2 Applications

Example 1 Given positive integers b and $n \geq 2$, let L be the set of nonnegative integer sequences λ satisfying

$$(nb - b + 1)\lambda_{\pi(n)} \geq b(\lambda_{\pi(1)} + \cdots + \lambda_{\pi(n-1)}) \quad \text{for all } \pi \in S_n.$$

The case $n = 2$, $b = 1$ is the Putnam problem (3). Here $a = [-b, -b, \ldots, -b, nb - b + 1]$, so by Theorem 1,

$$F(q) = \frac{\sum_{\pi \in S_n} \prod_{j \in D_\pi} q^{j+jbn}}{(1 - q^n) \prod_{j=1}^{n-1}(1 - q^{j+jbn})} = \frac{\sum_{\pi \in S_n} (q^{1+bn})^{\mathrm{maj}(\pi)}}{(1 - q^n) \prod_{j=1}^{n-1}(1 - q^{j+jbn})}.$$

By (4), the numerator is just $[n]_{q^{1+bn}}!$ and simplifying gives

$$F(q) = \frac{1 - q^{n(nb+1)}}{(1 - q^n)(1 - q^{nb+1})^n}.$$

This generating function was discovered by Andrews, Paule, and Riese and a complete parametrization was proved in [4] using partition analysis.

Example 2 Given positive integers b and $n \geq 2$, let L be the set of nonnegative integer sequences λ satisfying

$$b(\lambda_{\pi(2)} + \cdots + \lambda_{\pi(n-1)}) \geq (nb - b - 1)\lambda_{\pi(1)} \quad \text{for all } \pi \in S_n.$$

The case $n = 3$, $b = 1$ is the integer-sided triangle problem (2) and the case $n = 2$, $b = 2$ is the Putnam problem (3). Here $a = [-(nb - b - 1), b, b, \ldots, b]$, so by Theorem 1,

$$F(q) = \frac{\sum_{\pi \in S_n} \prod_{j \in D_\pi} q^{(bn-1)(n-j)}}{(1 - q^n) \prod_{j=1}^{n-1}(1 - q^{(bn-1)(n-j)})} = \frac{\sum_{\pi \in S_n}(q^{1-bn})^{\mathrm{maj}(\pi)}(q^{n(bn-1)})^{\mathrm{des}(\pi)}}{(1 - q^n) \prod_{j=1}^{n-1}(1 - q^{(bn-1)(n-j)})}.$$

By (7), the numerator is

$$C_n\left(q^{n(bn-1)}, q^{1-bn}\right) = \frac{(q^{n(bn-1)}; q^{1-bn})_n(-q)^{(1-bn)n}}{(1 - q^{1-bn})^n}.$$

Simplifying further and dividing by the denominator gives

$$F(q) = \frac{1 - q^{n(nb-1)}}{(1 - q^n)(1 - q^{nb-1})^n}.$$

This generating function was also originally proved by Andrews, Paule, and Riese in [4].

Example 3 Given positive integers b and $n \geq 2$, let L be the set of nonnegative integer sequences $\lambda = (\lambda_1, \ldots, \lambda_n)$ satisfying the constraints

$$(b + 1)\lambda_{\pi(n)} \geq b\lambda_{\pi(1)} \quad \text{for all } \pi \in S_n.$$

The case $n = 2$, $b = 1$ is the Putnam problem (3). Here $a = [-b, 0, 0, \ldots, 0, b + 1]$ so by Theorem 1,

$$F(q) = \frac{\sum_{\pi \in S_n} \prod_{j \in D_\pi} q^{j+bn}}{(1 - q^n) \prod_{j=1}^{n-1}(1 - q^{j+bn})} = \frac{\sum_{\pi \in S_n} q^{\mathrm{maj}(\pi)}(q^{bn})^{\mathrm{des}(\pi)}}{(1 - q^n) \prod_{j=1}^{n-1}(1 - q^{j+bn})}.$$

By (5), the numerator is

$$C_n(q^{bn}, q) = \frac{(q^{bn}; q)_{n+1}}{(1 - q)^n} \sum_{i=0}^{n} \binom{n}{i} \frac{(-q^i)}{1 - q^{bn+i}}.$$

Combining with the denominator and simplifying gives

$$F(q) = \frac{(1 - q^{bn})(1 - q^{bn+n})}{(1 - q^n)(1 - q)^n} \sum_{i=0}^{n} \binom{n}{i} \frac{(-q^i)}{1 - q^{bn+i}}.$$

Example 4 Given positive integers $k \leq \ell$, and $n \geq 3$, let $m = k + \ell - 1$ and let L be the set of nonnegative integer sequences λ satisfying

$$k\lambda_{\pi(n-1)} + \ell\lambda_{\pi(n)} \geq m\lambda_{\pi(1)} \quad \text{for all } \pi \in S_n.$$

The case $n = 3$ and $k = \ell = 1$ is the integer-sided triangles (2). Here $a = [-m, 0, 0, \ldots, 0, k, \ell]$, so by Theorem 1,

$$F(q) = \frac{\sum_{\pi \in S_n} \prod_{j \in D_\pi, j \neq n-1} q^{j+mn} \prod_{j \in D_\pi, j = n-1} q^{j+mn-nk}}{(1 - q^n)(1 - q^{n\ell-1}) \prod_{j=1}^{n-2}(1 - q^{j+bn})}.$$

Recall from (8) that $C_n^{(1)}$ is the joint distribution of des and maj over all permutations with no descent in position $n - 1$. Then in $F(q)$, we can split the sum over $\pi \in S_n$ into two sums according to whether or not $i \in D_\pi$. We get that the numerator can be written as:

$$C_n^{(1)}(q^{nm}, q) + q^{-nk}(C_n(q^{nm}, q) - C_n^{(1)}(q^{nm}, q)).$$

Using (8) and combining with the denominator gives (eventually)

$$F(q) = \frac{C_n(q^{nm}, q)(1 - q^{n\ell-1}) - C_{n-1}(q^{nm}, q)nq^{nm+n-1}(1 - q^{-nk})}{(1 - q^n)(1 - q^{n\ell-1})(q^{nm+1}; q)_{n-1}}.$$

Details appear in [8].

2.3 Efficient enumeration of symmetrically constrained compositions

We can compute the generating function $F(q)$ for compositions satisfying the $n!$ constraints

$$\sum_{i=1}^{n} a_i \lambda_{\pi(i)} \geq 0 \quad \text{for all } \pi \in S_n$$

via Theorem 1. The denominator is given explicitly, but the numerator is a sum of $n!$ terms. However, regardless of the values of the a_i, the numerator of $F(q)$, when simplified, is a polynomial with at most 2^{n-1} terms (one for each possible descent set).

Let $u_1, u_2, \ldots$ be arbitrary and define polynomials G_n by

$$G_n = \sum_{\pi \in S_n} \prod_{i \in D_\pi} u_i.$$

We can compute G_n in the following way: Let

$$G_n^{(i)} = \sum_{\pi \in \mathfrak{S}_n^{(i)}} \prod_{i \in D_\pi} u_i,$$

where $\mathfrak{S}_n^{(i)}$ is the set of all permutations $\pi \in S_n$ that end with i. Then $G_n = G_{n+1}^{(n+1)}$.

A permutation π in $\mathfrak{S}_n^{(i)}$ can be obtained uniquely from some permutation $\bar{\pi}$ in S_{n-1} by replacing each $j \geq i$ with $j + 1$ and then appending i at the end. The descent set of π will be the same as the descent set of $\bar{\pi}$ if the last entry of $\bar{\pi}$ is less than i

and the descent set of π will be $D_{\bar{\pi}} \cup \{n-1\}$ if the last entry of $\bar{\pi}$ is greater than or equal to i. Thus we have the recurrence

$$G_n^{(i)} = \sum_{j=1}^{i-1} G_{n-1}^{(j)} + u_{n-1} \sum_{j=i}^{n-1} G_{n-1}^{(j)}$$

with the initial condition $G_1^{(1)} = 1$. We can simplify this a bit to get "Algorithm G":

$$G_n^{(i)} = G_n^{(i-1)} + (1 - u_{n-1}) G_{n-1}^{(i-1)} \quad \text{for } i > 1,$$

with $G_n^{(1)} = u_{n-1} \sum_{j=1}^{n-1} G_{n-1}^{(j)}$.

Now, to compute the numerator of $F(q)$ in Theorem 1 using Algorithm G, simply set $u_i = q^{i-n(a_1+\cdots+a_i)}$ for $1 \le i < n$ and compute $G_{n+1}^{(n+1)}$.

If we use dynamic programming to implement the recurrence of Algorithm G, (e.g. "option remember" in Maple), then to compute $G_n = G_{n+1}^{(n+1)}$, at most $O(n^2)$ polynomials are computed. However, we must consider the time required to compute them. In order to compute one of the $G_k^{(i)}$, essentially we only need to add two polynomials. It is fair to assume that the time is proportional to the number of terms in the polynomials times the logarithm of the coefficient magnitude. So, overall, the time (and number of terms) grows roughly like 2^n in the dimension n, but logarithmically in the coefficient size, which is considered polynomial time in fixed dimension. In practice, we found that we could compute $F(q)$ for arbitrary a with $\sum a_i = 1$ within seconds for $n \le 11$, in about 10 seconds for $n = 12$ and in less than a minute up to $n = 15$, using a naive implementation in Maple on a tablet PC running Windows XP.

For comparison, there are existing software packages that, when given a collection of linear inequalities, produce the generating function for the integer points in the solution set. These packages include the Omega Package [3], Xin's speed-up of Omega [15], and LattE macchiato [10, 13]. We used these programs to compute symmetrically constrained compositions in n dimensions, by giving as input the $n!$ inequalities. The computation became infeasible when $n \ge 4$ for the Omega package and Xin's program. LattE was able to handle examples for $n = 5$ in under 10 seconds and $n = 6$ in under an hour.

Thus exploiting the symmetry via Theorem 1 and Algorithm G makes a huge difference in what we can compute.

3 The general case

3.1 A general version of the main theorem

We remove the requirement that $\sum a_i = 1$ and enumerate compositions $\lambda = (\lambda_1, \lambda_2, \ldots, \lambda_n) \in \mathbb{Z}_{\ge 0}^n$ that satisfy the $n!$ constraints

$$a_1 \lambda_{\pi(1)} + a_2 \lambda_{\pi(2)} + \cdots + a_n \lambda_{\pi(n)} \ge 0 \quad \text{for all } \pi \in S_n \tag{10}$$

via the generating function $F(z) = \sum_{\lambda} z^{\lambda}$.

Theorem 2 *Given integers $a_1 \leq a_2 \leq \cdots \leq a_n$, with $a_1 + a_2 + \cdots + a_j \leq 0$ for $1 \leq j \leq n - 1$ and $a_1 + a_2 + \cdots + a_n \geq 1$, define the vectors $A_1, A_2, \ldots, A_n \in \mathbb{Z}^n$ as the columns of the matrix*

$$
\begin{bmatrix}
a_2 + \cdots + a_n & a_3 + \cdots + a_n & a_4 + \cdots + a_n & \cdots & a_n & 1 \\
-a_1 & a_3 + \cdots + a_n & a_4 + \cdots + a_n & & a_n & 1 \\
-a_1 & -a_1 - a_2 & a_4 + \cdots + a_n & & a_n & 1 \\
-a_1 & -a_1 - a_2 & -a_1 - a_2 - a_3 & & a_n & 1 \\
\vdots & \vdots & \vdots & & \vdots & \vdots \\
-a_1 & -a_1 - a_2 & -a_1 - a_2 - a_3 & & a_n & 1 \\
-a_1 & -a_1 - a_2 & -a_1 - a_2 - a_3 & \cdots & -a_1 - \cdots - a_{n-1} & 1
\end{bmatrix}
$$

and let

$$
\mathcal{P} := \sum_{j=1}^{n} [0, 1) A_j = \left\{ \sum_{i=1}^{n} c_i A_j : 0 \leq c_i < 1 \right\}.
$$

Then

$$
F(z) = \sum_{p \in \mathcal{P} \cap \mathbb{Z}^n} \sum_{\pi \in S_n} \frac{z_\pi^p \prod_{i \in D_\pi, \, p_i = p_{i+1}} z_\pi^{A_i}}{\prod_{j=1}^{n} (1 - z_\pi^{A_i})},
$$

where we take the product over all descent positions i of π for which the ith and the $(i + 1)$st coordinate of p are the same.

If, for some i, d divides every coordinate of A_i, we can replace A_i by A_i/d in Theorem 2 and thereby reduce the number of lattice points in $\mathcal{P}$ by a factor of d.

Proof The start of our proof is similar to that of Theorem 1, except that we find it advantageous to view the compositions satisfying (10) as integer points in the simplicial cone

$$
K := \left\{ x = (x_1, x_2, \ldots, x_n) \in \mathbb{R}^n_{\geq 0} : \sum_{j=1}^{n} a_j x_{\pi(j)} \geq 0 \text{ for all } \pi \in S_n \right\}.
$$

A *simplicial cone* is a subset of $\mathbb{R}^n$ of the form $\{y \in \mathbb{R}^n \mid My \leq b\}$ where M is a nonsingular real matrix and $b \in \mathbb{R}^n$. From this point of view,

$$
F(z) = \sum_{\lambda \in K \cap \mathbb{Z}^d} z^\lambda.
$$

The setup now continues in analogy with the proof of Theorem 1. Like there, it suffices to study

$$
\tilde{K}_\pi := \left\{ x \in \mathbb{R}^n : \begin{array}{l} x_1 \geq x_2 \geq \cdots \geq x_n \geq 0 \text{ and } x_j > x_{j+1} \text{ if } j \in D_\pi \\ a_1 x_1 + a_2 x_2 + \cdots + a_n x_n \geq 0 \end{array} \right\}
$$

and the associated generating function $G_\pi(z) := \sum_{\lambda \in \tilde{K}_\pi \cap \mathbb{Z}^d} z^\lambda$; then

$$F(z) = \sum_{\pi \in S_n} G_\pi(z_\pi).$$

First, we study the cone K_{Id}. The constraints of K_{Id} are given by the system

$$
\begin{bmatrix}
1 & -1 & & & & \\
 & 1 & -1 & & & \\
 & & & \ddots & & \\
 & & & & 1 & -1 \\
a_1 & a_2 & a_3 & \cdots & a_{n-1} & a_n
\end{bmatrix}
x \geq 0,
\tag{11}
$$

and the inverse of the matrix on the left of (11) is

$$
\frac{1}{\sum_{j=1}^n a_j}
\begin{bmatrix}
a_2 + \cdots + a_n & a_3 + \cdots + a_n & a_4 + \cdots + a_n & \cdots & a_n & 1 \\
-a_1 & a_3 + \cdots + a_n & a_4 + \cdots + a_n & & a_n & 1 \\
-a_1 & -a_1 - a_2 & a_4 + \cdots + a_n & & a_n & 1 \\
-a_1 & -a_1 - a_2 & -a_1 - a_2 - a_3 & & a_n & 1 \\
\vdots & \vdots & \vdots & & \vdots & \vdots \\
-a_1 & -a_1 - a_2 & -a_1 - a_2 - a_3 & & a_n & 1 \\
-a_1 & -a_1 - a_2 & -a_1 - a_2 - a_3 & \cdots & -a_1 - \cdots - a_{n-1} & 1
\end{bmatrix}.
$$

The conditions on $a_1, a_2, \ldots, a_n$ guarantee that the inverse exists and that K_{Id} is a cone in $\mathbb{R}^n_{\geq 0}$. Thus the columns $A_1, A_2, \ldots, A_n$ of this matrix form a set of generators of K_{Id} and by an easy tiling argument (see, e.g., [5, Chap. 3])

$$
K_{\mathrm{Id}} \cap \mathbb{Z}_n = \left\{ p + \sum_{j=1}^n c_j A_j : p \in \mathcal{P}, \ c \in \mathbb{Z}^n_{\geq 0} \right\};
\tag{12}
$$

in other words,

$$
G_{\mathrm{Id}}(z) = \frac{\sum_{p \in \mathcal{P} \cap \mathbb{Z}^n} z^p}{\prod_{j=1}^n (1 - z_\pi^{A_j})}.
$$

Before turning to $\tilde{K}_\pi$, note that the generators A_j have a special form: we have (writing $A_{i,j}$ for the ith entry of A_j)

$$
A_{i,j} =
\begin{cases}
1 & \text{if } j = n, \text{ else,} \\
-(a_1 + a_2 + \cdots + a_j) & \text{if } i > j, \text{ else.} \\
a_{j+1} + a_{j+2} + \cdots + a_n)
\end{cases}
$$

Therefore, since $\sum_{j=1}^n a_j \geq 1$,

$$
A_{j,j} > A_{j+1,j} \quad \text{for } 1 \leq j < n \quad \text{and} \quad A_{i,j} = A_{i+1,j} \quad \text{for } j \neq i.
$$

Thus, if $p \in K_{\mathrm{Id}} \cap \mathbb{Z}^n$ satisfies $p_j = p_{j+1}$, then for any $c \in \mathbb{Z}^n_{\geq 0}$, if

$$r = (p + A_j) + \sum_{i=1}^{n} c_i A_i$$

then

$$r_j > r_{j+1}.$$

Now, what about $\widetilde{K}_\pi$? It contains all points $y \in K_{\mathrm{Id}}$ except those y with $y_i = y_{i+1}$ for some $i \in D_\pi$. By (12), every $y \in K_{\mathrm{Id}}$ has a unique representation as $y = p + \sum_{j=1}^{n} c_j A_j$ for some $c \in \mathbb{Z}^n_{\geq 0}$. Thus by the previous paragraph, $y_i = y_{i+1}$ iff both $p_i = p_{i+1}$ and $c_i = 0$. Now, in the same way as in (12),

$$\widetilde{K}_\pi \cap \mathbb{Z}^n = \left\{ p + \sum_{j=1}^{n} c_j A_j : p \in \mathcal{P}, \ c \in \mathbb{Z}^n_{\geq 0}, \right.$$

$$\left. \text{and if } j \in D_\pi \text{ and } p_j = p_{j+1} \text{ then } c_j > 0 \right\}$$

$$= \left\{ p + \sum_{j=1}^{n} c_j A_j + \sum_{j \in D_\pi, \ p_j = p_{j+1}} A_j : p \in \mathcal{P}, \ c \in \mathbb{Z}^n_{\geq 0} \right\}.$$

Thus

$$G_\pi(z) = \sum_{p \in \mathcal{P} \cap \mathbb{Z}^n} \frac{z^p \prod_{j \in D_\pi, \ p_j = p_{j+1}} z_\pi^{A_i}}{\prod_{j=1}^{n} (1 - z_\pi^{A_i})}. \qquad \square$$

In the special case of Theorem 1, $\sum_{j=1}^{n} a_j = 1$ and the origin is the only lattice point in $\mathcal{P}$.

3.2 Efficient computation for the general case

Give $a_1 \leq \cdots \leq a_n$ with $\sum_{j=1}^{n} a_j \geq 1$, once we find the generators $A_1, A_2, \ldots, A_n$, and the lattice points in $\mathcal{P}$, we can again use Algorithm G to efficiently compute $F(q)$: By Theorem 2, setting $z = (q, q, \ldots, q)$,

$$F(q) = \sum_{p \in \mathcal{P} \cap \mathbb{Z}^n} q^{|p|} \frac{\sum_{\pi \in S_n} \prod_{i \in D_\pi, p_i = p_{i+1}} q^{|A_i|}}{\prod_{j=1}^{n} (1 - q^{|A_i|})},$$

where $|x| = x_1 + \cdots + x_n$ for an n-dimensional vector x.

The denominator is easy. To find the numerator, for each point $p \in \mathcal{P} \cap \mathbb{Z}^n$, set

$$u_i = \begin{cases} q^{|A_i|} & \text{if } p_i = p_{i+1}, \\ 1 & \text{otherwise} \end{cases}$$

and then compute $G_{n+1}^{(n+1)}$.

Now the running time also depends on $|\mathcal{P} \cap \mathbb{Z}^n|$. This can grow linearly with the magnitude of the entries (rather than the logarithm of the magnitude), even in fixed dimension. However, when $|\mathcal{P} \cap \mathbb{Z}^n|$ is of moderate size, this computation method can be quite effective.

References

1. Andrews, G.E.: A note on partitions and triangles with integer sides. Am. Math. Mon. **86**(6), 477–478 (1979)
2. Andrews, G.E.: MacMahon's partition analysis. II. Fundamental theorems. Ann. Comb. **4**(3–4), 327–338 (2000)
3. Andrews, G.E., Paule, P., Riese, A.: MacMahon's partition analysis: the Omega package. Eur. J. Comb. **22**(7), 887–904 (2001)
4. Andrews, G.E., Paule, P., Riese, A.: MacMahon's partition analysis. VII. Constrained compositions. In: q-series with Applications to Combinatorics, Number Theory, and Physics, Urbana, IL, 2000. Contemp. Math., vol. 291, pp. 11–27. Am. Math. Soc., Providence (2001)
5. Beck, M., Robins, S.: Computing the Continuous Discretely: Lattice-point Enumeration in Polyhedra. Undergraduate Texts in Mathematics. Springer, New York (2007)
6. Carlitz, L.: q-Bernoulli and Eulerian numbers. Trans. Am. Math. Soc. **76**, 332–350 (1954)
7. Carlitz, L.: A combinatorial property of q-Eulerian numbers. Am. Math. Mon. **82**, 51–54 (1975)
8. Corteel, S., Gessel, I.M., Savage, C.D., Wilf, H.S.: The joint distribution of descent and major index over restricted sets of permutations. Ann. Comb. **11**(3–4), 375–386 (2007)
9. Corteel, S., Savage, C.D., Wilf, H.S.: A note on partitions and compositions defined by inequalities. Integers **5**(1), A24 (2005), 11 pp. (electronic)
10. De Loera, J.A., Hemmecke, R., Tauzer, J., Yoshida, R.: Effective lattice point counting in rational convex polytopes. J. Symb. Comput. **38**(4), 1273–1302 (2004)
11. Jordan, J.H., Walch, R., Wisner, R.J.: Triangles with integer sides. Am. Math. Mon. **86**(8), 686–689 (1979)
12. Klosinski, L.F., Alexanderson, G.L., Larson, L.C.: The sixtieth William Lowell Putnam mathematical competition. Am. Math. Mon. **107**(8), 721–732 (2000)
13. Köppe, M.: A primal Barvinok algorithm based on irrational decompositions. SIAM J. Discrete Math. **21**(1), 220–236 (2007) (electronic)
14. Stanley, R.P.: Enumerative Combinatorics, vol. 1. Cambridge Studies in Advanced Mathematics, vol. 49. Cambridge University Press, Cambridge (1997). Corrected reprint of the 1986 original
15. Xin, G.: A fast algorithm for MacMahon's partition analysis. Electron. J. Comb. **11**(1) (2004). Research Paper 58, 20 pp. (electronic)

Ramanujan J (2010) 23: 371–396
DOI 10.1007/s11139-010-9280-z

Bentley's conjecture on popularity toplist turnover under random copying

**Kimmo Eriksson · Fredrik Jansson ·
Jonas Sjöstrand**

Dedicated to George Andrews for his 70th birthday

Received: 30 June 2009 / Accepted: 15 October 2010 / Published online: 28 October 2010
© Springer Science+Business Media, LLC 2010

Abstract Bentley et al. studied the turnover rate in popularity toplists in a 'random copying' model of cultural evolution. Based on simulations of a model with population size N, list length ℓ and invention rate μ, they conjectured a remarkably simple formula for the turnover rate: $\ell\sqrt{\mu}$. Here we study an overlapping generations version of the random copying model, which can be interpreted as a random walk on the integer partitions of the population size. In this model we show that the conjectured formula, after a slight correction, holds asymptotically.

Keywords Toplists · Random walk · Integer partitions · Moran model · Popularity distribution

Mathematics Subject Classification (2000) Primary 60G50 · 05A17

This research was supported by the CULTAPTATION project (European Commission contract FP6-2004-NEST-PATH-043434) and the Swedish Research Council.

K. Eriksson (✉) · F. Jansson · J. Sjöstrand
School of Communication, Culture and Communication, Mälardalen University, 721 23 Västerås, Sweden
e-mail: kimmo.eriksson@mdh.se

F. Jansson
e-mail: fredrik.jansson@mdh.se

J. Sjöstrand
e-mail: jonas.sjostrand@mdh.se

K. Eriksson · F. Jansson · J. Sjöstrand
Centre for the Study of Cultural Evolution at Stockholm University, Stockholm, Sweden

1 Introduction

A pervasive phenomenon in modern culture are toplists like Top 100 Baby Names or the Billboard Top 200 Pop Chart. Mathematics is no exception; indeed, the present paper was partly inspired by Andrews and Berndt's paper on the *Top Ten Most Fascinating Formulas in Ramanujan's Lost Notebook* reporting the outcome of a popularity vote on these formulas among experts in the field [1]. Andrews and Berndt admit that their list would very likely look different in another week, and point to the corresponding phenomenon in pop charts: "it is the fate of popular songs to lose their popularity and fade off the charts" (p. 19). Here we will be concerned precisely with this phenomenon of turnover of toplists. By the *turnover rate* we will mean the number of entries that have gone off the list after a given time.

Our aim is to prove (a modified version of) a remarkable conjecture on turnover rates found by Bentley et al. [3]. These authors analyzed empirical data on the turnover of toplists of various things: baby names, pop albums, and dog breeds. Their data allowed them to study varying list lengths (like Top 10, Top 100, etc.), which yielded intriguing results: The turnover rate seemed to be approximately *proportional to the list length* and largely *independent of the underlying population size*.

In order to find a theoretical explanation for this empirical finding, Bentley et al. then simulated cultural evolution of the popularity distribution of cultural variants (say, pop songs) under a simple random copying model. Each individual of a new generation is assumed to copy the favorite song of a randomly drawn individual of the previous generation, but with a small probability μ the individual instead invents a new song. Simulations of this model gave results consistent with the empirical data, and the authors observed the following pattern (without any attempt at analytical verification), which we will refer to as Bentley's conjecture.

Conjecture 1 (Bentley's conjecture [3]) *Under the above random copying model with N individuals per generation, list length ℓ, and invention rate μ, the expected turnover rate of the toplist is very close to*

$$\ell\sqrt{\mu} \qquad (1)$$

for small μ and sufficiently large N compared to ℓ.

Bentley et al. made no attempt to pin down more precisely the assumptions and the result. Our aim in this paper is to come up with, and prove, a precise formulation of the result. First, we urge the reader to take a moment to appreciate the problem, since it seems to us to be quite novel. As we will discuss at the end, there are a number of well-known stochastic processes that can be interpreted as generating popularity distributions, but we have never before seen considered the question of how often the most popular elements get replaced. If Bentley's conjecture is to be believed, then such questions may have very nice answers.

1.1 Comparison of two random copying models

We will now briefly discuss two extensively studied random copying models from mathematical population genetics, called the *infinite alleles Wright–Fisher model* and the *infinite alleles Moran model*. For all details, we refer to the book by Ewens [6].

Fig. 1 Results of simulations of Bentley's model (*dotted curve*) and the IAM model (*solid curve*), for varying list length ℓ, with fixed population size $N = 4000$ and invention rate $\mu = 0.02$. The simulation first runs for a number of steps so that a stationary distribution is approached. The average turnover rate is then computed as the mean of the turnover rate in the following 200 generations

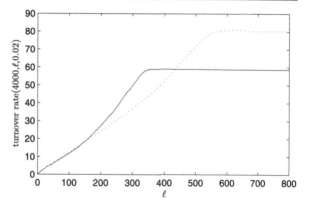

Bentley's model is equivalent to the infinite alleles Wright–Fisher model. In the present paper we will instead work with the infinite alleles Moran model (henceforth IAM). The IAM model differs from Bentley's model only in the respect that generations are overlapping: Each timestep sees the death of a randomly chosen individual and the birth of a new individual who either inherits the variant of a randomly chosen parent or, with probability μ, invents a new variant.

As described in [6], the two models often give pretty similar results (see also simulations in [8]). The IAM model seems on the whole to be more amenable to exact analysis. For instance, an exact expression for the stationary distribution is known for the IAM, whereas the same expression holds asymptotically for the infinite alleles Wright–Fisher model.

Simulations show that the two models also seem to behave similarly with respect to toplist turnover per generation, if we in the IAM model define a generation as N time steps. Figure 1 shows the turnover rate per generation for varying list length ℓ. Clearly, there are three regimes for both models. In the first regime (for short toplists), the two models give similar turnover rates with what looks like a linear dependence on ℓ. The second regime has a slightly convex shape. In the third regime, the turnover rate is constant.

Both Bentley's conjecture and the asymptotic analysis in the present paper apply to the first regime, which we will look into more carefully shortly. The third regime is trivially explained: When the list length is greater than the number of existing songs, then every newly invented song enters the list. Hence, the expected turnover rate in this regime is $N\mu$ in Bentley's model (in the figure, $N\mu = 80$), whereas it is somewhat lower in the IAM model as a song may cease to exist in the same generation it was invented, thus not contributing to the turnover. The second regime awaits closer investigation.

Figure 2 shows the first regime for the IAM model, with varying list length and four different population sizes. We see that, in line with Bentley's conjecture, the turnover rate is roughly independent of N and roughly linear in ℓ. A least squares fit of the four curves yields the linear expression 0.126ℓ, with $r^2 = 0.997$.

Finally, Bentley's conjecture says that the turnover rate shall be proportional to the square root of the invention rate μ. Figure 3 shows that this indeed seems to be

Fig. 2 Results of simulations of the IAM model, for varying ℓ, four different values of N, and constant $\mu = 0.02$. Average turnover rate is computed as the mean turnover rate in 70 generations

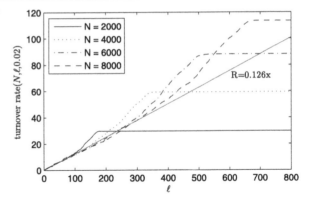

Fig. 3 Results of simulations of the IAM model, with $N = 4000$ and $\mu \in [0, 1]$. The *curve* shows the mean turnover rate divided by list length, taken over all list lengths $\ell \in \{1, 2, \ldots, 100\}$

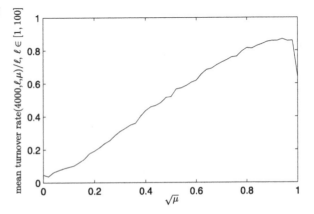

the case as long as μ is not too close to 1. A least squares fit for the part of the curve where $\sqrt{\mu} \leq 0.8$ yields the linear expression $1.037\sqrt{\mu}$ for the turnover rate, with $r^2 = 0.996$.

1.2 Outline of result and approach

The above simulation results indicate that Bentley's conjecture applies to the IAM model as well. For this model, Strimling et al. [8] recently obtained an expression for the expected number of variants of popularity k under the stationary distribution. Denoting this number by f_k, they proved that

$$f_k = \mu N \frac{(1 - \mu)^{k-1}}{k} \prod_{i=1}^{k-1} \frac{N - i}{N - i - 1 + i\mu}. \tag{2}$$

This formula will be the starting point for our analysis.

It can be helpful to observe that the popularity distribution of cultural variants can be viewed as an *integer partition* of the population size N. When we talk about a *song* that has k *votes*, one can think about a particular row of length k of the Young diagram. Thus, the IAM model of cultural evolution can be interpreted as a random walk on the set of Young diagrams of N squares, where each step involves the death of one random square and the birth of a new square, either through doubling of a random square or creation of a new row of a single square (and then possibly some reordering of the rows so that they are kept in order of decreasing length). In another paper [5] we explicitly use this interpretation to study limit shapes of integer partitions that evolve according to the IMA model (and related stochastic processes). In this paper we will instead use the terminology of songs and votes. The following key concepts will also be important.

- The *popularity* of a song is the number of votes for that song.
- The *toplist* is the set of the ℓ most popular songs. (This is not well-defined if two or more songs have the same popularity, but then we can use any rule to decide which of them should be included on the toplist as it does not matter for our analysis.)
- A *generation* consists of N time steps.
- Let $\alpha > 0$ be some constant. The *turnover rate* is the number of songs that are on the toplist at a given time t_0 but not α generations later. (We will only be interested in the expected turnover rate which is independent of t_0 since we assume a stationary distribution.)

As it turns out, Bentley's conjecture needs to be modified slightly. We will prove the following result:

Theorem 1 *Let $\alpha > 0$ be any constant, suppose N, ℓ, and μ satisfy Assumptions 1, 2 and 3 in Sect. 2, and let ψ be defined as in Assumption 2. Then, under the stationary distribution of the IAM model, the expected turnover rate in α generations (i.e., αN timesteps) is*

$$\sim \sqrt{\psi \alpha / \pi} \cdot \ell \cdot \sqrt{\mu \ln(N/\ell)}.$$

The paper is organized as follows. First we define the notation we will use, and present the assumptions needed for the theorem (Sect. 2). We then derive some basic results (like expectations and variance) about the number of songs of a given popularity (Sect. 3), after which we proceed to examine what popularity it takes to qualify on the toplist (Sect. 4). In the most technical part, we analyze the random process describing how the popularity of a song changes over time (Sect. 5). We then integrate the previous results into a proof of the main theorem (Sect. 6). We conclude by a brief discussion of future directions of research.

2 Notation and assumptions

The input to our problem consists of the three variables N, ℓ and μ and the constant α. To make the notation simple and clear in the following sections, we will think of the variables N, ℓ and μ, and functions of those, as depending on a single free variable ω. Thus, when we write e.g. $N/\ell \to \infty$ we mean that $N_\omega/\ell_\omega \to \infty$ when $\omega \to \infty$. However, the dependence on ω will always be invisible as we will drop the index and write e.g. N instead of N_ω.

We will use conventional ordo notation and the symbols $\ll$, $\sim$ and $\lesssim$.

- $A \ll B$ means that $A/B \to 0$.
- $A \sim B$ means that $A/B \to 1$.
- $A \lesssim B$ means that, for any $\varepsilon > 0$, eventually $|A/B| < 1 + \varepsilon$.

Our result requires that the variables N, ℓ and μ satisfy three assumptions that restrain their asymptotical behavior.

Assumption 1 $N/\ell \to \infty$.

Assumption 2 $\psi := 1 - \lim \frac{\ln(\mu^{-1})}{\ln(N/\ell)}$ exists, and $0 < \psi < 1$.

Assumption 3 $\mu\ell/\sqrt{\ln(N/\ell)} \to \infty$.

These assumptions easily imply the following basic asymptotical properties of our variables:

Corollary 1 *The following holds.*

$$N \to \infty, \tag{3}$$

$$\ell \to \infty, \tag{4}$$

$$\mu \ln(N/\ell) \to 0. \tag{5}$$

Proof We have

$$\frac{\ln((\mu \ln(N/\ell))^{-1})}{\ln(N/\ell)} = \frac{\ln(\mu^{-1})}{\ln(N/\ell)} - \frac{\ln \ln(N/\ell)}{\ln(N/\ell)},$$

which tends to $1 - \psi$ by Assumptions 1 and 2. Thus, $\mu \ln(N/\ell) \to 0$. Clearly, this implies that $\mu \to 0$ which together with Assumption 3 yields that $\ell \to \infty$. Finally, by Assumption 1 we get $N \to \infty$. $\square$

Finally, we will always assume that the IAM model has reached a stationary distribution (cf. [6, 8]).

3 The number of songs of a given popularity

For $1 \le m \le N$, let X_m be the number of songs with popularity m (assuming stationary distribution), which is a random variable. From (2) we already have an exact expression for the expected value $f_m = E(X_m)$. In this section we will examine $\text{Var}(X_m)$ and $x_{m,n} := E(X_{m,n})$, where $X_{m,n}$ denotes the number of ordered pairs (A, B) of distinct songs $A \neq B$ such that A has popularity m and B has popularity n, for $1 \le m, n \le N$.

Starting with $x_{m,n}$, first note that

$$X_{m,n} = \begin{cases} X_m X_n & \text{if } m \neq n, \\ X_m(X_m - 1) & \text{if } m = n. \end{cases}$$

Lemma 1

$$x_{m,n} \leq f_m f_n.$$

Proof For any $1 \leq i, m \leq N$, let $P(i \mapsto m)$ denote the probability that a song with popularity i will have popularity m at the next time step. For $1 \leq i, j, m, n \leq N$, define $B_{m,n;i,j}$ and $c_{m,n}$ as follows (when applicable).

$$B_{m,n;m,n} := P(m \mapsto m) + P(n \mapsto n) - 1$$

$$= 1 - \frac{m+n}{N} - a\frac{m+n}{N-1} + a\frac{m(2m-1) + n(2n-1)}{N(N-1)},$$

$$B_{m,n;m+1,n} := P(m+1 \mapsto m) = \frac{m+1}{N} - a\frac{m(m+1)}{N(N-1)},$$

$$B_{m,n;m,n+1} := P(n+1 \mapsto n) = \frac{n+1}{N} - a\frac{n(n+1)}{N(N-1)},$$

$$B_{m,n;m-1,n} := P(m-1 \mapsto m) = a\frac{m-1}{N-1} - a\frac{(m-1)^2}{N(N-1)},$$

$$B_{m,n;m,n-1} := P(n-1 \mapsto n) = a\frac{n-1}{N-1} - a\frac{(n-1)^2}{N(N-1)},$$

$$c_{m,n} := \mu(f_n\delta_{m,1} + f_m\delta_{n,1}).$$

All other $B_{m,n;i,j}$ are set to zero. Here, $a := 1 - \mu$, and $\delta_{m,1}$ is 1 if $m = 1$ and zero otherwise.

Since f_m is the expected value of X_m at the stationary distribution, we must have

$$f_m = \mu\delta_{m,1} + P(m+1 \mapsto m)f_{m+1} + P(m-1 \mapsto m)f_{m-1} + P(m \mapsto m)f_m \quad (6)$$

$$= \mu\delta_{m,1} + B_{m,n;m+1,n}f_{m+1} + B_{m,n;m-1,n}f_{m-1} + P(m \mapsto m)f_m, \quad (7)$$

and, similarly,

$$f_n = \mu\delta_{n,1} + P(n+1 \mapsto n)f_{n+1} + P(n-1 \mapsto n)f_{n-1} + P(n \mapsto n)f_n \quad (8)$$

$$= \mu\delta_{n,1} + B_{m,n;m,n+1}f_{n+1} + B_{m,n;m,n-1}f_{n-1} + P(n \mapsto n)f_n. \quad (9)$$

Multiplying (7) by f_n and (9) by f_m and then adding the resulting equations yields

$$2f_m f_n = c_{m,n} + \left(\sum_{1 \leq i,j \leq N} B_{m,n;i,j} f_i f_j \right) + f_m f_n,$$

and after subtracting $f_m f_n$ we obtain

$$f_m f_n = c_{m,n} + \sum_{1 \leq i,j \leq N} B_{m,n;i,j} f_i f_j. \quad (10)$$

Now, for $1 \le i, j, m, n \le N$, define $R_{m,n;i,j}$ and $d_{m,n}$ as follows (when applicable).

$$R_{m,n;m,n} := -2mn\frac{a}{N(N-1)},$$

$$R_{m,n;m+1,n} := n(m+1)\frac{a}{N(N-1)},$$

$$R_{m,n;m,n+1} := m(n+1)\frac{a}{N(N-1)},$$

$$R_{m,n;m-1,n} := n(m-1)\frac{a}{N(N-1)},$$

$$R_{m,n;m,n-1} := m(n-1)b\frac{a}{N(N-1)},$$

$$R_{m,n;m-1,n+1} := -(m-1)(n+1)\frac{a}{N(N-1)},$$

$$R_{m,n;m+1,n-1} := -(n-1)(m+1)\frac{a}{N(N-1)},$$

$$d_{m,n} := \frac{\mu}{N}\left((nf_n - (n+1)f_{n+1})\delta_{m,1} + (mf_m - (m+1)f_{m+1})\delta_{n,1}\right).$$

All other $R_{m,n;i,j}$ are set to zero. (Note that $f_{N+1} = 0$ by definition.)

Also, for any $1 \le i, j, m, n \le N$ such that $i + j \le N$, let $P((i,j) \mapsto (m,n))$ denote the probability that two distinct songs with popularity i and j will have popularity m and n, respectively, at the next time step. It is not difficult to check that, for any $1 \le i, j, m, n \le N$ such that $i + j \le N$, in fact

$$B_{m,n;i,j} - R_{m,n;i,j} = P((i,j) \mapsto (m,n)). \tag{11}$$

Now, let us drop the assumption of stationary distribution for a while and instead see what happens if we start with one single song with popularity N at time 0. Let $x_{m,n}^{(t)}$ denote the expected number of ordered pairs (A, B) of songs at time t such that A has popularity m and B has popularity n. It follows from (11) that

$$x_{m,n}^{(t+1)} = c_{m,n} - d_{m,n} + \sum_{1 \le i,j \le N} (B_{m,n;i,j} - R_{m,n;i,j})x_{i,j}^{(t)}. \tag{12}$$

Let $r_{m,n}^{(t)} := f_m f_n - x_{m,n}^{(t)}$. Subtracting (12) from (10) yields

$$r_{m,n}^{(t+1)} = d_{m,n} + \sum_{1 \le i,j \le N} R_{m,n;i,j} f_i f_j + \sum_{1 \le i,j \le N} (B_{m,n;i,j} - R_{m,n;i,j})r_{i,j}^{(t)}. \tag{13}$$

At time 0 there is only one song, so $x_{m,n}^{(0)} = 0$ and thus $r_{m,n}^{(0)} = f_m f_n > 0$ for all m and n. Suppose $r_{i,j}^{(t)} \ge 0$ and let us show that $r_{i,j}^{(t+1)} \ge 0$.

First, note that $r_{m,n}^{(t+1)} = f_m f_n > 0$ automatically when $m + n > N$, so in the following we will assume that $m + n \le N$. If $m + n$ is strictly less than N, $B_{m,n;i,j} - R_{m,n;i,j}$ either vanishes or equals $P((i,j) \mapsto (m,n)) \ge 0$. When $m + n = N$ it can easily be checked that $B_{m,n;i,j} - R_{m,n;i,j} \geq 0$ assuming that $\mu N \geq 1$ which follows from Assumptions 1 and 2. Thus, the last sum in (13) is nonnegative. Using again that $\mu N \geq 1$, we have $kf_k > (k+1)f_{k+1}$ for all k, which implies that $d_{m,n} \geq 0$. It remains to examine the sum $\sum_{1 \leq i,j \leq N} R_{m,n;i,j} f_i f_j$. But this sum can be written as

$$
\sum_{1 \leq i,j \leq N} R_{m,n;i,j} f_i f_j = \frac{a}{N(N-1)} \big[\big((m-1)f_{m-1} - mf_m\big)\big(nf_n - (n+1)f_{n+1}\big)
$$

$$
+ \big((n-1)f_{n-1} - nf_n\big)\big(mf_m - (m+1)f_{m+1}\big)\big], \tag{14}
$$

which is nonnegative if $m, n > 1$. If $m = 1$ or $n = 1$, then we must check that $d_{m,n} + \sum_{1 \leq i,j \leq N} R_{m,n;i,j} f_i f_j \geq 0$, and this is true if $\mu N \geq 1$.

Thus, we have proved by induction that $r_{m,n}^{(t)} \geq 0$ for all $1 \leq m, n \leq N$ and all $t \geq 0$. Since the process approaches the unique stationary distribution, it follows that $f_m f_n - x_{m,n} \geq 0$ for all m and n. $\qquad\square$

Proposition 1

$$
\mathrm{Var}(X_m) \leq f_m.
$$

Proof By Lemma 1 we have

$$
\mathrm{Var}(X_m) = E\big(X_m^2\big) - E\big(X_m\big)^2 = E\big(X_m(X_m - 1)\big) + E(X_m) - E(X_m)^2
$$

$$
\leq f_m^2 + E(X_m) - E(X_m)^2 = f_m. \qquad\square
$$

Proposition 2

$$
\mathrm{Var}\left(\sum_{k=m}^{N} X_k\right) \leq \sum_{k=m}^{N} f_k.
$$

Proof

$$
\mathrm{Var}\left(\sum_{k=m}^{N} X_k\right) = \sum_{m \leq i,j \leq N} \big(E(X_i X_j) - E(X_i)E(X_j)\big) \tag{15}
$$

$$
= \sum_{m \leq i,j \leq N} (x_{i,j} - f_i f_j) + \sum_{k=m}^{N} f_k \tag{16}
$$

$$
\leq \sum_{k=m}^{N} f_k, \tag{17}
$$

where the last inequality follows from Lemma 1. $\qquad\square$

4 What does it take to qualify on the toplist?

In this section we will analyze the asymptotics of the number of votes that are needed to qualify on the toplist. To this end we will shortly define and examine two related variables, $\hat{K}$ and $\hat{k}$. First, we need a couple of lemmas.

Lemma 2 *For any m,*

$$f_m \lesssim \mu N (1 - \mu)^m / m,$$

and for $m = O(\frac{\ln(N/\ell)}{\mu})$,

$$f_m \sim \mu N (1 - \mu)^m / m.$$

Proof By (2),

$$f_m = \mu N \frac{(1 - \mu)^{m-1}}{m} \prod_{i=1}^{m-1} \frac{N - i}{N - 1 - (1 - \mu)i}$$

so it suffices to show that the product

$$H := \prod_{i=1}^{m-1} \frac{N - i}{N - 1 - (1 - \mu)i}$$

is asymptotically smaller than 1, and that it tends to 1 if $m = O(\frac{\ln(N/\ell)}{\mu})$. It is easily verified that $\frac{N-i}{N-1-(1-\mu)i}$ is a decreasing function of i and that it is smaller than 1 if $i > 1/\mu$. Thus we obtain upper and lower bounds for H as follows.

$$H \leq \left(\frac{N - 1}{N - 2 + \mu} \right)^{\lceil 1/\mu \rceil} = \left(1 + \frac{2 - \mu}{N - 2 + \mu} \right)^{\lceil 1/\mu \rceil} \sim \exp\left(\frac{2}{\mu N} \right) \to 1$$

by Assumptions 1 and 3, and

$$H \geq \left(\frac{N - (m - 1)}{N - 1 - (1 - \mu)(m - 1)} \right)^{m-1} = \left(1 + \frac{1 - \mu(m - 1)}{N - 1 - (1 - \mu)(m - 1)} \right)^{m-1}$$

$$\sim \exp\left(\frac{(1 - \mu(m - 1))(m - 1)}{N - 1 - (1 - \mu)(m - 1)} \right),$$

which tends to 1 if $m = O(\frac{\ln(N/\ell)}{\mu})$ since

$$\frac{(\ln(N/\ell))^2}{\mu N} = \frac{1}{\mu \ell} \cdot \frac{(\ln(N/\ell))^2}{N/\ell} = o(\mu^{-1} \ell^{-1}) \to 0$$

by Assumption 3. □

For $1 \leq m \leq N$, let $S_m = \sum_{k=m}^{N} X_k$ be the number of songs with popularity at least m.

Lemma 3 *If $\mu^{-1} \ll m = O(\frac{\ln(N/\ell)}{\mu})$ then*

$$E(S_m) = \sum_{k=m}^{N} f_k \sim f_m/\mu.$$

Proof Let ϕ be an integer variable such that $\mu^{-1} \ll \phi \ll m$. First we divide the sum into two terms:

$$\sum_{k=m}^{N} f_k = f_m \left(\overbrace{\sum_{k=m}^{m+\phi} f_k/f_m}^{A} + \overbrace{\sum_{k=m+\phi+1}^{N} f_k/f_m}^{B} \right).$$

By Lemma 2 and the assumption on ϕ, we have

$$A \sim \sum_{k=m}^{m+\phi} \frac{m}{k}(1-\mu)^{k-m} \sim \sum_{k=m}^{m+\phi}(1-\mu)^{k-m} = \frac{1-(1-\mu)^{\phi+1}}{\mu} \sim \frac{1-e^{-\phi\mu}}{\mu} \sim \frac{1}{\mu}.$$

Once again by Lemma 2 and the assumption on ϕ, we have

$$B \lesssim \sum_{k=m+\phi+1}^{N} \frac{m}{k}(1-\mu)^{k-m}$$

$$< \sum_{k=m+\phi+1}^{\infty} (1-\mu)^{k-m} = \frac{(1-\mu)^{\phi+1}}{\mu} \sim \frac{e^{-\phi\mu}}{\mu} = o(1/\mu). \qquad (18)$$

$\square$

Let $\hat{K}$ be the popularity of the ℓth most popular song, i.e. $\hat{K}$ is the largest integer such that $S_{\hat{K}} \geq \ell$. In other words, $\hat{K}$ is the number of votes needed to qualify on the toplist. In order to estimate $\hat{K}$ we will study the related measure $\hat{k}$, defined as the largest integer such that $E(S_{\hat{k}}) \geq \ell$. Below we first determine the asymptotics of $\hat{k}$ (Proposition 3). We will then compute the probability of a large deviation of $\hat{K}$ from $\hat{k}$ (Proposition 5).

Proposition 3 *We have*

$$\hat{k} \sim \psi \frac{\ln(N/\ell)}{\mu}.$$

Proof By Lemma 3, it suffices to show that $f_{(\psi+\varepsilon)\mu^{-1}\ln(N/\ell)} \ll \mu\ell$ and $f_{(\psi-\varepsilon)\mu^{-1}\ln(N/\ell)} \gg \mu\ell$ for any (sufficiently small) fixed $\varepsilon > 0$.
By Lemma 2, we have

$$\frac{f_{(\psi\pm\varepsilon)\mu^{-1}\ln(N/\ell)}}{\mu\ell} \sim \frac{1}{\mu\ell} \frac{\mu N(1-\mu)^{(\psi\pm\varepsilon)\mu^{-1}\ln(N/\ell)}}{(\psi\pm\varepsilon)\mu^{-1}\ln(N/\ell)} =: A_{\pm\varepsilon}.$$

Taking the logarithm yields

$$
\begin{aligned}
\ln A_{\pm\varepsilon} &= -\ln(\mu^{-1}) + \ln(N/\ell) + (\psi \pm \varepsilon)\mu^{-1}\ln(N/\ell)\ln(1-\mu) \\
&\quad - \ln(\psi \pm \varepsilon) - \ln\ln(N/\ell) \\
&= -\ln(\mu^{-1}) + \ln(N/\ell) - (1+o(1))(\psi \pm \varepsilon)\ln(N/\ell) \\
&\quad - \ln(\psi \pm \varepsilon) - \ln\ln(N/\ell) \\
&= \{\text{Assumption 2}\} \\
&= -(1 - \psi + o(1))\ln(N/\ell) + \ln(N/\ell) \\
&\quad - (1+o(1))(\psi \pm \varepsilon)\ln(N/\ell) - \ln(\psi \pm \varepsilon) - \ln\ln(N/\ell) \\
&= (\mp\varepsilon + o(1))\ln(N/\ell) - \ln(\psi \pm \varepsilon) - \ln\ln(N/\ell) \to \mp\infty.
\end{aligned}
$$

Thus, $A_{+\varepsilon}$ tends to zero and $A_{-\varepsilon}$ tends to infinity. $\qquad\square$

Proposition 4

$$
f_{\hat{k}+o(\mu^{-1})} \sim \mu\ell.
$$

Proof From Lemma 3 and the definition of $\hat{k}$ it follows that $f_{\hat{k}} \sim \mu\ell$. Then Lemma 2 and Proposition 3 tell us that

$$
f_{\hat{k}+o(\mu^{-1})} \sim f_{\hat{k}} \frac{\hat{k}}{\hat{k} + o(\mu^{-1})}(1-\mu)^{o(\mu^{-1})} \sim f_{\hat{k}}. \qquad\square
$$

Proposition 5 *Suppose* $0 < \rho = o(\mu^{-1})$. *Then the following holds:*

$$
P(|\hat{K} - \hat{k}| > \rho) \lesssim 2\rho^{-2}\mu^{-2}\ell^{-1}.
$$

Proof Without loss of generality, we can assume that ρ is an integer. It follows from the definition of $\hat{K}$ that

$$
P(\hat{K} - \hat{k} > \rho) \le P(S_{\hat{k}+\rho} > \ell)
$$

and

$$
P(\hat{k} - \hat{K} > \rho) \le P(S_{\hat{k}-\rho} < \ell).
$$

Due to the assumption on ρ, Proposition 4 tells us that

$$
E(S_{\hat{k}\pm\rho}) = E(S_{\hat{k}}) \mp \sum_{i=1}^{\rho} f_{\hat{k}\pm i} \sim \ell \mp \rho\mu\ell,
$$

and we also have

$$\text{Var}(S_{\hat{k}\pm\rho}) \leq \{\text{Proposition 2}\} \leq E(S_{\hat{k}\pm\rho})$$

$$\sim \{\text{Lemma 3}\} \sim f_{\hat{k}\pm\rho}/\mu \sim \{\text{Proposition 4}\} \sim \ell.$$

Finally, combining the observations above, we obtain

$$P(S_{\hat{k}\pm\rho} \gtrless \ell) \leq P(|S_{\hat{k}\pm\rho} - E(S_{\hat{k}\pm\rho})| > |\ell - E(S_{\hat{k}\pm\rho})|)$$

$$\leq \{\text{Chebyshev's inequality}\}$$

$$\leq \frac{\text{Var}(S_{\hat{k}\pm\rho})}{|\ell - E(S_{\hat{k}\pm\rho})|^2}$$

$$\sim \frac{\ell}{(\rho\mu\ell)^2}. \qquad \square$$

5 How the popularity of a song changes over time

Define the probabilities

$$p^{\text{left}}(k) := \frac{k}{N}\left(1 - (1 - \mu)\frac{k-1}{N-1}\right)$$

and

$$p^{\text{right}}(k) := \left(1 - \frac{k}{N}\right)(1 - \mu)\frac{k}{N-1}$$

and $p^{\text{stay}}(k) := 1 - p^{\text{left}}(k) - p^{\text{right}}(k)$. It follows from the definition of the IAM model that $p^{\text{left}}(k)$ and $p^{\text{right}}(k)$ are the probabilities for a song with k votes to lose and gain a vote, respectively, in the next time step; $p^{\text{stay}}(k)$ is the probability that the number of votes for this song is not affected in this step.

Given a positive integer $\bar{k} \leq N$, define a random integer sequence $(K_t^{(\bar{k})})_{t=0}^\infty$ as follows. Put $K_0^{(\bar{k})} := \bar{k}$ and, assuming that $K_0^{(\bar{k})}, K_1^{(\bar{k})}, \ldots, K_{t-1}^{(\bar{k})}$ have already been defined, let $K_t^{(\bar{k})} := K_{t-1}^{(\bar{k})} - 1$ with probability $p^{\text{left}}(K_{t-1}^{(\bar{k})})$ and $K_t^{(\bar{k})} := K_{t-1}^{(\bar{k})} + 1$ with probability $p^{\text{right}}(K_{t-1}^{(\bar{k})})$; otherwise, $K_t^{(\bar{k})} := K_{t-1}^{(\bar{k})}$. Thus, the sequence $(K_t^{(\bar{k})})_{t=0}^\infty$ describes the evolution of the popularity of a song that has $\bar{k}$ votes from the beginning.

We will be interested in assessing the evolution of the random process after α generations, that is, αN time steps. In order to get a grip on this, we will define and examine three other random integer sequences derived from $(K_t^{(\bar{k})})_{t=0}^\infty$. We begin by a brief overview, saving the details until later. For $t = 1, 2, \ldots$:

- $U_t^{(\bar{k})}$ is basically $K_t^{(\bar{k})} - K_{t-1}^{(\bar{k})}$, but adjusted in such a way that $P(U_t^{(\bar{k})} = -1) = p^{\text{left}}(\bar{k})$ and $P(U_t^{(\bar{k})} = 1) = p^{\text{right}}(\bar{k})$;

- $V_t^{(\bar{k})}$ is the said adjustment, i.e., $V_t^{(\bar{k})} := K_t^{(\bar{k})} - K_{t-1}^{(\bar{k})} - U_t^{(\bar{k})}$;
- $V_t^{(\bar{k},\delta)}$ is basically $V_t^{(\bar{k})}$, but adjusted to zero if $|K_{t-1}^{(\bar{k})} - \bar{k}| > \delta$.

For notational convenience, we define symbols for the sums of the first αN elements of these sequences: $\tilde{U}^{(\bar{k})} := \sum_{t=1}^{\alpha N} U_t^{(\bar{k})}$, $\tilde{V}^{(\bar{k})} := \sum_{t=1}^{\alpha N} V_t^{(\bar{k})}$ and $\tilde{V}^{(\bar{k},\delta)} := \sum_{t=1}^{\alpha N} V_t^{(\bar{k},\delta)}$.

With this notation, we can express the total change over α generations to the popularity of a song that starts with $\bar{k}$ votes as

$$K_{\alpha N}^{(\bar{k})} - \bar{k} = \tilde{U}^{(\bar{k})} + \tilde{V}^{(\bar{k})}.$$

The purpose of this is that $\tilde{U}^{(\bar{k})}$ approximates the total change by assuming that the probabilities for going left and right are constant, and $\tilde{V}^{(\bar{k})}$ is the total adjustment one must make to that approximation.

Our aim in this section will be to prove that it is unlikely that the total change is large (Proposition 6), and unlikely that the total adjustment is large (Proposition 7). To achieve this, we must first examine the sequences and then their sums.

5.1 The $U_t^{(\bar{k})}$ sequence

Define the random integer sequence $(U_t^{(\bar{k})})_{t=1}^{\infty}$ as follows.

- If $K_t^{(\bar{k})} < K_{t-1}^{(\bar{k})}$, then we put $U_t^{(\bar{k})} = -1$ with probability $\min\{1, p^{\text{left}}(\bar{k})/p^{\text{left}}(K_{t-1}^{(\bar{k})})\}$, otherwise $U_t^{(\bar{k})} = 0$.
- If $K_t^{(\bar{k})} > K_{t-1}^{(\bar{k})}$, then we put $U_t^{(\bar{k})} = 1$ with probability $\min\{1, p^{\text{right}}(\bar{k})/p^{\text{right}}(K_{t-1}^{(\bar{k})})\}$, otherwise $U_t^{(\bar{k})} = 0$.
- If $K_t^{(\bar{k})} = K_{t-1}^{(\bar{k})}$, then we put $U_t^{(\bar{k})} = -1$ with probability

$$\max\left\{0, \frac{p^{\text{left}}(\bar{k}) - p^{\text{left}}(K_{t-1}^{(\bar{k})})}{p^{\text{stay}}(K_{t-1}^{(\bar{k})})}\right\},$$

$U_t^{(\bar{k})} = 1$ with probability

$$\max\left\{0, \frac{p^{\text{right}}(\bar{k}) - p^{\text{right}}(K_{t-1}^{(\bar{k})})}{p^{\text{stay}}(K_{t-1}^{(\bar{k})})}\right\},$$

and $U_t^{(\bar{k})} = 0$ otherwise.

In this way, $U_1^{(\bar{k})}, U_2^{(\bar{k})}, \ldots$ become independent identically distributed random variables, with $P(U_t^{(\bar{k})} = -1) = p^{\text{left}}(\bar{k})$ and $P(U_t^{(\bar{k})} = 1) = p^{\text{right}}(\bar{k})$.

5.2 The $V_t^{(\bar{k})}$ sequence

Define the random integer sequence $(V_t^{(\bar{k})})_{t=1}^{\infty}$ by $V_t^{(\bar{k})} := K_t^{(\bar{k})} - K_{t-1}^{(\bar{k})} - U_t^{(\bar{k})}$. Observe that

$$P\big(V_t^{(\bar{k})} = -1 \mid K_{t-1}^{(\bar{k})} = k\big)$$

$$= \max\{0, p^{\mathrm{left}}(k) - p^{\mathrm{left}}(\bar{k})\} + \max\{0, p^{\mathrm{right}}(\bar{k}) - p^{\mathrm{right}}(k)\}$$

$$= \max\left\{0, \frac{k - \bar{k}}{N}\left(1 - (1-\mu)\frac{\bar{k} + k - 1}{N-1}\right)\right\}$$

$$+ \max\left\{0, \left(1 - \frac{\bar{k} + k}{N}\right)(1-\mu)\frac{\bar{k} - k}{N-1}\right\}$$

$$\leq 2|k - \bar{k}|/N \tag{19}$$

and

$$P\big(V_t^{(\bar{k})} = +1 \mid K_{t-1}^{(\bar{k})} = k\big)$$

$$= \max\{0, p^{\mathrm{right}}(k) - p^{\mathrm{right}}(\bar{k})\} + \max\{0, p^{\mathrm{left}}(\bar{k}) - p^{\mathrm{left}}(k)\}$$

$$= \max\left\{0, \left(1 - \frac{\bar{k} + k}{N}\right)(1-\mu)\frac{k - \bar{k}}{N-1}\right\}$$

$$+ \max\left\{0, \frac{\bar{k} - k}{N}\left(1 - (1-\mu)\frac{\bar{k} + k - 1}{N-1}\right)\right\}$$

$$\leq 2|k - \bar{k}|/N. \tag{20}$$

We will also need that

$$P\big(V_t^{(\bar{k})} = +1 \mid K_{t-1}^{(\bar{k})} = k\big) - P\big(V_t^{(\bar{k})} = -1 \mid K_{t-1}^{(\bar{k})} = k\big)$$

$$= \big(p^{\mathrm{right}}(k) - p^{\mathrm{right}}(\bar{k})\big) - \big(p^{\mathrm{left}}(k) - p^{\mathrm{left}}(\bar{k})\big) = \mu(\hat{k} - k)/N. \tag{21}$$

5.3 The $V_t^{(\bar{k},\delta)}$ sequence

For any positive integer δ, define the random integer sequence $(V_t^{(\bar{k},\delta)})_{t=1}^{\infty}$ by $V_t^{(\bar{k},\delta)} = V_t^{(\bar{k})}$ if $|K_{t-1}^{(\bar{k})} - \bar{k}| \leq \delta$ and $V_t^{(\bar{k},\delta)} = 0$ otherwise. From (19), (20) and (21) it follows that

$$P\big(V_t^{(\bar{k},\delta)} = -1\big) \leq 2\delta/N, \tag{22}$$

$$P\big(V_t^{(\bar{k},\delta)} = +1\big) \leq 2\delta/N, \tag{23}$$

$$\left|E\left(V_t^{(\bar{k},\delta)}\right)\right| \le \mu\delta/N, \tag{24}$$

$$\left|E\left(\left(V_t^{(\bar{k},\delta)}\right)^2\right)\right| \le 4\delta/N. \tag{25}$$

5.4 Analysis of $\tilde{V}^{(\bar{k},\delta)}$

The purpose of defining $\tilde{V}^{(\bar{k},\delta)}$ is that it is easier to analyze than $\tilde{V}^{(\bar{k})}$. Before showing that it is unlikely that $\tilde{V}^{(\bar{k},\delta)}$ is large, we shall find bounds for its first and second momentum.

From (24), we obtain

$$\left|E\left(\tilde{V}^{(\bar{k},\delta)}\right)\right| \le \sum_{t=1}^{\alpha N}\left|E\left(V_t^{(\bar{k},\delta)}\right)\right| \le \alpha\mu\delta. \tag{26}$$

It follows from (21) that $|E(V_t^{(\bar{k},\delta)}|\text{"event"})| \le \mu\delta/N$ for any "event" that takes place at an earlier point in time than t. Using this and (22), (23) and (24), we conclude that, if $1 \le s < t$,

$$\left|E\left(V_s^{(\bar{k},\delta)}V_t^{(\bar{k},\delta)}\right)\right| \le P\left(V_s^{(\bar{k},\delta)} = 1\right) \cdot \left|E\left(V_t^{(\bar{k},\delta)}|V_s^{(\bar{k},\delta)} = 1\right)\right|$$

$$+ P\left(V_s^{(\bar{k},\delta)} = -1\right) \cdot \left|E\left(V_t^{(\bar{k},\delta)}|V_s^{(\bar{k},\delta)} = -1\right)\right|$$

$$\le \frac{2\delta}{N} \cdot \frac{\mu\delta}{N} + \frac{2\delta}{N} \cdot \frac{\mu\delta}{N}$$

$$= 4\mu\delta^2/N^2.$$

Combining this with (25), we obtain

$$E\left(\left(\tilde{V}^{(\bar{k},\delta)}\right)^2\right) = \sum_{t=1}^{\alpha N} E\left(\left(V_t^{(\bar{k},\delta)}\right)^2\right) + 2\sum_{1 \le s < t \le \alpha N} E\left(V_s^{(\bar{k},\delta)}V_t^{(\bar{k},\delta)}\right) \tag{27}$$

$$\le 4\alpha\delta + 4\alpha^2\delta^2\mu. \tag{28}$$

Lemma 4 *Suppose $\bar{k} = O(\hat{k})$ and $0 < \delta = O(\hat{k})$. Then, the following holds for $d > 0$:*

$$P\left(\left|\tilde{V}^{(\bar{k},\delta)}\right| > d\right) = O\left(\delta d^{-2}(1 + \delta\mu)\right).$$

Proof If $d = O(\mu\delta)$, then we would have $\delta d^{-2}(1 + \delta\mu) \to \infty$ by the assumption on δ together with Corollary 1. Thus, without loss of generality, we may assume that $d \gg \mu\delta$ and hence $d > |E(\tilde{V}^{(\bar{k},\delta)})|$ by (26). Then the following holds.

$$P\left(\left|\tilde{V}^{(\bar{k},\delta)}\right| > d\right) \le P\left(\left|\tilde{V}^{(\bar{k},\delta)} - E\left(\tilde{V}^{(\bar{k},\delta)}\right)\right| > d - E\left(\tilde{V}^{(\bar{k},\delta)}\right)\right) \tag{29}$$

$$\le \{\text{Chebyshev's inequality}\} \le \frac{\text{Var}(\tilde{V}^{(\bar{k},\delta)})}{(d - E(\tilde{V}^{(\bar{k},\delta)}))^2} \tag{30}$$

$$\leq \frac{E((\tilde{V}^{(\bar{k},\delta)})^2)}{(d - E(\tilde{V}^{(\bar{k},\delta)}))^2} = \{(26) \text{ and } (28)\} \tag{31}$$

$$= \frac{O(\delta + \delta^2\mu)}{(d - O(\mu\delta))^2} = \{\text{the assumption that } \delta = O(\hat{k})\} \tag{32}$$

$$= O\big(\delta d^{-2}(1 + \delta\mu)\big). \tag{33}$$

$\square$

5.5 Analysis of $\tilde{U}^{(\bar{k})}$

It is much easier to deal with the expected value and variance of $\tilde{U}^{(\bar{k})}$. Since

$$E\big(U_t^{(\bar{k})}\big) = p^{\text{right}}(\bar{k}) - p^{\text{left}}(\bar{k}) = -\mu\bar{k}/N \tag{34}$$

and

$$\text{Var}\big(U_t^{(\bar{k})}\big) = E\big((U_t^{(\bar{k})})^2\big) - E\big(U_t^{(\bar{k})}\big)^2 = p^{\text{right}}(\bar{k}) + p^{\text{left}}(\bar{k}) - \frac{\mu^2\bar{k}^2}{N^2} \tag{35}$$

$$= \frac{\bar{k}}{N}\left(2 - \mu - 2(1-\mu)\frac{\bar{k}-1}{N-1}\right) - \frac{\mu^2\bar{k}^2}{N^2} < \frac{2\bar{k}}{N}, \tag{36}$$

we get

$$E\big(\tilde{U}^{(\bar{k})}\big) = \alpha N E\big(U_t^{(\bar{k})}\big) = -\alpha\mu\bar{k} \tag{37}$$

and, since $U_1^{(\bar{k})}, U_2^{(\bar{k})}, \ldots$ are independent,

$$\text{Var}\big(\tilde{U}^{(\bar{k})}\big) = \alpha N \text{Var}\big(U_t^{(\bar{k})}\big) < 2\alpha\bar{k}. \tag{38}$$

Lemma 5 *If $\bar{k} = O(\hat{k})$, then, for $d > 0$,*

$$P\big(|\tilde{U}^{(\bar{k})}| > d\big) = O\big(d^{-2}\hat{k}\big).$$

Proof If $d = O(\mu\hat{k})$ we would have $d^{-2}\hat{k} \to \infty$ by Corollary 1. Thus, without loss of generality, we may assume that $d \gg \mu\hat{k}$ and hence $d > |E(\tilde{U}^{(\bar{k})})|$ by (37). Then the following holds.

$$P\big(|\tilde{U}^{(\bar{k})}| > d\big) \leq P\big(|\tilde{U}^{(\bar{k})} - E(\tilde{U}^{(\bar{k})})| > d - E(\tilde{U}^{(\bar{k})})\big) \tag{39}$$

$$\leq \{\text{Chebyshev's inequality}\} \leq \frac{\text{Var}(\tilde{U}^{(\bar{k})})}{(d - E(\tilde{U}^{(\bar{k})}))^2} \tag{40}$$

$$\leq \frac{E((\tilde{U}^{(\bar{k})})^2)}{(d - E(\tilde{U}^{(\bar{k})}))^2} = \{(37) \text{ and } (38)\} \tag{41}$$

$$= \frac{O(\bar{k})}{(d - \alpha\mu\bar{k})^2} = O(d^{-2}\hat{k}). \tag{42}$$

$\square$

5.6 Analysis of $\tilde{V}^{(\bar{k})}$ and $\tilde{U}^{(\bar{k})} + \tilde{V}^{(\bar{k})}$

In order to prove our desired propositions about the improbability of large values of $\tilde{V}^{(\bar{k})}$ and $\tilde{U}^{(\bar{k})} + \tilde{V}^{(\bar{k})}$, we need two additional lemmas. First, we study $\tilde{U}^{(\bar{k})} + \tilde{V}^{(\bar{k},\delta)}$ instead.

Lemma 6 *If* $\bar{k} = O(\hat{k})$ *and* $0 < \delta = O(\hat{k})$, *then the following holds*:

$$P\big(\big|\tilde{U}^{(\bar{k})} + \tilde{V}^{(\bar{k},\delta)}\big| > \delta/2\big) = O\big(\delta^{-2}\hat{k} + \mu\big).$$

Proof Obviously,

$$P\big(\big|\tilde{U}^{(\bar{k})} + \tilde{V}^{(\bar{k},\delta)}\big| > \delta/2\big) \le P\big(\big|\tilde{U}^{(\bar{k})}\big| > \delta/4\big) + P\big(\big|\tilde{V}^{(\bar{k},\delta)}\big| > \delta/4\big).$$

The first term, $P(|\tilde{U}^{(\bar{k})}| > \delta/4)$, is $O(\delta^{-2}\hat{k})$ by Lemma 5. The other term, $P(|\tilde{V}^{(\bar{k},\delta)}| > \delta/4)$, is $O(\delta^{-1} + \mu)$ by Lemma 4 with $d = \delta/4$, and $O(\delta^{-1} + \mu)$ is $O((\delta^{-2}\hat{k} + \mu)$ by the assumption that $\delta = O(\hat{k})$. $\square$

Next we must show that it is unlikely that $\tilde{V}^{(\bar{k},\delta)}$ differs from $\tilde{V}^{(\bar{k})}$.

Lemma 7 *If* $\bar{k} = O(\hat{k})$ *and* $0 < \delta = O(\hat{k})$, *then*

$$P\big(\exists t \le \alpha N : V_t^{(\bar{k},\delta)} \ne V_t^{(\bar{k})}\big) = O\big(\delta^{-2}\hat{k} + \mu\big).$$

Proof If there is a $t \le \alpha N$ such that $V_t^{(\bar{k},\delta)} \ne V_t^{(\bar{k})}$, then clearly $|K_{t-1} - \bar{k}| > \delta$ and there must exist a $T \le \alpha N$ such that $|\sum_{\tau=1}^{T}(U_\tau^{(\bar{k})} + V_\tau^{(\bar{k},\delta)})| = \lfloor\delta + 1\rfloor$. Then, by Lemma 6, the probability that $|\sum_{\tau=T+1}^{\alpha N}(U_\tau^{(\bar{k})} + V_\tau^{(\bar{k},\delta)})| \le \delta/2$ is $1 - O(\delta^{-2}\hat{k} + \mu)$. But if this probable event happens, then we must have $|\sum_{\tau=1}^{\alpha N}(U_\tau^{(\bar{k})} + V_\tau^{(\bar{k},\delta)})| > \delta/2$ which, again by Lemma 6, happens with probability $O(\delta^{-2}\hat{k} + \mu)$. Thus, the event that there is a $t \le \alpha N$ such that $V_t^{(\bar{k},\delta)} \ne V_t^{(\bar{k})}$ happens with probability $O(\delta^{-2}\hat{k} + \mu)$. $\square$

We are now in a position to derive the desired propositions.

Proposition 6 *If* $\bar{k} = O(\hat{k})$ *and* $0 < d = O(\hat{k})$ *then*

$$P\big(\big|\tilde{U}^{(\bar{k})} + \tilde{V}^{(\bar{k})}\big| > d\big) = O\big(d^{-2}\hat{k} + \mu\big).$$

Proof This follows directly from Lemmas 6 and 7 with $\delta = 2d$. $\square$

Proposition 7 *If $\bar{k} = O(\hat{k})$, then, for $d > 0$,*

$$P(|\tilde{V}^{(\bar{k})}| > d) = O(\mu^{-1}d^{-2} + \mu \ln(N/\ell)).$$

Proof With $\delta = \mu^{-1}$, Lemmas 4 and 7 yield

$$P(|\tilde{V}^{(\bar{k})}| > d) = O(\delta d^{-2}(1 + \delta\mu)) + O(\delta^{-2}\hat{k} + \mu) = O(\mu^{-1}d^{-2} + \mu \ln(N/\ell)).$$

$\square$

6 The proof of the main theorem

In this section we will use our previous achievements to finally prove Theorem 1. We will need the classical Berry–Esseen Theorem, which says how well the distribution of a sum of i.i.d. random variables is approximated by a normal distribution.

Theorem 2 (The Berry–Esseen Theorem) *Let $W_1, W_2, \ldots$ be independent and identically distributed random variables with $E(W_i) = 0$, $E(W_i^2) = \sigma^2 > 0$ and $E(|W_i|^3) = \tau < \infty$. Then, for any x and n,*

$$\left| P(W_1 + \cdots + W_n \le x) - \Phi(x/(\sigma\sqrt{n})) \right| \le \frac{\tau}{\sigma^3\sqrt{n}}.$$

At any point in time, let the *pseudolist* be the set of songs whose popularity is greater than or equal to $\hat{k}$.

Proposition 8 *The expected number L of songs that are on the pseudolist at a time t_0 but not at time $t_0 + \alpha N$ is $\sim \sqrt{\psi\alpha/\pi} \cdot \ell \cdot \sqrt{\mu \ln(N/\ell)}$.*

Proof Consider a song that has popularity $\bar{k} \ge \hat{k}$ at time t_0, and define the random sequence $(K_t^{(\bar{k})})_{t=0}^{\infty}$ by letting $K_t^{(\bar{k})}$ be the popularity of the song at time $t_0 + t$. (Note that this definition is statistically equivalent to the definition of $K_t^{(\bar{k})}$ given in Sect. 5.) Let $Q_{\bar{k}} := P(K_{\alpha N}^{(\bar{k})} < \hat{k})$ be the probability that the song has left the pseudolist after αN time steps.

Introduce a variable v that tends to infinity but slowly enough so that $v\mu \ln(N/\ell) \to 0$ and $v = o(\sqrt{\ln(N/\ell)})$. (In fact, $v = o(\sqrt{\ln(N/\ell)})$ implies $v\mu \ln(N/\ell) \to 0$ via Assumption 2 but that does not matter.) We can divide L into three terms as follows.

$$L = \sum_{\bar{k}=\hat{k}}^{N} f_{\bar{k}} Q_{\bar{k}} = \overbrace{\sum_{\bar{k}=\hat{k}}^{\lfloor\hat{k}+v\sqrt{\hat{k}}\rfloor} f_{\bar{k}} Q_{\bar{k}}}^{L_1} + \overbrace{\sum_{\bar{k}=\lfloor\hat{k}+v\sqrt{\hat{k}}\rfloor+1}^{\lfloor\hat{k}/\psi\rfloor} f_{\bar{k}} Q_{\bar{k}}}^{L_2} + \overbrace{\sum_{\bar{k}=\lfloor\hat{k}/\psi\rfloor+1}^{N} f_{\bar{k}} Q_{\bar{k}}}^{L_3}.$$

First, we will deal with the term L_1, so assume that $\bar{k} \in [\hat{k}, \hat{k} + v\sqrt{\hat{k}}]$. Let $d = \mu^{-1/2}(\ln(N/\ell))^{1/4}$ and define three events A_1, A_2, and B as follows.

$$A_1 \iff \bar{k} + \tilde{U}^{(\bar{k})} < \hat{k} - d, \tag{43}$$

$$A_2 \iff \bar{k} + \tilde{U}^{(\bar{k})} < \hat{k} + d, \tag{44}$$

$$B \iff |\tilde{V}^{(\bar{k})}| > d. \tag{45}$$

We have the implications

$$A_1 \text{ and not } B \implies K_{\alpha N}^{(\bar{k})} < \hat{k} \implies A_2 \text{ or } B$$

and hence the inequalities

$$P(A_1) - P(B) \le Q_{\bar{k}} \le P(A_2) + P(B). \tag{46}$$

By Proposition 7, we have $P(B) = O(\mu^{-1}d^{-2} + \mu\ln(N/\ell))$ which is $o(v^{-1})$ with our choice of d and v.

Let

$$W_t := U_t^{(\bar{k})} - E\left(U_t^{(\bar{k})}\right) = \{(34)\} = U_t^{(\bar{k})} + \frac{\mu\bar{k}}{N}$$

and define

$$\sigma^2 := E\left(W_t^2\right) = \text{Var}\left(U_t^{(\bar{k})}\right) = \{(36)\}$$

$$= \frac{\bar{k}}{N}\left(2 - \mu - 2(1-\mu)\frac{\bar{k}-1}{N-1}\right) - \frac{\mu^2\bar{k}^2}{N^2} \sim \frac{2\bar{k}}{N}$$

and

$$\tau := E\left(|W_t|^3\right)$$

$$\le E\left(|(U_t^{(\bar{k})})^3|\right) + 3E\left((U_t^{(\bar{k})})^2\right)\left|E(U_t^{(\bar{k})})\right| + 3E\left(|U_t^{(\bar{k})}|\right)E(U_t^{(\bar{k})})^2 + \left|E(U_t^{(\bar{k})})^3\right|$$

$$\sim \frac{2\bar{k}}{N}.$$

The Berry–Esseen Theorem now yields that, for any s,

$$\left|P(\tilde{U}^{(\bar{k})} \le s) - \Phi\left((s + \alpha\mu\bar{k})/(\sigma\sqrt{\alpha N})\right)\right| \lesssim \frac{1}{\sqrt{2\alpha\bar{k}}},$$

which implies, for $i = 1, 2$, that

$$\left|P(A_i) - \Phi\left((\hat{k} - \bar{k} + (-1)^i d + \alpha\mu\bar{k})/(\sigma\sqrt{\alpha N})\right)\right| = O(\bar{k}^{-1/2}). \tag{47}$$

Since $d = \mu^{-1/2}(\ln(N/\ell))^{1/4}$ and $\sigma^2 \sim 2\bar{k}/N = \Theta(\hat{k}/N)$, Corollary 1 yields that

$$\frac{d + \alpha\mu\bar{k}}{\sigma\sqrt{\alpha N}} \to 0.$$

Combining this with (46) and (47), we obtain

$$\left| Q_{\bar{k}} - \Phi\left((1 + o(1)) \frac{\hat{k} - \bar{k}}{\sqrt{2\alpha\hat{k}}} + o(1) \right) \right| = O(\hat{k}^{-1/2}) + P(B) = o(v^{-1}),$$

where the ordos converge uniformly over the interval $\bar{k} \in [\hat{k}, \hat{k} + v\sqrt{\hat{k}}]$. Summation over this interval yields

$$\left| \sum_{\bar{k}=\hat{k}}^{\lfloor \hat{k}+v\sqrt{\hat{k}} \rfloor} Q_{\bar{k}} - \sum_{\bar{k}=\hat{k}}^{\lfloor \hat{k}+v\sqrt{\hat{k}} \rfloor} \Phi\left((1 + o(1)) \frac{\hat{k} - \bar{k}}{\sqrt{2\alpha\hat{k}}} + o(1) \right) \right| = v\sqrt{\hat{k}} \cdot o(v^{-1}) = o(\sqrt{\hat{k}}).$$

By Proposition 4, $f_{\bar{k}} \sim \mu\ell$, so

$$\frac{L_1}{\mu\ell} \sim o(\sqrt{\hat{k}}) + \sum_{\bar{k}=\hat{k}}^{\lfloor \hat{k}+v\sqrt{\hat{k}} \rfloor} \Phi\left((1 + o(1)) \frac{\hat{k} - \bar{k}}{\sqrt{2\alpha\hat{k}}} + o(1) \right) \tag{48}$$

$$\sim o(\sqrt{\hat{k}}) + \int_{\bar{k}=\hat{k}}^{\hat{k}+v\sqrt{\hat{k}}} \Phi\left((1 + o(1)) \frac{\hat{k} - \bar{k}}{\sqrt{2\alpha\hat{k}}} + o(1) \right) d\bar{k} \tag{49}$$

$$= \left\{ x := \frac{\hat{k} - \bar{k}}{\sqrt{2\alpha\hat{k}}} \right\} \tag{50}$$

$$= o(\sqrt{\hat{k}}) + \sqrt{2\alpha\hat{k}} \int_{x=-v/\sqrt{2\alpha}}^{0} \Phi\left((1 + o(1))x + o(1) \right) dx \tag{51}$$

$$\sim o(\sqrt{\hat{k}}) + \sqrt{2\alpha\hat{k}} \int_{x=-\infty}^{0} \Phi(x) \, dx \tag{52}$$

$$\sim \frac{\sqrt{\alpha\hat{k}}}{\sqrt{\pi}}. \tag{53}$$

Next, we will deal with the term L_2.

$$L_2 < f_{\hat{k}} \sum_{\bar{k}=\lfloor \hat{k}+v\sqrt{\hat{k}} \rfloor}^{\lfloor \hat{k}/\psi \rfloor} Q_{\bar{k}} = \{\text{Propositions 4 and 6}\} \tag{54}$$

$$= \mu\ell \sum_{\bar{k}=\lfloor \hat{k}+v\sqrt{\hat{k}} \rfloor}^{\lfloor \hat{k}/\psi \rfloor} O\left((\bar{k} - \hat{k})^{-2}\hat{k} + \mu \right) \tag{55}$$

$$= \mu\ell \cdot O\left(\int_{x=v\sqrt{\hat{k}}}^{(\psi^{-1}-1)\hat{k}} (x^{-2}\hat{k} + \mu) \, dx \right) \tag{56}$$

$$= \mu\ell \cdot O\big(\hat{k}\big(v^{-1}\hat{k}^{-1/2} - \big(\psi^{-1} - 1\big)^{-1}\hat{k}^{-1}\big) + \mu\big(\big(\psi^{-1} - 1\big)\hat{k} - v\sqrt{\hat{k}}\big)\big) \quad (57)$$

$$= \mu\ell \cdot \big(o\big(\sqrt{\hat{k}}\big) + \ln(N/\ell)\big) = \{\text{Corollary 1}\} \quad (58)$$

$$= \mu\ell \cdot o\big(\sqrt{\hat{k}}\big). \quad (59)$$

The last term, L_3, is small simply because there are very few songs that are really popular:

$$\frac{L_3}{\ell\sqrt{\mu}} \leq \frac{1}{\ell\sqrt{\mu}} \sum_{\bar{k}=\lfloor \hat{k}/\psi \rfloor + 1}^{N} f_{\bar{k}} \sim \{\text{Lemma 3}\}$$

$$\sim \ell^{-1}\mu^{-3/2} f_{\lfloor \hat{k}/\psi \rfloor + 1} \sim \{\text{Lemma 2}\}$$

$$\sim \frac{\psi N(1-\mu)^{\hat{k}/\psi}}{\ell\sqrt{\mu} \cdot \hat{k}} \sim \{\text{Proposition 3}\}$$

$$\sim \frac{\sqrt{\mu} \cdot N(1-\mu)^{(1+o(1))\mu^{-1}\ln(N/\ell)}}{\ell\ln(N/\ell)} =: A.$$

Taking the logarithm yields

$$\ln A = -\frac{1}{2}\ln\big(\mu^{-1}\big) + \ln(N/\ell) + \big(1 + o(1)\big)\mu^{-1}\ln(N/\ell)\ln(1-\mu) - \ln\ln(N/\ell)$$

$$= -\frac{1}{2}\ln\big(\mu^{-1}\big) + \ln(N/\ell) - \big(1 + o(1)\big)\ln(N/\ell) - \ln\ln(N/\ell)$$

$$= \{\text{Assumption 2}\}$$

$$= -\bigg(\frac{1}{2} + o(1)\bigg)\ln(N/\ell) + \ln(N/\ell) - \big(1 + o(1)\big)\ln(N/\ell) - \ln\ln(N/\ell)$$

$$= -\bigg(\frac{1}{2} + o(1)\bigg)\ln(N/\ell) \to -\infty.$$

Thus, $A \to 0$ and $L_3 = o(\ell\sqrt{\mu})$. □

In the end we are interested in the actual toplist rather than the pseudolist. The following proposition gives an upper bound for the difference between these lists.

Proposition 9 *Let S be the number of songs that are on exactly one of the toplist and the pseudolist. Then, the expected value $E(S) = O(\ell\sqrt{\mu} + \mu^{-1/2}\ln\mu^{-1})$.*

Proof Let $\langle \hat{k}, \hat{K} \rangle := [\hat{k}, \hat{K}] \cup [\hat{K}, \hat{k}]$ denote the set of real numbers (inclusively) between $\hat{k}$ and $\hat{K}$.

First, we overestimate S like this:

$$S \le \overbrace{\sum_{\substack{k \in \langle \hat{k}, \hat{K} \rangle \\ |k - \hat{k}| \le \mu^{-1/2}}} X_k}^{A} + \overbrace{\sum_{\substack{k \in \langle \hat{k}, \hat{K} \rangle \\ \mu^{-1/2} < |k - \hat{k}| \le \mu^{-3/4}}} X_k}^{B}$$

$$+ \overbrace{1_{\hat{K} > \hat{k} + \mu^{-3/4}} \sum_{\hat{k} < k \le N} X_k}^{C} + \overbrace{1_{\hat{K} < \hat{k} - \mu^{-3/4}} \ell}^{D}, \tag{60}$$

where $1_{\text{"event"}}$ is an indicator variable for "event". From Proposition 4 we recall that $E(X_k) = f_k \sim \mu\ell$ if $|k - \hat{k}| = o(\mu^{-1})$. Thus

$$E(A) \le E\left(\sum_{|k - \hat{k}| < \mu^{-1/2}} X_k \right) = \sum_{|k - \hat{k}| < \mu^{-1/2}} f_k \sim \sum_{|k - \hat{k}| < \mu^{-1/2}} \mu\ell = O\left(\ell\sqrt{\mu}\right).$$

Now we will deal with the second sum, B, which can be written

$$B = \sum_{\mu^{-1/2} < |k - \hat{k}| \le \mu^{-3/4}} X_k 1_{k \in \langle \hat{k}, \hat{K} \rangle}.$$

By the Cauchy–Schwartz inequality,

$$E(B) = \sum_{\mu^{-1/2} < |k - \hat{k}| \le \mu^{-3/4}} \left(f_k P\left(k \in \langle \hat{k}, \hat{K} \rangle \right) + \mathrm{Cov}(X_k, 1_{k \in \langle \hat{k}, \hat{K} \rangle}) \right)$$

$$\le \sum_{\mu^{-1/2} < |k - \hat{k}| \le \mu^{-3/4}} \left(f_k P\left(k \in \langle \hat{k}, \hat{K} \rangle \right) + \sqrt{\mathrm{Var}(X_k)\,\mathrm{Var}(1_{k \in \langle \hat{k}, \hat{K} \rangle})} \right).$$

By Proposition 1, $\mathrm{Var}(X_k) \le f_k$, so we obtain

$$E(B) \le \sum_{\mu^{-1/2} < |k - \hat{k}| \le \mu^{-3/4}} \left(f_k P\left(k \in \langle \hat{k}, \hat{K} \rangle \right) + \sqrt{f_k P\left(k \in \langle \hat{k}, \hat{K} \rangle \right)} \right) \tag{61}$$

$$\sim \sum_{\mu^{-1/2} < |k - \hat{k}| \le \mu^{-3/4}} \left(\mu\ell P\left(k \in \langle \hat{k}, \hat{K} \rangle \right) + \sqrt{\mu\ell P\left(k \in \langle \hat{k}, \hat{K} \rangle \right)} \right). \tag{62}$$

Proposition 5 with $\rho = |k - \hat{k}|$ yields

$$P\left(k \in \langle \hat{k}, \hat{K} \rangle \right) \le P\left(|\hat{K} - \hat{k}| \ge |k - \hat{k}| \right) \lesssim 2|k - \hat{k}|^{-2} \mu^{-2} \ell^{-1},$$

and we obtain

$$E(B) \lesssim 2 \sum_{\mu^{-1/2} < |k-\hat{k}| \leq \mu^{-3/4}} \left(\mu^{-1} |k - \hat{k}|^{-2} + \mu^{-1/2} |k - \hat{k}| \right)$$

$$\leq 4 \sum_{i=\lceil \mu^{-1/2} \rceil}^{\lfloor \mu^{-3/4} \rfloor} \left(\mu^{-1} i^{-2} + \mu^{-1/2} i^{-1} \right).$$

In the last summand, the second term is larger than the first term over the summation interval, so

$$E(B) \lesssim 8\mu^{-1/2} \sum_{i=\lceil \mu^{-1/2} \rceil}^{\lfloor \mu^{-3/4} \rfloor} i^{-1}$$

$$= O\left(\mu^{-1/2} \ln \mu^{-1} \right).$$

By the Cauchy–Schwartz inequality,

$$E(C) = E(1_{\hat{K} > \hat{k} + \mu^{-3/4}}) E\left(\sum_{\hat{k} < k \leq N} X_k \right) + \mathrm{Cov}\left(1_{\hat{K} > \hat{k} + \mu^{-3/4}}, \sum_{\hat{k} < k \leq N} X_k \right)$$

$$\leq P(\hat{K} > \hat{k} + \mu^{-3/4}) E\left(\sum_{\hat{k} < k \leq N} X_k \right)$$

$$+ \sqrt{P(\hat{K} > \hat{k} + \mu^{-3/4}) \mathrm{Var}\left(\sum_{\hat{k} < k \leq N} X_k \right)}.$$

By definition of $\hat{k}$ we have $E(\sum_{\hat{k} < k \leq N} X_k) < \ell$, so by Proposition 2, $\mathrm{Var}(\sum_{\hat{k} < k \leq N} X_k) < \ell$. Proposition 5 with $\rho = \mu^{-3/4}$ yields $P(\hat{K} > \hat{k} + \mu^{-3/4}) = O(\mu^{-1/2} \ell^{-1})$. Thus,

$$E(C) = O\left(\mu^{-1/2} \ell^{-1} \ell + \sqrt{\mu^{-1/2} \ell^{-1} \ell} \right) = O\left(\mu^{-1/2} \right).$$

Finally, by Proposition 5, we get $E(D) = P(\hat{K} < \hat{k} - \mu^{-3/4}) \ell = O(\mu^{-1/2})$. □

Now, at last, we are ready to prove Theorem 1.

Proof of Theorem 1 By Proposition 8 it suffices to show that the expected number $E(S)$ of songs that leave either the toplist or the pseudolist without leaving the other one, is much smaller than $\ell \sqrt{\mu} \ln N/\ell$. Proposition 9 tells us that $E(S) = O(\ell \sqrt{\mu} + \mu^{-1/2} \ln \mu^{-1})$ and the first term $\ell \sqrt{\mu}$ is clearly small enough. The second term is

$$\mu^{-1/2}\ln\mu^{-1} = \frac{\ln\mu^{-1}}{\ln(N/\ell)}\mu^{-1/2}\ln(N/\ell) \sim \{\text{Assumption 2}\} \tag{63}$$

$$\sim (1 - \psi)\mu^{-1/2}\ln(N/\ell) < \mu^{-1/2}\sqrt{\ln(N/\ell)}\sqrt{\ln(N/\ell)} \tag{64}$$

$$\ll \{\text{Assumption 3}\} \ll \mu^{-1/2}\mu\ell\sqrt{\ln(N/\ell)}. \tag{65}$$

$\square$

7 Discussion

Bentley et al. [3] conjectured a simple expression for the turnover rate of popularity toplists in a random copying model with nonoverlapping generations (the infinite alleles Wright–Fisher model). In this paper we instead studied the overlapping generations version, known as the infinite alleles Moran model. We first showed by simulations that the toplist turnover rate seems to behave in the same way for the two models (for the appropriate regime of short lists compared to the population size). We then proved an asymptotic formula for the turnover rate, which modifies the conjectured formula by a factor $\sqrt{\ln(N/\ell)}$. In other words, the turnover rate is not perfectly independent of the population size N, but the dependence will not be noticeable in data unless one considers truly huge variations of N.

It is interesting that the two models behave so similarly with respect to toplist turnover. It is worth investigating how robust this behavior is to other reasonable changes of the model. For instance, there may be various forms of biases to the random copying, as discussed by Boyd and Richerson [4]. For instance, some pop songs may actually be better than others in some sense that makes them more likely to be voted for. Boyd and Richerson also discuss frequency-dependent biases, that would make already popular songs more (or less) likely to be voted for.

Another type of change is to let the population be increasing rather than fixed. For instance, if we remove the death step from the IAM model, it becomes equivalent to economist Herbert Simon's famous model of urban growth [7], for which toplist turnover would certainly be a relevant aspect to study. The book of Andrews and Eriksson [2] discusses a couple of other important random growth processes on Young diagrams for which the same question could be asked.

The toplist turnover problem seems to be novel. We envisage that a broader mathematical investigation of it may be fruitful.

References

1. Andrews, G.E., Berndt, B.C.: Your hit parade: The top ten most fascinating formulas in Ramanujan's lost notebook. Not. Am. Math. Soc. **55**, 18–30 (2008)
2. Andrews, G.E., Eriksson, K.: Integer Partitions. Cambridge University Press, Cambridge (2004)
3. Bentley, R.A., Lipo, C.P., Herzog, H.A., Hahn, M.W.: Regular rates of popular culture change reflect random copying. Evol. Human Behav. **28**, 151–158 (2007)
4. Boyd, R., Richerson, P.J.: Culture and the Evolutionary Process. University of Chicago Press, Chicago (1985)

5. Eriksson, K., Sjöstrand, J.: Limiting shapes of birth-and-death processes on Young diagrams. Adv. Appl. Math. (in press)
6. Ewens, W.J.: Mathematical Population Genetics, 2nd edn. Springer, New York (2004)
7. Simon, H.: On a class of skew distribution functions. Biometrika **42**, 425–440 (1955)
8. Strimling, P., Sjöstrand, J., Eriksson, K., Enquist, M.: Accumulation of independent cultural traits. Theor. Popul. Biol. **76**, 77–83 (2009)

Ramanujan J (2010) 23: 397–407
DOI 10.1007/s11139-010-9239-0

Log-convexity properties of Schur functions and generalized hypergeometric functions of matrix argument

Donald St.P. Richards

Dedicated to George Andrews on the occasion of this 70-th birthday

Received: 17 August 2009 / Accepted: 16 April 2010 / Published online: 1 September 2010
© Springer Science+Business Media, LLC 2010

Abstract We establish a positivity property for the difference of products of certain Schur functions, $s_\lambda(x)$, where λ varies over a fundamental Weyl chamber in $\mathbb{R}^n$ and x belongs to the positive orthant in $\mathbb{R}^n$. Further, we generalize that result to the difference of certain products of arbitrary numbers of Schur functions. We also derive a log-convexity property of the generalized hypergeometric functions of two Hermitian matrix arguments, and we show how that result may be extended to derive higher-order log-convexity properties.

Keywords Finite distributive lattice · FKG inequality · Generalized hypergeometric function of matrix argument · Log-supermodular · Monomial-positivity · Partition · Multivariate total positivity · Schur function · Schur-positivity · Sylvester's formula · Symmetric function · Young tableaux · Zonal polynomial

Mathematics Subject Classification (2000) Primary 05E05 · 33C67 · Secondary 05A17 · 15A15 · 60E15

1 Introduction

In this paper, we derive some positivity properties of the Schur functions and the generalized hypergeometric functions of matrix argument.

The Schur functions, s_λ, are indexed by vectors $\lambda \in \mathbb{R}^n$. For the case in which these vectors are partitions, the functions s_λ are polynomials and collectively form a linear basis of Λ, the ring of symmetric functions in the variables $x_1, x_2, \ldots \in \mathbb{R}$. The functions s_λ are ubiquitous, arising in the algebra of symmetric functions [15]; in

Research supported in part by National Science Foundation grants DMS-0705210.

D.St.P. Richards (✉)
Department of Statistics, Penn State University, University Park, PA 16802, USA
e-mail: richards@stat.psu.edu

combinatorics, in the theory of semi-standard Young tableaux [18]; and in the representation theory of GL(n, $\mathbb{C}$), the complex general linear group [7, 15]. In this paper, we will make use of the interpretation in mathematical statistics and in the theory of total positivity of the Schur functions s_λ as the "complex zonal polynomials" [5, 9], or as spherical functions in harmonic analysis on spaces of positive definite Hermitian matrices [5–7].

For $\lambda = (\lambda_1, \ldots, \lambda_n)$ and $\mu = (\mu_1, \ldots, \mu_n)$ in $\mathbb{R}^n$ define the *least upper bound*,

$$\lambda \vee \mu := \big(\max(\lambda_1, \mu_1), \ldots, \max(\lambda_n, \mu_n)\big), \tag{1.1}$$

and the *greatest lower bound*,

$$\lambda \wedge \mu := \big(\min(\lambda_1, \mu_1), \ldots, \min(\lambda_n, \mu_n)\big). \tag{1.2}$$

These operations induce on $\mathbb{R}^n$ a partial order, called the *product order*, and $\mathbb{R}^n$ then becomes a distributive lattice.

For $x = (x_1, \ldots, x_n) \in \mathbb{C}^n$, define

$$\Delta_\lambda(x) = \det\big(x_i^{\lambda_j + n - j}\big), \tag{1.3}$$

the $n \times n$ determinant with (i, j)th entry $x_i^{\lambda_j + n - j}$. For the case in which $\lambda = 0$, the zero vector, it is well-known that (1.3) reduces to the Vandermonde determinant,

$$\Delta_0(x) := \det\big(x_i^{n-j}\big) = \prod_{i < j}(x_i - x_j). \tag{1.4}$$

The *Schur function* indexed by the vector λ is defined as

$$s_\lambda(x) = \frac{\Delta_\lambda(x)}{\Delta_0(x)} \equiv \frac{\det(x_i^{\lambda_j + n - j})}{\prod_{i<j}(x_i - x_j)}, \tag{1.5}$$

with the convention that L'Hospital's formula is applied whenever there are equalities among $x_1, \ldots, x_n$. For the case in which λ is a partition, it is well-known that this definition of s_λ is equivalent to the combinatorial definition in terms of Young tableaux.

Let $\mathcal{W} = \{(\lambda_1, \ldots, \lambda_n) \in \mathbb{R}^n : \lambda_1 \geq \cdots \geq \lambda_n\}$, a fundamental Weyl chamber in $\mathbb{R}^n$, and denote by $\mathbb{R}_+$ the positive real line. Then our first aim is to establish by analytical methods the following log-convexity property of the Schur functions.

Theorem 1.1 *Let* $\lambda, \mu \in \mathcal{W}$ *and* $x \in \mathbb{R}_+^n$. *Then,*

$$s_{\lambda \vee \mu}(x) s_{\lambda \wedge \mu}(x) - s_\lambda(x) s_\mu(x) \geq 0.$$

As a consequence of this result, we will also derive in Theorem 2.1 an extension that involves any finite number of vectors drawn from $\mathcal{W}$.

A symmetric function p in the variables $x_1, \ldots, x_n \in \mathbb{R}$ is called *monomial positive* if, on expressing p as a linear combination of monomials in $x_1, \ldots, x_n$, each monomial term has a nonnegative coefficient. Similarly, p is *Schur positive* if in the expansion of p as a linear combination of Schur functions each term has a nonnegative coefficient.

Recall that $\lambda \in \mathcal{W}$ is a *partition* if $\lambda_1, \ldots, \lambda_n$ are nonnegative integers. Corresponding to each partition λ is a unique *Young diagram* [15, 18]. In particular, given partitions λ and μ, the partitions corresponding to the union and intersection, respectively, of the Young diagrams of λ and μ are given by (1.1) and (1.2).

In some recent articles Lam et al. [12, 13] have established some interesting positivity properties of the Schur functions. A consequence of the work of Lam et al. [12, 13] is the following result.

Theorem 1.2 (Lam et al. [12, 13]) *For partitions λ and μ, the symmetric function $s_{\lambda \vee \mu} s_{\lambda \wedge \mu} - s_\lambda s_\mu$ is monomial positive and Schur positive.*

For $x_1, \ldots, x_n \in \mathbb{R}_+$, the positive real line, it is well-known that $s_\lambda(x_1, \ldots, x_n) \geq 0$, so Theorem 1.2 implies Theorem 1.1 for the case in which λ and μ are partitions. Nevertheless, the two results do not seem to be directly comparable, for Theorem 1.1 applies to all $\lambda, \mu \in \mathcal{W}$, but proves positivity only, whereas Theorem 1.2 applies to partitions λ, μ only, but proves monomial and Schur positivity. Moreover, the markedly different methods by which the two results are established further underscores their incomparability.

The second aim of this paper is to derive log-convexity properties of the (generalized) hypergeometric functions of two Hermitian matrix arguments [5, 9]. These functions are infinite series of "complex" zonal polynomials with coefficients extending the classical rising factorial, and they generalize the classical hypergeometric functions in many ways. For suitably chosen scalars $a_1, \ldots, a_p, b_1, \ldots, b_q \in \mathbb{C}$, denote by

$$_pF_q\left(\begin{matrix} a_1, \ldots, a_p \\ b_1, \ldots, b_q \end{matrix}; x, y\right)$$

the hypergeometric function of $n \times n$ Hermitian matrix arguments, x and y. By applying properties of the Schur functions established in Sect. 2 together with application of the FKG inequality [4, 16], we shall establish the positivity of certain differences of products of these hypergeometric functions. By means of the generalized FKG inequalities in [16], we obtain results in which these products involve as many as five $_pF_q$ functions of two Hermitian matrix arguments.

2 Log-convexity properties of the Schur functions

Let $\mathcal{D} \subseteq \mathbb{R}^n$ (we assume for the sake of definiteness that $\mathcal{D}$ is sufficiently smooth, say, a C^∞ manifold). A function $\phi : \mathcal{D} \to [0, \infty)$ is called *multivariate totally positive of order 2* (MTP$_2$) on $\mathcal{D}$ if

$$\phi(\lambda \vee \mu)\phi(\lambda \wedge \mu) \geq \phi(\lambda)\phi(\mu) \tag{2.1}$$

for all $\lambda, \mu \in \mathcal{D}$. We remark that the "MTP$_2$" terminology is common in mathematical statistics and parts of probability theory [11, 16]; in combinatorics [1, p. 83], game theory, and economics [19, p. 64], the MTP$_2$ concept is known as *log-supermodularity*; and in mathematical physics and related areas of probability [2, 4, 8], it is known as the *FKG condition*.

There is an analytical characterization of the class of positive, sufficiently smooth MTP$_2$ functions [2, 8, 11, 14]: Suppose that $\phi : \mathcal{D} \to \mathbb{R}_+$ is in $C^2(\mathcal{D})$; then ϕ is MTP$_2$ on $\mathcal{D}$ if and only if

$$\frac{\partial^2}{\partial \lambda_i \partial \lambda_j} \log \phi(\lambda) \geq 0 \tag{2.2}$$

on $\mathcal{D}$ for all $1 \leq i \neq j \leq n$.

As an application of this characterization, it follows that if $\phi_1, \phi_2 : \mathcal{D} \to \mathbb{R}_+$ both are in $C^2(\mathcal{D})$ and are MTP$_2$ on $\mathcal{D}$ then the product $\phi_1 \phi_2$ also is MTP$_2$ on $\mathcal{D}$. This result holds, more generally, for all MTP$_2$ functions, a result which is established by Karlin and Rinott [11] by algebraic methods.

By means of the characterization (2.2), we now establish Theorem 1.1 *via* analytical methods; in so doing, we write $s_\lambda(x)$ simply as s_λ.

Proof of Theorem 1.1 To ascertain the sign of $\partial^2 \log s_\lambda / \partial \lambda_i \partial \lambda_j$, $\lambda \in \mathcal{W}$, for any $i \neq j$, it suffices to take $(i, j) = (1, 2)$, for all other cases are resolved in the same way. Let $\alpha_j = \lambda_j + n - j$, $1 \leq j \leq n$, so that $\alpha_1 > \cdots > \alpha_n$, set $\alpha = (\alpha_1, \ldots, \alpha_n)$, and define

$$\widetilde{\Delta}_\alpha(x) = \det\left(x_i^{\alpha_j}\right).$$

By (1.3)–(1.5), we have

$$\Delta_\lambda^2 \cdot \frac{\partial^2}{\partial \lambda_1 \lambda_2} \log s_\lambda = \Delta_\lambda \cdot \frac{\partial^2}{\partial \lambda_1 \lambda_2} \Delta_\lambda - \frac{\partial}{\partial \lambda_1} \Delta_\lambda \cdot \frac{\partial}{\partial \lambda_2} \Delta_\lambda$$

$$= \widetilde{\Delta}_\alpha \cdot \frac{\partial^2}{\partial \alpha_1 \alpha_2} \widetilde{\Delta}_\alpha - \frac{\partial}{\partial \alpha_1} \widetilde{\Delta}_\alpha \cdot \frac{\partial}{\partial \alpha_2} \widetilde{\Delta}_\alpha. \tag{2.3}$$

To simplify this difference of product terms, we proceed as in [17, Sect. 3], applying a result derived by Karlin [10, p. 7, (0.16)] from Sylvester's formula for compound determinants: For column vectors $a, b, f_1, \ldots, f_{n-2} \in \mathbb{R}^n$, define the determinant

$$D(a, b, f_1, \ldots, f_{n-2}) = \det(a, b, f_1, \ldots, f_{n-2}).$$

Then, for $a_1, a_2, b_1, b_2 \in \mathbb{R}^n$,

$$\begin{vmatrix} D(a_1, b_1, f_1, \ldots, f_{n-2}) & D(a_1, b_2, f_1, \ldots, f_{n-2}) \\ D(a_2, b_1, f_1, \ldots, f_{n-2}) & D(a_2, b_2, f_1, \ldots, f_{n-2}) \end{vmatrix}$$
$$= D(a_1, a_2, f_1, \ldots, f_{n-2}) \cdot D(b_1, b_2, f_1, \ldots, f_{n-2}). \tag{2.4}$$

For $u \in \mathbb{R}_+$ and $v \in \mathbb{R}$, define the kernel $K : \mathbb{R}_+ \times \mathbb{R} \to \mathbb{R}_+$ by

$$K(u, v) = u^v; \tag{2.5}$$

set

$$a_1 = \begin{pmatrix} K(x_1, \alpha_1) \\ \vdots \\ K(x_n, \alpha_1) \end{pmatrix}, \qquad a_2 = \frac{\partial}{\partial \alpha_1} \begin{pmatrix} K(x_1, \alpha_1) \\ \vdots \\ K(x_n, \alpha_1) \end{pmatrix},$$

$$b_1 = \begin{pmatrix} K(x_1, \alpha_2) \\ \vdots \\ K(x_n, \alpha_2) \end{pmatrix}, \qquad b_2 = \frac{\partial}{\partial \alpha_2} \begin{pmatrix} K(x_1, \alpha_2) \\ \vdots \\ K(x_n, \alpha_2) \end{pmatrix};$$

and, for $j = 1, \ldots, n - 2$, set

$$f_j = \begin{pmatrix} K(x_1, \alpha_j) \\ \vdots \\ K(x_n, \alpha_j) \end{pmatrix}.$$

Then

$$D(a_1, b_1, f_1, \ldots, f_{n-2}) = \tilde{\Delta}_\alpha(x),$$

$$D(a_1, b_2, f_1, \ldots, f_{n-2}) = \frac{\partial}{\partial \alpha_2} \tilde{\Delta}_\alpha(x),$$

$$D(a_2, b_1, f_1, \ldots, f_{n-2}) = \frac{\partial}{\partial \alpha_1} \tilde{\Delta}_\alpha(x),$$

and

$$D(a_2, b_2, f_1, \ldots, f_{n-2}) = \frac{\partial^2}{\partial \alpha_1 \partial \alpha_2} \tilde{\Delta}_\alpha(x).$$

Substituting these results into (2.3) and (2.4), we obtain

$$\Delta_\lambda^2 \cdot \frac{\partial^2}{\partial \lambda_1 \lambda_2} \log s_\lambda = \tilde{\Delta}_\alpha \cdot \frac{\partial^2}{\partial \alpha_1 \alpha_2} \tilde{\Delta}_\alpha - \frac{\partial}{\partial \alpha_1} \tilde{\Delta}_\alpha \cdot \frac{\partial}{\partial \alpha_2} \tilde{\Delta}_\alpha$$

$$= \begin{vmatrix} D(a_1, b_1, f_1, \ldots, f_{n-2}) & D(a_1, b_2, f_1, \ldots, f_{n-2}) \\ D(a_2, b_1, f_1, \ldots, f_{n-2}) & D(a_2, b_2, f_1, \ldots, f_{n-2}) \end{vmatrix}$$

$$= D(a_1, a_2, f_1, \ldots, f_{n-2}) \cdot D(b_1, b_2, f_1, \ldots, f_{n-2}). \quad (2.6)$$

Next,

$$D(a_1, a_2, f_1, \ldots, f_{n-2})$$

$$= \begin{vmatrix} K(x_1, \alpha_1) & \frac{\partial}{\partial \alpha_1} K(x_1, \alpha_1) & K(x_1, \alpha_3) & \cdots & K(x_1, \alpha_n) \\ \vdots & \vdots & \vdots & \cdots & \vdots \\ K(x_n, \alpha_1) & \frac{\partial}{\partial \alpha_1} K(x_n, \alpha_1) & K(x_n, \alpha_3) & \cdots & K(x_n, \alpha_n) \end{vmatrix}$$

$$\equiv \lim_{\alpha_2 \to \alpha_1} \frac{\det(K(x_i, \alpha_j))}{\alpha_2 - \alpha_1}.$$

It is well-known [10] that the kernel K in (2.5) is strictly totally positive of order infinity (STP$_\infty$), i.e., $\det(K(x_i, \alpha_j)) > 0$ for all $x_1 > \cdots > x_n > 0, \alpha_1 > \cdots > \alpha_n$, and all $n \geq 1$. Hence, in the above limit, the numerator is positive and the denominator is negative; therefore, $D(a_1, a_2, f_1, \ldots, f_{n-2}) \leq 0$. (We remark that the limit can be shown to be negative by utilizing the fact [10, pp. 16] that the kernel K also is extended totally positive of order infinity (ETP$_\infty$); however, this result is not needed for our purposes.)

Similarly, we deduce that $D(b_1, b_2, f_1, \ldots, f_{n-2}) \leq 0$, and then it follows from (2.6) that for $\lambda \in \mathcal{W}$ and $x_1 > \cdots > x_n > 0$,

$$\frac{\partial^2}{\partial \lambda_i \partial \lambda_j} \log s_\lambda(x) > 0, \tag{2.7}$$

$i \neq j$. By the characterization (2.2) of functions that are positive, MTP$_2$, and C^2, we deduce that if $\lambda, \mu \in \mathcal{W}$ and $x_1 > \cdots > x_n > 0$ then

$$s_{\lambda \vee \mu}(x) s_{\lambda \wedge \mu}(x) - s_\lambda(x) s_\mu(x) \geq 0. \tag{2.8}$$

As a limiting case, we find that positivity also holds in (2.7) if $x_1 \geq \cdots \geq x_n$; moreover, because the function $s_\lambda(x)$ is symmetric in $x_1, \ldots, x_n$, then the condition $x_1 \geq \cdots \geq x_n$ can be dispensed with. Consequently, (2.8) holds for all $x \in \mathbb{R}^n_+$. $\square$

Extending Theorem 1.1, we have the following result.

Theorem 2.1 *Let p be a positive integer and, for $\kappa_1, \ldots, \kappa_p \in \mathcal{W}$, define*

$$v_j = \bigvee_{1 \leq i_1 < \cdots < i_j \leq p} (\kappa_{i_1} \wedge \cdots \wedge \kappa_{i_j}),$$

$j = 1, \ldots, p$. Then, for $x \in \mathbb{R}^n_+$ and any positive integer q, where $1 \leq q \leq p$,

$$\prod_{j=q}^{p} \left(s_{v_j}(x) \right)^{\binom{j-1}{q-1}} - \prod_{1 \leq i_1 < \cdots < i_q \leq p} s_{\kappa_{i_1} \wedge \cdots \wedge \kappa_{i_q}}(x) \geq 0. \tag{2.9}$$

Moreover, these inequalities remain valid if the operations $\vee$ and $\wedge$ are interchanged; that is, if

$$\widetilde{v}_j = \bigwedge_{1 \leq i_1 < \cdots < i_j \leq p} (\kappa_{i_1} \vee \cdots \vee \kappa_{i_j}),$$

$j = 1, \ldots, p$, then, for $x \in \mathbb{R}^n_+$ and any positive integer q, where $1 \leq q \leq p$,

$$\prod_{j=q}^{p} \left(s_{\widetilde{v}_j}(x) \right)^{\binom{j-1}{q-1}} - \prod_{1 \leq i_1 < \cdots < i_q \leq p} s_{\kappa_{i_1} \vee \cdots \vee \kappa_{i_q}}(x) \geq 0. \tag{2.10}$$

Proof By Theorem 1.1, $s_{\lambda \vee \mu}(x)s_{\lambda \wedge \mu}(x) \geq s_\lambda(x)s_\mu(x)$; equivalently,

$$\log s_{\lambda \vee \mu}(x) + \log s_{\lambda \wedge \mu}(x) \geq \log s_\lambda(x) + \log s_\mu(x),$$

so the function $\lambda \mapsto -\log s_\lambda(x)$ is *subadditive* in the sense defined by Fan [3]. On applying the theorem of Fan, we obtain the inequality,

$$\sum_{j=q}^{p} \binom{j-1}{q-1} \log \big(s_{\nu_j}(x) \big) \geq \sum_{1 \leq i_1 < \cdots < i_q \leq p} \log s_{\kappa_{i_1} \wedge \cdots \wedge \kappa_{i_q}}(x),$$

which yields (2.9) by exponentiation.

Finally, the dual result (2.10) is obtained similarly.   □

The results in Theorem 2.1 also raise the combinatorial problem of whether the symmetric functions in (2.9) and (2.10) are monomial positive or Schur positive.

3 Log-convexity properties of the hypergeometric functions of matrix argument

In this section, we apply Theorem 1.1 and the FKG inequality to derive log-convexity properties of the generalized hypergeometric functions of two Hermitian matrix arguments.

Let $\phi : \mathbb{R}^n \to \mathbb{R}$ be a MTP$_2$ probability density function; let $f : \mathbb{R}^n \to \mathbb{R}$; and denote by $\mathbb{E}(f)$ the expectation, whenever it exists, of f with respect to the probability distribution corresponding to ϕ. A function $f : \mathbb{R}^n \to \mathbb{R}$ is called *increasing* if $f(x)$ is monotone increasing in each coordinate of the argument $x \in \mathbb{R}^n$.

We recall the FKG inequality [4, 16]: If f and g are increasing functions on $\mathbb{R}^n$ then

$$\mathbb{E}(fg) - \mathbb{E}(f)\mathbb{E}(g) \geq 0. \tag{3.1}$$

Let $\lambda = (\lambda_1, \ldots, \lambda_n)$ be a partition. The *length*, $\ell(\lambda)$, of λ is the largest integer l such that $\lambda_l > 0$; and the *weight* of λ is $|\lambda| := \lambda_1 + \cdots + \lambda_n$. We denote by $\mathcal{P}_n$ the set of all partitions λ of length $\ell(\lambda) \leq n$.

To apply the FKG inequality, we construct a MTP$_2$ probability density function on $\mathcal{P}_n$, and then we deduce certain hypergeometric function inequalities by a careful choice of the functions f and g in (3.1).

Let $\mathcal{H}_n$ denote the space of $n \times n$ Hermitian matrices, and $\mathcal{H}_n^+$ denote the positive definite matrices in $\mathcal{H}_n$. Whenever $x \in \mathcal{H}_n$, with eigenvalues $x_1, \ldots, x_n$, we write $s_\lambda(x) \equiv s_\lambda(x_1, \ldots, x_n)$, an interpretation that is based on the interpretation of the Schur functions as "class functions" in the representation theory of $GL(n, \mathbb{C})$.

We shall construct positive coefficients c_λ, $\lambda \in \mathcal{P}_n$, such that c_λ is MTP$_2$ in λ; and then, for fixed $x \in \mathcal{H}_n^+$, we form the function,

$$\phi(\lambda) = \frac{c_\lambda s_\lambda(x)}{\sum_{\mu \in \mathcal{P}_n} c_\mu s_\mu(x)}, \tag{3.2}$$

$\lambda \in \mathcal{P}_n$. Since $x \in \mathcal{H}_n^+$ then $s_\lambda(x) > 0$, and hence $\phi(\lambda) > 0$, $\lambda \in \mathcal{P}_n$. Therefore, subject to the absolute convergence of the denominator in (3.2), ϕ is a probability density function on $\mathcal{P}_n$. Moreover, because the class of MTP$_2$ functions is closed under pointwise multiplication [11], we shall deduce from the MTP$_2$ nature of the coefficients c_λ and the functions $s_\lambda(x)$ that ϕ also is MTP$_2$.

For $a \in \mathbb{C}$ define the *classical shifted factorial*, $(a)_k = a(a+1)\cdots(a+k-1)$, $k = 0, 1, 2, \ldots$ and, for $\lambda \in \mathcal{P}_n$, define the *partitional shifted factorial*,

$$(a)_\lambda = \prod_{j=1}^{n}(a-j+1)_{\lambda_j}.$$

Lemma 3.1 *Let $a_1, \ldots, a_p, b_1, \ldots, b_q > n-1$, $y \in \mathcal{H}_n^+$, and define*

$$c_\lambda = \frac{(a_1)_\lambda \cdots (a_p)_\lambda}{(b_1)_\lambda \cdots (b_q)_\lambda (n)_\lambda} s_\lambda(y), \tag{3.3}$$

$\lambda \in \mathcal{P}_n$. *Then, c_λ is MTP$_2$ in λ.*

Proof If $a > n-1$ then the function $\lambda \mapsto (a)_\lambda$, $\lambda \in \mathcal{P}_n$ clearly is positive. Moreover, $(a)_\lambda$ is MTP$_2$, for it is straightforward to verify that $(a)_{\lambda \vee \mu}(a)_{\lambda \wedge \mu} \equiv (a)_\lambda(a)_\mu$ for all $\lambda, \mu \in \mathcal{P}_n$, so that (2.1) is satisfied as an equality. Hence, for $a_1, \ldots, a_p, b_1, \ldots, b_q > n-1$, the function

$$\lambda \mapsto \frac{(a_1)_\lambda \cdots (a_p)_\lambda}{(b_1)_\lambda \cdots (b_q)_\lambda (n)_\lambda},$$

$\lambda \in \mathcal{P}_n$, also is positive and MTP$_2$.

Next, $s_\lambda(y) > 0$ because $y \in \mathcal{H}_n^+$. Also, as we have observed in Sect. 2, $s_\lambda(y)$ is MTP$_2$ in λ. Again because the product of MTP$_2$ functions remains MTP$_2$, it follows that c_λ is positive and is MTP$_2$ in λ. $\square$

Let us now draw from [5, 9, 15] the definition and some properties of the (generalized) hypergeometric functions of two Hermitian matrix arguments.

For each partition $\lambda = (\lambda_1, \ldots, \lambda_n) \in \mathcal{P}_n$, let Z_λ denote the *zonal polynomial* indexed by λ. The zonal polynomials are spherical functions on $\mathcal{H}_n$ and, up to a constant multiple, Z_λ coincides with the Schur functions: For $x \in \mathcal{H}_n$ with eigenvalues $x_1, \ldots, x_n$, the zonal polynomial is given by the explicit formula,

$$Z_\lambda(x) = \omega_\lambda s_\lambda(x), \tag{3.4}$$

where $s_\lambda(x) \equiv s_\lambda(x_1, \ldots, x_n)$ is given explicitly by the ratio formula (1.5), and

$$\omega_\lambda = |\lambda|! \frac{\prod_{j<k}(\lambda_j - \lambda_k - j + k)}{\prod_{j=1}^{n}(\lambda_j + n - j)!}. \tag{3.5}$$

It is well-known, and follows as a limiting case of the formula (1.5), that $d_\lambda :=$ $s_\lambda(1,\ldots,1)$ is given explicitly by the *Weyl dimension formula*,

$$d_\lambda = \frac{\prod_{j<k}(\lambda_j - \lambda_k - j + k)}{\prod_{j=1}^n (j-1)!}. \tag{3.6}$$

By (3.4),

$$Z_\lambda(I_n) = \omega_\lambda d_\lambda, \tag{3.7}$$

and it can also be deduced from (3.5) and (3.6) that

$$\omega_\lambda = \frac{|\lambda|! d_\lambda}{(n)_\lambda}. \tag{3.8}$$

Although these results are well-known to have profound significance in the representation theory of $GL(n, \mathbb{C})$, that subject will not play a role in our results.

Suppose that $a_1, \ldots, a_p, b_1, \ldots, b_q \in \mathbb{C}$ are such that, for $1 \le i \le q$ and $1 \le j \le n$, none of the numbers $-b_i + j - 1$ is a nonnegative integer. For $x, y \in \mathcal{H}_n$, the *hypergeometric function of two Hermitian matrix arguments* is defined as the series

$$_pF_q\left(\begin{matrix} a_1, \ldots, a_p \\ b_1, \ldots, b_q \end{matrix}; x, y\right) = \sum_\lambda \frac{(a_1)_\lambda \cdots (a_p)_\lambda}{(b_1)_\lambda \cdots (b_q)_\lambda} \frac{Z_\lambda(x) Z_\lambda(y)}{|\lambda|! Z_\lambda(I_n)}, \tag{3.9}$$

where the sum is over all partitions λ of all nonnegative integers.

For $x \in \mathcal{H}_n$, let $\|x\| = \max\{|x_j| : j = 1, \ldots, n\}$ where $x_1, \ldots, x_n$ are the eigenvalues of x. Then the convergence properties of the series (3.9) are given in [5, Theorem 4.1]: If $p \le q$ then the series (3.9) converges absolutely for all $x, y \in \mathcal{H}_n$; if $p = q + 1$ then (3.9) converges absolutely if $\|x\| \cdot \|y\| < 1$ and diverges if $\|x\| \cdot \|y\| > 1$; and if $p > q + 1$ then (3.9) diverges unless it terminates.

We can now establish a log-convexity property for the $_pF_q$ functions in (3.9).

Theorem 3.2 *Let* $x, y \in \mathcal{H}_n^+$, $p < q$, *and* $a_1, \ldots, a_{p+1}, b_1, \ldots, b_{q+1} > n - 1$. *Then,*

$$_{p+1}F_{q+1}\left(\begin{matrix} a_1, \ldots, a_{p+1} \\ b_1, \ldots, b_{q+1} \end{matrix}; x, y\right) {}_pF_q\left(\begin{matrix} a_1, \ldots, a_p \\ b_1, \ldots, b_q \end{matrix}; x, y\right)$$

$$\le {}_{p+1}F_q\left(\begin{matrix} a_1, \ldots, a_{p+1} \\ b_1, \ldots, b_q \end{matrix}; x, y\right) {}_pF_{q+1}\left(\begin{matrix} a_1, \ldots, a_p \\ b_1, \ldots, b_{q+1} \end{matrix}; x, y\right).$$

If $p = q$ *then the same result holds under the additional assumption that* $\|x\| \cdot \|y\| < 1$.

Proof By (3.4), (3.7), and (3.8),

$$\frac{Z_\lambda(x) Z_\lambda(y)}{|\lambda|! Z_\lambda(I_n)} = \frac{s_\lambda(x) s_\lambda(y)}{(n)_\lambda};$$

therefore, with the coefficients c_λ as defined in (3.3), we have

$$\sum_{\lambda \in \mathcal{P}_n} c_\lambda s_\lambda(x) = {}_pF_q \begin{pmatrix} a_1, \ldots, a_p \\ b_1, \ldots, b_q \end{pmatrix}; x, y \end{pmatrix}.$$

For $a_{p+1}, b_{q+1} > n-1$, set $f(\lambda) = (a_{p+1})_\lambda$ and $g(\lambda) = 1/(b_{q+1})_\lambda$; then f is increasing, and g is *decreasing*, on $\mathcal{P}_n$. On applying the FKG inequality (3.1), we obtain

$$\left(\sum_{\lambda \in \mathcal{P}_n} c_\lambda s_\lambda(x) f(\lambda) g(\lambda) \right) \cdot \left(\sum_{\lambda \in \mathcal{P}_n} c_\lambda s_\lambda(x) \right)$$

$$\leq \left(\sum_{\lambda \in \mathcal{P}_n} c_\lambda s_\lambda(x) f(\lambda) \right) \cdot \left(\sum_{\lambda \in \mathcal{P}_n} c_\lambda s_\lambda(x) g(\lambda) \right).$$

Substituting (3.3) for c_λ, we find that each sum reduces to a hypergeometric function of matrix argument, and then we obtain the desired result. □

In [16], a class of generalizations of the FKG inequality with to up to five functions is provided. In the case of three functions, it is proved [16, Theorem 1.1] that if ϕ is a MTP_2 density on $\mathbb{R}^n$ and f, g, and h are nonnegative, increasing functions on $\mathbb{R}^n$ then

$$2\mathbb{E}(fgh) - \left[\mathbb{E}(fg)\mathbb{E}(h) + \mathbb{E}(fh)\mathbb{E}(g) + \mathbb{E}(f)\mathbb{E}(gh) \right]$$

$$+ \mathbb{E}(f)\mathbb{E}(g)\mathbb{E}(h) \geq 0, \qquad\qquad (3.10)$$

and analogous generalizations are given for four and five functions. By proceeding as before, choosing c_λ as in (3.3), and choosing each of $f(\lambda)$, $g(\lambda)$, and $h(\lambda)$ as a partitional shifted factorial or its inverse, we can obtain inequalities that extend Theorem 3.2.

References

1. Alon, N., Spencer, J.H.: The Probabilistic Method, 3rd edn. Wiley, Hoboken (2008)
2. Battle, G.A., Rosen, L.: The FKG inequality for the Yukawa quantum field theory. J. Stat. Phys. **22**, 123–192 (1980)
3. Fan, K.: An inequality for subadditive functions on a distributive lattice, with application to determinantal inequalities. Linear Algebra Appl. **1**, 33–38 (1968)
4. Fortuin, C.M., Kasteleyn, P.W., Ginibre, J.: Correlation inequalities on some partially ordered sets. Commun. Math. Phys. **22**, 89–103 (1971)
5. Gross, K.I., Richards, D.St.P.: Total positivity, spherical series, and hypergeometric functions of matrix argument. J. Approx. Theory **59**, 224–246 (1987)
6. Gross, K.I., Richards, D.St.P.: Total positivity, finite reflection groups, and a formula of Harish-Chandra. J. Approx. Theory **82**, 60–87 (1995)
7. Helgason, S.: Groups and Geometric Analysis. Academic Press, New York (1984)
8. Herbst, I., Pitt, L.: Diffusion equation techniques in stochastic monotonicity and positive correlations. Probab. Theory Relat. Fields **87**, 275–312 (1991)
9. James, A.T.: Distributions of matrix variates and latent roots derived from normal samples. Ann. Math. Stat. **35**, 475–501 (1964)
10. Karlin, S.: Total Positivity. Stanford University Press, Stanford (1968)

11. Karlin, S., Rinott, Y.: Classes of orderings of measures and related correlation inequalities. I. Multivariate totally positive distributions. J. Multivar. Anal. **10**, 467–498 (1980)
12. Lam, T., Postnikov, A., Pylyavskyy, P.: Schur positivity and Schur log-concavity. Am. J. Math. **129**, 1611–1622 (2007)
13. Lam, T., Pylyavskyy, P.: Cell transfer and monomial positivity. J. Algebr. Comb. **26**, 209–224 (2007)
14. Lorentz, G.G.: An inequality for rearrangements. Am. Math. Mon. **60**, 176–179 (1953)
15. Macdonald, I.G.: Symmetric Functions and Hall Polynomials, 2nd edn. Oxford University Press, New York (1995)
16. Richards, D.St.P.: Algebraic methods toward higher-order probability inequalities, II. Ann. Probab. **32**, 1509–1544 (2004)
17. Richards, D.St.P.: Total positivity properties of generalized hypergeometric functions of matrix argument. J. Stat. Phys. **116**, 225–231 (2004)
18. Stanley, R.P.: Enumerative Combinatorics, vol. 2. Cambridge University Press, Cambridge (1999)
19. Topkis, D.M.: Supermodularity and Complementarity. Princeton University Press, Princeton (1998)

Ramanujan J (2010) 23: 409–416
DOI 10.1007/s11139-010-9283-9

Infinite families of strange partition congruences for broken 2-diamonds

Peter Paule · Silviu Radu

Dedicated to our friend George E. Andrews on the occasion of his 70th birthday

Received: 18 December 2009 / Accepted: 26 October 2010 / Published online: 12 November 2010
© Springer Science+Business Media, LLC 2010

Abstract In 2007, George E. Andrews and Peter Paule (Acta Arithmetica 126:281–294, 2007) introduced a new class of combinatorial objects called broken k-diamonds. Their generating functions connect to modular forms and give rise to a variety of partition congruences. In 2008, Song Heng Chan proved the first infinite family of congruences when $k = 2$. In this note, we present two non-standard infinite families of broken 2-diamond congruences derived from work of Oliver Atkin and Morris Newman. In addition, four conjectures related to $k = 3$ and $k = 5$ are stated.

Keywords Partition congruences · Partition analysis · Partition diamonds

Mathematics Subject Classification (2000) Primary 05A17 · 11P83 · Secondary 05A15

1 Introduction

A combinatorial study guided by MacMahon's Partition Analysis led G.E. Andrews and P. Paule [1] to the construction of a new class of directed graphs called broken k-diamonds. These objects were constructed in such a way that the generating functions of their counting sequences $(\Delta_k(n))_{n\geq 0}$ are closely related to modular forms,

P. Paule was partially supported by grant P2016-N18 of the Austrian Science Funds FWF.
S. Radu was supported by DK grant W1214-DK6 of the Austrian Science Funds FWF.

P. Paule (✉) · S. Radu
Research Institute for Symbolic Computation (RISC), Johannes Kepler University, 4040 Linz, Austria
e-mail: Peter.Paule@risc.uni-linz.ac.at

S. Radu
e-mail: Silviu.Radu@risc.uni-linz.ac.at

namely:

$$\sum_{n=0}^{\infty} \Delta_k(n)q^n = \prod_{n=1}^{\infty} \frac{(1-q^{2n})(1-q^{(2k+1)n})}{(1-q^n)^3(1-q^{(4k+2)n})}$$
$$= q^{(k+1)/12} \frac{\eta(2\tau)\eta((2k+1)\tau)}{\eta(\tau)^3 \eta((4k+2)\tau)}, \quad k \geq 1,$$

where we recall the Dedekind eta function

$$\eta(\tau) := q^{\frac{1}{24}} \prod_{n=1}^{\infty} (1-q^n) \quad (q = e^{2\pi i \tau}).$$

This modular aspect, in turn, led to various arithmetic theorems and conjectures. For example, M. Hirschhorn and J. Sellers [5] supplied a proof of

$$\Delta_2(10n+2) \equiv 0 \pmod 2, \quad n \geq 0, \tag{1}$$

which is Conjecture 1 of [1]. In addition, they observed and proved the congruences

$$\Delta_1(4n+2) \equiv 0 \pmod 2, \quad n \geq 0,$$
$$\Delta_1(4n+3) \equiv 0 \pmod 2, \quad n \geq 0,$$

and

$$\Delta_2(10n+6) \equiv 0 \pmod 2, \quad n \geq 0, \tag{2}$$

the latter being a class-mate of (1). The first parametrized families of congruences were given by S.H. Chan [4]; namely for $\alpha \geq 1$,

$$\Delta_2(5^{\alpha+1}n + \lambda_\alpha) \equiv 0 \pmod 5, \quad n \geq 0, \tag{3}$$

and

$$\Delta_2(5^{\alpha+1}n + \mu_\alpha) \equiv 0 \pmod 5, \quad n \geq 0, \tag{4}$$

where λ_α and μ_α are the smallest positive integer solutions to

$$4\lambda_\alpha \equiv 1 \pmod{11 \cdot 5^\alpha} \quad \text{and} \quad 4\mu_\alpha \equiv 1 \pmod{19 \cdot 5^\alpha},$$

respectively. For example, $\alpha = 1$ gives the congruences

$$\Delta_2(25n+14) \equiv \Delta_2(25n+24) \equiv 0 \pmod 5, \quad n \geq 0; \tag{5}$$

the first one was stated as Conjecture 2 in [1].

Based on numerical experiments the authors of [1] wrote: "The following observations about congruences suggest strongly that there are undoubtedly a myriad of partition congruences for $\Delta_k(n)$. This list is only to indicate the tip of the iceberg." This note tries to continue along this line by presenting further evidence of the rich arithmetical structure of broken k-diamonds. Our parametrized families of congruences are different from those of Ramanujan type for ordinary partitions; in this sense our attribute "strange" has to be understood. For example, inspired by Atkin [3] and Atkin and O'Brien [2], in Sect. 2 we shall prove (Theorem 2.5) that if p is a prime such that $p \equiv 13 \pmod{20}$ or $p \equiv 17 \pmod{20}$, then

$$\Delta_2\left((5n+4)p - \frac{p-1}{4}\right) \equiv 0 \pmod{5} \tag{6}$$

for all nonnegative integers n such that $20n + 15 \not\equiv 0 \pmod{p}$. Also in Sect. 2, inspired by M. Newman we shall prove (Lemma 2.10) that for all nonnegative integers k:

$$\Delta_2\left(4 \cdot 29^k - \frac{29^k - 1}{4}\right) \equiv k+1 \pmod{5}. \tag{7}$$

As a consequence, the entries of the sequence $(\Delta_2(n))_{n \geq 0}$ visit all residue classes modulo 5 infinitely often. In Sect. 3, we present some concluding remarks and four open problems.

2 Strange partition congruences

The set of natural numbers is supposed to include 0, i.e., we have $\mathbb{N} = \{0, 1, \ldots\}$ and $\mathbb{N}^* = \{1, 2, \ldots\}$. Writing $f(q) \equiv g(q) \pmod{n}$ for power series $f(q)$ and $g(q)$ as usually means that the coefficient sequences agree modulo n. Finally, it will be convenient to introduce the following convention: For a coefficient sequence $(a(n))_{n \geq 0}$ of a power series $\sum_{n \geq 0} a(n)q^n$ we extend the domain of the argument to the rational numbers by defining

$$a(r) := 0 \quad \text{if } r \in \mathbb{Q} \backslash \mathbb{N}.$$

Lemma 2.1

$$\sum_{n=0}^{\infty} \Delta_2(5n+4)q^n \equiv \prod_{n=1}^{\infty} \frac{(1-q^{2n})^{12}}{(1-q^n)^6} \pmod{5}.$$

Proof This follows immediately from S.H. Chan's result [4, (3.6)],

$$\sum_{n=0}^{\infty} \Delta_2(5n-1)q^n \equiv q \prod_{n=1}^{\infty} \frac{(1-q^{10n})^4(1-q^n)^4}{(1-q^{5n})^2(1-q^{2n})^8} \pmod{5},$$

by utilizing the fact that $1 - q^{5m} \equiv (1-q^m)^5 \pmod{5}$. $\qquad \square$

Defining

$$\sum_{n=0}^{\infty} c(n)q^n := \prod_{n=1}^{\infty} \frac{(1-q^{2n})^{12}}{(1-q^n)^6},$$

we recall a special case of Newman's Theorem 3 in [6]:

Lemma 2.2 *For each prime p with $p \equiv 1 \pmod{4}$ there exists an integer $x(p)$ such that for all $n \in \mathbb{N}$:*

$$c\left(np + 3\frac{p-1}{4}\right) + p^2 c\left(\frac{n}{p} - 3\frac{p-1}{4p}\right) = x(p)c(n). \tag{8}$$

Lemma 2.2, in view of $c(N) \equiv \Delta_2(5N + 4) \pmod 5$, implies that for each prime p, $p \equiv 1 \pmod 4$, and all $n \in \mathbb{N}$:

$$\Delta_2\left((5n + 4)p - \frac{p-1}{4}\right) \equiv c\left(np + 3\frac{p-1}{4}\right)$$

$$\equiv x(p)\Delta_2(5n + 4)$$

$$- p^2 \Delta_2\left(\frac{5n+4}{p} + \frac{p-1}{4p}\right) \pmod 5. \tag{9}$$

Setting $n = 0$ and noting that $\Delta_2(4) \equiv c(0) = 1$, we obtain

$$\Delta_2\left(4p - \frac{p-1}{4}\right) \equiv x(p) - p^2 \Delta_2\left(\frac{p+15}{4p}\right) \pmod 5.$$

Noting that

$$p \neq 5 \quad \Rightarrow \quad \frac{p+15}{4p} \notin \mathbb{N} \quad \Rightarrow \quad \Delta_2\left(\frac{p+15}{4p}\right) = 0,$$

one obtains

Corollary 2.3 *For all primes p with $p \equiv 1 \pmod 4$ we have*

$$\Delta_2\left((5n + 4)p - \frac{p-1}{4}\right) \equiv \Delta_2\left(4p - \frac{p-1}{4}\right)\Delta_2(5n + 4) \pmod 5 \tag{10}$$

for all $n \in \mathbb{N}$ such that $20n + 15 \not\equiv 0 \pmod p$.

Lemma 2.4 *For all primes p with $p \equiv 13$ or $p \equiv 17 \pmod{20}$ we have*

$$\Delta_2\left(4p - (p-1)/4\right) \equiv 0 \pmod 5.$$

Proof By (5), we know that $\Delta_2(4p - (p-1)/4) \equiv 0 \pmod 5$ if

$$4p - (p-1)/4 \equiv 14 \pmod{25}, \tag{11}$$

or

$$4p - (p-1)/4 \equiv 24 \pmod{25}. \tag{12}$$

The statement follows by verifying (11) and (12) for $p \equiv 17 \pmod{20}$ and $p \equiv 13 \pmod{20}$, respectively. $\qquad\square$

Finally Corollary 2.3 and Lemma 2.4 imply

Theorem 2.5 *For any prime p with $p \equiv 13 \pmod{20}$ or $p \equiv 17 \pmod{20}$ we have*

$$\Delta_2\big(p(5n+4) - (p-1)/4\big) \equiv 0 \pmod{5},$$

for all $n \in \mathbb{N}$ such that $20n + 15 \not\equiv 0 \pmod{p}$.

Considering such n where $n = pk$, we see that the condition $20n + 15 \not\equiv 0 \pmod{p}$ holds for all nonnegative integers k if $p \neq 3, 5$. Hence we have

$$\Delta_2\big(p(5pk+4) - (p-1)/4\big) \equiv 0 \pmod{5}, \tag{13}$$

whenever $p \equiv 13, 17 \pmod{20}$ and k a nonnegative integer.

For example, the primes $13, 17, 37, 53, 73, 97$ are either congruent 13 or 17 modulo 20. Hence by (13):

$$\Delta_2\big(5 \cdot 13^2 n + 49\big) \equiv 0 \pmod{5},$$
$$\Delta_2\big(5 \cdot 17^2 n + 64\big) \equiv 0 \pmod{5},$$
$$\Delta_2\big(5 \cdot 37^2 n + 139\big) \equiv 0 \pmod{5},$$
$$\Delta_2\big(5 \cdot 53^2 n + 199\big) \equiv 0 \pmod{5},$$
$$\Delta_2\big(5 \cdot 73^2 n + 274\big) \equiv 0 \pmod{5},$$

and

$$\Delta_2\big(5 \cdot 97^2 n + 364\big) \equiv 0 \pmod{5}$$

for all $n \in \mathbb{N}$.

We also like to mention that the special case (9) of Newman's Lemma 2.2 implies [4, Eq. 3.5], namely:

Lemma 2.6 *For $k \in \mathbb{N}^*$,*

$$\Delta_2(5n+4) \equiv \Delta_2\left(5^k n + \frac{1+3 \cdot 5^k}{4}\right) \pmod{5}, \quad n \geq 0. \tag{14}$$

Proof Setting $p = 5$ in (9) gives

$$\Delta_2(25n+19) \equiv x(5)\Delta_2(5n+4) \pmod{5}. \tag{15}$$

Since $\Delta_2(19) = 85606 \equiv 1 \pmod{5}$ and $\Delta_2(4) \equiv 1 \pmod{5}$, we obtain from (15) that $x(5) \equiv 1 \pmod{5}$. Hence (14) is true for $k = 2$. Next assume that (14) is true for $2 \leq k < N$. From this we conclude correctness for $k = N$ as follows. By the induction hypothesis, we have for $v \in \mathbb{N}$:

$$\Delta_2(5(5v+3)+4) \equiv \Delta_2\left(5^{N-1}(5v+3) + \frac{1+3 \cdot 5^{N-1}}{4}\right) \pmod{5},$$

which is equivalent to

$$\Delta_2(25v+19) \equiv \Delta_2\left(5^N v + \frac{1+3 \cdot 5^N}{4}\right) \pmod{5}.$$

Next we apply (15) to conclude that $\Delta_2(5v + 4) \equiv \Delta_2(25v + 19)$ (mod 5), and the proof is finished. $\qquad\qquad\qquad\qquad\qquad\qquad\qquad\qquad\qquad\qquad\qquad$ □

Motivated by Newman's work [7], in combination with (9) we obtain interesting congruences for primes with $p \equiv 1$ (mod 4) and $x(p) \not\equiv 0$ (mod 5). For this purpose, we introduce the following definition:

Definition 2.7 For $n \in \mathbb{Z}$ and p a prime with $p \equiv 1$ (mod 4) we define a map $a_{p,n}$: $\mathbb{Z} \rightarrow \mathbb{Z}$ by

$$a_{p,n}(k) := \Delta_2\big(p^k(5n + 4) - (p^k - 1)/4\big).$$

The next proposition is a straightforward verification; it will be used to prove Lemma 2.9.

Proposition 2.8 *For $n \in \mathbb{Z}$ and p a prime with $p \equiv 1$ (mod 4) we have*

$$a_{p,pn+3(p-1)/4}(k) = a_{p,n}(k + 1), \quad k \in \mathbb{Z}.$$

Lemma 2.9 *For $n \in \mathbb{Z}$ and p a prime with $p \equiv 1$ (mod 4) we have*

$$a_{p,n}(k + 2) - x(p)a_{p,n}(k + 1) + p^2 a_{p,n}(k) \equiv 0 \quad (\text{mod } 5), \quad k \geq -1, \qquad (16)$$

where $x(p)$ is as in (9).

Proof By (9), we have for all $n \in \mathbb{Z}$ and all primes such that $p \equiv 1$ (mod 4):

$$a_{p,n}(1) - x(p)a_{p,n}(0) + p^2 a_{p,n}(-1) \equiv 0 \quad (\text{mod } 5). \qquad (17)$$

So (16) holds for $k = -1$. Proceeding by induction assume that (16) holds for all $k > N \geq -1$. To prove (16) for $k = N$, apply to

$$a_{p,n}(N + 1) - x(p)a_{p,n}(N) + p^2 a_{p,n}(N - 1) \equiv 0 \quad (\text{mod } 5), \qquad (18)$$

which is (16) with $k = N - 1$, the transformation $n \mapsto pn + 3(p - 1)/4$. Using Proposition 2.8 completes the proof of Lemma 2.9. $\qquad\qquad\qquad\qquad\qquad$ □

Finally, we consider the special choice $p = 29$ with $p \equiv 1$ (mod 4). One computes

$$\Delta_2\big(4p - (p - 1)/4\big) = 339953476833877 \equiv 2 \quad (\text{mod } 5).$$

We find that $x(29) = 2$ (mod 5). For the choice $p = 29$ and $n = 0$, Lemma 2.9 turns into

$$a_{29,0}(k + 2) - 2a_{29,0}(k + 1) + a_{29,0}(k) \equiv 0 \quad (\text{mod } 5), \quad k \in \mathbb{Z}.$$

This congruence, viewed as an integer recurrence in k for $k \geq -1$, has the general solution $c_1 k + c_0$ with $c_0, c_1 \in \mathbb{Z}$. From $a_{29,0}(-1) = 0$ and $a_{29,0}(0) = \Delta_2(4) \equiv 1$ (mod 5) we obtain the particular solution $k + 1$ with $c_0 = c_1 = 1$. Thus we have proven statement (7), namely:

Lemma 2.10 *For $k \in \mathbb{N}$:*

$$a_{29,0}(k) = \Delta_2\left(29^k \cdot 4 - (29^k - 1)/4\right) \equiv k + 1 \pmod{5}.$$

3 Some conjectures

Newman's Theorem 3 from [6], implying Lemma 2.2 as a special case, played a crucial role in this note. Concerning broken diamond congruences it seems that its scope of applications exceeds the Δ_2 case by far. To illustrate this point, we pose some conjectures that involve analogous congruences to the ones we presented. Let

$$\sigma_3(n) = \sum_{d\mid n} d^3, \quad n \in \mathbb{N}^*.$$

Let

$$E_4(q) = 1 + 240 \sum_{n=1}^{\infty} \sigma_3(n) q^n$$

be the Eisenstein series of weight 4 for the full modular group. Numerical computations show strong evidence for the following four conjectures to be true.

Conjecture 3.1

$$\prod_{n=1}^{\infty}\left(1 - q^n\right)^4\left(1 - q^{2n}\right)^6 \equiv 6 \sum_{n=0}^{\infty} \Delta_3(7n + 5) q^n \pmod{7}.$$

Conjecture 3.2

$$\Delta_3\left(7^3 n + 82\right) \equiv \Delta_3\left(7^3 n + 278\right) \equiv \Delta_3\left(7^3 n + 327\right) \equiv 0 \pmod{7}, \quad n \in \mathbb{N}.$$

Conjecture 3.3

$$E_4\left(q^2\right) \prod_{n=1}^{\infty}\left(1 - q^n\right)^8\left(1 - q^{2n}\right)^2 \equiv 8 \sum_{n=0}^{\infty} \Delta_5(11n + 6) q^n \pmod{11}.$$

Conjecture 3.4 *Let $\sum_{n=0}^{\infty} c(n) q^n := E_4(q^2) \prod_{n=1}^{\infty}(1 - q^n)^8(1 - q^{2n})^2$. Then for every prime p with $p \equiv 1 \pmod{4}$ there exists an integer $y(p)$ such that*

$$c\left(pn + \frac{p-1}{2}\right) + p^8 c\left(\frac{n - (p-1)/2}{p}\right) = y(p)c(n)$$

for all $n \in \mathbb{N}$.

Let

$$\sum_{n=0}^{\infty} b(n) q^n := \prod_{n=1}^{\infty}\left(1 - q^n\right)^4\left(1 - q^{2n}\right)^6.$$

We find again in the list of Newman [6, p. 486] that for all primes p such that $p \equiv 1 \pmod{12}$ there exists an integer $z(p)$ such that

$$b\left(np + \frac{2(p-1)}{3}\right) + p^4 b\left(\frac{n - 2(p-1)/3}{p}\right) = z(p)b(n) \tag{19}$$

for all $n \in \mathbb{N}$. In particular, equation (19), together with Conjecture 3.1, implies an identity analogous to (9). Similarly, Conjecture 3.3, together with Conjecture 3.4, also implies an identity analogous to (9), which lead to some generalizations of the results of this paper. Conjecture 3.2 is analogous to (5). We also tried to find a congruence similar to (5) for Δ_5 but failed. However, when $z(p) \equiv 0 \pmod{7}$ in (19), we obtain congruences modulo 7 for Δ_3. Similarly, when $y(p) \equiv 0 \pmod{11}$ in Conjecture 3.4, we obtain congruences modulo 11 for Δ_5.

References

1. Andrews, G.E., Paule, P.: MacMahon's partition analysis XI: broken diamonds and modular forms. Acta Arith. **126**, 281–294 (2007)
2. Atkin, A.O.L.: Multiplicative congruence properties and density problems for $p(n)$. In: Proceedings of London Mathematical Society, pp. 563–576 (1968)
3. Atkin, A.O.L., O'Brien, J.N.: Some properties of $p(n)$ and $c(n)$ modulo powers of 13. Trans. Am. Math. Soc. **126**, 442–459 (1967)
4. Chan, S.H.: Some congruences for Andrews–Paule's broken 2-diamond partitions. Discrete Math. **308**, 5735–5741 (2008)
5. Hirschhorn, M.D., Sellers, J.A.: On recent congruence results of Andrews and Paule. Bull. Aust. Math. Soc. **75**, 121–126 (2007)
6. Newman, M.: Modular forms whose coefficients possess multiplicative properties. Ann. Math. **70**, 478–489 (1959)
7. Newman, M.: Modulo m and divisibility properties of the partition function. Trans. Am. Math. Soc. **97**, 225–236 (1960)

Ramanujan J (2010) 23: 417–431
DOI 10.1007/s11139-010-9251-4

On the Andrews–Schur proof
of the Rogers–Ramanujan identities

Hei-Chi Chan

Dedicated to George Andrews for his 70th birthday

Received: 6 May 2010 / Accepted: 28 May 2010 / Published online: 31 August 2010
© Springer Science+Business Media, LLC 2010

Abstract In this article, we study one of Andrews' proofs of the Rogers–Ramanujan identities published in 1970. His proof inspires connections to some famous formulas discovered by Ramanujan. During the course of study, we discovered identities such as

$$\sum_{n\geq 0} \frac{q^{n^2}}{(q;q)_n} = \frac{1}{\sqrt{5}}\left(\beta \prod_{n=1}^{\infty} \frac{1}{1+\alpha q^{n/5}+q^{2n/5}} - \alpha \prod_{n=1}^{\infty} \frac{1}{1+\beta q^{n/5}+q^{2n/5}}\right),$$

where $\beta = -1/\alpha$ is the Golden Ratio.

Keywords Andrews–Schur proof · Rogers–Ramanujan identities

Mathematics Subject Classification (2000) 05A15 · 05A30 · 05A40

1 The Andrews–Schur's proof

Over the years Professor Andrews made many important contributions in the mathematics related to the Rogers–Ramanujan identities: for $|q| < 1$,

$$\sum_{n\geq 0} \frac{q^{n^2+an}}{(1-q)(1-q^2)\cdots(1-q^n)} = \prod_{n=1}^{\infty} \frac{1}{(1-q^{5n-1-a})(1-q^{5n-4+a})}, \qquad (1)$$

where $a = 0, 1$. For introductions, see [3, 4, 9, 11]. We hope, therefore, that he will be delighted by this article, in which we look at one of his proofs of these amazing identities.

H.-C. Chan (✉)
University of Illinois at Springfield, Springfield, IL 62703-5407, USA
e-mail: chan.hei-chi@uis.edu

One way of proving (1) is to start with proving a finite version of it, namely:

Theorem 1 (Andrews [2], Schur [32]) *If $a = 0, 1$, then*

$$\sum_{j \geq 0} q^{j^2+aj} \begin{bmatrix} n-j \\ j \end{bmatrix} = \sum_{j=-\infty}^{\infty} (-1)^j q^{\frac{j(5j+1)}{2}-2aj} \begin{bmatrix} n+a \\ \lfloor \frac{n+3a-5j}{2} \rfloor \end{bmatrix}, \tag{2}$$

where

$$\begin{bmatrix} A \\ B \end{bmatrix} = \begin{cases} 0 & \text{if } B < 0 \text{ or } B > A, \\ \frac{(1-q^A)(1-q^{A-1})\cdots(1-q^{A-B+1})}{(1-q^B)(1-q^{B-1})\cdots(1-q)} & \text{otherwise} \end{cases}$$

are the usual q-binomial numbers (or the Gaussian polynomials), and $\lfloor x \rfloor$ means the largest integer $\leq x$.

First, we note that (2) reduces to (1) as $n \to \infty$ because of the following property of the q-binomial numbers (see, e.g., [3, 9]): for fixed m, m_1 and m_2, with $R > S$ positive,

$$\lim_{N \to \infty} \begin{bmatrix} N \\ m \end{bmatrix} = \frac{1}{(q;q)_m}, \tag{3}$$

$$\lim_{N \to \infty} \begin{bmatrix} RN+m_1 \\ SN+m_2 \end{bmatrix} = \frac{1}{(q;q)_\infty}. \tag{4}$$

Here, we follow the customary q-product notation: $(a;q)_0 := 1$ and, for $n \geq 1$,

$$(a;q)_n := \prod_{k=0}^{n-1} (1 - aq^k).$$

With (3) and (4), (1) follows by a simple application of the Jacobi Triple Identity. This trick of "finitization" is very powerful and can be used to prove many generalizations of (1); see A. Sills' excellent paper [33].

I. Schur studied the polynomials involved in Theorem 1 in his second proof of the Rogers–Ramanujan identities and established the above theorem through some complicated methods. Years later, Andrews gave a new and simpler proof of the theorem. Andrews' proof also forms the basis for many exciting generalizations in the past several decades.

Andrews' idea of proving Theorem 1 is as follows. Let $E_{n+1}(a)$ be the left-hand side of (2), and $D_{n+1}(a)$ be the right-hand side. If we can show that they satisfy the same difference equation, i.e.,

$$F_{n+1}(a) = F_n(a) + q^{n+a-1}F_{n-1}(a) \tag{5}$$

with the same initial values (indeed, $F_1(a) = F_2(a) = 1$), then we are done: $E_n(a) = D_n(a)$, and Theorem 1 is established.

It turns out, by using a well-known property of the q-binomial numbers, namely,

$$\begin{bmatrix} n \\ m \end{bmatrix} = \begin{bmatrix} n-1 \\ m \end{bmatrix} + \begin{bmatrix} n-1 \\ m-1 \end{bmatrix} q^{n-m}, \tag{6}$$

$$= \begin{bmatrix} n-1 \\ m \end{bmatrix} q^m + \begin{bmatrix} n-1 \\ m-1 \end{bmatrix} \tag{7}$$

(e.g., see [3, 9]), it is quite easy to show that $E_n(a)$ (the left-hand side of (2)) satisfies the required difference equation. When coming to show that $D_n(a)$ satisfies the same equation, the situation is very different, even though one still uses (6) and (7). Beginners may be bewildered by the mysterious cancelations and regrouping of various terms involved. Therefore it might be instructive to look at the verification of a numerical example, which illustrates the strategy used in the general proof.

Consider the case $a = 0$. In this example, we write $D_n(0)$ as D_n. Say, we want to show

$$D_{10} = D_9 + q^8 D_8. \tag{8}$$

First, we write down the difference $D_{10} - D_9$ (and we need (6) and (7)):

$$D_{10} - D_9 = \begin{bmatrix} 8 \\ 8 \end{bmatrix} q^9 - \begin{bmatrix} 8 \\ 7 \end{bmatrix} q^9 + \begin{bmatrix} 8 \\ 3 \end{bmatrix} q^5 - \begin{bmatrix} 8 \\ 2 \end{bmatrix} q^5. \tag{9}$$

Likewise, we use the definition of D_8 to write down

$$q^8 D_8 = \begin{bmatrix} 7 \\ 8 \end{bmatrix} q^{17} - \begin{bmatrix} 7 \\ 6 \end{bmatrix} q^{10} + \begin{bmatrix} 7 \\ 3 \end{bmatrix} q^8 - \begin{bmatrix} 7 \\ 1 \end{bmatrix} q^{11} \tag{10}$$

(note: we added the first term, which is zero, for bookkeeping reason). Now, (9) and (10) do not "look" the same! Thankfully, (6) and (7) will come to our rescue. Let us subtract (10) from (9). First, we pair up terms as follows. As both (9) and (10) have four terms, we pair up the first terms from both equations, then we pair up the second terms, etc. Next, we use (6) and (7) for subtraction. For example, the difference of the third terms gives

$$\begin{bmatrix} 8 \\ 3 \end{bmatrix} q^5 - \begin{bmatrix} 7 \\ 3 \end{bmatrix} q^8 = \begin{bmatrix} 7 \\ 2 \end{bmatrix} q^5,$$

where we have used (7) to expand $\begin{bmatrix} 8 \\ 3 \end{bmatrix}$. After all these are done, we have

$$D_{10} - D_9 - q^8 D_8 = \begin{bmatrix} 7 \\ 7 \end{bmatrix} q^9 - \begin{bmatrix} 7 \\ 7 \end{bmatrix} q^9 + \begin{bmatrix} 7 \\ 2 \end{bmatrix} q^5 - \begin{bmatrix} 7 \\ 2 \end{bmatrix} q^5 = 0,$$

with the cancelations clearly displayed.

In Andrews' paper [2], this program of proving the difference equation for $D_n(a)$ is implemented by considering separately the case for even n and the case for odd n. A logical question is: can we do the same thing without this separate consideration?

The answer is affirmative. In [18]—which was motivated by a feature article by Andrews [5]—the present author shows how this can be done. It turns out that a crucial ingredient needed is a trick that Andrews used in another paper around the same time [1], in which he studied Fibonacci numbers (i.e., $q \to 1$ in (5)). The key is to decompose $D_n(a)$ as a sum of five terms. Precisely (cf. the proof of Lemma 1 in [18]):

Lemma 1

$$D_{n+1}(a) = \frac{1}{5} \sum_{j=0}^{4} \left(\sum_{\sigma=-\infty}^{\infty} (-1)^\sigma q^{f_a(\sigma)} \left[\begin{array}{c} n+a \\ \lfloor \frac{n+3a-\sigma}{2} \rfloor \end{array} \right] \zeta^{j\sigma} \right)$$

$$:= \frac{1}{5} \sum_{j=0}^{4} C_{n+1}(\zeta^j, a). \tag{11}$$

Here, $\zeta := e^{i(2\pi/5)}$ and $f_a(\sigma) = \frac{\sigma(\sigma+1)}{10} - \frac{2}{5}a\sigma$.

Proof First, we observe that $-1 = (-1)^5$, and so $D_{n+1}(a)$ can be written as (cf. the right-hand side of (2) for its definition)

$$D_{n+1}(a) = \sum_{\sigma \equiv 0 \,(\mathrm{mod}\,5)} (-1)^\sigma q^{f_a(\sigma)} \left[\begin{array}{c} n+a \\ \lfloor \frac{n+3a-\sigma}{2} \rfloor \end{array} \right]. \tag{12}$$

Next, we want to remove the mod 5 restriction on the sum. Observe that

$$\frac{1}{5} \sum_{j=0}^{4} \zeta^{j\sigma} = \begin{cases} 1 & \text{if } \sigma \equiv 0 \ (\mathrm{mod}\,5), \\ 0 & \text{otherwise.} \end{cases} \tag{13}$$

This allows us to write (12) as

$$D_{n+1}(a) = \frac{1}{5} \sum_{j=0}^{4} \left(\sum_{\sigma=-\infty}^{\infty} (-1)^\sigma q^{f_a(\sigma)} \left[\begin{array}{c} n+a \\ \lfloor \frac{n+3a-\sigma}{2} \rfloor \end{array} \right] \zeta^{j\sigma} \right),$$

which is the desired result.   □

Next, we need to find the difference equation satisfied by $C_{n+1}(z, a)$ (here, z is not necessarily a power of ζ). Following [18], we can show that

$$C_{n+1}(z, a) = C_n(z, a) + q^{n+a-1} C_{n-1}(z, a) + (1 - z^{-5}) \Delta_{n+1}(z, a). \tag{14}$$

Here, the explicit form of $\Delta_{n+1}(z, a)$ is irrelevant. It is just the collection of all the "unwanted" terms. It turns out that the form of $C_n(z, a)$ allows us to prove (14) without considering separately the case of even n and the case of odd n. For details, see [18].

With this understood, we can prove the difference equation for $D_n(a)$. Plug in $z = \zeta^k$ ($k = 0, 1, \ldots, 4$) in (14). Note that these values of z make the Δ term decouple from the rest because of the factor $1 - z^{-5}$. Sum the $C_{n+1}(\zeta^k, a)$ over $k = 0, \ldots, 4$ and divide the sum by 5. This gives

$$\left(\frac{1}{5} \sum_{k=0}^{4} C_{n+1}(\zeta^k, a) \right) = \left(\frac{1}{5} \sum_{k=0}^{4} C_n(\zeta^k, a) \right) + q^{n+a-1} \left(\frac{1}{5} \sum_{k=0}^{4} C_{n-1}(\zeta^k, a) \right),$$

which, of course, is $D_{n+1}(a) = D_n(a) + q^{n+a-1} D_{n-1}(a)$, and we are done.

What else can we do with the decomposition in Lemma 1? It turns out that this decomposition is quite rewarding: $C_n(z, a)$ is easier to play with, and its properties are closely tied with several famous formulas discovered by Ramanujan, as we will see below.

In the next section, we set up some notation and prove several properties for $C_n(z, a)$. The formula displayed in the abstract can be found at the end of Sect. 2. In Sect. 3, we turn to four formulas discovered by Ramanujan.

2 More notation and some preliminary results

First, we start with several definitions that will be used repeatedly:

- *Two constants α and β*

  We define $\alpha := \frac{1-\sqrt{5}}{2}$ and $\beta := \frac{1+\sqrt{5}}{2}$. So β is the Golden Ratio, and α is the negative of the reciprocal of the Golden Ratio.
- *The function f*

$$f(-q) := (q; q)_\infty.$$

- *The function $J(x)$*

$$J(x) := \prod_{n=1}^{\infty} \left(1 + xq^{n/5} + q^{2n/5} \right). \tag{15}$$

- *The $G(q)$ and $H(q)$ functions*

  These functions are the standard notation for the right-hand side of the Rogers–Ramanujan identities:

$$G(q) := \frac{1}{(q; q^5)_\infty (q^4; q^5)_\infty}, \qquad H(q) := \frac{1}{(q^2; q^5)_\infty (q^3; q^5)_\infty}.$$

Note:

$$D_\infty(a) = \begin{cases} H(q) & \text{if } a = 1, \\ G(q) & \text{if } a = 0. \end{cases} \tag{16}$$

An easy consequence is

$$G(q)H(q) = \frac{f(-q^5)}{f(-q)}. \tag{17}$$

- *The Rogers–Ramanujan continued fraction*

  For $|q| < 1$, define

$$R(q) := \frac{q^{1/5}}{1} + \frac{q}{1} + \frac{q^2}{1} + \frac{q^3}{1} + \cdots$$

$$= q^{1/5} \frac{H(q)}{G(q)}. \tag{18}$$

The first line is the definition of $R(q)$. The second equality is proven by L. Rogers in 1894 [31].

Next, we recall two results (Lemmas 2 and 3) from [22].

Lemma 2 *With the above notation,*

$$C_\infty(1, 0) = 0, \tag{19}$$

$$C_\infty(\zeta, 0) = \overline{C_\infty(\zeta^4, 0)} = \frac{f(-q^{1/5})}{f(-q)} \left(1 - \frac{1}{\zeta}\right) J(\alpha), \tag{20}$$

$$C_\infty(\zeta^2, 0) = \overline{C_\infty(\zeta^3, 0)} = \frac{f(-q^{1/5})}{f(-q)} \left(1 - \frac{1}{\zeta^2}\right) J(\beta), \tag{21}$$

$$C_\infty(1, 1) = 0, \tag{22}$$

$$C_\infty(\zeta, 1) = \overline{C_\infty(\zeta^4, 1)} = \frac{f(-q^{1/5})}{f(-q)} \cdot \frac{\zeta(\zeta - 1)}{q^{1/5}} J(\alpha), \tag{23}$$

$$C_\infty(\zeta^2, 1) = \overline{C_\infty(\zeta^3, 1)} = \frac{f(-q^{1/5})}{f(-q)} \cdot \frac{\zeta^2(\zeta^2 - 1)}{q^{1/5}} J(\beta). \tag{24}$$

Here, $\overline{w}$ denotes the complex conjugate of w.

Proof Our proof relies on Jacobi's Triple Product Identity (see, e.g., [3, 6, 9, 13])

$$\sum_{n=-\infty}^{\infty} (-z)^n Q^{n^2} = \prod_{j=1}^{\infty} (1 - Q^{2j})(1 - zQ^{2j-1})(1 - z^{-1}Q^{2j-1}). \tag{25}$$

From the definition of $C_{n+1}(\zeta^k, a)$ in Lemma 1 we have

$$C_\infty(\zeta^k, a) = \frac{\sum_{n=-\infty}^{\infty} (-1)^n q^{f_a(n)} \zeta^{kn}}{(q; q)_\infty}$$

$$= \frac{\prod_{n=1}^{\infty} (1 - q^{n/5})(1 - \zeta^k q^{(n-2a)/5})(1 - \zeta^{-k} q^{(n+2a-1)/5})}{(q; q)_\infty}. \tag{26}$$

Note that we have used (4) for the first equality and (25) for the second one.

Considering $a = 0$ in (26), we have

$$C_\infty(\zeta^k, 0) = \frac{f(-q^{1/5})}{f(-q)} \prod_{n=1}^\infty \left(1 - \zeta^k q^{n/5}\right)\left(1 - \zeta^{-k} q^{(n-1)/5}\right)$$

$$= \frac{f(-q^{1/5})}{f(-q)} \left(1 - \zeta^{-k}\right) \prod_{n=1}^\infty \left(1 - \zeta^k q^{n/5}\right)\left(1 - \zeta^{-k} q^{n/5}\right)$$

$$= \frac{f(-q^{1/5})}{f(-q)} \left(1 - \zeta^{-k}\right) \prod_{n=1}^\infty \left(1 - 2\cos k\theta \, q^{n/5} + q^{2n/5}\right), \qquad (27)$$

where $\theta := 2\pi/5$. Note that the last line implies (19); i.e., $C_\infty(1, 0) = 0$, as $1 - \zeta^0 = 0$.

For (20)–(21), we apply the known facts

$$\cos\theta = \cos 4\theta = -\alpha/2,$$

$$\cos 2\theta = \cos 3\theta = -\beta/2$$

to (27) and obtain

$$C_\infty(\zeta, 0) = \frac{f(-q^{1/5})}{f(-q)} \left(1 - \zeta^{-1}\right) \prod_{n=1}^\infty \left(1 + \alpha q^{n/5} + q^{2n/5}\right),$$

$$C_\infty(\zeta^2, 0) = \frac{f(-q^{1/5})}{f(-q)} \left(1 - \zeta^{-2}\right) \prod_{n=1}^\infty \left(1 + \beta q^{n/5} + q^{2n/5}\right),$$

$$C_\infty(\zeta^3, 0) = \frac{f(-q^{1/5})}{f(-q)} \left(1 - \zeta^{-3}\right) \prod_{n=1}^\infty \left(1 + \beta q^{n/5} + q^{2n/5}\right),$$

$$C_\infty(\zeta^4, 0) = \frac{f(-q^{1/5})}{f(-q)} \left(1 - \zeta^{-4}\right) \prod_{n=1}^\infty \left(1 + \alpha q^{n/5} + q^{2n/5}\right).$$

These, with the definition of $J(x)$ in (15) and $\overline{\zeta^{-k}} = \zeta^{-(5-k)}$, give (20)–(21). The proof of (22)–(24) is similar and will be omitted. $\qquad\square$

Lemma 3

$$G(q) = \frac{f(-q^{1/5})}{\sqrt{5}f(-q)} \left(\beta J(\beta) - \alpha J(\alpha)\right), \qquad (28)$$

$$q^{1/5} H(q) = \frac{f(-q^{1/5})}{\sqrt{5}f(-q)} \left(J(\beta) - J(\alpha)\right). \qquad (29)$$

Proof The proof of (28) goes as follows. As $G(q) = D_\infty(0)$ (cf. (16)), by using Lemmas 1 and 2, we have

$$G(q) = \frac{2\,\mathrm{Re}\,C_\infty(\zeta^1,0) + 2\,\mathrm{Re}\,C_\infty(\zeta^2,0)}{5}$$

$$= \frac{2f(-q^{1/5})}{5f(-q)}\left(J(\alpha)\,\mathrm{Re}\left(1 - \frac{1}{\zeta}\right) + J(\beta)\,\mathrm{Re}\left(1 - \frac{1}{\zeta^2}\right)\right)$$

$$= \frac{2f(-q^{1/5})}{5f(-q)}\left(\left(-\frac{\sqrt{5}}{2}\alpha\right)J(\alpha) + \left(\frac{\sqrt{5}}{2}\beta\right)J(\beta)\right)$$

$$= \frac{f(-q^{1/5})}{\sqrt{5}f(-q)}\left(\beta J(\beta) - \alpha J(\alpha)\right).$$

This gives (28). The proof of (29) follows the same method and will be omitted. $\square$

An easy calculation shows that

$$J(\alpha)J(\beta) = \prod_{n=1}^{\infty}\left(1 + q^{n/5} + q^{2n/5} + q^{3n/5} + q^{4n/5}\right) = \frac{f(-q)}{f(-q^{1/5})}. \tag{30}$$

This, with Lemma 3, gives

$$G(q) = \frac{1}{\sqrt{5}}\left(\frac{\beta}{J(\alpha)} - \frac{\alpha}{J(\beta)}\right),$$

$$H(q) = \frac{1}{q^{1/5}\sqrt{5}}\left(\frac{1}{J(\alpha)} - \frac{1}{J(\beta)}\right),$$

or, by using (1):

Corollary 1

$$\sum_{n\geq 0}\frac{q^{n^2}}{(q;q)_n} = \frac{1}{\sqrt{5}}\left(\beta\prod_{n=1}^{\infty}\frac{1}{1 + \alpha q^{n/5} + q^{2n/5}} - \alpha\prod_{n=1}^{\infty}\frac{1}{1 + \beta q^{n/5} + q^{2n/5}}\right),$$

$$\sum_{n\geq 0}\frac{q^{n^2+n}}{(q;q)_n} = \frac{1}{q^{1/5}\sqrt{5}}\left(\prod_{n=1}^{\infty}\frac{1}{1 + \alpha q^{n/5} + q^{2n/5}} - \prod_{n=1}^{\infty}\frac{1}{1 + \beta q^{n/5} + q^{2n/5}}\right).$$

Note that these identities closely resemble two remarkable identities discovered by H.H. Chan, S.H. Chan, and Z.-G. Liu in a recent paper [21]:

Theorem 2 (H.H. Chan, S.H. Chan, and Z.-G. Liu)

$$q^{1/10}\sqrt{\frac{5f(-q^5)}{f(-q)}}\,R(q) = \prod_{n=1}^{\infty}\frac{1}{1 + \alpha q^{n/5} + q^{2n/5}} - \prod_{n=1}^{\infty}\frac{1}{1 + \beta q^{n/5} + q^{2n/5}},$$

$$q^{1/10}\sqrt{\frac{5f(-q^5)}{f(-q)}\frac{1}{R(q)}} = \beta\prod_{n=1}^{\infty}\frac{1}{1+\alpha q^{n/5}+q^{2n/5}} - \alpha\prod_{n=1}^{\infty}\frac{1}{1+\beta q^{n/5}+q^{2n/5}}.$$

H.H. Chan et al. used elementary trigonometric sums and the Jacobi theta function θ_1 to prove these wonderful results.

Before we conclude this section, let us rewrite Lemma 3 as follows, as this will be very handy for the next section (again, (30) is used):

Lemma 4

$$G(q) - \alpha q^{1/5}H(q) = \frac{1}{J(\alpha)}, \tag{31}$$

$$G(q) - \beta q^{1/5}H(q) = \frac{1}{J(\beta)}. \tag{32}$$

3 Applications

In this section, we sketch the proofs of four famous formulas (Theorems 3 to 6) by using Lemma 4. It should be said from the beginning that the proofs below are essentially the same as the existing proofs of these formulas (hence we are not claiming any originality). Our aim is to show that these four identities can all be derived by Lemma 4, and, hopefully, this may be of pedagogical interest.

3.1 A factorization theorem in the Lost Notebook

Theorem 3 (Entry 1.4.1 (p. 206)) *Let* $t = R(q)$. *Then*

$$\frac{1}{\sqrt{t}} - \alpha\sqrt{t} = \frac{1}{q^{1/10}J(\alpha)}\sqrt{\frac{f(-q)}{f(-q^5)}}, \tag{33}$$

$$\frac{1}{\sqrt{t}} - \beta\sqrt{t} = \frac{1}{q^{1/10}J(\beta)}\sqrt{\frac{f(-q)}{f(-q^5)}}. \tag{34}$$

Idea of Proof This is Lemma 4 in disguise. For (33), divide (31) by $\sqrt{q^{1/5}G(q)H(q)}$ $= \sqrt{q^{1/5}f(-q^5)/f(-q)}$ (cf. (17)) and recall that $t = q^{1/5}H(q)/G(q)$. This gives the desired result. $\qquad\square$

Remarks

• This is a "factorization" theorem: if we multiply (33) and (34), we have

$$\frac{1}{R(q)} - 1 - R(q) = \frac{f(-q^{1/5})}{q^{1/5}f(-q^5)}.$$

This equation gives many important applications, including the evaluation of $R(e^{-2\pi})$, one of the results that convinced G. Hardy that Ramanujan was a "mathematician of the highest class" (cf. [27], p. 9). See also [7, 11].

- Berndt, H.H. Chan, S.-S. Huang, S.-Y. Kang, J. Sohn, and S.H. Son gave a complete proof of Theorem 3 in [14]. See also [7, pp. 21–24]. These authors derived Theorem 3 from their Corollary 4.6 in their paper [14]. In fact, Lemma 4 above is equivalent to their Corollary 4.6. Berndt et al. derived their Corollary 4.6 by using

$$f(U_1, V_1) = \sum_{r=0}^{n-1} U_r f\left(\frac{U_{n+r}}{U_r}, \frac{V_{n-r}}{U_r}\right). \tag{35}$$

Here, with $|ab| < 1$,

$$f(a, b) := \sum_{n=-\infty}^{\infty} a^{n(n+1)/2} b^{n(n-1)/2},$$

where $U_n := a^{n(n+1)/2} b^{n(n-1)/2}$ and $V_n := a^{n(n-1)/2} b^{n(n+1)/2}$. For a proof, see [11, p. 48, Entry 31]. For more applications of (35), see, e.g., [16, 17]. Note also that Lemmas 1 and 2 above are in the same spirit of (35). We hope that our presentation indicates how Andrews' proof [2] motivates the decomposition in Lemma 1.
- In the same paper mentioned above (after Corollary 1), H.H. Chan, S.H. Chan, and Z.-G. Liu [21] give a new proof of Theorem 3.
- Finally, the same idea sketched here can also be used to prove two identities involving the cubic continued fraction, see [19].

3.2 A "difficult and deep" identity involving the Rogers–Ramanujan continued fraction

Theorem 4 *Let $u = R(q)$ and $v = R(q^5)$. Then*

$$u^5 = v\frac{1 - 2v + 4v^2 - 3v^3 + v^4}{1 + 3v + 4v^2 + 2v^3 + v^4}. \tag{36}$$

Idea of Proof Write

$$G_5 := G(q^5), \qquad H_5 := H(q^5)$$

and

$$W(z, q) := G_5 - zq H_5.$$

Note that $W(z, q)$ is motivated by the left-hand sides of the equations in Lemma 4 (with $q \to q^5$). Consider the product (recall that ζ is the 5th root of unity):

$$P(q) := \left(\frac{W(\beta, \zeta q) W(\beta, \zeta^4 q)}{W(\beta, \zeta^2 q) W(\beta, \zeta^3 q)}\right)\left(\frac{W(\alpha, \zeta^2 q) W(\alpha, \zeta^3 q)}{W(\alpha, \zeta q) W(\alpha, \zeta^4 q)}\right)$$

$$= \frac{G_5^4 - 2q G_5^3 H_5 + 4q^2 G_5^2 H_5^2 - 3q^3 G_5 H_5^3 + q^4 H_5^4}{G_5^4 + 3q G_5^3 H_5 + 4q^2 G_5^2 H_5^2 + 2q^3 G_5 H_5^3 + q^4 H_5^4}$$

$$= \frac{1 - 2v + 4v^2 - 3v^3 + v^4}{1 + 3v + 4v^2 + 2v^3 + v^4}.$$

We get almost the right-hand side of (36). To obtain the rest, we do the following. Replace $q^{1/5}$ by $\zeta^k q$ in Lemma 4. This produces $W(\alpha, \zeta^k q)$ and $W(\beta, \zeta^k q)$. Apply the resulting equations to evaluate $P(q)$ defined above. After some tedious calculations, one will arrive at $P(q) = u^5/v$. □

Remarks

- Again, this is one of the formulas that Hardy thinks they are "obviously both difficult and deep" (see [27, p. 9]).
- For references on Theorem 4, see [7, 12]. See also a recent paper of Gugg [26]. In fact, the sketch of proof above is motivated by Gugg's paper. For hints for the "tedious calculations" mentioned above, readers are encouraged to read Gugg's paper, especially Lemmas 4.1 to 4.3, in which he sorts out the patterns behind many infinite products involved.

3.3 An identity in the Lost Notebook that deeply impacts the theory of partition

Theorem 5 *Recall that* $\zeta = e^{2\pi i/5}$. *Then*

$$\frac{(q;q)_\infty}{(\zeta q;q)_\infty(\zeta^{-1}q;q)_\infty} = A(q^5) - q(\zeta + \zeta^{-1})^2 B(q^5)$$
$$+ q^2(\zeta^2 + \zeta^{-2})C(q^5) - q^3(\zeta + \zeta^{-1})D(q^5), \quad (37)$$

where

$$A(q) = f(-q^5)\frac{G^2(q)}{H(q)},$$

$$B(q) = f(-q^5)G(q),$$

$$C(q) = f(-q^5)H(q),$$

$$D(q) = f(-q^5)\frac{H^2(q)}{G(q)}.$$

Idea of Proof Consider the ("uneven") product

$$(G(q) - \alpha q^{1/5}H(q))^2(G(q) - \beta q^{1/5}H(q))$$
$$= \left(\frac{f(-q^{1/5})}{f(-q)}\right)^3 J(\alpha)J(\beta)^2 \quad \text{(by Lemma 4)}$$
$$= \frac{f(-q^{1/5})}{f(-q)} \cdot \frac{1}{J(\alpha)} \quad \text{(by (30))}$$
$$= \frac{f(-q^{1/5})}{f(-q)} \cdot \prod_{n=1}^{\infty} \frac{1}{(1 - \zeta q^{n/5})(1 - \zeta^{-1}q^{n/5})}.$$

Divide this equation by $G(q)H(q) = f(-q^5)/f(-q)$ (i.e., (17)). Expand the product $(G - \alpha q^{1/5} H)^2 (G - \beta q^{1/5} H)$ and rearrange terms. This gives (write $G := G(q)$ and $H := H(q)$)

$$
\frac{f(-q^{1/5})}{\prod_{n=1}^{\infty}(1 - \zeta q^{n/5})(1 - \zeta^{-1} q^{n/5})}
$$

$$
= f(-q^5)\frac{G^3 + (\zeta + \zeta^{-1} - 1)q^{1/5}G^2 H - (\zeta + \zeta^{-1} + 1)q^{2/5}GH^2}{GH}
$$

$$
\times \frac{-(\zeta + \zeta^{-1})q^{1/5}H^3}{GH}
$$

$$
= A(q) + (\zeta + \zeta^{-1} - 1)q^{1/5}B(q) - (\zeta + \zeta^{-1} + 1)q^{2/5}C(q)
$$

$$
- (\zeta + \zeta^{-1})q^{1/5}D(q).
$$

Replacing q by q^5 gives the desired result, as $\zeta + \zeta^{-1} - 1 = -(\zeta + \zeta^{-1})^2$ and $\zeta + \zeta^{-1} + 1 = -(\zeta^2 + \zeta^{-2})$. $\square$

Remarks

- This identity was first proven by Garvan [25]. Our presentation is a streamlined version of his proof. Garvan used (37) to give a new proof of the famous Ramanujan's congruences. Later, Ekin [24] and Berndt, H.H. Chan, S.H. Chan, and W.C. Liaw [15] gave further proofs. Garvan [25] and Andrews and Garvan [10] later used (37) to discover the notion of "crank"—a conjectured statistic (by F. Dyson) that provides a combinatorial interpretation for Ramanujan's congruences. For this amazing development, we refer readers to an excellent article [8] on the top ten most fascinating formulas in the lost notebook.
- An exercise for interested readers: consider the "uneven" product the product $(G - \alpha q^{1/5} H)(G - \beta q^{1/5} H)^2$ to derive an accompanying formula for (37).

3.4 Ramanujan's "most beautiful identity"

Theorem 6 *Let $p(n)$ be the number of partitions of n, defined by $\sum_{n \geq 0} p(n)q^n :=$ $\prod_{n \geq 1}(1 - q^n)^{-1}$. Then*

$$
\sum_{n=0}^{\infty} p(5n + 4)q^n = 5\frac{(q^5; q^5)_{\infty}^5}{(q; q)_{\infty}^6}. \tag{38}
$$

Idea of Proof Consider the product

$$
T := \frac{\prod_{i=1}^{4}(G(q^5) - \alpha q \zeta^i H(q^5))(G(q^5) - \beta q \zeta^i H(q^5))}{q^4 G^4(q^5) H^4(q^5)}
$$

(note: i starts from 1, not 0). As in Theorem 4, define $v = R(q^5)$. Expand the product in T (just as we did in Theorems 4 and 5). This gives

$$T = v^4 - v^3 + 2v^2 + 3v + 5 - \frac{3}{v} + \frac{2}{v^2} + \frac{1}{v^3} + \frac{1}{v^4}. \tag{39}$$

By using Lemma 4, T can be shown to be (again, after some algebra)

$$T = \frac{(q^5; q^5)_\infty^6}{q^5 (q^{25}; q^{25})_\infty^5} \left(\frac{q}{(q; q)_\infty} \right). \tag{40}$$

Substitute

$$\frac{q}{(q; q)_\infty} = \sum_{n \geq 0} p(n) q^{n+1}$$

into (40). Note that

$$v = R(q^5) = q \frac{(q^5; q^{25})_\infty (q^{20}; q^{25})_\infty}{(q^{10}; q^{25})_\infty (q^{15}; q^{25})_\infty} = q \left(\sum_{n \geq 0} v_n q^{5n} \right). \tag{41}$$

The overall q in the power series expansion of v is crucial. Finally, from both expressions of T (i.e., (39) and (40)) extract the terms that involve only the power of $5n$ (and one needs the observation in (41)). This gives

$$\sum_{n \geq 0} p(5n + 4) q^{5(n+1)} = 5q^5 \frac{(q^{25}; q^{25})_\infty^5}{(q^5; q^5)_\infty^6}.$$

Scaling q to $q^{1/5}$ in the last equation gives the desired formula. $\qquad\square$

Remarks

- Both Hardy and MacMahon considered (38) as Ramanujan's "Most Beautiful Identity" (cf. [30], p. xxxv). For introductions, see the wonderful books of [3, 6, 13, 23]. See also Hirschhorn's beautiful papers [28, 29].
- Note that our proof here is the same as the one using the quotient $(R^{-5} - 11 - R^5)/(R^{-1} - 1 - R)$ (why?). Readers can read more about this approach in Berndt's excellent book [13].
- Note that there is an analogue of the "most beautiful identity" associated with the cubic continued fraction [20] (see Cao's new proof [17], in which he used various dissection identities). Define $a(n)$ by

$$\sum_{n=0}^{\infty} a(n) q^n := \frac{1}{(q; q)_\infty (q^2; q^2)_\infty}.$$

We have the following:

Theorem 7 (H.-C. Chan)

$$\sum_{n=0}^{\infty} a(3n+2)q^n = 3\frac{(q^3;q^3)_{\infty}^3(q^6;q^6)_{\infty}^3}{(q;q)_{\infty}^4(q^2;q^2)_{\infty}^4}.$$

Acknowledgement   Professor Andrews, thank you for your wonderful work on q-series, in particular the Rogers–Ramanujan identities, without which this work will never see the light of day. Happy birthday to you again!

References

1. Andrews, G.E.: Some formulae for the Fibonacci sequence with generalizations. Fibonacci Q. **7**, 113–130 (1969)
2. Andrews, G.E.: A polynomial identity which implies the Rogers–Ramanujan identities. Scr. Math. **28**, 297–305 (1970)
3. Andrews, G.E.: In: The Theory of Partitions. Encycl. Math. and Its Appl., vol. 2, Rota, G.-C. (ed.) Addison-Wesley, Reading (1976). Reissued: Cambridge University Press, Cambridge (1998)
4. Andrews, G.E.: q-Series: Their Development and Application in Analysis, Number Theory, Combinatorics, Physics and Computer Algebra. C.B.M.S. Regional Conference Series in Math., vol. 66. Am. Math. Soc., Providence (1986)
5. Andrews, G.E.: Fibonacci numbers and the Rogers–Ramanujan identities. Fibonacci Q. **42**, 3–19 (2004)
6. Andrews, G.E., Askey, R., Roy, R.: Special Functions. Cambridge University Press, Cambridge (1999)
7. Andrews, G.E., Berndt, B.C.: Ramanujan's Lost Notebook, Part I. Springer, New York (2005)
8. Andrews, G.E., Berndt, B.C.: Your hit parade—the top ten most fascinating formulas from Ramanujan's lost notebook. Not. Am. Math. Soc. **55**, 1830 (2008)
9. Andrews, G.E., Eriksson, K.: Integer Partitions. Cambridge University Press, Cambridge (2004)
10. Andrews, G.E., Garvan, F.G.: Dyson's crank of a partition. Bull. Am. Math. Soc. **18**, 167–171 (1988)
11. Berndt, B.C.: Ramanujan's Notebooks, Part III. Springer, New York (1991)
12. Berndt, B.C.: Ramanujan's Notebooks, Part V. Springer, New York (1998)
13. Berndt, B.C.: Number Theory in the Spirit of Ramanujan. Am. Math. Soc., Providence (2004)
14. Berndt, B.C., Chan, H.H., Huang, S.-S., Kang, S.-K., Sohn, J., Son, S.H.: The Rogers–Ramanujan continued fraction. J. Comput. Appl. Math. **105**, 9–24 (1999)
15. Berndt, B.C., Chan, H.H., Chan, S.H., Liaw, W.C.: Cranks and dissections in Ramanujan's lost notebook. J. Comb. Theory, Ser. A **109**, 91–120 (2005)
16. Berndt, B.C., Choi, G., Choi, Y.-S., Hahn, H., Yeap, B.P., Yee, A.J., Yesilyurt, H., Yi, J.: Ramanujan's forty identities for the Rogers–Ramanujan functions (with). Mem. Am. Math. Soc. **880**, 188 (2007)
17. Cao, Z.: On Somos' dissection identities. J. Math. Anal. Appl. **365**, 659–667 (2010)
18. Chan, H.-C.: From Andrews' formula for the Fibonacci numbers to the Rogers–Ramanujan identities. Fibonacci Q. **45**, 221–239 (2007)
19. Chan, H.-C.: A new proof for two identities involving the Ramanujan's cubic continued fraction. Ramanujan J. **21**, 173–180 (2010)
20. Chan, H.-C.: Ramanujan's cubic continued fraction and a generalization of his "most beautiful identity." Int. J. Number Theory **6**(3), 673–680 (2010)
21. Chan, H.H., Chan, S.H., Liu, Z.-G.: The Rogers–Ramanujan continued fraction and a new Eisenstein series identity. J. Number Theory **129**, 1786–1797 (2009)
22. Chan, H.-C., Ebbing, S.: Factorization theorems for the Rogers–Ramanujan continued fraction. In: The Lost Notebook (preprint)
23. Chu, W., Di Claudio, L.: Classical partition identities and basic hypergeometric series. Università degli Study di Lecce, Lecce (2004)
24. Ekin, A.B.: Some properties of partitions in terms of crank. Trans. Am. Math. Soc. **352**, 2145–2156 (2000)
25. Garvan, F.: New combinatorial interpretations of Ramanujan's partition congruences mod 5, 7, and 11. Trans. Am. Math. Soc. **305**, 47–77 (1988)

26. Gugg, C.: A new proof of Ramanujan's modular equation relating $R(q)$ with $R(q^5)$. Ramanujan J. **20**, 163–177 (2009)
27. Hardy, G.: Ramanujan. AMS Chelsea, Rhode Island (2002)
28. Hirschhorn, M.D.: An identity of Ramanujan, and applications. In: q-Series from a Contemporary Perspective. Contemporary Mathematics, vol. 254, pp. 229–234. Am. Math. Soc., Providence (2000)
29. Hirschhorn, M.D.: Ramanujan's "Most Beautiful Identity". Aust. Math. Soc. Gaz. **31**, 259–262 (2005)
30. Ramanujan, S.: Collected Papers. Cambridge University Press, Cambridge (1927). Reprinted by Chelsea, New York (1962); reprinted by Am. Math. Soc., Providence (2000)
31. Rogers, L.J.: Second memoir on the expansion of certain infinite products. Proc. Lond. Math. Soc. **25**, 318–343 (1894)
32. Schur, I.: Ein Beitrag zur Additiven Zahlentheorie Sitzungsber. Akad. Wiss. Berlin, Phys.-Math. Kl., 302–321 (1917)
33. Sills, A.V.: Finite Rogers–Ramanujan type identities. Electron. J. Comb. **10**, R13 (2003)

Ramanujan J (2010) 22: 101–117
DOI 10.1007/s11139-009-9197-6

Congruences modulo powers of 2 for a certain partition function

Hei-Chi Chan · Shaun Cooper

Dedicated to George Andrews for his 70th birthday

Received: 31 March 2009 / Accepted: 12 August 2009 / Published online: 1 January 2010
© Springer Science+Business Media, LLC 2009

Abstract We study the divisibility properties of the coefficients $c(n)$ defined by

$$\prod_{n=1}^{\infty} \frac{1}{(1-q^n)^2(1-q^{3n})^2} = \sum_{n=0}^{\infty} c(n)q^n.$$

An analogue of Ramanujan's partition congruences is obtained for certain coefficients $c(n)$ modulo powers of 2. Furthermore, an analogue of the identity that Hardy regarded as Ramanujan's most beautiful is proved.

Keywords Ramanujan-type congruences · Partition function

Mathematics Subject Classification (2000) Primary 05A15 · 05A30 · 05A40

1 Introduction

Let $p(n)$ denote the number of partitions of n, defined by

$$\prod_{n=1}^{\infty} \frac{1}{1-q^n} = \sum_{n=0}^{\infty} p(n)q^n.$$

H.-C. Chan (✉)
University of Illinois at Springfield, Springfield, IL 62703-5407, USA
e-mail: hchan1@uis.edu

S. Cooper
Institute of Information and Mathematical Sciences, Massey University, Private Bag 102904,
North Shore Mail Centre, Auckland, New Zealand
e-mail: s.cooper@massey.ac.nz

Ramanujan conjectured [10, 11, pp. 210–213] and later proved [12, pp. 135–177] that if $j \geq 1$ and if δ_j is the reciprocal modulo 5^j of 24, then

$$p(5^j n + \delta_j) \equiv 0 \pmod{5^j}. \tag{1.1}$$

A commentary on Ramanujan's work [12, pp. 135–177] and a comprehensive list of references to work of other authors has been given by Berndt and Ono [3]. Perhaps the simplest proof of (1.1), relying only on classical identities of Euler and Jacobi, is due to Hirschhorn and Hunt [7].

Recently [5], an analogue of (1.1) for the numbers $a(n)$ defined by

$$\prod_{n=1}^{\infty} \frac{1}{(1-q^n)(1-q^{2n})} = \sum_{n=0}^{\infty} a(n)q^n$$

was obtained:

$$a(3^j n + d_j) \equiv 0 \pmod{3^{j+e(j)}}, \tag{1.2}$$

where d_j is the reciprocal modulo 3^j of 8, and $e(n) = 1$ or 0 according to whether n is even or odd, respectively. For subsequent studies involving the partition function $a(n)$, see the works of Kim [8] and of Sinick [13].

In this work, we shall study the numbers $c(n)$ defined by

$$\prod_{n=1}^{\infty} \frac{1}{(1-q^n)^2(1-q^{3n})^2}$$

$$= \sum_{n=0}^{\infty} c(n)q^n = 1 + 2q + 5q^2 + 12q^3 + 24q^4 + 46q^5 + 90q^6 + 160q^7 + \cdots.$$

The main goal is to prove the following analogue of (1.1) and (1.2):

Theorem 1.1 *For any nonnegative integer n and any positive integer j,*

$$c(2n+1) \equiv 0 \pmod{2}, \tag{1.3}$$

$$c\left(2^{2j}n + \frac{1}{3}(2 \times 4^j + 1)\right) \equiv 0 \pmod{2^{j+1}}, \quad and \tag{1.4}$$

$$c\left(2^{2j+1}n + \frac{1}{3}(5 \times 4^j + 1)\right) \equiv 0 \pmod{2^{j+4}}. \tag{1.5}$$

The paper is organized as follows. In Sect. 2, three key lemmas are presented. In Sect. 3, generating functions for each of the sequences in (1.3)–(1.5) are obtained. In Sect. 4, the divisibility properties of the coefficients in the generating functions are analyzed, and Theorem 1.1 is proved as a consequence. We also deduce the congruences

$$c(8n+3) \equiv 6c(2n+1) \pmod{2^8}$$

and

$$c(16n + 11) \equiv 198c(4n + 3) \pmod{2^{10}}.$$

The strategy of the proof of Theorem 1.1 follows that of Atkin [2] and of Hirschhorn and Hunt [7]. For wonderful introductions to Atkin's proof, see [1, 9].

2 Preliminary results

For a positive integer j, let

$$E_j = \prod_{\ell=1}^{\infty} (1 - q^{j\ell}),$$

where $|q| < 1$. Let

$$z = z(q) = q \prod_{j=1}^{\infty} \frac{(1 - q^{12j-10})^2 (1 - q^{12j-2})^2}{(1 - q^{12j-8})^2 (1 - q^{12j-4})^2} = q \frac{E_2^2 E_{12}^4}{E_4^4 E_6^2}.$$

The starting point in our analysis is the following:

Lemma 2.1

$$\frac{1}{z} - 2 - 3z = \frac{1}{q} \frac{E_1^2 E_3^2}{E_4^2 E_{12}^2},$$

$$\frac{1}{z^2} - 10 + 9z^2 = \frac{1}{q^2} \frac{E_2^6 E_6^6}{E_4^6 E_{12}^6},$$

$$\frac{1}{z} + 2 - 3z = \frac{1}{q} \frac{E_2^6 E_6^6}{E_4^4 E_{12}^4} \times \frac{1}{E_1^2 E_3^2}.$$

Proof The first two results were proved in [6], and the third follows from these by division. □

Let

$$G = G(q) = \frac{1}{q} \frac{E_2^6 E_6^6}{E_4^4 E_{12}^4} \times \frac{1}{E_1^2 E_3^2},$$

$$g = g(q) = q \frac{E_2^6 E_6^6}{E_1^6 E_3^6}, \quad \text{and}$$

$$u = u(q) = \frac{1}{q^2} \frac{E_2^6 E_6^6}{E_4^6 E_{12}^6}.$$

The following simple relationships will be used frequently.

Lemma 2.2

$$G(q) = \frac{1}{z} + 2 - 3z,$$

$$u(q) = \frac{1}{z^2} - 10 + 9z^2,$$

$$\frac{1}{g(q^2)} = u(q).$$

Proof These are immediate from the definitions of G, g, and u and from the results in Lemma 2.1. □

The results in the next lemma will turn out to be useful.

Lemma 2.3

$$g(q) = g^2(q^2) G^3(q),$$

$$\frac{q}{E_1^2 E_3^2} = \frac{g(q^2)}{E_4^2 E_{12}^2} G(q), \quad and$$

$$G^2(q) = 4G(q) + u(q).$$

Proof The first two results follow immediately from the definitions of G, g, and u, and we omit the details. By Lemma 2.2 we have

$$G(q)\big(G(q) - 4\big) = \left(\frac{1}{z} + 2 - 3z\right)\left(\frac{1}{z} - 2 - 3z\right) = \frac{1}{z^2} - 10 + 9z^2 = u(q),$$

and this proves the third result. □

Let $f(q) = \sum_{n=-\infty}^{\infty} f_n q^n$ in the annulus $0 < |q| < r$, where r is a positive number or $+\infty$. Define three operators H, V, and U by

$$Hf = \sum_{n=-\infty}^{\infty} f_{2n} q^{2n},$$

$$Vf = \sum_{n=-\infty}^{\infty} f_n q^{n/2},$$

$$Uf = V(Hf) = \sum_{n=-\infty}^{\infty} f_{2n} q^n.$$

3 Generating functions

In this section, generating functions are obtained for the sequences in Theorem 1.1.

The identity described by Hardy as Ramanujan's most beautiful is

$$\sum_{n=0}^{\infty} p(5n+4)q^n = 5 \prod_{n=1}^{\infty} \frac{(1-q^{5n})^5}{(1-q^n)^6}.$$

Analogous to this is the result, proved recently in [4]:

$$\sum_{n=0}^{\infty} a(3n+2)q^n = 3 \prod_{n=1}^{\infty} \frac{(1-q^{3n})^3(1-q^{6n})^3}{(1-q^n)^4(1-q^{2n})^4}.$$

The analogue of these results for the coefficients $c(n)$ is the following:

Theorem 3.1

$$\sum_{n=0}^{\infty} c(2n+1)q^{n+1} = \frac{2g(q)}{E_2^2 E_6^2} = 2q \prod_{n=1}^{\infty} \frac{(1-q^{2n})^4(1-q^{6n})^4}{(1-q^n)^6(1-q^{3n})^6}.$$

Proof The second equality is immediate from the definitions of $g(q)$ and E_j. It remains to prove the first equality. By the definition of the numbers $c(n)$ and Lemmas 2.2 and 2.3, we have

$$\sum_{n=0}^{\infty} c(n)q^{n+1} = \frac{q}{E_1^2 E_3^2}$$

$$= \frac{g(q^2)}{E_4^2 E_{12}^2} G(q)$$

$$= \frac{g(q^2)}{E_4^2 E_{12}^2} \left(\frac{1}{z} + 2 - 3z \right).$$

Now extract the terms with even powers of q. From the definition, the series expansion of z contains only odd powers of q. Hence,

$$\sum_{n=0}^{\infty} c(2n+1)q^{2n+2} = \frac{2g(q^2)}{E_4^2 E_{12}^2}.$$

Replace q^2 with q to complete the proof.   □

It will be necessary to examine the effect of the operator H on powers of $G(q)$. In preparation for this, we define an infinite matrix of integers $\{m(j,k)\}_{j,k \geq 1}$ by

(1) $m(1,1) = 2$ and $m(1,k) = 0$ for $k \geq 2$.
(2) $m(2,1) = 1$, $m(2,2) = 8$, and $m(2,k) = 0$ for $k \geq 3$.
(3) $m(j,1) = 0$ for $j \geq 3$.
(4) $m(j,k) = 4m(j-1,k-1) + m(j-2,k-1)$ for $j \geq 3$ and $k \geq 2$.

Using induction, it is easy to prove that

(5) $m(j,k) = 0$ for $k > j$.
(6) $m(j,k) = 0$ for $j > 2k$.
(7) $m(2k,k) = 1$ for $k \geq 1$.
(8) $m(j,j) = 2^{2j-1}$ for $j \geq 1$.

We omit the details. The first seven rows of m are given by

$$
\begin{pmatrix}
2 & 0 & 0 & 0 & 0 & 0 & 0 & 0 & \cdots \\
1 & 8 & 0 & 0 & 0 & 0 & 0 & 0 & \cdots \\
0 & 6 & 32 & 0 & 0 & 0 & 0 & 0 & \cdots \\
0 & 1 & 32 & 128 & 0 & 0 & 0 & 0 & \cdots \\
0 & 0 & 10 & 160 & 512 & 0 & 0 & 0 & \cdots \\
0 & 0 & 1 & 72 & 768 & 2048 & 0 & 0 & \cdots \\
0 & 0 & 0 & 14 & 448 & 3548 & 8192 & 0 & \cdots
\end{pmatrix}. \tag{3.1}
$$

Lemma 3.2 *Let the functions G and u and the operator H be as defined in Sect. 2, and let $m(j,k)$ be the integers defined above. Then*

$$H(G) = 2,$$

$$H(G^2) = u + 8,$$

$$H(G^3) = 6u + 32,$$

$$H(G^4) = u^2 + 32u + 128,$$

and in general

$$
H(G^j) = \sum_{k=1}^{j} m(j,k) u^{j-k} = \sum_{k=\lfloor (j+1)/2 \rfloor}^{j} m(j,k) u^{j-k}, \quad j = 1, 2, 3, \ldots. \tag{3.2}
$$

Proof The second equality in (3.2) is a restatement of property (6) of the numbers $m(j,k)$. It remains to prove the first equality, and for this, we use induction on j. From the definitions, z and u are odd and even functions of q, respectively, so

$$H(z) = 0 \quad \text{and} \quad H(u) = u.$$

Thus by Lemma 2.2 and linearity of H we have

$$H(G) = H\left(\frac{1}{z} + 2 - 3z\right) = 2,$$

and further by Lemma 2.3 we have

$$H(G^2) = H(4G + u) = 8 + u.$$

This proves the cases $j = 1$ and $j = 2$.

Now assume that $H(G^{j+1})$ and $H(G^j)$ are given by formula (3.2) for some positive integer j. Then by Lemma 2.3 we have

$$H(G^{j+2}) = 4H(G^{j+1}) + H(uG^j).$$

Since u is even, this becomes

$$H(G^{j+2}) = 4H(G^{j+1}) + uH(G^j).$$

Apply the induction hypothesis and properties (3), (4), and (8) of $m(j,k)$ to get

$$H(G^{j+2}) = 4\sum_{k=1}^{j+1} m(j+1,k)u^{j+1-k} + u\sum_{k=1}^{j} m(j,k)u^{j-k}$$

$$= 4m(j+1,j+1) + \sum_{k=2}^{j+1}(4m(j+1,k-1) + m(j,k-1))u^{j+2-k}$$

$$= m(j+2,j+2) + \sum_{k=2}^{j+1} m(j+2,k)u^{j+2-k}$$

$$= \sum_{k=1}^{j+2} m(j+2,k)u^{j+2-k}.$$

This completes the induction.   □

Remark 3.3 A standard calculation using generating functions yields

$$\sum_{j=1}^{\infty} H(G^j)t^j = \frac{(2+ut)t}{1-4t-ut^2};$$

however we shall not require this.

Let us define the other infinite matrix of integers $\{b(j,k)\}$ by

(9) $b(1,1) = 2$ and $b(1,k) = 0$ for $k \geq 2$;
(10)

$$b(j+1,k) = \begin{cases} \sum_{i=1}^{\infty} b(j,i)m(3i,i+k) & \text{if } j \text{ is odd,} \\ \sum_{i=1}^{\infty} b(j,i)m(3i+1,i+k) & \text{if } j \text{ is even.} \end{cases}$$

Although (10) involves an infinite series, only finitely many terms are nonzero. In fact,

(11)

$$b(j,k) = 0 \quad \text{if } \begin{cases} k > (2^{j+1}-1)/3 \text{ and } j \text{ is odd,} \\ k > (2^{j+1}-2)/3 \text{ and } j \text{ is even;} \end{cases}$$

and

(12)

$$b\big(j, (2^{j+1} - 1)/3\big) = 2^{4(2^j - 1) - 3j} \quad \text{if } j \text{ is odd,}$$

$$b\big(j, (2^{j+1} - 2)/3\big) = 2^{4(2^j - 1) - 3j} \quad \text{if } j \text{ is even.}$$

These results are easily proved by induction using properties of the numbers $m(j, k)$, and we omit the details. For future reference, the first three rows of b are given by

$$\begin{pmatrix} 2 & 0 & 0 & 0 & 0 & 0 & \cdots \\ 2^2 \times 3 & 2^6 & 0 & 0 & 0 & 0 & \cdots \\ 2^2 \times 3 & 2^8 \times 5 & 2^9 \times 59 & 2^{15} \times 7 & 2^{19} & 0 & \cdots \end{pmatrix}. \tag{3.3}$$

Theorem 3.4 *Let $b(j, k)$ be the numbers defined above. For $j \geq 1$, we have*

$$\sum_{n=0}^{\infty} c\left(2^{2j-1}n + \frac{1}{3}(2^{2j-1} + 1)\right)q^{n+1} = \frac{1}{E_2^2 E_6^2} \sum_{\ell=1}^{\frac{1}{3}(4^j - 1)} b(2j - 1, \ell)g^{\ell}(q) \tag{3.4}$$

and

$$\sum_{n=0}^{\infty} c\left(2^{2j}n + \frac{1}{3}(2^{2j+1} + 1)\right)q^{n+1} = \frac{1}{E_1^2 E_3^2} \sum_{\ell=1}^{\frac{2}{3}(4^j - 1)} b(2j, \ell)g^{\ell}(q). \tag{3.5}$$

Before giving the proof, we note the following examples:

$$\sum_{n=0}^{\infty} c(2n + 1)q^{n+1} = \frac{1}{E_2^2 E_6^2} \times 2g,$$

$$\sum_{n=0}^{\infty} c(4n + 3)q^{n+1} = \frac{1}{E_1^2 E_3^2} \times (12g + 64g^2),$$

$$\sum_{n=0}^{\infty} c(8n + 3)q^{n+1} = \frac{1}{E_2^2 E_6^2}$$

$$\times \left(12g + 1280g^2 + 30208g^3 + 229376g^4 + 524288g^5\right),$$

where $g = g(q)$.

Proof of Lemma 3.4 We use induction on j, passing alternately between (3.4) and (3.5). To begin, note that (3.4) is true for $j = 1$ by Theorem 3.1.

Step 1. Suppose that (3.4) is true for a positive integer j. Applying the operator U to both sides gives

$$\sum_{n=0}^{\infty} c\left(2^{2j-1}(2n + 1) + \frac{1}{3}(2^{2j-1} + 1)\right)q^{n+1}$$

$$= U\left(\frac{1}{E_2^2 E_6^2} \sum_{k=1}^{\frac{1}{3}(4^j-1)} b(2j-1,k)g^k(q)\right)$$

$$= \frac{1}{E_1^2 E_3^2} \sum_{k=1}^{\frac{1}{3}(4^j-1)} b(2j-1,k)U\left(g^k(q)\right). \tag{3.6}$$

By Lemmas 2.3, 3.2, and 2.2, we have

$$U\left(g^k(q)\right) = U\left(g^{2k}(q^2)G^{3k}(q)\right)$$

$$= g^{2k}(q)U\left(G^{3k}(q)\right)$$

$$= g^{2k}(q)V\left(H\left(G^{3k}(q)\right)\right)$$

$$= g^{2k}(q)V\left(\sum_{\ell=\lfloor(3k+1)/2\rfloor}^{3k} m(3k,\ell)u^{3k-\ell}(q)\right)$$

$$= \sum_{\ell=\lfloor(3k+1)/2\rfloor}^{3k} m(3k,\ell)g^{\ell-k}(q)$$

$$= \sum_{\ell=\lfloor(k+1)/2\rfloor}^{2k} m(3k,\ell+k)g^{\ell}(q). \tag{3.7}$$

Combining (3.6) and (3.7) gives

$$\sum_{n=0}^{\infty} c\left(2^{2j}n + \frac{1}{3}(2^{2j+1}+1)\right)q^{n+1}$$

$$= \frac{1}{E_1^2 E_3^2} \sum_{k=1}^{\frac{1}{3}(4^j-1)} \sum_{\ell=\lfloor(k+1)/2\rfloor}^{2k} b(2j-1,k)m(3k,\ell+k)g^{\ell}(q).$$

Interchanging the order of summation and using properties (5), (6), and (11) of the numbers m and b to extend the sums to all positive integers, we get

$$\sum_{n=0}^{\infty} c\left(2^{2j}n + \frac{1}{3}(2^{2j+1}+1)\right)q^{n+1}$$

$$= \frac{1}{E_1^2 E_3^2} \sum_{\ell=1}^{\infty} \left(\sum_{k=1}^{\infty} b(2j-1,k)m(3k,\ell+k)\right)g^{\ell}(q)$$

$$= \frac{1}{E_1^2 E_3^2} \sum_{\ell=1}^{\infty} b(2j,\ell)g^{\ell}(q).$$

Since $b(2j, \ell) = 0$ for $\ell > (2^{2j+1} - 2)/3$, we obtain (3.5). In summary, we have shown that if (3.4) is true for some positive integer j, then (3.5) is also true for that value of j.

Step 2. Suppose that (3.5) is true for some positive integer j. We will deduce identity (3.4) with $j + 1$ in place of j.

Multiplying (3.5) by q and applying the operator U, we get

$$\sum_{n=0}^{\infty} c\left(2^{2j}(2n) + \frac{1}{3}(2^{2j+1} + 1)\right)q^{n+1} = U\left(\frac{q}{E_1^2 E_3^2} \sum_{k=1}^{\frac{2}{3}(4^j-1)} b(2j, k)g^k(q)\right).$$

By the first two parts of Lemma 2.3, this is

$$\sum_{n=0}^{\infty} c\left(2^{2j+1}n + \frac{1}{3}(2^{2j+1} + 1)\right)q^{n+1}$$

$$= U\left(\frac{1}{E_4^2 E_{12}^2} \sum_{k=1}^{\frac{2}{3}(4^j-1)} b(2j, k)g^{2k+1}(q^2)G^{3k+1}(q)\right)$$

$$= \frac{1}{E_2^2 E_6^2} \sum_{k=1}^{\frac{2}{3}(4^j-1)} b(2j, k)g^{2k+1}(q)U\left(G^{3k+1}(q)\right). \tag{3.8}$$

By Lemmas 3.2 and 2.2, we have

$$U\left(G^{3k+1}(q)\right) = V\left(H\left(G^{3k+1}(q)\right)\right)$$

$$= V\left(\sum_{\ell=\lfloor(3k+2)/2\rfloor}^{3k+1} m(3k + 1, \ell)u^{3k+1-\ell}(q)\right)$$

$$= \sum_{\ell=\lfloor(k+2)/2\rfloor}^{2k+1} m(3k + 1, \ell + k)g^{\ell-1-2k}(q). \tag{3.9}$$

Now combine (3.8) and (3.9), interchange the order of summation, and extend both sums to all positive integers. The result is

$$\sum_{n=0}^{\infty} c\left(2^{2j+1}n + \frac{1}{3}(2^{2j+1} + 1)\right)q^{n+1}$$

$$= \frac{1}{E_2^2 E_6^2} \sum_{\ell=1}^{\infty} \left(\sum_{k=1}^{\infty} b(2j, k)m(3k + 1, \ell + k)\right)g^\ell(q)$$

$$= \frac{1}{E_2^2 E_6^2} \sum_{\ell=1}^{\infty} b(2j + 1, \ell)g^\ell(q).$$

Since $b(2j+1, \ell) = 0$ for $\ell > (4^{j+1} - 1)/3$, we have obtained (3.4) with j replaced with $j+1$. □

The next lemma will be useful in obtaining the final generating function.

Lemma 3.5 *Let the functions G and z and the operator H be as defined in Sect. 2. Then for any positive integer ℓ,*

$$q H\left(\frac{1}{q} G^{\ell}\right) = \sum_{k \text{ odd}} \binom{\ell}{k} 2^{\ell-k} \left(\frac{1}{z} - 3z\right)^k.$$

Proof By Lemma 2.2 and the binomial theorem, we have

$$q H\left(\frac{1}{q} G^{\ell}\right) = q H\left(\frac{1}{q}\left(\frac{1}{z} + 2 - 3z\right)^{\ell}\right)$$

$$= q H\left(\frac{1}{q} \sum_{k=0}^{\ell} \binom{\ell}{k} 2^{\ell-k} \left(\frac{1}{z} - 3z\right)^k\right).$$

Observe that $q H(\frac{1}{q} f(q))$ is the odd part of $f(q)$. Since z is an odd function, the operator $q H\frac{1}{q}$ picks out precisely the terms in the above sum that correspond to odd values of k. This completes the proof. □

For any positive odd integer k, let

$$f_k(q^2) = \frac{1}{q}\left(\frac{1}{z(q)} - 3z(q)\right)^k.$$

Observe that since $z(q)$ is an odd function and k is an odd integer, the function on the right-hand side is an even function of q.

Theorem 3.6 *For $j \geq 1$, we have*

$$\sum_{n=0}^{\infty} c\left(2^{2j+1} n + \frac{1}{3}(5 \times 4^j + 1)\right) q^{n+1}$$

$$= \frac{1}{E_2^2 E_6^2} \sum_{\ell=1}^{\frac{2}{3}(4^j-1)} \sum_{k \text{ odd}} b(2j, \ell) \binom{3\ell+1}{k} 2^{3\ell+1-k} g^{2\ell+1}(q) f_k(q).$$

Proof The proof is similar to Step 2 of the proof of Theorem 3.4. The difference is that we apply the operator U to (3.5), without first multiplying by q, to get

$$\sum_{n=0}^{\infty} c\left(2^{2j}(2n+1) + \frac{1}{3}(2^{2j+1} + 1)\right) q^{n+1} = U\left(\frac{1}{E_1^2 E_3^2} \sum_{\ell=1}^{\frac{2}{3}(4^j-1)} b(2j, \ell) g^{\ell}(q)\right).$$

By the first two parts of Lemma 2.3, this is

$$\sum_{n=0}^{\infty} c\left(2^{2j+1}n + \frac{1}{3}(5 \times 4^j + 1)\right)q^{n+1}$$

$$= U\left(\frac{1}{E_4^2 E_{12}^2} \sum_{\ell=1}^{\frac{2}{3}(4^j-1)} b(2j,\ell)g^{2\ell+1}(q^2) \times \frac{1}{q}G^{3\ell+1}(q)\right)$$

$$= \frac{1}{E_2^2 E_6^2} \sum_{\ell=1}^{\frac{2}{3}(4^j-1)} b(2j,\ell)g^{2\ell+1}(q)U\left(\frac{1}{q}G^{3\ell+1}(q)\right). \tag{3.10}$$

By Lemma 3.5 and the definition of f_k,

$$U\left(\frac{1}{q}G^{3\ell+1}(q)\right) = V\left(H\left(\frac{1}{q}G^{3\ell+1}(q)\right)\right)$$

$$= V\left(\frac{1}{q}\sum_{k \text{ odd}}\binom{3\ell+1}{k}2^{3\ell+1-k}\left(\frac{1}{z}-3z\right)^k\right)$$

$$= \sum_{k \text{ odd}}\binom{3\ell+1}{k}2^{3\ell+1-k}f_k(q). \tag{3.11}$$

Combining (3.10) and (3.11), we complete the proof. $\qquad\square$

4 Congruences

In this section, we shall consider the powers of 2 that divide the numbers $m(j,k)$ and $b(j,k)$. This information will be used to prove Theorem 1.1.

For a positive integer n, let $v_2(n)$ be the exact power of 2 that divides n, and define $v_2(0) = +\infty$.

Lemma 4.1 *For the numbers $m(j,k)$ defined in Sect. 3, we have*

$$v_2\big(m(j,k)\big) \geq N(j,k),$$

where

$$N(j+k,j) = 2(j-k)-1 \quad for\ k = 0,1,2,\ldots,j-1$$

and all other $N(j,k)$ are 0.

Proof The first seven rows of the matrix $N = \{N(j,k)\}_{j,k \geq 1}$ are given by

$$
\begin{pmatrix}
1 & 0 & 0 & 0 & 0 & 0 & 0 & 0 & \cdots \\
0 & 3 & 0 & 0 & 0 & 0 & 0 & 0 & \cdots \\
0 & 1 & 5 & 0 & 0 & 0 & 0 & 0 & \cdots \\
0 & 0 & 3 & 7 & 0 & 0 & 0 & 0 & \cdots \\
0 & 0 & 1 & 5 & 9 & 0 & 0 & 0 & \cdots \\
0 & 0 & 0 & 3 & 7 & 11 & 0 & 0 & \cdots \\
0 & 0 & 0 & 1 & 5 & 9 & 13 & 0 & \cdots
\end{pmatrix}. \tag{4.1}
$$

Comparing (3.1) and (4.1), if follows immediately that $v_2(m(j,k)) \geq N(j,k)$ for $1 \leq j \leq 7$. The general result follows by an easy induction proof using the defining properties (1)–(4) of $m(j,k)$. We omit the details. □

Lemma 4.2 *For the numbers $b(j,k)$ defined in Sect. 3, we have*

$$
v_2\big(b(2j-1,k)\big) \geq j + 3(k-1) \tag{4.2}
$$

and

$$
v_2\big(b(2j,k)\big) \geq j + 1 + 4(k-1) \tag{4.3}
$$

for any positive integers j and k. Equality holds in each case for $k = 1$.

Proof We shall use induction on j, passing alternately between (4.2) and (4.3). To begin with, note that from the values in (3.3), (4.2) is true for $j = 1, 2$, and (4.3) is true for $j = 1$, and equality holds in each instance for $k = 1$.

Step 1. Suppose that (4.2) is true for some positive integer j.

Consider the case $k = 1$. Then, by definition of the numbers b and property (6) of the numbers m,

$$
b(2j,1) = \sum_{i=1}^{\infty} b(2j-1,i)m(3i,i+1)
$$

$$
= 6b(2j-1,1) + b(2j-1,2).
$$

Now

$$
v_2\big(b(2j-1,1)\big) = j \quad \text{and} \quad v_2\big(b(2j-2,2)\big) \geq j+3,
$$

by hypothesis. It follows that $v_2(b(2j,1)) = j+1$.

Now suppose that $k \geq 2$. Then by the definition of the numbers b and properties (5) and (6) of the numbers m,

$$
v_2\big(b(2j,k)\big) = v_2\left(\sum_{i=1}^{\infty} b(2j-1,i)m(3i,i+k) \right)
$$

$$
= v_2\left(\sum_{\frac{k}{2} \leq i \leq 2k} b(2j-1,i)m(3i,i+k) \right)
$$

$$
\geq \min_{\frac{k}{2} \leq i \leq 2k} v_2\big(b(2j-1,i)m(3i,i+k)\big). \tag{4.4}
$$

By the induction hypothesis and Lemma 4.1,

$$\min_{\frac{k}{2} \le i \le 2k-1} v_2\big(b(2j-1,i)m(3i,i+k)\big)$$

$$\ge \min_{\frac{k}{2} \le i \le 2k-1} j + 3(i-1) + 2(2k-i) - 1$$

$$= \min_{\frac{k}{2} \le i \le 2k-1} j + i + 4(k-1)$$

$$\ge j + 1 + 4(k-1). \tag{4.5}$$

Moreover, by property (7) of the numbers m and the induction hypothesis,

$$v_2\big(b(2j-1,i)m(3i,i+k)\big)\big|_{i=2k} = v_2\big(b(2j-1,2k)m(6k,3k)\big)$$

$$= v_2\big(b(2j-1,2k)\big)$$

$$\ge j + 3(2k-1). \tag{4.6}$$

Using (4.5) and (4.6) in (4.4), we get

$$v_2\big(b(2j,k)\big) \ge \min\big\{j + 1 + 4(k-1), j + 3(2k-1)\big\}$$

$$= j + 1 + 4(k-1).$$

Thus, if (4.2) is true for a particular positive integer j, then (4.3) is also true for j.

Step 2. Suppose that (4.3) is true for some positive integer j.

Consider the case $k = 1$. Then, by the definition of the numbers b and properties (6) and (7) of the numbers m,

$$b(2j+1,1) = \sum_{i=1}^{\infty} b(2j,i)m(3i+1,i+1)$$

$$= b(2j,1),$$

and so

$$v_2\big(b(2j+1,1)\big) = v_2\big(b(2j,1)\big) = j + 1.$$

Now suppose that $k \ge 2$. Then by the definition of the numbers b and properties (5) and (6) of the numbers m,

$$v_2\big(b(2j+1,k)\big) = v_2\left(\sum_{i=1}^{\infty} b(2j,i)m(3i+1,i+k)\right)$$

$$= v_2\left(\sum_{\frac{k-1}{2} \le i \le 2k-1} b(2j,i)m(3i+1,i+k)\right)$$

$$\ge \min_{\frac{k-1}{2} \le i \le 2k-1} v_2\big(b(2j-1,i)m(3i,i+k)\big). \tag{4.7}$$

By the induction hypothesis and Lemma 4.1,

$$\min_{\frac{k-1}{2} \le i \le 2k-2} v_2\big(b(2j, i)m(3i + 1, i + k)\big)$$

$$\ge \min_{\frac{k-1}{2} \le i \le 2k-2} j + 1 + 4(i - 1) + 2(2k - i - 1) - 1$$

$$= \min_{\frac{k-1}{2} \le i \le 2k-2} j + 1 + 3(k - 1) + (k + 2i - 4)$$

$$\ge j + 1 + 3(k - 1). \tag{4.8}$$

Moreover, by property (7) of the numbers m and the induction hypothesis,

$$v_2\big(b(2j, i)m(3i + 1, i + k)\big)\big|_{i=2k-1}$$

$$= v_2\big(b(2j, 2k - 1)m(6k - 2, 3k - 1)\big)$$

$$= v_2\big(b(2j, 2k - 1)\big)$$

$$\ge j + 1 + 4(2k - 3). \tag{4.9}$$

Using (4.8) and (4.9) in (4.7), we get

$$v_2\big(b(2j, k)\big) \ge \min\{j + 1 + 3(k - 1), j + 1 + 4(2k - 3)\}$$

$$= j + 1 + 3(k - 1).$$

Thus, if (4.3) is true for a particular positive integer j, then (4.2) is true for the value $j + 1$.

This completes the proof by induction. □

We are now ready to prove Theorem 1.1.

Proof of (1.3) This is immediate from Theorem 3.1. □

Proof of (1.4) This is immediate from identity (3.5) in Theorem 3.4 and Lemma 4.2. □

Proof of (1.5) Consider the right-hand side of the identity in Theorem 3.6. The $\ell = 1$ term is

$$\frac{g^3(q)}{E_2^2 E_6^2} b(2j, 1)\left(\binom{4}{1} 8 f_1(q) + \binom{4}{3} 2 f_3(q)\right),$$

which clearly contains a factor of $8b(2j, 1)$. Lemma 4.2 implies that this factor is divisible by $8 \times 2^{j+1} = 2^{j+4}$.

Each of the remaining terms $\ell \ge 2$ contains a factor $b(2j, \ell)$, and by Lemma 4.2,

$$v_2\big(b(2j, \ell)\big) \ge j + 1 + 4(\ell - 1) \ge j + 5.$$

It follows that every term in the q-expansion of the identity in Theorem 4.2 is divisible by 2^{j+4}, and this completes the proof. □

We end with two examples that are analogues of [7, Theorem 5.2].

Theorem 4.3 *For $n \geq 0$,*

$$c(8n + 3) \equiv 6c(2n + 1) \pmod{2^8}$$

and

$$c(16n + 11) \equiv 198c(4n + 3) \pmod{2^{10}}.$$

Proof By the examples listed in Theorem 3.4,

$$\sum_{n=0}^{\infty} \left(c(8n + 3) - 6c(2n + 1) \right) q^{n+1}$$

$$= \frac{2^8}{E_2^2 E_6^2} \left(5g^2 + 118g^3 + 896g^4 + 2048g^5 \right),$$

and the first result follows. The second result follows similarly from Theorem 3.4, using the values

$$b(4, 1) = 1352, \quad \text{and} \quad b(4, 2) = 865664,$$

which have been computed using (10) and the fact that

$$v_2\big(b(4, k)\big) \geq 11 \quad \text{for } k \geq 3,$$

which holds by Lemma 4.2. □

Acknowledgements We would like to thank Professors Bruce Berndt and Michael Hirschhorn for their encouragement and interest in the present work. Part of this work was reported by one of us (HCC) at the Combinatory Analysis 2008: Partitions, q-series, and Applications. He would like to thank the organizers for organizing such a wonderful conference.

References

1. Andrews, G.E.: In: Rota, G.-C. (ed.) The Theory of Partitions. Encycl. Math. and Its Appl., vol. 2. Addison-Wesley, Reading (1976). (Reissued: Cambridge University Press, Cambridge, 1998)
2. Atkin, A.O.L.: Proof of a conjecture of Ramanujan. Glasg. Math. J. **8**, 14–32 (1967)
3. Berndt, B.C., Ono, K.: Ramanujan's unpublished manuscript on the partition and tau functions with proofs and commentary. Sémin. Lothar. Comb. **42**, 63 (1999)
4. Chan, H.-C.: Ramanujan's cubic continued fraction and a generalization of his "most beautiful identity". Int. J. Number Theory (to appear)
5. Chan, H.-C.: Ramanujan's cubic continued fraction and Ramanujan type congruences for a certain partition function. Int. J. Number Theory (to appear)
6. Cooper, S.: Series and iterations for $1/\pi$. Acta Arith. (to appear)
7. Hirschhorn, M.D., Hunt, D.C.: A simple proof of the Ramanujan conjecture for powers of 5. J. Reine Angew. Math. **326**, 1–17 (1981)
8. Kim, B.: A crank analog on a certain kind of partition function arising from the cubic continued fraction. Preprint

9. Knopp, M.I.: Modular Functions in Analytic Number Theory. Markham Publ. Co., Chicago (1970)
10. Ramanujan, S.: Some properties of $p(n)$, the number of partitions of n. Proc. Camb. Philos. Soc. **19**, 207–210 (1919)
11. Ramanujan, S.: Collected Papers. AMS-Chelsea, Providence (2000)
12. Ramanujan, S.: The Lost Notebook and Other Unpublished Papers. Narosa, New Delhi (1988)
13. Sinick, J.: Ramanujan congruences for a class of eta quotients. Int. J. Number Theory (to appear)

Ramanujan J (2011) 25:319–342
DOI 10.1007/s11139-011-9310-5

New multiple $_6\psi_6$ summation formulas and related conjectures

Vyacheslav P. Spiridonov · S. Ole Warnaar

Dedicated to George Andrews on the occasion of his 70th birthday

Received: 7 August 2009 / Accepted: 5 April 2011 / Published online: 27 May 2011
© Springer Science+Business Media, LLC 2011

Abstract Three new summation formulas for $_6\psi_6$ bilateral basic hypergeometric series attached to root systems are presented. Remarkably, two of these formulae, labelled by the A_{2n-1} and A_{2n} root systems, can be reduced to multiple $_6\phi_5$ sums generalising the well-known van Diejen sum. This latter sum serves as the weight-function normalisation for the BC_n q-Racah polynomials of van Diejen and Stokman. Two $_8\phi_7$-level extensions of the multiple $_6\phi_5$ sums, as well as their elliptic analogues, are conjectured. This opens up the prospect of constructing novel A-type extensions of the Koornwinder–Macdonald theory.

Keywords Basic hypergeometric series · Elliptic hypergeometric series · Root systems · Orthogonal polynomials

Mathematics Subject Classification (2000) 05E05 · 33D52 · 33D67

1 Introduction

Bailey's $_6\psi_6$ summation formula [5]

$$\sum_{k=-\infty}^{\infty} \frac{1-aq^{2k}}{1-a} \frac{(b,c,d,e)_k}{(aq/b,aq/c,aq/d,aq/e)_k} \left(\frac{a^2q}{bcde}\right)^k$$
$$= \frac{(aq/bc,aq/bd,aq/be,aq/cd,aq/ce,aq/de,aq,q/a,q)_\infty}{(aq/b,aq/c,aq/d,aq/e,q/b,q/c,q/d,q/e,a^2q/bcde)_\infty}, \quad (1.1)$$

Work supported by the Australian Research Council and Russian Foundation for Basic Research, grant no. 08-01-00392.

V.P. Spiridonov
Laboratory of Theoretical Physics, JINR, Dubna, Moscow Region 141980, Russia

S.O. Warnaar (✉)
School of Mathematics and Physics, The University of Queensland, Brisbane, QLD 4072, Australia
e-mail: o.warnaar@maths.uq.edu.au

where

$$|q| < 1 \quad \text{and} \quad \left| \frac{a^2 q}{bcde} \right| < 1,$$

is one of the most impressive identities in the theory of bilateral basic hypergeometric series. (Readers unfamiliar with q-series notation are referred to the next section.) Throughout his long and distinguished career George Andrews has been a master in extracting combinatorial information from identities such as (1.1). In a paper on $_6\psi_6$ summations dedicated to George it seems appropriate to highlight one of his applications of (1.1) pertaining to two of his favourite subjects, partition theory and the mathematical discoveries of Ramanujan.

Let $p(n)$ be the number of integer partitions of n. Then one of Ramanujan's celebrated congruences states that

$$p(5n + 4) \equiv 0 \quad (\text{mod } 5).$$

For example, $p(4) = 5$, $p(9) = 30$, $p(14) = 135$ and so on. Ramanujan proved his congruence by establishing the beautiful identity

$$\sum_{n=0}^{\infty} p(5n + 4)q^n = 5 \frac{(q^5; q^5)_\infty^5}{(q)_\infty^6}. \tag{1.2}$$

As noted by Andrews in his SIAM review *Applications of basic hypergeometric functions* [2], identity (1.2) readily follows from (1.1) (for some of the details, see also [6]). Indeed, after replacing

$$(a, b, c, d, e, q) \mapsto \left(q^4, q, q, q^3, q^3, q^5 \right)$$

in Bailey's $_6\psi_6$ sum and carrying out some standard manipulations, one obtains

$$\sum_{n=1}^{\infty} \left(\frac{n}{5} \right) \frac{q^n}{(1 - q^n)^2} = q \frac{(q^5; q^5)_\infty^5}{(q)_\infty} \tag{1.3a}$$

$$= (q^5; q^5)_\infty^5 \sum_{n=0}^{\infty} p(n)q^{n+1}, \tag{1.3b}$$

where $\left(\frac{n}{p} \right)$, for p an odd prime, is the Legendre symbol [17]. For m a positive integer, let the Hecke operator U_m act on formal power series in q as

$$U_m \sum_{n=0}^{\infty} a_n q^n = \sum_{n=0}^{\infty} a_{nm} q^n.$$

Since $\left(\frac{n}{p} \right) = 0$ when n is a multiple of p,

$$U_p \sum_{n=1}^{\infty} \left(\frac{n}{p}\right) \frac{q^n}{(1-q^n)^2} = U_p \sum_{n,m=1}^{\infty} \left(\frac{n}{p}\right) mq^{nm}$$

$$= p \sum_{n,m=1}^{\infty} \left(\frac{n}{p}\right) mq^{nm} = p \sum_{n=1}^{\infty} \left(\frac{n}{p}\right) \frac{q^n}{(1-q^n)^2}.$$

Acting with U_5 on (1.3b) thus gives

$$(q)_{\infty}^5 \sum_{n=0}^{\infty} p(5n+4)q^{n+1} = 5 \sum_{n=1}^{\infty} \left(\frac{n}{5}\right) \frac{q^n}{(1-q^n)^2}.$$

By (1.3a) this proves (1.2).

Another important consequence of the $_6\psi_6$ sum pointed out by Andrews in his SIAM review is the Jacobi triple product identity. Specifically, taking the limit $b, c, d, e \to \infty$ in (1.1) and replacing a by z yields[1]

$$\sum_{k=-\infty}^{\infty} (-z)^k q^{\binom{k}{2}} = (z, q/z, q)_{\infty}, \quad z \neq 0. \tag{1.4}$$

In the landmark paper [20], Macdonald generalised the triple product identity to all (reduced irreducible) affine root systems. For example, for the affine root system of type A_{n-1}, he proved that[2]

$$\sum_{\substack{\lambda \in \mathbb{Z}^n \\ |\lambda|=0}} \prod_{i=1}^{n} z_i^{n\lambda_i} q^{n\binom{\lambda_i}{2}+i\lambda_i} \prod_{1 \le i < j \le n} \left(1 - q^{\lambda_i - \lambda_j} z_i/z_j\right)$$

$$= (q)_{\infty}^{n-1} \prod_{1 \le i < j \le n} (z_i/z_j, qz_j/z_i)_{\infty} \tag{1.5}$$

for $z_1, \ldots, z_n \neq 0$.

In view of the above limit reducing (1.1) to (1.4), it is a natural question to ask for a generalisation of (1.5) and other Macdonald identities to multiple $_6\psi_6$ Bailey sums. To a large extent this question was settled by Gustafson [10–12], who proved four $_6\psi_6$ sums corresponding to the affine root systems of type A_{n-1}, C_n, $B_n^\vee$ and G_2. By taking various limits, these four identities yield *all* of the infinite families of Macdonald identities, corresponding to A_{n-1}, B_n, $B_n^\vee$ C_n, $C_n^\vee$, D_n and BC_n, as well as the Macdonald identity for G_2. For example, if for $z \in (\mathbb{C}^*)^n$ and $\lambda \in \mathbb{Z}^n$,

$$\Delta(z) = \prod_{1 \le i < j \le n} (z_i - z_j) \quad \text{and} \quad \Delta(zq^\lambda) = \prod_{1 \le i < j \le n} (z_i q^{\lambda_i} - z_j q^{\lambda_j}), \tag{1.6}$$

---

[1] Incidentally, Bailey himself obtained the triple product identity by specialising $b = a^{1/2}$, $c = -a^{1/2}$, then taking the $d, e \to \infty$ limit, and finally replacing $(a, q) \mapsto (zq^{-1/2}, q^{1/2})$.

[2] Equation (1.5) is a particular form of the A_{n-1} Macdonald identity first stated in [22].

then Gustafson's A_{n-1} $_6\psi_6$ sum [10] reads

$$\sum_{\substack{\lambda \in \mathbb{Z}^n \\ |\lambda|=0}} \frac{\Delta(zq^\lambda)}{\Delta(z)} \prod_{i,j=1}^n \frac{(z_j/b_i)_{\lambda_j}}{(qa_iz_j)_{\lambda_j}} = \frac{(qAZ, qB/Z)_\infty}{(q, qAB)_\infty} \prod_{i,j=1}^n \frac{(qa_ib_j, qz_i/z_j)_\infty}{(qa_iz_j, qb_i/z_j)_\infty}, \quad (1.7)$$

where $A = a_1 \cdots a_n$, $B = b_1 \cdots b_n$, $Z = z_1 \cdots z_n$,

$$|q| < 1 \quad \text{and} \quad |qAB| < 1.$$

Letting $a_i, b_i \to 0$ for $1 \le i \le n$, one recovers the Macdonald identity (1.5).

Multiple $_6\psi_6$ summations are not only important in relation to the Macdonald identities but also have a close connection to q-beta-integrals on root systems, which in turn play a role in the theory of multivariable orthogonal polynomials. The integral counterpart of (1.7), for example, is Gustafson's A_{n-1} integral [14]

$$\int_{\mathbb{T}^{n-1}} \prod_{1 \le i < j \le n} (z_i/z_j, z_j/z_i)_\infty \prod_{j=1}^n \frac{(ABz_j)_\infty}{\prod_{i=1}^{n+1}(a_iz_j)_\infty \prod_{i=1}^n (b_i/z_j)_\infty} \frac{dz_1}{z_1} \cdots \frac{dz_{n-1}}{z_{n-1}}$$

$$= \frac{n!(2\pi i)^{n-1} \prod_{i=1}^{n+1}(AB/a_i)_\infty \prod_{i=1}^n (Ab_i)_\infty}{(q)_\infty^{n-1}(B)_\infty \prod_{i=1}^{n+1}(A/a_i)_\infty \prod_{i=1}^{n+1} \prod_{j=1}^n (a_ib_j)_\infty}.$$

Here $\mathbb{T}$ is the positively oriented unit circle, $z_1 \cdots z_n = 1$, $A = a_1 \cdots a_{n+1}$, $B = b_1 \cdots b_n$, where $a_1, \ldots, a_{n+1}, b_1, \ldots, b_n \in \mathbb{C}$ such that $|a_1|, \ldots, |b_n| < 1$.

There are a number of known multiple beta integrals for which no corresponding $_6\psi_6$ summation has ever been found. Filling this gap in the literature has been the initial motivation for this paper, and in Sects. 4 and 5 three new multiple $_6\psi_6$ sums are stated. The consequences of these results go far beyond the completion of the classification of $_6\psi_6$ summations. Indeed, as it turns out, two of our new summations corresponding to A_{2n-1} and A_{2n} have rather unexpected consequences for multivariable orthogonal polynomials. In particular, both these $_6\psi_6$ summations can be reduced to new multiple $_6\phi_5$ sums. Conjecturally, these $_6\phi_5$ sums can be interpreted as weight-function normalisations for some yet-to-be found generalisations of the BC_n q-Racah polynomials. Moreover, by conjecturing elliptic extensions of the $_6\phi_5$ sums we are led to speculate on the existence of A_{2n-1} and A_{2n} elliptic generalisation of the entire BC_n-Koornwinder–Macdonald theory.

2 Notation

In this section we collect some standard notation from the theory of basic hypergeometric series and partitions.

Throughout this paper we view the base q either as a formal variable or as a fixed complex number such that $|q| < 1$. Then the q-shifted factorials $(a)_\infty$ and $(a)_n$ (for $n \in \mathbb{Z}$) are defined as

$$(a)_\infty = (a; q)_\infty := \prod_{k=0}^\infty (1 - aq^k)$$

and

$$(a)_n = (a; q)_n = \frac{(a)_\infty}{(aq^n)_\infty}.$$

Note that

$$(a)_n = \prod_{k=0}^{n-1} (1 - aq^k)$$

for n a nonnegative integer, and

$$(a)_{-n} = \frac{(-q/a)^n}{(q/a)_n} q^{\binom{n}{2}}$$

for all $n \in \mathbb{Z}$. Hence $1/(q^m)_n = 0$ unless $n \geq -m$. We also adopt the usual condensed notations

$$(a_1, \ldots, a_k)_m = (a_1, \ldots, a_k; q)_m := \prod_{i=1}^{k} (a_i)_m$$

and

$$\left(z^\pm\right)_m = \left(z, z^{-1}\right)_m, \qquad \left(z^{\pm n}\right)_m = \left(z^n, z^{-n}\right)_m,$$
$$\left(z^\pm w^\pm\right)_m = \left(zw, zw^{-1}, z^{-1}w, z^{-1}w^{-1}\right)_m,$$

where $m \in \mathbb{Z} \cup \{\infty\}$.

Because we are dealing with series on root systems, we need some notation pertaining to integer sequences, and for $\lambda = (\lambda_1, \lambda_2, \ldots)$ a finite sequence of integers, we set

$$|\lambda| = \sum_{i \geq 1} \lambda_i,$$

$$n(\lambda) = \sum_{i \geq 1} (i - 1)\lambda_i$$

and

$$(a)_\lambda = (a; q, t)_\lambda := \prod_{i \geq 1} \left(at^{1-i}\right)_{\lambda_i}.$$

In a few instances we also use

$$(a)_{(N^m)} = (a; q, t)_{(N^m)} := \prod_{i=1}^{m} \left(at^{1-i}\right)_N.$$

We already defined the Vandermonde product $\Delta(z)$ in (1.6). Subsequently we also need the analogous product for the (classical) root system of type C_n, and for $z \in \mathbb{C}^n$, we define

$$\Delta^+(z) = \prod_{i=1}^n \left(1 - z_i^2\right) \prod_{1 \le i < j \le n} (z_i - z_j)(1 - z_i z_j).$$

Finally we need some notation concerning partitions. All our partitions will have at most n parts, and we define

$$\Lambda = \left\{ \lambda \in \mathbb{Z}^n : \lambda_1 \ge \lambda_2 \ge \cdots \ge \lambda_n \ge 0 \right\}$$

and

$$\Lambda_N = \{\lambda \in \Lambda : \lambda_1 \le N\}.$$

As usual we identify a partition with its diagram or Ferrers graph [3, 21]. Given $\lambda, \mu \in \Lambda$, we say that μ is contained in λ, denoted $\mu \subseteq \lambda$, if $\mu_i \le \lambda_i$ for all $1 \le i \le n$. In other words, μ is contained in λ if the graph of μ is a subset of the graph of λ. If $\mu \subseteq \lambda$ we also use the more customary $|\lambda - \mu|$ instead of $|\lambda| - |\mu|$. We write $\mu \preccurlyeq \lambda$ if $\mu \subseteq \lambda$ and the graphs of λ and μ differ by at most one square in each column. (In the terminology of [21], $\mu \preccurlyeq \lambda$ if the skew shape $\lambda - \mu$ is a horizontal strip.) Note that $\mu \preccurlyeq \lambda$ if and only if we have the interlacing property

$$\lambda_1 \ge \mu_1 \ge \lambda_2 \ge \cdots \ge \mu_{n-1} \ge \lambda_n \ge \mu_n \ge 0.$$

3 Some known $_6\psi_6$ summations

Our derivations of new multiple $_6\psi_6$ summations rely on a technique employed by van Diejen in his proof of Theorem 3.3 below [33]. This method essentially coincides with the one used by Gustafson for proving generalised beta integrals [13], and by Anderson for proving the Selberg integral [1, 4]. At its core is a clever sequential use of existing multiple $_6\psi_6$ summations, and for our purposes the following three known summations are crucial.

Recall that throughout this paper it is assumed that $|q| < 1$.

Theorem 3.1 (Gustafson's type I C_n $_6\psi_6$ sum [11]) *For $a_1, \ldots, a_{2n+2}, z_1, \ldots, z_n \in \mathbb{C}^*$,*

$$\sum_{\lambda \in \mathbb{Z}^n} (qA)^{|\lambda|} \frac{\Delta^+(zq^\lambda)}{\Delta^+(z)} \prod_{i=1}^{2n+2} \prod_{j=1}^n \frac{(z_j/a_i)_{\lambda_j}}{(qa_i z_j)_{\lambda_j}}$$

$$= \frac{(q)_\infty^n}{(qA)_\infty} \prod_{1 \le i < j \le 2n+2} (qa_i a_j)_\infty \prod_{j=1}^n \frac{(qz_j^{\pm 2})_\infty}{\prod_{i=1}^{2n+2}(qa_i z_j^\pm)_\infty} \prod_{1 \le i < j \le n} (qz_i^\pm z_j^\pm)_\infty, \quad (3.1)$$

where $A = a_1 \cdots a_{2n+2}$ and $|qA| < 1$.

We remark that in the limit $a_i \to 0$ for $1 \le i \le 2n+2$ this yields the Macdonald identity of type C_n, in the limit $a_i \to 0$ for $1 \le i \le 2n+1$ and $a_{2n+2} \mapsto -1$ it yields the Macdonald identity of type BC_n, and in the limit $a_i \to 0$ for $1 \le i \le 2n$ and $a_{2n+1} \mapsto -1$, $a_{2n+2} \mapsto -q^{-1/2}$ it yields the Macdonald identity of type $C_n^\vee$. Similar comments apply to the other $_6\psi_6$ summations listed below.

Theorem 3.2 (Gustafson's type I $B_n^\vee$ $_6\psi_6$ sum [11]) *For $a_1, \ldots, a_{2n}, z_1, \ldots, z_n \in \mathbb{C}^*$ and $\sigma = 0, 1$,*

$$\sum_{\substack{\lambda \in \mathbb{Z}^n \\ |\lambda| \equiv \sigma \pmod 2}} (-A)^{|\lambda|} \frac{\Delta^+(zq^\lambda)}{\Delta^+(z)} \prod_{i=1}^{2n} \prod_{j=1}^{n} \frac{(z_j/a_i)_{\lambda_j}}{(qa_i z_j)_{\lambda_j}}$$

$$= \frac{(q)_\infty^{n-1}(q^2; q^2)_\infty}{(-A)_\infty} \prod_{i=1}^{2n} (qa_i^2; q^2)_\infty \prod_{1 \le i < j \le 2n} (qa_i a_j)_\infty$$

$$\times \prod_{j=1}^{n} \frac{(q^2 z_j^{\pm 2}; q^2)_\infty}{\prod_{i=1}^{2n}(qa_i z_j^{\pm})_\infty} \prod_{1 \le i < j \le n} (q z_i^{\pm} z_j^{\pm})_\infty, \tag{3.2}$$

where $A = a_1 \cdots a_{2n}$ and $|A| < 1$.

This result in not entirely independent of the C_n $_6\psi_6$ sum; setting $a_{2n+1} = q^{-1/2}$ and $a_{2n+2} = -q^{-1/2}$ in the latter, we obtain the former summed over σ.

Theorem 3.3 (van Diejen's type II C_n $_6\psi_6$ sum [33]) *For $a_1, \ldots, a_4, t, z_1, \ldots, z_n \in \mathbb{C}^*$,*

$$\sum_{\lambda \in \mathbb{Z}^n} (qt^{2-2n} A)^{|\lambda|} \left(\frac{t^2}{q}\right)^{n(\lambda)} \frac{\Delta^+(zq^\lambda)}{\Delta^+(z)} \prod_{i=1}^{4} \prod_{j=1}^{n} \frac{(z_j/a_i)_{\lambda_j}}{(qa_i z_j)_{\lambda_j}}$$

$$\times \prod_{1 \le i < j \le n} \frac{(tz_i z_j)_{\lambda_i + \lambda_j}}{(qt^{-1} z_i z_j)_{\lambda_i + \lambda_j}} \frac{(tz_i z_j^{-1})_{\lambda_i - \lambda_j}}{(qt^{-1} z_i z_j^{-1})_{\lambda_i - \lambda_j}}$$

$$= \prod_{j=1}^{n} \frac{(q, qt^{-j}, qz_j^{\pm 2})_\infty \prod_{1 \le k < l \le 4}(qt^{1-j} a_k a_l)_\infty}{(qt^{-1}, qt^{2-j-n} A)_\infty \prod_{i=1}^{4}(qa_i z_j^{\pm})_\infty} \prod_{1 \le i < j \le n} \frac{(qz_i^{\pm} z_j^{\pm})_\infty}{(qt^{-1} z_i^{\pm} z_j^{\pm})_\infty}, \tag{3.3}$$

where $A = a_1 \cdots a_4$ and $\max\{|qt^{2-2n}A|, |q^{2-n}A|\} < 1$.

Other multiple $_6\psi_6$ summations that appear not to be amenable to the methods employed in this paper may be found in [18, 27, 28].

In the above we have made reference not only to the root system attached to each $_6\psi_6$ sum, but also to its type. In type I hypergeometric sums (or type I multiple beta integrals) the number of free parameters (such as the a_i and b_j) is of the form $2n + m$ where m is a constant and n (or $n - 1$) the rank of the root system. The sums (1.5), (3.1), (3.2) are all of type I. In type II sums or integrals the number of free parameters is assumed to be independent of the rank of the underlying root system. The sum (3.3) is an example of a type II hypergeometric sum. From the point of view of orthogonal polynomial theory, type II sums and integrals are by far the most important. Koornwinder's multivariable Askey–Wilson polynomials, for example, depend on 5 free variables, and the corresponding orthogonality measure is determined by a type II q-beta integral, see e.g., [19, 24, 25, 34] for more details.

4 Type II $B_n^\vee$ $_6\psi_6$ sum

As a warm up to the much more important results of the next section, we prove a type II variant of Gustafson's $B_n^\vee$ sum.

Theorem 4.1 *For* $a_1, a_2, t, z_1, \ldots, z_n \in \mathbb{C}^*$ *and* $\sigma = 0, 1$,

$$
\sum_{\substack{\lambda \in \mathbb{Z}^n \\ |\lambda| \equiv \sigma \pmod 2}} \left(-t^{2-2n} A\right)^{|\lambda|} \left(\frac{t^2}{q}\right)^{n(\lambda)} \frac{\Delta^+(zq^\lambda)}{\Delta^+(z)} \prod_{i=1}^{2} \prod_{j=1}^{n} \frac{(z_j/a_i)_{\lambda_j}}{(qa_i z_j)_{\lambda_j}}
$$

$$
\times \prod_{1 \le i < j \le n} \frac{(tz_i z_j)_{\lambda_i + \lambda_j}}{(qt^{-1} z_i z_j)_{\lambda_i + \lambda_j}} \frac{(tz_i z_j^{-1})_{\lambda_i - \lambda_j}}{(qt^{-1} z_i z_j^{-1})_{\lambda_i - \lambda_j}}
$$

$$
= \frac{1}{2} \prod_{j=1}^{n} \left(\frac{(q, qt^{-j}, qt^{1-j} A, -t^{1-j})_\infty (q^2 z_j^{\pm 2}; q^2)_\infty}{(qt^{-1}, -t^{2-j-n} A)_\infty} \prod_{i=1}^{2} \frac{(qt^{2-2j} a_i^2; q^2)_\infty}{(qa_i z_j^{\pm})_\infty} \right)
$$

$$
\times \prod_{1 \le i < j \le n} \frac{(qz_i^\pm z_j^\pm)_\infty}{(qt^{-1} z_i^\pm z_j^\pm)_\infty}, \tag{4.1}
$$

where $A = a_1 a_2$ *such that* $\max\{|t^{2-2n} A|, |q^{1-n} A|\} < 1$.

The above identity bears the same relation to (3.3) as (3.2) to (3.1). That is, if we set $a_{2n+1} = q^{-1/2}$ and $a_{2n+2} = -q^{-1/2}$ in (3.3), we obtain (4.1) summed over σ.

Before proving the theorem we list a number of easy consequences. First of all we note that both the $B_n^\vee$ and D_n ($n \ge 2$) Macdonald identities [20] follow from (4.1); the $B_n^\vee$ case is obtained if we let $1/a_1, 1/a_2, t \to \infty$ and take $\sigma = 0$, and the D_n case is obtained if we let $t \to \infty$ and take $a_1 = -a_2 = 1$ and $\sigma = 0$.

A collection of rather curious variations of (some of) the Macdonald identities arises if we take Theorems 3.3 and 4.1 and consider the $t \to q$ limit. Depending on the choice of the a_i this yields the following five infinite families, which we, perhaps somewhat misleadingly, label B_n, $B_n^\vee$ C_n, $C_n^\vee$ and BC_n based on the corresponding Macdonald identities (obtained by replacing the $t \to q$ limit by a $t \to \infty$ limit in the proofs).

For $z = (z_1, \ldots, z_n)$, let

$$
\Xi\left(zq^\lambda\right) = \prod_{1 \le i < j \le n} \left(1 - z_i z_j^{-1} q^{\lambda_i - \lambda_j}\right)^2 \left(1 - z_i z_j q^{\lambda_i + \lambda_j}\right)^2.
$$

Corollary 4.2 (B_n identity) *For* $z_1, \ldots, z_n \in \mathbb{C}^*$,

$$\sum_{\substack{\lambda \in \mathbb{Z}^n \\ |\lambda| \equiv 0 \ (\mathrm{mod}\ 2)}} \Xi\left(zq^\lambda\right) \prod_{i=1}^{n} z_i^{\lambda_i} q^{\binom{\lambda_i}{2}+(2i-2n+1)\lambda_i} \left(1 - z_i q^{\lambda_i}\right)$$
$$= 2^{n-1} n! \prod_{i=1}^{n} z_i^{2(n-i)} (q^{-1}; q^{-1})_{2i-2} (z_i, z_i^{-1}q, q)_\infty (z_i^{\pm 2}q; q^2)_\infty.$$

Proof In (4.1) let $a_1 \to 0$, $t \to q$ and choose $a_2 = -1$, $\sigma = 0$. $\hspace{2cm}$ □

Corollary 4.3 ($B_n^\vee$ identity) *For* $z_1, \ldots, z_n \in \mathbb{C}^*$,

$$\sum_{\substack{\lambda \in \mathbb{Z}^n \\ |\lambda| \equiv 0 \ (\mathrm{mod}\ 2)}} \Xi\left(zq^\lambda\right) \prod_{i=1}^{n} z_i^{2\lambda_i} q^{2\binom{\lambda_i}{2}+2(i-n)\lambda_i} \left(1 - z_i^2 q^{2\lambda_i}\right)$$
$$= 2^{n-1} n! \prod_{i=1}^{n} z_i^{2(n-i)} (q^{-2}; q^{-2})_{i-1} (z_i^2, z_i^{-2}q^2, q^2; q^2)_\infty.$$

Proof In (4.1) let $a_1, a_2 \to 0$, $t \to q$ and choose $\sigma = 0$. $\hspace{2cm}$ □

Corollary 4.4 (C_n identity) *For* $z_1, \ldots, z_n \in \mathbb{C}^*$,

$$\sum_{\lambda \in \mathbb{Z}^n} \Xi\left(zq^\lambda\right) \prod_{i=1}^{n} z_i^{4\lambda_i} q^{4\binom{\lambda_i}{2}+(2i-2n+1)\lambda_i} \left(1 - z_i^2 q^{2\lambda_i}\right)$$
$$= n! \prod_{i=1}^{n} z_i^{2(n-i)} (q^{-1}; q^{-1})_{i-1} (z_i^2, z_i^{-2}q, q)_\infty.$$

Proof In (3.3) let $a_1, \ldots, a_4 \to 0$ and $t \to q$. $\hspace{2cm}$ □

Corollary 4.5 ($C_n^\vee$ identity) *For* $z_1, \ldots, z_n \in \mathbb{C}^*$,

$$\sum_{\lambda \in \mathbb{Z}^n} \Xi\left(zq^\lambda\right) \prod_{i=1}^{n} z_i^{2\lambda_i} q^{2\binom{\lambda_i}{2}+(2i-2n+1/2)\lambda_i} \left(1 - z_j q^{\lambda_i}\right)$$
$$= n! \prod_{i=1}^{n} z_i^{2(n-i)} (q^{-1/2}; q^{-1/2})_{2i-2} (z_i, z_i^{-1}q^{1/2}, q^{1/2}; q^{1/2})_\infty.$$

Proof In (3.3) let $a_1, a_2 \to 0$, $t \to q$ and choose $a_3 = -1$, $a_4 = -q^{-1/2}$. $\hspace{1cm}$ □

Corollary 4.6 (BC_n identity) *For* $z_1, \ldots, z_n \in \mathbb{C}^*$,

$$\sum_{\lambda \in \mathbb{Z}^n} \Xi\left(zq^\lambda\right) \prod_{i=1}^{n} z_i^{3\lambda_i} q^{3\binom{\lambda_i}{2}+(2i-2n+1)\lambda_i} \left(1 - z_i q^{\lambda_i}\right)$$
$$= n! \prod_{i=1}^{n} z_i^{2(n-i)} (q^{-1}; q^{-1})_{i-1} (z_i, z_i^{-1}q, q)_\infty (qz_i^{\pm 2}; q^2)_\infty.$$

Proof In (3.3) let $a_1, \ldots, a_3 \to 0$, $t \to q$ and set $a_4 = -1$. □

For later comparison, we give one further special case of Theorem 4.1. For $\lambda \in \Lambda$, let $\Delta_\lambda(a) = \Delta_\lambda(a; q, t)$ be defined as

$$
\Delta_\lambda(a) = \prod_{i=1}^{n} \frac{1 - at^{2-2i}q^{2\lambda_i}}{1 - at^{2-2i}} \prod_{1 \le i < j \le n} \frac{1 - t^{j-i}q^{\lambda_i - \lambda_j}}{1 - t^{j-i}} \frac{1 - at^{2-i-j}q^{\lambda_i + \lambda_j}}{1 - at^{2-i-j}}
$$

$$
\times \prod_{1 \le i < j \le n} \frac{(at^{3-i-j})_{\lambda_i + \lambda_j}}{(aqt^{1-i-j})_{\lambda_i + \lambda_j}} \frac{(t^{j-i+1})_{\lambda_i - \lambda_j}}{(qt^{j-i-1})_{\lambda_i - \lambda_j}}. \tag{4.2}
$$

Then the $a_1 = a^{-1/2}t^{n-1}$, $a_2 = a^{1/2}b$ and $z_i = a^{1/2}t^{1-i}$ ($1 \le i \le n$) specialisation of (4.1) boils down to

$$
\sum_{\substack{\lambda \in \Lambda \\ |\lambda| \equiv \sigma \pmod 2}} \Delta_\lambda(a) \frac{(at^{1-n}, 1/b)_\lambda}{(qt^{n-1}, abq)_\lambda} \left(-bt^{1-n}\right)^{|\lambda|} t^{2n(\lambda)}
$$

$$
= \frac{1}{2} \prod_{i=1}^{n} \frac{(aqt^{1-i}, -t^{1-i})_\infty}{(abqt^{1-i}, -bt^{1-i})_\infty} \frac{(ab^2qt^{2-2i}; q^2)_\infty}{(aqt^{2-2i}; q^2)_\infty}. \tag{4.3}
$$

Proof of Theorem 4.1 Making the substitutions

$$
a_i \mapsto \begin{cases} uy_i q^{\mu_i}, & \text{for } 1 \le i \le n, \\ u(y_i q^{\mu_i})^{-1}, & \text{for } n+1 \le i \le 2n, \end{cases}
$$

in (3.2) yields

$$
\sum_{\substack{\lambda \in \mathbb{Z}^n \\ |\lambda| \equiv \sigma \pmod 2}} (-u^{2n})^{|\lambda|} \frac{\Delta^+(zq^\lambda)}{\Delta^+(z)} \prod_{i,j=1}^{n} \frac{(z_i y_j^\pm / u)_{\lambda_i}}{(quz_i y_j^\pm)_{\lambda_i}} \frac{((q^{\lambda_i} z_i)^\pm y_j / u)_{\mu_j}}{(qu(q^{\lambda_i} z_i)^\pm y_j)_{\mu_j}}
$$

$$
= \frac{(q)_\infty^{n-1}(q^2; q^2)_\infty(qu^2)_\infty^n}{(-u^{2n})_\infty} \prod_{i=1}^{n}(qu^2 y_i^{\pm 2}; q^2)_\infty \prod_{1 \le i < j \le n}(qu^2 y_i^\pm y_j^\pm)_\infty
$$

$$
\times \prod_{i=1}^{n}(q^2 z_i^{\pm 2}; q^2)_\infty \prod_{1 \le i < j \le n}(q z_i^\pm z_j^\pm)_\infty \prod_{i,j=1}^{n} \frac{1}{(quy_i^\pm z_j^\pm)_\infty}
$$

$$
\times \left(-\frac{u^{2(n-1)}}{q}\right)^{|\mu|} \left(\frac{1}{qu^4}\right)^{n(\mu)} \prod_{i=1}^{n} \frac{(qy_i^2/u^2; q^2)_{\mu_i}}{(qu^2 y_i^2; q^2)_{\mu_i}}
$$

$$
\times \prod_{1 \le i < j \le n} \frac{(y_i y_j / u^2)_{\mu_i + \mu_j}(y_i / y_j u^2)_{\mu_i - \mu_j}}{(qu^2 y_i y_j)_{\mu_i + \mu_j}(qu^2 y_i / y_j)_{\mu_i - \mu_j}}.
$$

If we multiply this by

$$\left(qb_1b_2u^{2n}\right)^{|\mu|}\frac{\Delta^+(yq^\mu)}{\Delta(y)}\prod_{i=1}^{2}\prod_{j=1}^{n}\frac{(y_j/b_i)_{\mu_j}}{(qb_iy_j)_{\mu_j}},$$

where $y=(y_1,\ldots,y_n)$, and note that

$$\prod_{i=1}^{n}\frac{(qu^{-2}y_i^2;q^2)_{\mu_i}}{(qu^2y_i^2;q^2)_{\mu_i}}=\prod_{i=1}^{n}\frac{(q^{1/2}u^{-1}y_i,-q^{1/2}u^{-1}y_i)_{\mu_i}}{(q^{1/2}uy_i,-q^{1/2}uy_i)_{\mu_i}},$$

then the left can be summed over μ by (3.1), and the right can be summed over μ by (3.3). The resulting identity corresponds to the claim with $(a_1,a_2,t)\mapsto(ub_1,ub_2,1/u^2)$.

The above application of (3.1), (3.2) and (3.3) is only valid provided that

$$|t|>1,\qquad\left|qt^{1-n}A\right|<1,\qquad\left|q^{1-n}A\right|<1,\quad\text{and}\quad\left|t^{2-2n}A\right|<1,$$

but the first two conditions may be dropped by analytic continuation. $\qquad\square$

5 Type II A_{2n-1} and A_{2n} $_6\psi_6$ summations

Our next two results, which are the series counterparts of q-beta integrals of Gustafson [15], are new $_6\psi_6$ summations for the root systems A_{2n-1} and A_{2n}. Both are much deeper than Theorem 4.1 and a lot more intricate to prove. As alluded to in the introduction, they have some rather surprising consequences, to be discussed in the next section. Because of some intricate convergence issues, which we failed to completely settle, the A_{2n-1} and A_{2n} sums are stated as Claims instead of fully fledged Theorems.

Claim 5.1 (Type II A_{2n-1} $_6\psi_6$ sum) *For $a_1,a_2,b_1,b_2,z_1,\ldots,z_{2n}\in\mathbb{C}^*$,*

$$\sum_{\substack{\lambda\in\mathbb{Z}^{2n}\\|\lambda|=0}}\frac{\Delta(zq^\lambda)}{\Delta(z)}\prod_{1\le i<j\le 2n}\frac{(z_iz_j/t)_{\lambda_i+\lambda_j}}{(qtz_iz_j)_{\lambda_i+\lambda_j}}\prod_{i=1}^{2}\prod_{j=1}^{2n}\frac{(z_j/b_i)_{\lambda_j}}{(qa_iz_j)_{\lambda_j}}$$

$$=(q)_\infty^{2n-1}\left(qt^nZ,qt^n/Z,qa_1a_2t^{n-1}Z,qb_1b_2t^{n-1}/Z\right)_\infty$$

$$\times\prod_{i=1}^{n-1}(qt^{2i},qa_1a_2t^{2i-1},qb_1b_2t^{2i-1})_\infty\prod_{i=1}^{n}\frac{\prod_{j,k=1}^{2}(qa_jb_kt^{2i-2})_\infty}{(qa_1a_2b_1b_2t^{2i+2n-4})_\infty}$$

$$\times\prod_{i=1}^{2}\prod_{j=1}^{2n}\frac{1}{(qa_iz_j,qb_i/z_j)_\infty}\prod_{1\le i<j\le 2n}\frac{(qz_i/z_j,qz_j/z_i)_\infty}{(qtz_iz_j,qt/z_iz_j)_\infty},$$

where $Z=z_1\cdots z_{2n}$ and $|qa_1a_2b_1b_2t^{4n-4}|<1$.

Claim 5.2 (Type II A$_{2n}$ $_6\psi_6$ sum) *For $a_1, a_2, b_1, b_2, z_1, \ldots, z_{2n+1} \in \mathbb{C}^*$,*

$$
\sum_{\substack{\lambda \in \mathbb{Z}^{2n+1} \\ |\lambda|=0}} \frac{\Delta(zq^\lambda)}{\Delta(z)} \prod_{1 \le i < j \le 2n+1} \frac{(z_i z_j/t)_{\lambda_i+\lambda_j}}{(qt z_i z_j)_{\lambda_i+\lambda_j}} \prod_{i=1}^{2} \prod_{j=1}^{2n+1} \frac{(z_j/b_i)_{\lambda_j}}{(qa_i z_j)_{\lambda_j}}
$$

$$
= (q)_\infty^{2n} \prod_{i=1}^{2} (qa_i t^n Z, qb_i t^n/Z)_\infty
$$

$$
\times \prod_{i=1}^{n} \left[\frac{(qt^{2i}, qa_1 a_2 t^{2i-1}, qb_1 b_2 t^{2i-1})_\infty \prod_{j,k=1}^{2}(qa_j b_k t^{2i-2})_\infty}{(qa_1 a_2 b_1 b_2 t^{2i+2n-2})_\infty} \right]
$$

$$
\times \prod_{i=1}^{2} \prod_{j=1}^{2n+1} \frac{1}{(qa_i z_j, qb_i/z_j)_\infty} \prod_{1 \le i < j \le 2n+1} \frac{(qz_i/z_j, qz_j/z_i)_\infty}{(qt z_i z_j, qt/z_i z_j)_\infty},
$$

where $Z = z_1 \cdots z_{2n+1}$ and $|qa_1 a_2 b_1 b_2 t^{4n-2}| < 1$.

Proof We first combine the two claims into one statement, for which we give a formal proof.

For $a_1, a_2, b_1, b_2, z_1, \ldots, z_n \in \mathbb{C}^*$,

$$
\sum_{\substack{\lambda \in \mathbb{Z}^n \\ |\lambda|=0}} \frac{\Delta(zq^\lambda)}{\Delta(z)} \prod_{1 \le i < j \le n} \frac{(z_i z_j/t)_{\lambda_i+\lambda_j}}{(qt z_i z_j)_{\lambda_i+\lambda_j}} \prod_{i=1}^{2} \prod_{j=1}^{n} \frac{(z_j/b_i)_{\lambda_j}}{(qa_i z_j)_{\lambda_j}}
$$

$$
= (q)_\infty^{n-1} \left(qAt^m Z, qBt^m/Z, q\hat{A}t^{n-m-1} Z, q\hat{B}t^{n-m-1}/Z \right)_\infty
$$

$$
\times \prod_{i=1}^{n-m-1} (qt^{2i}, qa_1 a_2 t^{2i-1}, qb_1 b_2 t^{2i-1})_\infty \prod_{i=1}^{m} \frac{\prod_{j,k=1}^{2}(qa_j b_k t^{2i-2})_\infty}{(qa_1 a_2 b_1 b_2 t^{2i+2n-2m-4})_\infty}
$$

$$
\times \prod_{i=1}^{2} \prod_{j=1}^{n} \frac{1}{(qa_i z_j, qb_i/z_j)_\infty} \prod_{1 \le i < j \le n} \frac{(qz_i/z_j, qz_j/z_i)_\infty}{(qt z_i z_j, qt/z_i z_j)_\infty}, \tag{5.1}
$$

where $m = \lfloor n/2 \rfloor$, $Z = z_1 \cdots z_n$, $|qa_1 a_2 b_1 b_2 t^{2n-4}| < 1$ and

$$
(A, \hat{A}, B, \hat{B}) = \begin{cases} (1, a_1 a_2, 1, b_1 b_2), & \text{for } n \text{ even,} \\ (a_1, a_2, b_1, b_2), & \text{for } n \text{ odd.} \end{cases}
$$

To prove this we set $n = 2m + k$ where $k = 0, 1$ in (1.7), and simultaneously replace

$$
a_i \mapsto t w_i q^{\nu_i}, \qquad a_{i+m} \mapsto t(w_i q^{\nu_i})^{-1} \quad \text{for } 1 \le i \le m,
$$

$$
b_i \mapsto s^{-1} y_i q^{\mu_i}, \qquad b_{i+m} \mapsto s^{-1}(y_i q^{\mu_i})^{-1} \quad \text{for } 1 \le i \le m
$$

and

$$a_{2i+m} \mapsto a_i, \qquad b_{2i+m} \mapsto b_i \quad \text{for } 1 \leq i \leq k.$$

After some elementary manipulations this yields

$$\sum_{\substack{\lambda \in \mathbb{Z}^n \\ |\lambda|=0}} \frac{\Delta(zq^\lambda)}{\Delta(z)} \prod_{j=1}^n \left[\prod_{i=1}^m \frac{(sy_i^{\pm} z_j)_{\lambda_j}}{(qtw_i^{\pm} z_j)_{\lambda_j}} \prod_{i=1}^k \frac{(z_j/b_i)_{\lambda_j}}{(qa_i z_j)_{\lambda_j}} \right]$$

$$\times \prod_{i=1}^n \prod_{j=1}^m \frac{(sq^{\lambda_i} z_i y_j)_{\mu_j}}{(q^{1-\lambda_i} y_j/sz_i)_{\mu_j}} \frac{(q^{-\lambda_i} w_j/tz_i)_{\nu_j}}{(q^{\lambda_i+1} tz_i w_j)_{\nu_j}}$$

$$= \frac{(qt^{2m} AZ, qB/s^{2m} Z)_\infty}{(q, q(t/s)^{2m} AB)_\infty}$$

$$\times \prod_{j=1}^n \left[\prod_{i=1}^n (qz_i/z_j)_\infty \prod_{i=1}^m \frac{1}{(qy_i^{\pm}/sz_j, qtw_i^{\pm} z_j)_\infty} \prod_{i=1}^k \frac{1}{(qb_i/z_j, qa_i z_j)_\infty} \right]$$

$$\times \prod_{i=1}^m \left[\prod_{j=1}^m (qty_i^{\pm} w_j^{\pm}/s)_\infty \prod_{j=1}^k (qy_i^{\pm} a_j/s, qtw_i^{\pm} b_j)_\infty \right] \prod_{i,j=1}^k (qa_i b_j)_\infty$$

$$\times (t^{2m} AZ)^{|\mu|} \left(\frac{B}{s^{2m} Z} \right)^{|\nu|} \prod_{i,j=1}^m \frac{(q^{\nu_i} sw_i y_j/t, q^{-\nu_i} sy_j/tw_i)_{\mu_j}}{(q^{1-\nu_i} ty_j/sw_i, q^{\nu_i+1} tw_i y_j/s)_{\mu_j}}$$

$$\times \frac{(sy_i^{\pm} w_j/t)_{\nu_j}}{(qty_i^{\pm} w_j/s)_{\nu_j}} \prod_{i=1}^k \prod_{j=1}^m \frac{(sy_j/a_i)_{\mu_j}}{(qa_i y_j/s)_{\mu_j}} \frac{(w_j/tb_i)_{\nu_j}}{(qtb_i w_j)_{\nu_j}},$$

where $A = a_1 \cdots a_k$ and $B = b_1 \cdots b_k$. We multiply the above equation by

$$\left(\frac{q\hat{B}}{s^{n+k-2} Z} \right)^{|\mu|} \frac{\Delta^+(yq^\mu)}{\Delta^+(y)} \prod_{i=1}^m \prod_{j=k+1}^2 \frac{(y_i/sb_j)_{\mu_i}}{(qsy_i b_j)_{\mu_i}}$$

$$\times \left(qt^{n+k-2} \hat{A} Z \right)^{|\nu|} \frac{\Delta^+(wq^\nu)}{\Delta^+(w)} \prod_{i=1}^m \prod_{j=k+1}^2 \frac{(tw_i a_j)_{\nu_i}}{(qw_i a_j/t)_{\nu_i}},$$

where $y = (y_1, \ldots, y_m)$, $w = (w_1, \ldots, w_m)$, $\hat{A} = a_{k+1} \cdots a_2$ and $\hat{B} = b_{k+1} \cdots b_2$, and sum both sides over $\mu, \nu \in \mathbb{Z}^m$.

Now we change the order of summations $\sum_{\mu, \nu \in \mathbb{Z}^m}$ and $\sum_{\lambda \in \mathbb{Z}^n, |\lambda|=0}$ in the triple sum on the left-hand side. Then the sums over μ and ν on the left can be evaluated with the help of (3.1) with $m \mapsto n$. Evaluating in the same way the resulting sum over

μ on the right-hand side using (3.1), we arrive at the formula

$$\sum_{\substack{\lambda\in\mathbb{Z}^n \\ |\lambda|=0}} \frac{\Delta(zq^\lambda)}{\Delta(z)} \prod_{1\le i<j\le n} \frac{(s^2 z_i z_j)_{\lambda_i+\lambda_j}}{(qt^2 z_i z_j)_{\lambda_i+\lambda_j}} \prod_{i=1}^2\prod_{j=1}^n \frac{(z_j/b_i)_{\lambda_j}}{(qa_i z_j)_{\lambda_j}}$$

$$= \frac{(qt^{2m}AZ, qB/s^{2m}Z, q\hat{B}/s^{n+k-2}Z, qt^{n+k-2}\hat{A}Z)_\infty (qt^2/s^2)_\infty^m}{(q, q(t/s)^{2m}AB, qt^{2m}A\hat{B}/s^{n+k-2})_\infty (q)_\infty^m}$$

$$\times \prod_{i=1}^m \frac{1}{(qw_i^{\pm 2})_\infty} \prod_{1\le i<j\le m} \frac{(qt^2 w_i^\pm w_j^\pm/s^2)_\infty}{(qw_i^\pm w_j^\pm)_\infty}$$

$$\times \prod_{j=1}^m \left[\prod_{i=1}^k (qta_i w_j^\pm/s^2)_\infty \prod_{i=k+1}^2 (qa_i w_j^\pm/t)_\infty \prod_{i=1}^2 (qtw_j^\mp b_i)_\infty\right]$$

$$\times \prod_{1\le i<j\le n} \frac{1}{(q/s^2 z_i z_j, qt^2 z_i z_j)_\infty} \prod_{j=1}^n \left[\prod_{i=1}^n (qz_i/z_j)_\infty \prod_{i=1}^2 \frac{1}{(qb_i/z_j, qa_i z_j)_\infty}\right]$$

$$\times \prod_{k+1\le i<j\le 2} \frac{1}{(qa_i a_j/t^2)_\infty} \prod_{1\le i<j\le k} (qa_i a_j/s^2)_\infty \prod_{i=1}^k\prod_{j=1}^2 (qa_i b_j)_\infty$$

$$\times \sum_{\nu\in\mathbb{Z}^m} (q(t/s)^{4m+2k-4} a_1 a_2 b_1 b_2)^{|\nu|} \left(\frac{s^4}{qt^4}\right)^{n(\nu)} \frac{\Delta^+(wq^\nu)}{\Delta^+(w)}$$

$$\times \prod_{1\le i<j\le m} \frac{(s^2 w_i w_j/t^2)_{\nu_i+\nu_j} (s^2 w_i/t^2 w_j)_{\nu_i-\nu_j}}{(qt^2 w_i w_j/s^2)_{\nu_i+\nu_j} (qt^2 w_i/s^2 w_j)_{\nu_i-\nu_j}}$$

$$\times \prod_{j=1}^m \left[\prod_{i=1}^2 \frac{(w_j/tb_i)_{\nu_j}}{(qtb_i w_j)_{\nu_j}} \prod_{i=1}^k \frac{(s^2 w_j/a_i t)_{\nu_j}}{(qta_i w_j/s^2)_{\nu_j}} \prod_{i=k+1}^2 \frac{(tw_j/a_i)_{\nu_i}}{(qa_i w_j/t)_{\nu_i}}\right].$$

Summing over ν on the right using (3.3) with $n \mapsto m$, formula (5.1) follows upon the substitution $(s^2, t^2) \mapsto (1/t, t)$.

For the convergence of the initial triple sum on the left (and, so, convergence of the series in (5.1)) and of the double sum on the right, the following conditions on the parameters are required:

$$|q(t/s)^{2m}AB| < 1, \qquad |qt^{n+k-2}\hat{A}Z| < 1, \qquad |qs^{2-n-k}\hat{B}/Z| < 1,$$

$$|qt^{2m}s^{2-n-k}A\hat{B}| < 1, \qquad |q^{2-n}(s/t)^{2n}a_1 a_2 b_1 b_2| < 1,$$

$$|q(t/s)^{2n-4}a_1 a_2 b_1 b_2| < 1.$$

By analytic continuation these may be relaxed to yield (after the rescaling $(s^2, t^2) \mapsto (1/t, t)$) the condition $|qt^{2n-4}a_1 a_2 b_1 b_2| \le 1$ imposed on (5.1). $\qquad\square$

Unfortunately the above is only a formal proof of (5.1). Although all of the series used converge, we need to take caution since they do not converge absolutely, so that the interchange of the summations $\sum_{\mu,\nu\in\mathbb{Z}^m}$ with $\sum_{\lambda\in\mathbb{Z}^n,|\lambda|=0}$ is not justified. It thus remains to be proved rigorously that both series converge to the same function. Indeed, on the left we are looking at evaluating a triple sum of the form

$$\sum_{\substack{\mu,\nu\in\mathbb{Z}^m \\ \lambda\in\mathbb{Z}^n \\ |\lambda|=0}} f_{\lambda\mu\nu},$$

but, as pointed out to us by the anonymous referee, neither

$$\sum_{\substack{\lambda\in\mathbb{Z}^n \\ |\lambda|=0}}\sum_{\mu\in\mathbb{Z}^m} f_{\lambda\mu\nu} \quad \text{nor} \quad \sum_{\substack{\lambda\in\mathbb{Z}^n \\ |\lambda|=0}}\sum_{\nu\in\mathbb{Z}^m} f_{\lambda\mu\nu}$$

converge (for more details, see the appendix in [29]). We do believe it should be possible to give meaning even to these divergent series along the lines described in [16]. That is, one should replace the formal series by an appropriate analytical function which generates formally the corresponding series. It is well known that the summation formulas allow an analytical continuation of functions to the region of parameters where the series representations diverge. If one finds an appropriate analytic continuation of our formal manipulations with the series, then this could lead to a rigorous justification of formula (5.1).

6 New type II $_6\phi_5$ summations

We begin this section by reviewing some results from [24, 26, 33, 35–37]. Making the specialisations

$$(a_1, a_2, a_3, a_4) = \left(t^{n-1}/a^{1/2}, a^{1/2}/b, a^{1/2}/c, a^{1/2}q^N\right)$$

and

$$z_i = a^{1/2}t^{1-i} \quad \text{for } 1 \leq i \leq n$$

in (3.3) yields van Diejen's type II C_n $_6\phi_5$ summation [33]

$$\sum_{\lambda\in\Lambda_N} \Delta_\lambda(a)\frac{(at^{1-n}, b, c, q^{-N})_\lambda}{(qt^{n-1}, aq/b, aq/c, aq^{N+1})_\lambda}\left(\frac{aq^{N+1}t^{1-n}}{bc}\right)^{|\lambda|}t^{2n(\lambda)}$$

$$= \frac{(aq, aq/bc)_{(N^n)}}{(aq/b, aq/c)_{(N^n)}}, \tag{6.1}$$

where $\Delta_\lambda(a)$ is defined in (4.2). This sum, the $N \to \infty$ limit of which should be compared with (4.3), can be interpreted as the weight-function normalisation for the BC_n q-Racah polynomials of van Diejen and Stokman [36] as follows. Let $\mathcal{H}^{BC_n}$ denote the space of BC_n-symmetric Laurent polynomials (that is, the space of Laurent polynomials symmetric under $W = \mathfrak{S}_n \ltimes (\mathbb{Z}_2)^n$, where $\mathfrak{S}_n$ acts by permuting the variables and $\mathbb{Z}_2$ acts by inversion in the sense of $x \mapsto 1/x$). $\mathcal{H}^{BC_n}$ is spanned by $\{m_\lambda : \lambda \in \Lambda\}$ with m_λ a monomial symmetric function on BC_n:

$$m_\lambda(z) = \sum_\alpha z_1^{\alpha_1} \cdots z_n^{\alpha_n},$$

where the sum is over all distinct signed permutations α of λ. Now define the restricted space $\mathcal{H}_N^{qR}$ as

$$\mathcal{H}_N^{qR} = \text{Span}\{m_\lambda : \lambda \in \Lambda_N\},$$

and for $t, t_0, \ldots, t_3$ such that $t_0 t_i t^{n-1} = q^{-N}$ for a fixed $i \in \{1, 2, 3\}$, let the q-Racah weight function be given by

$$\Delta^{qR}(\lambda) = \Delta_\lambda\left(t_0^2 t^{2n-2}\right)\left(\frac{qt^{2-2n}}{t_0 t_1 t_2 t_3}\right)^{|\lambda|} t^{2n(\lambda)} \prod_{r=0}^{3} \frac{(t_0 t_r t^{n-1})_\lambda}{(qt_0 t^{n-1}/t_r)_\lambda}.$$

Note that $\Delta^{qR}(\lambda)$ is exactly the summand of (6.1) if we identify $a = t_0^2 t^{2n-2}$ and

$$\{b, c, q^{-N}\} = \{t_0 t_1 t^{n-1}, t_0 t_2 t^{n-1}, t_0 t_3 t^{n-1}\}.$$

With the above notation we can define a bilinear form

$$\langle f, g \rangle_N^{qR} = \sum_{\lambda \in \Lambda_N} \Delta^{qR}(\lambda) f\left(t_0 \langle \lambda \rangle\right) g\left(t_0 \langle \lambda \rangle\right)$$

for $f, g \in \mathcal{H}_N^{qR}$ and

$$\langle \lambda \rangle = \left(q^{\lambda_1} t^{n-1}, q^{\lambda_2} t^{n-2}, \ldots, q^{\lambda_n}\right)$$

a spectral vector. The BC_n q-Racah (or discrete BC_n Askey–Wilson) polynomials p_λ are the unique monic polynomials in $\mathcal{H}_N^{qR}$ such that

$$\langle p_\lambda, p_\mu \rangle_N^{qR} = 0, \quad \lambda \neq \mu, \ \lambda, \mu \in \Lambda_N.$$

van Diejen's identity (6.1) may now be put in the equivalent form

$$\langle 1, 1 \rangle_N^{qR} = \prod_{i=1}^{n} \frac{(qt_0^2 t^{2n-i-1}, qt^{1-i}/t_1 t_2, qt^{1-i}/t_1 t_3, qt^{1-i}/t_2 t_3)_\infty}{(qt_0 t^{n-i}/t_1, qt_0 t^{n-i}/t_2, qt_0 t^{n-i}/t_3, qt^{2-i-n}/t_0 t_1 t_2 t_3)_\infty}.$$

In [37] this sum was conjectured to generalise to the elliptic level.[3] To state this now ex-conjecture, we adopt all the notation for q-shifted factorials introduced in Sect. 2 but with $(a)_n$ representing the elliptic shifted factorial [9, 31]:

$$(a)_n = (a; q, p)_n := \prod_{k=0}^{n-1} \theta\left(aq^k\right), \tag{6.2}$$

---

[3]For a recent review of elliptic hypergeometric functions, see [31].

where

$$\theta(x) = \theta(x; p) := (x, p/x; p)_\infty \quad \text{for } |p| < 1.$$

Also defining the elliptic analogue of (4.2) as

$$\Delta_\lambda(a) = \Delta_\lambda(a; q, t; p)$$

$$= \prod_{i=1}^{n} \frac{\theta(at^{2-2i}q^{2\lambda_i})}{\theta(at^{2-2i})} \prod_{1\le i<j\le n} \frac{\theta(t^{j-i}q^{\lambda_i-\lambda_j})}{\theta(t^{j-i})} \frac{\theta(at^{2-i-j}q^{\lambda_i+\lambda_j})}{\theta(at^{2-i-j})}$$

$$\times \prod_{1\le i<j\le n} \frac{(at^{3-i-j})_{\lambda_i+\lambda_j}}{(aqt^{1-i-j})_{\lambda_i+\lambda_j}} \frac{(t^{j-i+1})_{\lambda_i-\lambda_j}}{(qt^{j-i-1})_{\lambda_i-\lambda_j}}, \tag{6.3}$$

the elliptic generalisation of van Diejen's sum is [37]

$$\sum_{\lambda \in \Lambda_N} \Delta_\lambda(a) \frac{(at^{1-n}, b, c, d, e, q^{-N})_\lambda}{(qt^{n-1}, aq/b, aq/c, aq/d, aq/e, aq^{N+1})_\lambda} q^{|\lambda|} t^{2n(\lambda)}$$

$$= \frac{(aq, aq/bc, aq/bd, aq/cd)_{(N^n)}}{(aq/b, aq/c, aq/d, aq/bcd)_{(N^n)}}, \tag{6.4}$$

provided that $bcdet^{n-1} = a^2q^{N+1}$. For $n = 1$ this is Frenkel–Turaev's elliptic extension of the Jackson sum [8], which was shown in [32] to serve as a normalisation condition of the weight function for a family of elliptic biorthogonal rational functions with discrete arguments (for their continuous analogues, see [30]).

For $p = 0$ (and general n), identity (6.4) was first proved in [35] using residue calculus on Gustafson's type II C_n q-Selberg integral. In its full generality (6.4) was proved by Rosengren in [26]. Subsequently Rains [24] and Coskun and Gustafson [7] not only generalised (6.4) to allow for more general partitions than (N^n), but also connected it to the theory of BC_n abelian functions generalising the Koornwinder polynomials (which include the above-discussed q-Racah polynomials) and Macdonald interpolation polynomials to the elliptic level.

After these preliminaries we turn to Claim 5.1. If we make the simultaneous substitutions

$$z_{2i} \mapsto t^{1-2i}(t/a)^{1/2} \quad \text{for } 1 \le i \le n-1,$$

$$z_{2i-1} \mapsto t^{2i-2}(a/t)^{1/2} \quad \text{for } 1 \le i \le n,$$

$$b_1 \mapsto q^N(a/t)^{1/2},$$

$$b_2 \mapsto (a/t)^{1/2}/b,$$

$$a_1 \mapsto t^{2-2n}(t/\hat{a})^{1/2},$$

$$a_2 \mapsto a(t/a)^{1/2}/c,$$

$$z_{2n} \mapsto (a/t)^{1/2}/\hat{a},$$

followed by $t^2 \mapsto 1/t$, the summand contains the term

$$\prod_{i=1}^{n-1} \frac{(1)_{\lambda_{2i}+\lambda_{2i+1}}}{(q)_{\lambda_{2i-1}+\lambda_{2i}}} \prod_{i=1}^{n-1} \frac{(q^{-N})_{\lambda_1}}{(q)_{\lambda_{2n-1}}}.$$

Hence it vanishes unless

$$N \geq \lambda_1 \geq -\lambda_2 \geq \lambda_3 \geq \cdots \geq -\lambda_{2n-2} \geq \lambda_{2n-1} \geq 0.$$

If we now relabel $\lambda_{2i} \mapsto -\mu_i$ for $1 \leq i \leq n-1$ followed by $\lambda_{2i-1} \mapsto \lambda_i$ for $1 \leq i \leq n$, we obtain the following new $_6\phi_5$ summation. For $m \leq n$ and $\lambda, \mu \in \Lambda$ such that $\mu_{m+1} = \cdots = \mu_n = 0$, let

$$\Delta_{\lambda\mu}^{nm}(a, \hat{a}) = \Delta_{\lambda\mu}^{nm}(a, \hat{a}; q, t)$$

$$:= \prod_{1 \leq i < j \leq n} \frac{1 - t^{j-i} q^{\lambda_i - \lambda_j}}{1 - t^{j-i}} \frac{(at^{3-i-j})_{\lambda_i + \lambda_j}}{(aqt^{2-i-j})_{\lambda_i + \lambda_j}}$$

$$\times \prod_{1 \leq i < j \leq m} \frac{1 - t^{j-i} q^{\mu_i - \mu_j}}{1 - t^{j-i}} \frac{(at^{2-i-j})_{\mu_i + \mu_j}}{(aqt^{1-i-j})_{\mu_i + \mu_j}}$$

$$\times \prod_{i=1}^{n} \prod_{j=1}^{m} \frac{1 - at^{2-i-j} q^{\lambda_i + \mu_j}}{1 - at^{2-i-j}} \frac{(t^{j-i+1})_{\lambda_i - \mu_j}}{(qt^{j-i})_{\lambda_i - \mu_j}}$$

$$\times \prod_{i=1}^{n} \frac{1 - \hat{a} t^{1-i} q^{\lambda_i + |\lambda - \mu|}}{1 - \hat{a} t^{1-i}} \frac{(at^{2-i}/\hat{a})_{\lambda_i - |\lambda - \mu|}}{(aqt^{1-i}/\hat{a})_{\lambda_i - |\lambda - \mu|}}$$

$$\times \prod_{i=1}^{m} \frac{1 - at^{1-i} q^{\mu_i - |\lambda - \mu|}/\hat{a}}{1 - at^{1-i}/\hat{a}} \frac{(\hat{a} t^{1-i})_{\mu_i + |\lambda - \mu|}}{(\hat{a} q t^{-i})_{\mu_i + |\lambda - \mu|}}.$$

Corollary 6.1 (Type II A_{2n-1} $_6\phi_5$ sum) *For N a nonnegative integer,*

$$\sum \Delta_{\lambda\mu}^{n,n-1}(a, \hat{a}) \frac{(b, q^{-N})_\lambda}{(aq/b, aq^{N+1})_\mu} \frac{(at^{1-n}, c)_\mu}{(qt^{n-1}, aq/c)_\lambda} \frac{(\hat{a} t^{1-n}, \hat{a} c/a)_{|\lambda - \mu|}}{(\hat{a} q/b, \hat{a} q^{N+1})_{|\lambda - \mu|}}$$

$$\times \left(\frac{aq^{N+1}}{bc} \right)^{|\lambda|} q^{n(\lambda) + n(\mu) - (n-1)|\mu|}$$

$$= \frac{(aq/bc)_{(N^n)}}{(aq/c)_{(N^n)}} \frac{(aq)_{(N^{n-1})}}{(aq/b)_{(N^{n-1})}} \frac{(\hat{a} q)_N}{(\hat{a} q/b)_N},$$

where the sum is over $\lambda, \mu \in \Lambda_N$ such that $\mu_n = 0$ and $\mu \preccurlyeq \lambda$.

Remarkably, the above identity contains van Diejen's $_6\phi_5$ sum as a special case, establishing

$$\text{type II } C_n \ _6\phi_5 \text{ sum} \quad \hookrightarrow \quad \text{type II } A_{2n-1} \ _6\phi_5 \text{ sum}.$$

Specifically, if we fix $\hat{a} = at^{1-n}$, then $\Delta_{\lambda\mu}^{n,n-1}(a,\hat{a})$ contains the factor

$$\frac{1}{(q)_{\lambda_n - |\lambda - \mu|}},$$

which implies that it vanishes unless $\mu_i = \lambda_i$ for $1 \le i \le n-1$. Assuming that μ is fixed in this manner (so that $|\lambda - \mu| = \lambda_n$), it takes a routine calculation to show that

$$\Delta_{\lambda\mu}^{n,n-1}(a, at^{1-n}) = \left(\frac{t}{q}\right)^{2n(\lambda)-(n-1)|\lambda|} \Delta_\lambda(a).$$

It is now easily seen that Corollary 6.1 reduces to (6.1).

In view of the above result and our previous discussion of elliptic hypergeometric series, it takes little imagination to make the following conjecture (which has been extensively checked for small values of n and N). Let $(a)_n$ again represent the elliptic shifted factorial (6.2), and for $m \le n$ and $\lambda, \mu \in \Lambda$ such that $\mu_m = \cdots = \mu_n = 0$, let

$$\Delta_{\lambda\mu}^{nm}(a, \hat{a}) = \Delta_{\lambda\mu}^{nm}(a, \hat{a}; q, t; p)$$

$$:= \prod_{1 \le i < j \le n} \frac{\theta(t^{j-i}q^{\lambda_i - \lambda_j})}{\theta(t^{j-i})} \frac{(at^{3-i-j})_{\lambda_i + \lambda_j}}{(aqt^{2-i-j})_{\lambda_i + \lambda_j}}$$

$$\times \prod_{1 \le i < j \le m} \frac{\theta(t^{j-i}q^{\mu_i - \mu_j})}{\theta(t^{j-i})} \frac{(at^{2-i-j})_{\mu_i + \mu_j}}{(aqt^{1-i-j})_{\mu_i + \mu_j}}$$

$$\times \prod_{i=1}^{n} \prod_{j=1}^{m} \frac{\theta(at^{2-i-j}q^{\lambda_i + \mu_j})}{\theta(at^{2-i-j})} \frac{(t^{j-i+1})_{\lambda_i - \mu_j}}{(qt^{j-i})_{\lambda_i - \mu_j}}$$

$$\times \prod_{i=1}^{n} \frac{\theta(\hat{a}t^{1-i}q^{\lambda_i + |\lambda - \mu|})}{\theta(\hat{a}t^{1-i})} \frac{(at^{2-i}/\hat{a})_{\lambda_i - |\lambda - \mu|}}{(aqt^{1-i}/\hat{a})_{\lambda_i - |\lambda - \mu|}}$$

$$\times \prod_{i=1}^{m} \frac{\theta(at^{1-i}q^{\mu_i - |\lambda - \mu|}/\hat{a})}{\theta(at^{1-i}/\hat{a})} \frac{(\hat{a}t^{1-i})_{\mu_i + |\lambda - \mu|}}{(\hat{a}qt^{-i})_{\mu_i + |\lambda - \mu|}}.$$

Use this to define the new type II elliptic hypergeometric series

$$V(a, \hat{a}; b_1, \ldots, b_{r+1}; c_1, \ldots, c_r)$$

$$= V(a, \hat{a}; b_1, \ldots, b_{r+1}; c_1, \ldots, c_r; q, t; p)$$

$$:= \sum \Delta_{\lambda\mu}^{n,n-1}(a, \hat{a}) \frac{(b_1, \ldots, b_{r+1})_\lambda}{(aq/b_1, \ldots, aq/b_{r+1})_\mu} \frac{(at^{1-n}, c_1, \ldots, c_r)_\mu}{(qt^{n-1}, aq/c_1, \ldots, aq/c_r)_\lambda}$$

$$\times \frac{(\hat{a}t^{1-n}, \hat{a}c_1/a, \ldots, \hat{a}c_r/a)_{|\lambda - \mu|}}{(\hat{a}q/b_1, \ldots, \hat{a}q/b_{r+1})_{|\lambda - \mu|}} (qt^{n-1})^{|\lambda|} q^{n(\lambda) + n(\mu) - (n-1)|\mu|},$$

where one of the b_i is of the form q^{-N}, the balancing condition

$$b_1 \cdots b_{r+1} c_1 \cdots c_r t^{n-1} = a^r q^{r-1}$$

holds, and where the sum is over $\lambda, \mu \in \Lambda_N$ such that $\mu_n = 0$ and $\mu \preccurlyeq \lambda$.

Conjecture 6.2 (Type II A_{2n-1} elliptic ${}_8\phi_7$ sum) *Assuming the balancing condition*

$$bcdet^{n-1} = a^2 q^{N+1},$$

we have

$$V\left(a, \hat{a}; b, c, q^{-N}; d, e\right)$$
$$= \frac{(aq/bd, aq/cd)_{(N^n)}}{(aq/d, aq/bcd)_{(N^n)}} \frac{(aq, aq/bc)_{(N^{n-1})}}{(aq/b, aq/c)_{(N^{n-1})}} \frac{(\hat{a}q, \hat{a}q/bc)_N}{(\hat{a}q/b, \hat{a}q/c)_N}. \tag{6.5}$$

For $n = 1$ this again corresponds to the elliptic Frenkel and Turaev sum [8], and for general n and $(a, \hat{a}) = (a, at^{1-n})$ it reduces to (6.4).

Conjecture 6.3 *Conjecture 6.2 follows from the elliptic type II A_{2n-1} beta integral of [30] by an appropriate residue calculus.*

The preceding manipulations can be repeated in the A_{2n} case. That is, if in Claim 5.2 we specialise

$$z_{2i} \mapsto t^{1-2i}(t/a)^{1/2}, \quad 1 \le i \le n,$$
$$z_{2i-1} \mapsto t^{2i-2}(a/t)^{1/2}, \quad 1 \le i \le n,$$
$$c_1 \mapsto q^N (a/t)^{1/2},$$
$$c_2 \mapsto t^{1-2n}(t/a)^{1/2},$$
$$d_1 \mapsto t(a/t)^{1/2}/b,$$
$$d_2 \mapsto t(a/t)^{1/2}/c,$$
$$z_{2n+1} \mapsto (a/t)^{1/2}/\hat{a},$$

and finally make the substitution $t^2 \mapsto 1/t$, we obtain by a similar reasoning as before the following companion of Corollary 6.1.

Corollary 6.4 (Type II A_{2n} ${}_6\phi_5$ sum) *For N a nonnegative integer,*

$$\sum \Delta_{\lambda\mu}^{nn}(a, \hat{a}) \frac{(at^{1-n}, q^{-N})_\lambda}{(qt^{n-1}, aq^{N+1})_\mu} \frac{(b, c)_\mu}{(aq/b, aq/c)_\lambda}$$
$$\times \frac{(\hat{a}b/a, \hat{a}c/a)_{|\lambda-\mu|}}{(\hat{a}qt^{n-1}/a, \hat{a}q^{N+1})_{|\lambda-\mu|}} \left(\frac{aq^{N+2}}{bct}\right)^{|\lambda|} q^{n(\lambda)+n(\mu)-n|\mu|}$$
$$= \frac{(aq/bct, aq)_{(N^n)}}{(aq/b, aq/c)_{(N^n)}} \frac{(\hat{a}q)_N}{(\hat{a}q/t^n)_N},$$

where the sum is over $\lambda, \mu \in \Lambda_N$ such that $\mu \preccurlyeq \lambda$.

Once again (6.1) arises through an appropriate specialisation, so that now

$$\text{type II } C_n \; _6\phi_5 \text{ sum} \quad \hookrightarrow \quad \text{type II } A_{2n} \; _6\phi_5 \text{ sum}.$$

To be more precise, $\Delta_{\lambda\mu}^{nn}(a/t, a)$ contains the factor $(1)_{\lambda_1 - |\lambda - \mu|}$. Since

$$\lambda_1 - |\lambda - \mu| = (\mu_1 - \lambda_2) + \cdots + (\mu_{n-1} - \lambda_n) + \mu_n$$

and $\mu_i \geq \lambda_{i+1}$ (recall that $\mu \preccurlyeq \lambda$), $\Delta_{\lambda\mu}^{nn}(a/t, a)$ vanishes unless $\mu_i = \lambda_{i+1}$ for $1 \leq i \leq n-1$ and $\mu_n = 0$. But for such μ,

$$\Delta_{\lambda\mu}^{nn}(a/t, a) = \left(\frac{t}{q}\right)^{2n(\lambda) - (n-1)|\lambda|} q^{|\lambda| - (n+1)\lambda_1} \frac{(qt^n, at^{1-n})_\lambda}{(qt^{n-1}, at^{-n})_\lambda} \Delta_\lambda(a).$$

It thus follows that Corollary 6.4 reduces to (6.1) after the substitution $(a, \hat{a}, b, c) \mapsto (a/t, a, b/t, c/t)$.

Conjecturally Corollary 6.4 again admits an elliptic generalisation. To state this, we define

$$V(a, \hat{a}; b_1, \ldots, b_r; c_1, \ldots, c_{r+1})$$

$$= V(a, \hat{a}; b_1, \ldots, b_r; c_1, \ldots, c_{r+1}; q, t; p)$$

$$:= \sum \Delta_{\lambda\mu}^{nn}(a, \hat{a}) \frac{(at^{1-n}, b_1, \ldots, b_r)_\lambda}{(qt^{n-1}, aq/b_1, \ldots, aq/b_r)_\mu} \frac{(c_1, \ldots, c_{r+1})_\mu}{(aq/c_1, \ldots, aq/c_{r+1})_\lambda}$$

$$\times \frac{(\hat{a}c_1/a, \ldots, \hat{a}c_{r+1}/a)_{|\lambda - \mu|}}{(\hat{a}qt^{n-1}/a, \hat{a}q/b_1, \ldots, \hat{a}q/b_r)_{|\lambda - \mu|}} (q^2 t^{n-1})^{|\lambda|} q^{n(\lambda) + n(\mu) - n|\mu|},$$

where one of the b_i is of the form q^{-N}, the balancing condition is

$$b_1 \cdots b_r c_1 \cdots c_{r+1} t^n = a^r q^{r-1},$$

and where the sum is over $\lambda, \mu \in \Lambda_N$ such that $\mu \preccurlyeq \lambda$.

Conjecture 6.5 (Type II elliptic A_{2n} $_8\phi_7$ sum) *Assuming the balancing condition*

$$bcdet^n = a^2 q^{N+1},$$

we have

$$V\left(a, \hat{a}; b, q^{-N}; c, d, e\right) = \frac{(aq, aq/bc, aq/bd, aq/cdt)_{(N^n)}}{(aq/b, aq/c, aq/d, aq/bcdt)_{(N^n)}} \frac{(\hat{a}q, \hat{a}q/bt^n)_N}{(\hat{a}/t^n, \hat{a}q/b)_N}.$$

If we substitute $(a, \hat{a}, c, d, e) \mapsto (a/t, a, c/t, d/t, e/t)$, this again simplifies to the elliptic C_n sum (6.4).

Conjecture 6.6 *Conjecture 6.5 follows from the elliptic type II A_{2n} beta integral of [30] by an appropriate residue calculus.*

The sum (6.4) serves as a normalisation of the weight function for Rains' BC_n abelian biorthogonal functions [24, 25] for specifically fixed discrete values of the arguments. It is natural to expect that our two V-function sums conjectured above have similar interpretation for more general biorthogonal functions attached to the root systems A_{2n-1} and A_{2n}. We hope to present a more detailed study of the orthogonal polynomials associated with the sums (6.1) and (6.4), and of the abelian biorthogonal functions based on the type II A_n elliptic beta integrals of [30] in future publications.

7 Further applications of A_{n-1} $_6\psi_6$ summations

In this last section we make some final remarks regarding A_{n-1} $_6\psi_6$ summations. Such summations contain a sum over $\lambda \in \mathbb{Z}^n$ subject to the restriction $|\lambda| = 0$. It is trivial to lift this restriction to $|\lambda| = M$ for $M \in \mathbb{Z}$ simply be replacing $\lambda_1 \mapsto \lambda_1 - M$ and $z_1 \mapsto z_1 q^M$. For example, in the case of (1.7) one obtains [10]

$$
\sum_{\substack{\lambda \in \mathbb{Z}^n \\ |\lambda| = M}} \frac{\Delta(zq^\lambda)}{\Delta(z)} \prod_{i,j=1}^n \frac{(z_j/b_i)_{\lambda_j}}{(qa_i z_j)_{\lambda_j}} = \frac{(Z/B)_M}{(qAZ)_M} \frac{(qAZ, qB/Z)_\infty}{(q, qAB)_\infty} \prod_{i,j=1}^n \frac{(qa_i b_j, qz_i/z_j)_\infty}{(qa_i z_j, qb_i/z_j)_\infty},
$$

where $A = a_1 \cdots a_n$, $B = b_1 \cdots b_n$, $Z = z_1 \cdots z_n$ and $|qAB| < 1$. This implies the following useful lemma [23].

Lemma 7.1 *Provided that both sides converge,*

$$
\sum_{\lambda \in \mathbb{Z}^n} f_{|\lambda|} \frac{\Delta(zq^\lambda)}{\Delta(z)} \prod_{i,j=1}^n \frac{(z_j/b_i)_{\lambda_j}}{(qa_i z_j)_{\lambda_j}}
$$

$$
= \frac{(qAZ, qB/Z)_\infty}{(q, qAB)_\infty} \prod_{i,j=1}^n \frac{(qa_i b_j, qz_i/z_j)_\infty}{(qa_i z_j, qb_i/z_j)_\infty} \sum_{M=-\infty}^\infty f_M \frac{(Z/B)_M}{(qAZ)_M}.
$$

For example, taking $f_k = t^k$, the sum on the right can be performed using Ramanujan's $_1\psi_1$ sum, resulting in a multivariable $_1\psi_1$ sum, see [10, 23].

In much the same way, Claims 5.1 and 5.2 imply the following lemma.

Lemma 7.2 *With the same notation as (5.1),*

$$
\sum_{\lambda \in \mathbb{Z}^n} f_{|\lambda|} \frac{\Delta(zq^\lambda)}{\Delta(z)} \prod_{1 \le i < j \le n} \frac{(z_i z_j/t)_{\lambda_i + \lambda_j}}{(qt z_i z_j)_{\lambda_i + \lambda_j}} \prod_{i=1}^2 \prod_{j=1}^n \frac{(z_j/b_i)_{\lambda_j}}{(qa_i z_j)_{\lambda_j}}
$$

$$
= (q)_\infty^{n-1} \left(qAt^m Z, qBt^m/Z, q\hat{A}t^{n-m-1}Z, q\hat{B}t^{n-m-1}/Z \right)_\infty
$$

$$
\times \prod_{i=1}^{n-m-1} \left(qt^{2i}, qa_1 a_2 t^{2i-1}, qb_1 b_2 t^{2i-1} \right)_\infty \prod_{i=1}^m \frac{\prod_{j,k=1}^2 (qa_j b_k t^{2i-2})_\infty}{(qa_1 a_2 b_1 b_2 t^{2i+2n-2m-4})_\infty}
$$

$$\times \prod_{i=1}^{2}\prod_{j=1}^{n} \frac{1}{(qa_iz_j, qb_i/z_j)_\infty} \prod_{1\le i<j\le n} \frac{(qz_i/z_j, qz_j/z_i)_\infty}{(qtz_iz_j, qt/z_iz_j)_\infty}$$

$$\times \sum_{M=-\infty}^{\infty} f_M \frac{(Zt^{-m}/B, t^{m-n+1}Z/\hat{B})_M}{(qAt^mZ, q\hat{A}t^{n-m-1}Z)_M},$$

provided that both sides converge.

For a number of different choices of f_k, the right-hand sides of the above two formulas become explicitly summable. We omit the details and instead refer the interested reader to [23] where applications of Lemma 7.1 are discussed.

Acknowledgements We thank George Andrews for his many beautiful contributions to mathematics, for his continued interest in our work, for fruitful collaborations and above all for his friendship and support. George, we wish you many more happy, healthy and productive years.

We also gratefully acknowledge the anonymous referee for pointing out several errors in an earlier version of this paper.

The first author is indebted to the University of Queensland for supporting his visit to Australia in February–March 2009 during which some of the results of this paper were obtained.

References

1. Anderson, G.W.: A short proof of Selberg's generalized beta formula. Forum Math. **3**, 415–417 (1991)
2. Andrews, G.E.: Applications of basic hypergeometric functions. SIAM Rev. **16**, 441–484 (1974)
3. Andrews, G.E.: The Theory of Partitions. Encyclopedia of Mathematics and Its Applications, vol. 2. Addison–Wesley, Reading (1976)
4. Andrews, G.E., Askey, R., Roy, R.: Special Functions. Encyclopedia of Mathematics and Its Applications, vol. 71. Cambridge University Press, Cambridge (1999)
5. Bailey, W.N.: Series of hypergeometric type which are infinite in both directions. Q. J. Math. **7**, 105–115 (1936)
6. Bailey, W.N.: A note on two of Ramanujan's formulae. Q. J. Math., Ser. (2) **3**, 29–31 (1952)
7. Coskun, H., Gustafson, R.A.: Well-poised Macdonald functions W_λ and Jackson coefficients ω_λ on BC_n. In: Jack, Hall–Littlewood and Macdonald Polynomials. Contemp. Math., vol. 417, pp. 127–155. AMS, Providence (2006)
8. Frenkel, I.B., Turaev, V.G.: Elliptic solutions of the Yang–Baxter equation and modular hypergeometric functions. In: The Arnold–Gelfand Mathematical Seminars, pp. 171–204. Birkhäuser, Boston (1997)
9. Gasper, G., Rahman, M.: Basic Hypergeometric Series, 2nd edn. Encyclopedia of Mathematics and Its Applications, vol. 96. Cambridge University Press, Cambridge (2004)
10. Gustafson, R.A.: Multilateral summation theorems for ordinary and basic hypergeometric series in $U(n)$. SIAM J. Math. Anal. **18**, 1576–1596 (1987)
11. Gustafson, R.A.: The Macdonald identities for affine root systems of classical type and hypergeometric series very-well-poised on semisimple Lie algebras. In: Thakare, N.K. (ed.) Ramanujan International Symposium on Analysis, pp. 187–224 (1989)
12. Gustafson, R.A.: A summation theorem for hypergeometric series very-well-poised on G_2. SIAM J. Math. Anal. **21**, 510–522 (1990)
13. Gustafson, R.A.: A generalization of Selberg's beta integral. Bull. Am. Math. Soc. (N.S.) **22**, 97–105 (1990)
14. Gustafson, R.A.: Some q-beta and Mellin–Barnes integrals with many parameters associated to the classical groups. SIAM J. Math. Anal. **23**, 525–551 (1992)
15. Gustafson, R.A.: Some q-beta integrals on $SU(n)$ and $Sp(n)$ that generalize the Askey–Wilson and Nassrallah–Rahman integrals. SIAM J. Math. Anal. **25**, 441–449 (1994)

16. Hardy, G.H.: Divergent Series. Clarendon, Oxford (1949)
17. Hardy, G.H., Wright, E.M.: An Introduction to the Theory of Numbers, 2nd edn. Oxford University Press, London (1980)
18. Ito, M.: A product formula for Jackson integral associated with the root system F_4. Ramanujan J. **6**, 279–293 (2002)
19. Koornwinder, T.H.: Askey–Wilson polynomials for root systems of type BC. In: Hypergeometric Functions on Domains of Positivity, Jack Polynomials, and Applications. Contemp. Math., vol. 138, pp. 189–204. AMS, Providence (1992)
20. Macdonald, I.G.: Affine root systems and Dedekind's η-function. Invent. Math. **15**, 91–143 (1972)
21. Macdonald, I.G.: Symmetric Functions and Hall Polynomials, 2nd edn. Oxford University Press, London (1995)
22. Milne, S.C.: An elementary proof of the Macdonald identities for $A_l^{(1)}$. Adv. Math. **57**, 34–70 (1985)
23. Milne, S.C., Schlosser, M.: A new A_n extension of Ramanujan's $_1\psi_1$ summation with applications to multilateral A_n series. Rocky Mt. J. Math. **32**, 759–792 (2002)
24. Rains, E.M.: BC_n-symmetric Abelian functions. Duke Math. J. **135**, 99–180 (2006)
25. Rains, E.M.: Transformations of elliptic hypergeometric integrals. Ann. Math. **171**, 169–243 (2010)
26. Rosengren, H.: A proof of a multivariable elliptic summation formula conjectured by Warnaar. In: q-Series with Applications to Combinatorics, Number Theory, and Physics. Contemp. Math., vol. 291, pp. 193–202. AMS, Providence (2001)
27. Schlosser, M.: Summation theorems for multidimensional basic hypergeometric series by determinant evaluations. Discrete Math. **210**, 151–169 (2000)
28. Schlosser, M.: A new multivariable $_6\psi_6$ summation formula. Ramanujan J. **17**, 305–319 (2008)
29. Schlosser, M.J.: Multilateral inversion of A_r, C_r, and D_r basic hypergeometric series. Ann. Comb. **13**, 341–363 (2009)
30. Spiridonov, V.P.: Theta hypergeometric integrals. Algebra Anal. **15**(6), 161–215 (2003). (St. Petersburg. Math. J. **15**, 929–967 (2004))
31. Spiridonov, V.P.: Essays on the theory of elliptic hypergeometric functions. Usp. Mat. Nauk **63**(3), 3–72 (2008). (Russ. Math. Surv. **63**, 405–472 (2008))
32. Spiridonov, V.P., Zhedanov, A.S.: Classical biorthogonal rational functions on elliptic grids. C. R. Math. Acad. Sci. **22**, 70–76 (2000)
33. van Diejen, J.F.: On certain multiple Bailey, Rogers and Dougall type summation formulas. Publ. Res. Inst. Math. Sci. **33**, 483–508 (1997)
34. van Diejen, J.F.: Properties of some families of hypergeometric orthogonal polynomials in several variables. Trans. Am. Math. Soc. **351**, 233–270 (1999)
35. van Diejen, J.F., Spiridonov, V.P.: An elliptic Macdonald–Morris conjecture and multiple modular hypergeometric sums. Math. Res. Lett. **7**, 729–746 (2000)
36. van Diejen, J.F., Stokman, J.V.: Multivariable q-Racah polynomials. Duke Math. J. **91**, 89–136 (1998)
37. Warnaar, S.O.: Summation and transformation formulas for elliptic hypergeometric series. Constr. Approx. **18**, 479–502 (2002)